Advanced Bio-Inspired and Energy Materials

Advanced Bio-Inspired and Energy Materials

Dr. Vikas Mittal
Editor and Lead Author

CWP
Central West Publishing

This edition has been published by Central West Publishing, Australia

For more information about the books published by Central West Publishing, please visit https://centralwestpublishing.com

Disclaimer
Every effort has been made by the publisher, editor and authors while preparing this book, however, no warranties are made regarding the accuracy and completeness of the content. The publisher, editor and authors disclaim without any limitation all warranties as well as any implied warranties about sales, along with fitness of the content for a particular purpose. Citation of any website and other information sources does not mean any endorsement from the publisher and authors. For ascertaining the suitability of the contents contained herein for a particular lab or commercial use, consultation with the subject expert is needed. In addition, while using the information and methods contained herein, the practitioners and researchers need to be mindful for their own safety, along with the safety of others, including the professional parties and premises for whom they have professional responsibility. To the fullest extent of law, the publisher, editor and authors are not liable in all circumstances (special, incidental, and consequential) for any injury and/or damage to persons and property, along with any potential loss of profit and other commercial damages due to the use of any methods, products, guidelines, procedures contained in the material herein.

NATIONAL LIBRARY OF AUSTRALIA

A catalogue record for this book is available from the National Library of Australia

ISBN (print): 978-1-922617-13-2

Contents

Preface XI

1. Biofuels: Classification, Generations and Feedstocks/ Techniques for Biofuels Generation 1

1.1 Introduction 1
1.2 Classification of Biofuels 2
1.2.1 First-generation Biofuels 2
1.2.2 Second-generation Biofuels 3
1.2.3 Third-generation Biofuels 4
1.3 Types of Liquid Biofuels 6
1.3.1 Biodiesel 6
1.3.2 Bioethanol 7
1.3.3 Butanol 8
1.4 Processes Involved for the Generation of Biodiesel 9
1.4.1 Alkali Catalytic Transesterification 10
1.4.2 Acid Catalytic Transesterification 10
1.4.3 Biocatalytic Transesterification 11
1.4.4 Non-catalytic Techniques 11
1.5 Different Feedstocks and Techniques for Producing Biofuels 13
1.6 Conclusions 22
References 22

2. Nanocellulose as Reinforcement in Polymer Nanocomposites 33

2.1 Introduction 33
2.2 Types of Nanocellulose 35
2.2.1 Micro-fibrillated Cellulose 35
2.2.2 Nano-crystalline Cellulose 39
2.2.3 Bacterial Nanocellulose 43
2.3 Use of Nanocellulose as a Reinforcement in Polymer Nanocomposites 45
2.3.1 MFC as Reinforcement in Polymer Nanocomposites 45
2.3.2 NCC as Reinforcement in Polymer Nanocomposites 47

2.3.3 BNC as Reinforcement in Polymer Nanocomposites 49
2.4 Summary 50
References 50

3. Bio-Inspired Nanomaterials 59

3.1 Introduction 59
3.2 Bio-Inspired Gold Nanomaterials 60
3.3 Bio-Inspired Silver Nanomaterials 67
3.4 Bio-Inspired Palladium Nanomaterials 72
3.5 Bio-Inspired Iron Oxide Nanomaterials 74
3.6 Bio-Inspired Zinc Oxide Nanomaterials 78
3.7 Bio-Inspired Tin Dioxide Nanomaterials 80
3.8 Bio-Inspired Titanium Dioxide Nanomaterials 83
3.9 Bio-Inspired Copper Nanomaterials 84
3.10 Bio-Inspired Nanomaterials Based on Other Metal Oxides 84
3.11 Bio-Inspired Silica Based Nanomaterials 85
3.12 Bio-Inspired Graphene Based Nanomaterials 86
3.13 Other Bio-Inspired Carbonaceous Nanomaterials 91
3.14 Bio-Inspired Polymer Nanomaterials 92
3.15 Conclusions 95
References 95

4. Polymers and Polymer Nanocomposites for Tissue Engineering and Artificial Joints 131

4.1 Introduction 131
4.2 Polymeric Materials for Tissue Engineering 131
4.2.1 Polymeric Hydrogels 133
4.2.2 Polymer Nanofibers 138
4.2.3 Conducting Polymers 141
4.2.4 Other Polymer Composites and Blends 142
4.3 Artificial Joints 146
4.3.1 UHMWPE 148
4.3.2 PVA Hydrogels Based Materials 156
4.4 Conclusions 158
References 158

5. Recent Advances in Nanocarbon-based Polymer Composite Materials as Membranes in Fuel Cell Applications **175**

5.1 Introduction 175
5.2 Types of Polymer/Nanocarbon Composite Membranes 178
5.2.1 Polymer/CNT Composite Membranes 178
5.2.2 Polymer/Functionalized CNTs Composite Membranes 183
5.2.3 Polymer/Graphene Oxide Composite Membranes 191
5.2.4 Polymer/Functionalized Graphene Oxide Membranes 195
5.3 Conclusion and Future Prospects 211
References 212

6. Energy Storage Devices **221**

6.1 Introduction 221
6.2 Supercapacitors 223
6.2.1 Graphene Based Supercapacitors 223
6.2.2 Activated Carbon Based Supercapacitors 228
6.2.3 Metal Based Supercapacitors 231
6.2.4 Polymer Based Supercapacitors 238
6.3 Lithium-Ion Batteries (LIBs) 242
6.3.1 Silicon Based LIB Anodes 242
6.3.2 Lithium Based LIB Electrodes 245
6.3.3 Metal Disulfide Based LIB Anodes 246
6.3.4 Metal Oxide Based LIB Anodes 248
6.3.5 Graphene Based LIB Anodes 250
6.3.6 Other Carbonaceous Materials Based LIB Anodes 253
6.4 Sodium-Ion Batteries (SIBs) 254
6.4.1 Graphene Based SIB Electrodes 254
6.4.2 Other Carbonaceous Materials Based SIB Electrodes 255
6.4.3 Metal Based SIB Electrodes 257

6.5 Lithium-Sulfur Batteries (LSBs) 257
 6.5.1 Graphene Based LSB Electrodes 257
 6.5.2 Carbon Nanotubes Based LSB Electrodes 258
 6.5.3 Other Carbonaceous Materials Based LSB Electrodes 259
 6.5.4 Polymers Based LSB Electrodes 260
6.6 Conclusions 260
References 260

7. Nano-catalysts: Advances in Synthesis and Applications in the Field of Energy 293

7.1 Introduction 293
7.2 Applications of Nano-catalysts in the Field of Energy 295
 7.2.1 Oil Refining 295
 7.2.2 Alternative/Renewable Energy 308
 7.2.3 Other Systems 317
7.3 Conclusion 319
References 320

8. Polymers and Copolymers Utilized in Photovoltaic Applications 329

8.1 Introduction 329
8.2 Factors Influencing Power Conversion Efficiency of Polymer Solar Cell Devices 329
8.3 Importance of Generating Low Band Gap Polymers 330
8.4 Various Polymers and Copolymers for Photovoltaic Devices 330
8.5 Conclusions 350
References 350

9. Thermally Phase Changing Materials: Synthesis, Properties and Applications 361

9.1 Introduction 361
9.2 Different Phase Change Materials (PCMs) 362
 9.2.1 Nanomaterial PCMs 362

9.2.2 Carbon PCMs 363
9.2.3 Metal Foam PCMs 366
9.3 Classification of PCMs Based on Morphology 367
9.3.1 Fibrous Structure 368
9.3.2 Porous Structure 369
9.4 Applications of PCMs 370
9.4.1 Energy Storage 370
9.4.2 Smart Buildings 371
9.4.3 Textile Applications 372
9.5 Conclusions 373
References 374

10. Thermally Conducting Polymer Nanocomposites: Synthesis, Properties and Applications 381

10.1 Introduction 381
10.2 Modeling of Thermal Conductivity 382
10.3 Measurement Techniques 386
10.4 Polymer and Fillers for Thermally Conducting Nanocomposites 388
10.5 Applications 392
10.5.1 Heat Exchange Materials 393
10.5.2 Materials for Harnessing Solar Energy 396
10.5.3 Batteries 398
10.5.4 Light-Emitting Diode Devices 399
10.5.5 Electronic Packaging 400
10.5.6 Energy Storage 403
10.6 Conclusion 406
References 406

Index 417

Preface

Materials science has attained tremendous advancements in the recent years owing to the refinements in synthetic methodologies and better understanding of the fundamental structure property correlations of different materials as well as tuning of the material behaviors based on need. As a result, a large number of advanced materials have been developed for specific applications. A few examples of these material categories include biofuels, bio-inspired materials, polymers for tissue engineering, fuel cell membranes, energy storage devices (such as supercapacitors, lithium ion batteries, etc.), nano-catalysts, photovoltaic materials, phase changing materials, etc. The book aims to present a variety of such advanced materials by providing both basic information as well as recent developments reported for these materials. All aspects related to synthesis, properties and applications of these materials have been reviewed.

Chapter 1 reviews the various generations of biofuels developed in the recent years, along with commenting on the challenges and outlook. In Chapter 2, substantial advancements and prospects of the nanocellulose-based materials are presented. Different preparation and characterization methods of the three types of nanocellulose, namely MFC, NCC and BNC, are also discussed. In Chapter 3, recent progresses in different strategies developed for the effective generation of various bio-inspired nanomaterials have been reviewed, along with focusing on properties and applications of the developed materials. Chapter 4 provides insights into the polymer and polymer nanocomposite systems developed for tissue engineering and artificial joints. Chapter 5 reviews the state-of-the-art research on composite membranes based on polymers and nanocarbon materials such as carbon nanotubes (CNTs) and graphene oxide (GO) used as proton exchange membranes (PEMs) for fuel cells. Chapter 6 focuses on some of the recent advancements in the development of the functional energy storage devices, such as supercapacitors, lithium-ion batteries, lithium-sulfur batteries, sodium-ion batteries, etc. Chapter 7 reviews the advancements in the synthesis and application of nanocatalysts in the field of energy. Chapter 8 presents a thorough compilation of recent studies reporting the design, synthesis and property evaluation of various polymers and copolymers for utilization in photovoltaic applications. Chapter 9 reviews the properties and applications of different phase changing materials. Chapter 10 discusses recent developments in enhancing thermal conductivity of polymer based nanocomposites. Various polymer and filler types as well as techniques used to evaluate the thermal conductivity and microstructure of thermally conducting polymer nanocomposites are discussed, followed by the application of these materials.

The support of all the chapter contributors is greatly acknowledged towards the successful accomplishment of the book. The book is dedicated to my family for their constant support.

Vikas MITTAL

1

Biofuels: Classification, Generations and Feedstocks/ Techniques for Biofuels Generation

1.1 Introduction

Numerous factors, such as increased global warming, high consumption of fossil fuels, environmental pollution and decreasing petroleum reserves, necessitate the need of energy production from environmentally friendly sources for fulfilling the demands of the industrialized world. Some alternative sources of energy like wind, solar, biofuels, etc., have exhibited promising results to replace the fossil fuels.

In this respect, biofuel like biodiesel is a promising replacement of petroleum fuel since it is non-toxic, renewable, biodegradable and emits less harmful gases like sulfur oxide [1-4]. Transesterification is the most employed technique for the production of biodiesel which involves a vegetable oil or animal fat, methanol or any other alcohol (ethanol and butanol) and a homogeneous or heterogeneous catalyst. Biodiesel can be utilized in the existing diesel engines without any modification requirement as the biofuels and fossil fuels have almost similar properties.

Currently, many European countries and United States employ biodiesel and bioethanol in different applications. Other nations like Malaysia also utilize the blend consisting of 95% petroleum diesel and 5% palm oil. In future, it is envisaged that the consumption of biodiesel will grow as an alternative to fossil fuels [5-9].

Higher cost of processing of biodiesel is still the main limitation for its extensive commercialization. There are numerous factors which decide the cost of biodiesel including storage terminal, purification process, raw materials, reactants, etc. Among these components, feedstock alone costs 80% of the complete processing cost of biodiesel. Therefore, it is viable to achieve a low-cost biodiesel by adapting to the economical feedstocks like waste fats and oils [3,10]. Additionally, there are other challenges associated with the usage of biofuels, which need to be overcome so as to realize the large scale

Naman Arora and Vikas Mittal
Khalifa University of Science and Technology, Abu Dhabi, UAE

usage of these advanced materials [11]. Significant research efforts have been devoted in this direction, and the various current challenges are envisaged to be resolved in the near future.

1.2 Classification of Biofuels

Biofuels are generally categorized as primary and secondary fuels. The sources of primary biofuels are wood chips, fuel-wood and pellets which are employed in an unprocessed form for the production of electricity as well as for cooking and heating purposes. Thus, primary biofuels are naturally existing unprocessed biomass which is consumed without making any modification in its natural form. On the other hand, biomass processing is performed for generating secondary biofuels. This type of biofuel can be obtained in the liquid, solid or gas form such as charcoal, biogas, bio-oil, bioethanol, hydrogen and biodiesel. Secondary biofuels are utilized in numerous industrial applications, transportation and other processing purposes. Secondary biofuels are further classified as first-, second- and third-generation biofuels based on the type of raw materials and production methodologies [12,13].

1.2.1 First-generation Biofuels

In this category, grains [14-17], sugars [18-24] and seeds are the primary sources for biofuel production. This type of biofuels involves simple processing for obtaining the final product. Ethanol is one of the most common examples of first-generation biofuels which is produced by the fermentation process of sugar derived from different plants and crops containing starch. Another popular example of first-generation biofuels is biodiesel which is prepared from vegetable oils of oleaginous plants by employing transesterification method. In various countries, first-generation biofuels are generated in remarkable commercial amounts. The feasibility of producing first-generation biofuels is, however, challenging due to strong contradiction with the supply of food [13]. Operating cost of producing these biofuels is also high owing to the competition with food demand associated with primary crops like oil-seeds, grains and sugars. These challenges signify the need for identifying the non-edible sources of biomass for obtaining first-generation biofuels, which can also help to maintain sustainable supply of food for the communities, especially in the poorer regions and countries.

1.2.2 Second-generation Biofuels

Thermochemical and biological processing based techniques usually produce the second-generation biofuels. These type of biofuels are extracted from non-eatable plant biomass and non-edible residual matter of different crops. Basically, specific biomass is generated for producing second-generation biofuels to fulfill the demand of energy which does not involve the edible crops. Classification of second-generation biofuels can be made on the basis of biochemical and thermochemical processes which are employed for converting biomass into biofuels. Based on the biochemical process, biofuels like bioethanol, biodiesel and butanol are generated. On the other hand, fuels like dimethyl ether, Fischer-Tropsch liquids (FTL) and methanol are generated by using the thermochemical process.

Larson [12] stated that the basic properties of feedstocks can lower the processing cost and have potential energy and environmental advantages for most of the second-generation biofuels. Thegarid *et al.* [25] reported the generation of second-generation biofuels where bio-oil was produced by using catalytic pyrolysis in catalytic cracking co-processing mode with vacuum gas oil. The study confirmed the possibility for the development of biofuels through co-processing in fluid catalytic cracking unit. Schenk *et al.* [26] reported microalgae based second-generation biofuels. The benefits of second-generation microalgal systems are that they generate a broad variety of feedstocks for producing bio-methane, biodiesel, bio-hydrogen and bioethanol. The important benefits associated with microalgal system are: a) it exhibits a higher photon conversion capability, b) this type of system can be harvested batch-wise almost throughout a year which imparts constant oil supply, c) this system is environment-friendly as it only requires limited amount of freshwater and only consumes waste water and salt water, d) it is capable of coupling carbon-dioxide-neutral fuel production with carbon-dioxide sequestration and e) this system generates biodegradable and toxicity-free biofuels. Process of harvesting and delivery of carbon-dioxide are the main limitations for the potential production of the biofuels. Dias *et al.* [27] analyzed the simulation results associated with different stages of coupling between first- and second-generation ethanol based setups using modified hydrolysis technology and pentoses fermentation. It was revealed that the merged first- and second-generation exhibited numerous benefits over independent and unmerged second-generation ethanol production setups.

1.2.3 Third-generation Biofuels

As discussed earlier, pursuing first-generation biofuels may result in immense pressure on the commercial food markets all around the world. Moreover, fuels associated with first-generation lead to the deterioration of the plant and water shortages. Some of the challenges associated with the first-generation biofuels were effectively resolved by the introduction of second-generation biofuels. However, the challenges of land usage and associated changes [28] still limit the application of biofuels associated with second-generation. Therefore, in order to overcome the limitations of first- and second-generation biofuels, a latest approach (third-generation biofuels) is aimed towards using microscopic organisms as feedstock instead of agricultural residues and vegetable oils. Biofuels associated with third-generation are extracted from microbes and microalgae which have proved to be a reliable source of energy to replace petroleum-derived fuels.

The consolidated bioprocessing (CBP) system is capable in reducing the cost of biofuels because of its simple feedstock processing and lower energy input needs. In this system, the production of cellulose, fermentation and hydrolysis of substrate are carried out in one process step through cellulolytic microorganism. Therefore, CBP system is an economical approach towards the production of third-generation biofuels [29]. Saidane-Bchir *et al.* [30] reported the production of third-generation bioethanol by using the isolated strain of microalgae which was extracted from the wastewater of a slaughterhouse. It was revealed that the strain of microalgae is a potential feedstock for producing third-generation bioethanol. Zhu *et al.* [31] reported the potential production of lipid content by using Trichosporon fermentans and subsequently, it was used for producing biodiesel. It was observed that in pre-treated waste molasses, Trichosporon fermentans evolved or developed effectively with a yield of obtained lipid content of 12.8 g/L in the molasses of 15% total sugar concentration. The study demonstrated that yeast is a promising source for the preparation of inexpensive microbial oil from the residues of agriculture based industries.

Samori *et al.* [32] reported the extraction of hydrocarbons from Botryococcus braunii by employing switchable-polarity solvents. The system of switchable solvent comprised of 1,8-diazabicyclo-[5.4.0]-undec-7-ene (DBU) and an alcohol. The authors examined two different systems of switchable-polarity solvents, DBU-octanol and DBU-ethanol, for extracting hydrocarbons from the samples of Botryococcus braunii. The system based on DBU-octanol exhibited better extraction efficiency as compared to conventional solvents like methanol, n-hexane, etc. Thus, the usage of switchable-polarity solvents represents a beneficial route for harvesting the algal biomass as it is environmentally friendly and lowers the overall cost of biodiesel production. El-Shimi *et al.* [33] developed biodiesel from Spirulina-Platensis microalgae by employing the acid catalyzed in-situ transesterification technique. The authors studied factors such as amount of methanol, acid catalyst concentration, stirring speed, temperature and time, which could immensely affect the yield and cost of biodiesel production.

Chng *et al.* [34] successfully produced bioethanol from lipid-extracted biomass of Scenedesmus dimorphus. In this work, biomass was directly introduced to simultaneous saccharification and fermentation, thereby, avoiding the expensive pre-treatment, reducing the possibility of contamination and decreasing the complexity of high sugar content. Abreu *et al.* [35] reported the strategy to decrease the cost of producing microalagal biomass, thus, developing cheaper biodiesel fuel. In this study, chlorella vulgaris was grown under both mixotrophic and photoautotrophic conditions. In order to grow chlorella vulgaris, cheese whey was utilized as a source of carbon. It was observed that mixotrophic microalgae evolved more rapidly as compared to photoautotrophic cells. Usage of industrial dairy waste (cheese whey) also contributed towards overcoming the environmental issues associated with the disposal of cheese whey in the diary industries.

In a recent study, Srinophakun *et al.* [36] reported the strategy to produce feedstock for the preparation of biofuels by cultivating Arthrobacter Ak19 in the presence of stress agent like potassium permanganate. Herein, glycerol was used as a source of pure carbon. Response surface design was employed for the optimization of the proportion of nutrients for enhancing the concentration of biomass. It was observed that with the usage of potassium permanganate (stress agent) at a concentration of 3.2 mg/L, 1.95 times higher content of lipid was obtained as compared to the lipid content achieved under normal conditions. In another study, Wang *et al.* [37] also reported the preparation of biodiesel from corn oil by using lipozyme TL IM catalyzed methanolysis. Similar to the other study, the authors reported that the production of biodiesel improved by employing response surface methodology.

1.3 Types of Liquid Biofuels

1.3.1 Biodiesel

Biodiesel is extracted from organic feedstock like waste animal fats or vegetable oils. On burning, biodiesel emits lower percentage of harmful gases as compared to petroleum-derived diesel. The usage of biodiesel does not directly or indirectly raise the level of carbon dioxide in the environment, thereby, contributing to the reduction of the greenhouse effect. Furthermore, biodiesel exhibits better properties as compared to petroleum-derived diesel in respect to the content of sulfur, aromatic content and flash point [38,39]. However, biodiesel cannot currently entirely replace the petroleum-derived diesel, though its importance is steadily increasing as reserves of petroleum are reducing with time. Both edible and non-edible vegetable oils have been employed to produce biodiesel. Some of the common examples of edible vegetable oils are corn and canola, whereas non-edible vegetable oils such as madhuca indica and Pongamia pinnata are commonly employed [40-42]. As biodiesel is usually seen as an effective alternative for petroleum-based diesel fuel, thus, it is critical to examine its key features discussed below.

Chemical structure: Petroleum-extracted diesel is comprised of combinations of a wide variety of hydrocarbons (C12-C25) made up of aromatics, napthenes and paraffins. Biodiesel is a mixture of a small variety of molecules which are esters of fatty acids of C12, C14, C16, C18 and C22. Ring structures like aromatic molecules are exhibited by diesel fuel, whereas straight chain hydrocarbon esters constitute the primary structure of a variety of biodiesels.

Antifoaming: It is convenient to fill pure biodiesel into the fuel tanks of vehicles as it exhibits superior anti-foaming property. As a consequence, foam generation is minimized, which avoids overflowing and leakage of the biodiesel out of the fuel tank. On the other hand, petroleum-extracted diesel has poor anti-foaming property.

Oxygen content: Petroleum-derived diesel does not hold any oxygen content. On the other hand, biodiesel usually exhibits 11% of oxygen content and exhibits properties such as conductivity, solvency and detergency. Due to this reason, biodiesel can also create linkages with metal surfaces (wettability), as it is polar in nature.

Cetane number: The cetane number of petroleum-extracted diesel and biodiesel is usually observed to be in the range of 40-52 and 45-70 respectively. Distribution pattern of fatty acids present in oils and fats decides the cetane number of biodiesel. High value of cetane number can be achieved by fatty acids which are longer and more saturated.

Conductivity: Biodiesel exhibits lower risk of fires and sparks induced by static charge as pure biodiesel has high conductivity, which exceeds 500 picoS/m.

Corrosion: Generally, corrosion occurs due to the presence of oxygen which is associated with the absorption of water. However, the wettability property of pure biodiesel opposes the corrosion process by decreasing the movement of oxygen molecules to the metal surface. Additionally, pure biodiesel does not contain any sulphur content and corrosive entities, which subsequently avoids the corrosion of copper and yellow metals.

Cold flow properties: In pure biodiesel, solidification process is very fast as biodiesel is a simple mixture of low number of constituents. On the other hand, the solidification process of petroleum-derived diesel is relatively slow as the crystallization temperatures of the constituents present in diesel have different values [43].

1.3.2 Bioethanol

Bioethanol results due to the utilization of renewable biomass for producing ethanol. Bioethanol is harmless, environmental-friendly and renewable fuel [44]. In 2006, 51 billion litres of bioethanol based on first-generation processes were produced globally. India and China produced 11% of the global generation of bioethanol in 2006. Recently, most of the countries have focused on the development of first-generation bioethanol. Blending of bioethanol and petrol has also been observed to exhibit complete burning of fuel inside the existing transportation vehicles. A litre of ethanol generates only 66% of energy produced by the same amount of petrol, however, ethanol exhibits higher octane number. Moreover, it demonstrates better performance when blended with petroleum-extracted fuel as compared to pure petrol. The combustion process of the fuel in engines has been observed to improve with the usage of ethanol due to the reduction in partially burned hydrocarbons and carbon monoxide emissions. An additional benefit of utilizing ethanol-petrol blends is decreased sulphur oxide emissions [45,46].

One of the simplest methods for producing bioethanol is the direct fermentation of biomass containing sugars into ethanol [47,48]. Sindhu *et al.* [49] reported the generation of bioethanol from bamboo (Dendrocalamus sp.) after dilute acid pre-treatment and enzymatic saccharification. In another study, Trivedi *et al.* [50] reported the generation of bioethanol from green seaweed Ulva fasciata. Kwon *et al.* [51] used spent coffee grounds as feedstock for the co-generation of bioethanol and biodiesel. In another recent study, Enquist-Newman *et al.* [52] proposed the potential approach for producing bioethanol from brown macroalgae sugars. Kumar *et al.* [53] reported the utilization of Gracilaria verrucosa as a feedstock for the generation of bioethanol. Borines *et al.* [54] used the macroalgae Sargassum *spp* as a potential renewable feedstock for the purpose of generating bioethanol. Shukla *et al.* [55] also developed the process for producing bioethanol by using waste algal biomass of Gracilaria verrucosa as a feedstock. Raghavi *et al.* [56] developed the sequential approach for generating bioethanol by utilizing sugarcane trash. In another study, Klein *et al.* [57] also demonstrated the microwave irradiation facilitated process for generating bioethanol by using the leaves of Ficus religiosa as feedstock.

1.3.3 Butanol

Butanol is 4-carbon alcohol ($C_4H_{10}O$) based biofuel [58]. It can be effectively blended with gasoline and other hydrocarbons and exhibits higher amount of heat energy as compared to ethanol, equivalent to 25% enhanced harvestable energy (BTU). Butanol has 110,000 BTUs per gallon which is almost equal to the BTUs of gasoline (115,000) [59]. Butanol is less corrosive than bioethanol, thereby, making it more likely to be transported or distributed by using existing pipelines and fuel stations. Butanol is less likely to evaporate as compared to ethanol and gasoline which makes it more secure. Acceptability of butanol-gasoline blend in conventional engines and generation of less volatile organic compound emissions are some of the benefits associated with the usage of butanol-gasoline blends [60]. The burning of butanol is cleaner than ethanol as it contains 22% oxygen [61]. Butanol is commonly observed to exist in the form of four isomers, i.e., n-butanol, 2-butanol, t-butanol and i-butanol referred to as normal-butanol, secondary-butanol, ter-butanol and iso-butanol, respectively. Same energy is exhibited by all isomers, however, different manufacturing techniques are used for these isomers [58]. Jiang *et al.* [62] studied the production of n-butanol from sugarcane juice by utilizing Clostridium acetobutylicum JB200, which is a mutant with high butanol tolerance and potential to produce high-titer n-butanol from glucose. Moradi *et al.* [63] applied alkali and acid pre-treatments for the improved generation of butanol from rice straw. Herein, Clostridium acetobutylicum bacterium was utilized to hydrolyze and ferment rice straw into acetone, butanol and ethanol. In another study, Du *et al.* [64] reported the generation of butanol from acid hydrolyzed corn fiber with Clostridium beijerinckii. Xin *et al.* [65] also developed the process of fermentation for generating butyl-butyrate and butanol by utilizing Clostridium sp. strain. Kong *et al.* [66] used gamma-valerolactone for hydrolyzing the sugarcane bagasse to generate bio-butanol. In another study, Tan *et al.* [67] reported the generation of butanol, ethanol and acetone from barley based dried distillers grains with solubles (DDGS).

1.4 Processes Involved for the Generation of Biodiesel

In order to generate biodiesel, transesterification reaction takes place between fats or oils and alcohol (ethanol or methanol). Generally, two methods are used for the transesterification process, i.e., catalytic transesterification and non-catalytic transesterification. The presence of catalyst is significant as alcohol has very low solubility in oil or fat. Thus, the main function of the catalyst is to improve alcohol solubility, thereby, enhancing the rate of reaction. Catalytic transesterification method involves base catalytic transesterification and acid catalytic transesterification techniques. Selecting the proper catalyst is a crucial factor for lowering the cost of biodiesel generation [68]. Quantity of free fatty acid (FFA) present in the oil is another factor which decides the selection of catalyst type employed for the reaction. In the case of low amount of FFAs present in the oil, base catalyzed esterification is suitable, whereas oils with high amount of FFAs must be processed by acid catalytic transesterification method [69]. Some examples of base catalysts are sodium hydroxide and potassium hydroxide, whereas acid catalysts include sulphuric acid and phosphoric acid. Supercritical alcohol transesterification and BIOX co-solvent transesterification are the methods associated with non-catalytic transesterification process.

Conventionally, homogeneous acid or base catalysts are employed for the process of transesterification to produce biodiesel. Heterogeneous catalytic transesterification is attaining much attention recently as the heterogeneous catalyst does not lead to the formation of soap [70]. Temperature also strongly influences the rate of transesterification reaction. As temperature increases, the rate of reaction also increases and the time of reaction decreases. Solid base and acid catalysts are utilized in heterogeneous catalytic transesterification process, where solid acid catalysts are considered to be less active as compared to solid base catalysts. Hydrotalcites, zeolites and alkaline earth metal oxides are some examples of solid base catalysts, whereas solid acid catalysts include tungstated zirconia, sulfated zirconia and Nafion-NR50. Solid base catalysts ensure reasonable cost for producing biodiesel and utilization of these catalysts in the form of fixed bed reactor set-up leads to the catalyst separation from the product, which have been transesterified [71]. In spite of exhibiting lower activity, solid acid catalysts are employed in various processes associated with industry due to the following benefits: a) no influence of the content of FFA, b) simultaneous esterification and transesterification [72], c) removal of purification stage while generating biodiesel [73], d) effective separation of catalyst from the end products and e) corrosion control [74]. Various catalytic and non-catalytic transesterification processes employed for the generation of biodiesel are explained below.

1.4.1 Alkali Catalytic Transesterification

The base catalytic transesterification process is usually quick, however, it can be easily influenced in contact with the free fatty acids (FFAs) present in oil or fat and water. Reactions of base catalysts with FFAs lead to the formation of soap and water. Formation of soap lowers the generation of alkyl esters and creates complications for separating glycerol and biodiesel as well as water washing as emulsion is formed. It has been noticed that methoxide catalysts exhibit high productivity as compared to hydroxide catalysts. Potassium based catalysts are also observed to be better than sodium based catalysts in terms of producing biodiesel [75].

1.4.2 Acid Catalytic Transesterification

The systems based on acid catalysts have a slower rate of reaction. In acid catalyzed reactions, free fatty acids are converted to esters. Soaps can also be transformed to esters as a pre-treatment step for feedstocks containing high content of free fatty acids [76]. Excess quantity of alcohol is required by the acid catalytic transesterification process, therefore, bigger distillation column and reactor are required as compared to alkali catalyzed process for generating same quantity of biodiesel. The acid catalyzed transesterification provides high conversion efficiency through the increased molar ratio of alcohol to oil, acid catalyst concentration, time of reaction and reaction temperature [77].

1.4.3 Biocatalytic Transesterification

Lipases are the most common examples of naturally available biocatalysts which have the potential for performing the transesterification reactions required for generating biodiesel. These lipases can be extracted from the various species of bacteria, i.e., Rhizomucor miehei, Candida Antarctica, Candida rugosa, etc. [78]. Enzymatic biocatalysts can be classified into two major categories: a) extracellular lipases and b) intracellular lipases. In case of extracellular lipases, enzymes are extracted from the microorganism broth. Mucor miehei, P. cepacia and R. oryzae are the common microorganisms utilized for extracting enzymes. Intracellular lipases are observed to remain inside the cell-producing walls, where enzyme is immobilized. Various methods are involved for the immobilization of lipase such as encapsulation, covalent bonding, entrapment, cross-linking and adsorption. The purpose of employing these methods is to enhance the stability of lipase for generating biodiesel. Among these methods, adsorption is widely utilized for the immobilization process [79]. Despite the high specificity and purity of enzymatic reactions, these reactions are very slow requiring a reaction time varying from 4 to 40 h at temperatures ranging between 35 and 45 °C.

Overall, biocatalytic transesterification process is promising for the generation of biodiesel, however, overcoming its high operational cost is necessary to make it viable for large-scale commercial production.

1.4.4 Non-catalytic Techniques

The catalytic processes for the generation of biodiesel involve various steps associated with ester purification and separation of unreacted reactants and catalysts. As an alternative, biodiesel generation from triglycerides by employing non-catalytic reactions also represents a simple approach. Supercritical alcohol and BIOX processes are two main non-catalytic transesterification methods for generating biodiesel.

Supercritical Alcohol Transesterification Process

Supercritical alcohol process is one of the non-catalytic methods employed for generating biodiesel in which temperature and high pressure are the main factors employed for carrying out the transesterification reaction in place of catalysts [80]. The triglycerides transesterification by employing supercritical ethanol and other alcohols (like methanol, butanol and propanol) has been observed to demonstrate effective output. The reaction is observed to be fast and the conversion efficiency increases from 50% to 95% for the initial 10 minutes with the temperature ranging between 250 and 400 °C. Supercritical methanol transesterification process has been designed in such a way that it can overcome the lag time of initial reaction which occurs due to the lower solubility of alcohol in triglyceride. Separate phases do not exist when the solvent is exposed to temperature and pressure exceeding the critical point.

Supercritical alcohol process is observed to be beneficial as there is no presence of catalyst, which makes the recovery of glycerol and purification of biodiesel relatively easier and environment-friendly [81].

BIOX Co-solvent Transesterification Process

BIOX co-solvent process has gained significant attention for generating biodiesel. In this process, both triglycerides and free fatty acids are transformed in a two-step, single phase and continuous process at atmospheric pressures and near ambient temperatures [43]. The poor solubility of alcohol in triglyceride causes slow rate of reaction which can be overcome by employing co-solvent process. Biox co-solvent process exhibit benefits as it utilizes inert and retrievable co-solvents in a single pass and fast reaction at ambient temperature and pressure. Similar to the supercritical alcohol process, the catalyst does not emerge in biodiesel as well as glycerol [82]. Almost all oils or fats such as crude vegetable oils, animal fats and waste cooking oils can be processed by using this technique.

1.5 Different Feedstocks and Techniques for Producing Biofuels

Zheng *et al.* [83] reported the preparation of biodiesel from unusable frying oil. In this study, the authors investigated the reaction kinetics of the acid catalyzed transesterification of oil in the presence of methanol for generating fatty acid methyl esters. It was observed that reaction kinetics followed pseudo first-order model. In another study, Demirbas and Demirbas [84] reported the significance of algae oil for potential utilization as a source for producing biodiesel. Algae are observed to grow faster than other plants like terrestrial crops and almost half of the weight of algae is oil. In addition, algae is easily available almost everywhere, for example in waste water, salt water, around the rivers or lakes, sewage, etc. Thus, no agricultural land and crops are required in order to obtain algae for biodiesel generation. Moreover, algae requires less energy for processing as compared to the energy provided by it. Overall, algae based biofuels are envisaged to be a better alternative to petroleum-derived fuels in the near future.

Leung and Guo [85] carried out the alkaline catalyzed transesterification of canola oil and waste frying oil, along with studying the properties of three catalysts (sodium methoxide, potassium hydroxide and sodium hydroxide). Content of ester, acid value and kinematic viscosity were the main factors, which controlled the properties of the biofuel generated using three catalysts. It was observed that sodium hydroxide was most beneficial among these catalysts due to its lower price and intermediate catalytic response. Another study was reported by Rashid and Anwar [86] for producing biodiesel by alkaline catalyzed transesterification of rapeseed oil. ASTM (American Society for Testing and Materials) and European EN standards confirmed that the as-prepared biodiesel from rapeseed oil exhibited the prescribed fuel properties.

Dube *et al.* [87] employed the two-phase membrane reactor for the production of biodiesel or fatty acid methyl esters from methanol and canola oil. The main aim of using the two-phase membrane reactor was to eliminate the unreacted canola oil from fatty acid methyl esters (FAME), thereby, producing the biodiesel of pure quality and shifting the reaction equilibrium to the product side. Berchmans and Hirata [88] developed the method for producing biodiesel from crude Jatropha curcas seed oil containing excessive percentage of free fatty acids. In this study, two stage transesterification was applied for improving the production of methyl ester. The first stage was associated with the usage of acid, which decreased the level of free fatty acids of Jatropha curcas seed oil to less than 1%. The second stage was transesterification with alkaline catalyst which provided 90% yield of methyl esters of fatty acids. In another study, Kouzu *et al.* [89] conducted the transesterification of soybean oil with methanol in the presence of solid base catalyst for the purpose of producing biodiesel. Selected solid base catalysts were calcium oxide, calcium hydroxide and calcium carbonate. The yield of fatty acid methyl ester was recorded as 93%, 12% and 0% respectively after 1 hour reaction time.

Liu *et al.* [90] conducted the transesterification of soybean oil with methanol in the presence of solid base catalyst (calcium methoxide) for the production of biodiesel. Miao and Wu [91] also introduced a new method integrating transesterification process and bio-engineering together for obtaining biodiesel from microalagal oil. Herein, high content of lipid was extracted from the heterotrophic growth of Chlorella protothecoides and n-hexane was used to extract high quantity of microalgal oil from C. protothecoides. It was noticed that the combined method was a successful approach for developing biodiesel from microalgal oil. Wei *et al.* [92] demonstrated the production of biofuels by introducing an approach for developing effective and proper utilization of cellulosic carbons. This strategy was implemented by combining the fermentation route of xylose and pentose as well as acetic acid reduction route into a Saccharomyces cerevisiae strain. With the combination of synthetic biological strategies and different heterologous pathways into a single microbial strain, potential effects were generated for supplementing the production of biofuels. The metabolic routes of simultaneous consumption of cellobiose, xylose and acetic acid are demonstrated in Figure 1.1.

Sarin *et al.* [93] studied the physico-chemical characteristics of Jatropha-Palm biodiesel blends. In this study, the authors discussed the requirement of blending of these two biodiesels as feedstock required for producing biodiesel must exhibit proper integration of saturated and unsaturated fatty compounds for obtaining substantial oxidation stability and low temperature properties. Biodiesel extracted from Jatropha exhibited inadequate oxidation stability and effective low temperature attributes. On the other hand, substantial oxidative stability and poor low temperature properties were exhibited by biodiesel extracted from Palm. Hence, Jatropha and Palm biodiesels were blended for obtaining biodiesel which exhibited substantial oxidation stability and low temperature properties. Additionally, the authors discussed the issues associated with the stability of

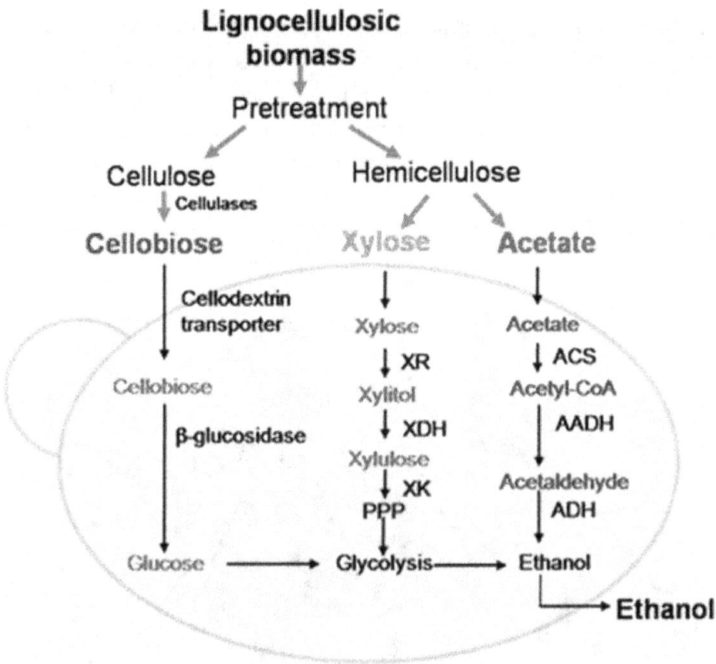

Figure 1.1 Schematic diagram for producing biofuel through simultaneous consumption of cellobiose, xylose, and acetic acid from lignocellulosic biomass by yeast. Reproduced from Reference 92 with permission from American Chemical Society.

biodiesel as it needs antioxidant for the storage purposes. According to EN 14112 specification, it is essential to add 200 mg/L of antioxidant into biodiesel for maintaining its stability, which is comparatively higher than the antioxidant concentration required in petroleum-extracted diesel. For reducing the dosage of antioxidant, Palm biodiesel was blended with Jatropha biodiesel in an appropriate proportion as Palm biodiesel does not require the presence of antioxidant for maintaining its own stability. It was noticed that dosage of antioxidant was decreased by 80-90% after Palm biodiesel was mixed with Jatropha biodiesel at a concentration of 20-40%. Also, the disadvantages associated with Palm biodiesel, such as inappropriate cloud point and pour point, were neutralized by Jatropha biodiesel. Thus, optimized integration of two biodiesels could lead to substantial oxidation stability and low temperature property. In another study, Pal *et al.* [94] synthesized a series of Mn-doped mesoporous ceria-silica composites and utilized these as heterogeneous and reusable catalysts for liquid phase transesterification of different esters (containing both long chain and short chain alcohols) in the presence of different alcohols like n-octanol, n-octanol, etc., for the generation of biodiesel under mild conditions. Transmission electron microscopy (TEM) revealed the hexagonal arrangement of pores in the composites. Scanning electron microscopy (SEM) also revealed that the particles sizes of these composites were in the range 300-400 nm (Figure 1.2).

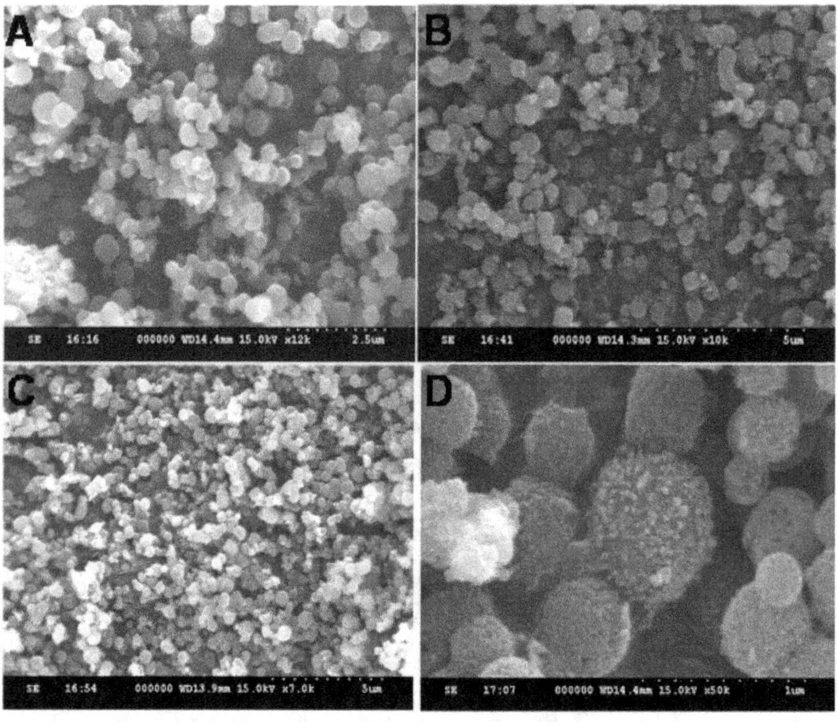

Figure 1.2 Scanning electron microscopy images of the Mn-doped ceria-silica composites. Reproduced from Reference 94 with permission from American Chemical Society.

Sahoo and Das [95] reported different methods of producing biodiesel from Polanga, Jatropha and Karanja oils. The authors employed one-step base catalyzed transesterification process for Karanja and Jatropha oils for the production of biodiesel, while a three-step acid base process was adopted for Polanga oil. In another study, Neumann *et al.* [96] introduced fully off-grid approach for producing cellulosic bioethanol which involves the integration of solar steam processing and solar distillation. Basically, three steps are followed for the development of cellulosic bioethanol: a) sugars are extracted from cellulosic feedstock, b) ethanol is obtained from the fermentation of sugars and c) distillation process is used for ethanol purification. In comparison, the conventional techniques involving extraction and distillation stages for producing bioethanol are associated with consumption of high energy, thereby, enhancing the cost. In this study, nanoparticle-activated solar steam was generated by the illumination of light-absorbing nanoparticles (dispersed in water) with sunlight. With the direct exposure of solar irradiation, ethanol could be obtained from cellulosic feedstock. The authors constructed the system based on nanoparticle-enabled solar steam generation where it was possible to deliver steam into a container of 7.5 L at the temperature ranging from 150-180°C for degrading the cellulosic chains of feedstock into monomeric sugars.

Schematic demonstration as well as digital images of the three-step procedure for generating solar biofuel are demonstrated in Figure 1.3.

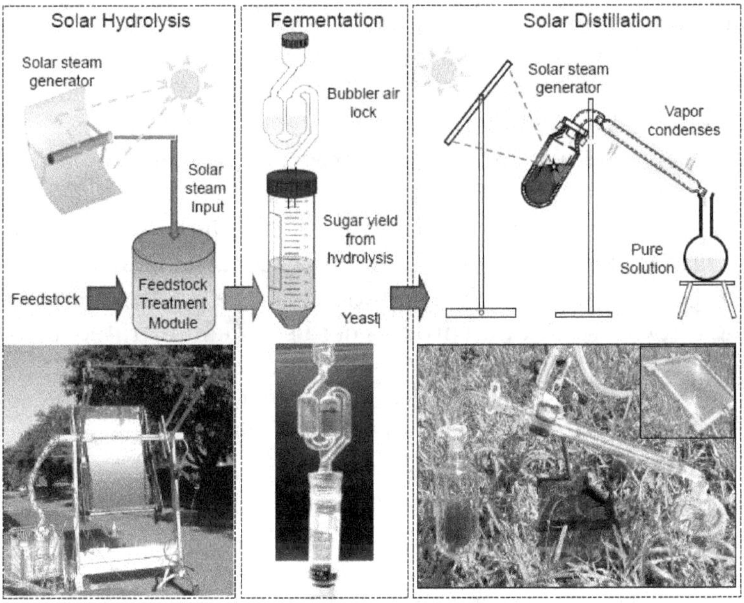

Figure 1.3 Schematic demonstration and digital images of three-step procedure for generating solar biofuel. Reproduced from Reference 96 with permission from American Chemical Society.

Mu *et al.* [97] investigated the integrated bioprocess coupling preparation of biodiesel by lipase with microbial production of 1,3-propanediol using a hollow fiber membrane. During the transesterification process, glycerol was released as a byproduct which could flow through the membrane and transformed into 1,3-propanediol by Klebsiella pneumonia. After the reaction time of 20 h, 84% of biodiesel was generated. It was also observed that productivity of 1,3-propanediol was 1.7 $g.L^{-1}.h^{-1}$ and the value of molar yield of 1,3-propanediol was observed as 0.47 $mol.mol^{-1}$. Glycerol inhibition was avoided by the integrated bioprocess, thereby, decreasing the cost of production and enhancing the yield of biodiesel and 1,3-propanediol. In another study, Yuan *et al.* [98] reported the development of biodiesel from waste rapeseed oil (feedstock). Response surface methodology was employed for optimizing the conditions in order to achieve the highest transformation to biodiesel. Gas chromatography/mass spectroscopy was used for analyzing the generated biodiesel and the as-prepared biodiesel was observed to contain six fatty acid methyl esters. Dorado *et al.* [99] also conducted the studies for optimizing the alkali catalyzed transesterification process of Brassica carinata oil. In order to obtain an economical procedure for generating biodiesel, the reaction was carried out with 30 min stirring of a mixture of oil/methanol (in a ratio of 1:4.6) and potassium hydroxide in the temperature range 20-45 °C. These conditions were observed to be suitable for the low-cost production of biodiesel. Gas chromatography analysis was carried out for determining the composition of fatty acid of B. carinata oil and its esters. The results disclosed that the content of free fatty acid increased due to the presence of erucic acid which inhibited the transformation of B. carinata oil into its methyl ester. Schematic representation of the biofuel generation through transesterification process has been demonstrated in Figure 1.4.

Johnson and Wen [100] reported the generation of biodiesel by applying the direct transesterification of algal biomass. Microalga Schizochytrium limacinum was utilized for the production of biodiesel as it generates high levels of biomass and total fatty acid. Biodiesel generated from direct transesterification process fulfilled the ASTM specifications, thereby, representing a viable biodiesel product. In another study, Alexander *et al.* [101] reported the generation of biofuels using fish oil. For this purpose, the authors carried out the partial hydrogenation of fish oil extracted fatty acid methyl esters (FAME) using a batch reactor at room temperature for < 15 minutes.

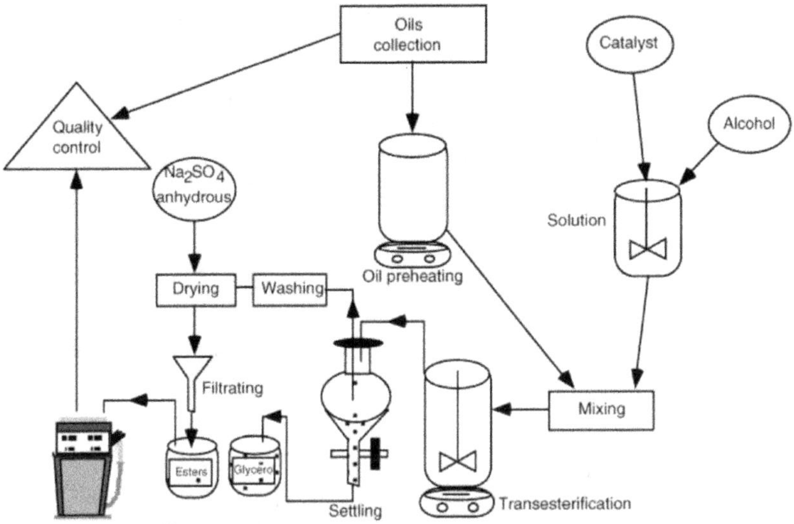

Figure 1.4 Schematic representation of generating biofuel through transesterification process. Reproduced from Reference 99 with permission from American Chemical Society.

Herein, selective hydrogenation of polyunsaturated fatty acids was carried out, whereas the content of saturated fatty acids remained at a constant level as compared to the feedstock. The as-prepared biofuels and their blends with petroleum-extracted fuel were examined for fuel characteristics as well as for EN 14214 and EN 590 specifications. The results suggested that the obtained biofuels were suitable for generating blends with petroleum-derived fuel and the blends fulfilled EN 590 specification. In a recent study, Phelan *et al.* [102] introduced terpene biosynthesis in Streptomyces cerevisiae for the generation of modified biofuel precursor bisabolene (Figure 1.5). Morshed *et al.* [103] also reported rubber seed oil as an effective feedstock for producing biodiesel. Techniques like cold percolation and mechanical press with and without solvent were employed for the extraction of oil from the rubber seeds. Biodiesel was produced by using the method involving three steps, i.e., oil saponification, soap acidification and esterification of free fatty acid. FAME was generated in the esterification step. The maximum productivity of free fatty acid from rubber seed oil was observed as 86%. Nuclear magnetic resonance (NMR) spectroscopy was also used to analyze the rubber seed oil and biodiesel samples, which confirmed the transformation of rubber seed oil into biodiesel.

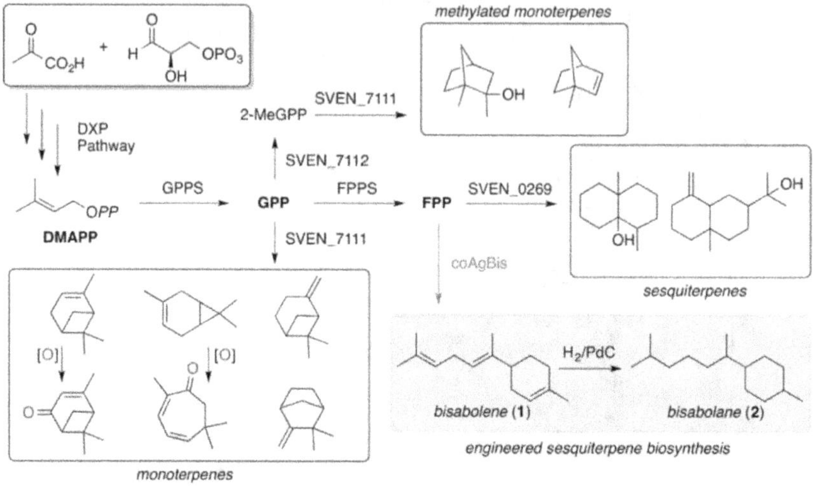

Figure 1.5 Biosynthetic routes to predominant isoprenoids produced by endogenous (black arrows) and heterologous (red arrow) terpene synthases in S. venezuelae. Reproduced from Reference 102 with permission from American Chemical Society.

In another study, Yang *et al.* [104] also reported the production of biodiesel utilizing rubber seed oil as feedstock. The authors developed poly(sodium acrylate) and sodium hydroxide (NaOH-NaPAA) integrated system acting as a water resistant catalyst, which further reduced the unfavorable effects of water content on the base catalyzed transesterification process for producing biodiesel. The results revealed that the modified catalyst system exhibited significant water resistance during transesterification process and maintained high catalytic activity. Ghadge and Raheman [105] reported optimized design for producing biodiesel from mahua oil. In the study, central composite rotatable pattern was utilized for examining the effects of quantity of methanol and acid as well as durability of reaction for reducing free fatty acids of mahua oil during its pre-treatment for the generation of biodiesel. Ge *et al.* [106] also reported an approach to recover microalgae from water for the generation of biofuel. In this study, modification of CO_2-switchable crystalline nanocellulose (CNC) was achieved with 1-(3-aminopropyl)-imidazole (APIm) so that it behaves as a reversible coagulant for the purpose of microalgae harvesting. Positively charged APIm-modified CNC exhibited good dispersion in the carbonated water as compared to unmodified CNC. On CO_2-treatment, the modified CNC was observed to exhibit substantial electrostatic interactions with negatively charged Chlorella vulgaris. Multiple mechanisms were integrated in order to harvest microalgae such as electrostatic interaction during CO_2 sparging stage, enmeshment during air sparging triggered interaction of APIm modified CNC with Chlorella vulgaris, followed by further coagulation, flocculation and sedimentation stages (Figure 1.6). This approach was significant as it involved biocompatible, readily available and natural components like air, CO_2 and cellulose for generating biofuel.

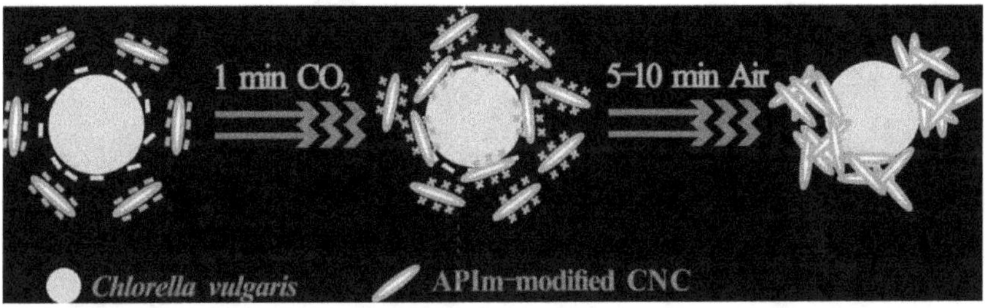

Figure 1.6 Schematic representation of the electrostatic attraction and enmeshment mechanisms involved in interactions between C. vulgaris and APIm-modified CNC. Reproduced from Reference 106 with permission from American Chemical Society.

Xie *et al.* [107] developed Guerbet-type process catalyzed using complexes of ruthenium pincer for generating biofuel from ethanol (Figure 1.7). In another study, Hums *et al.* [108] also performed the life cycle assessment of biodiesel obtained from grease trap waste. Basically, life cycle assessment was utilized to measure the quantity

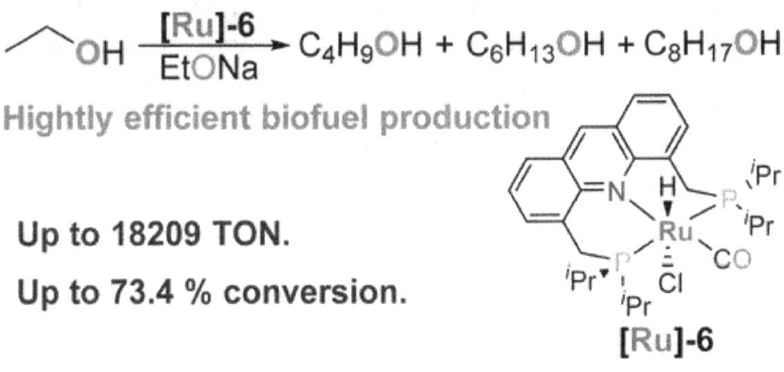

Figure 1.7 Representation of the Guerbet-type process. Reproduced from Reference 107 with permission from American Chemical Society.

of greenhouse gas emissions, necessity of fossil energy and air pollutant emissions for the process of grease trap waste extracted biodiesel. Herein, Monte Carlo simulations were performed for analyzing the sensitivity to the concentration of lipid present in grease trap waste. The assessment displayed complete life cycle of biofuel from the point of arrangement of feedstock (grease trap waste) to the combustion of biofuel (Figure 1.8).

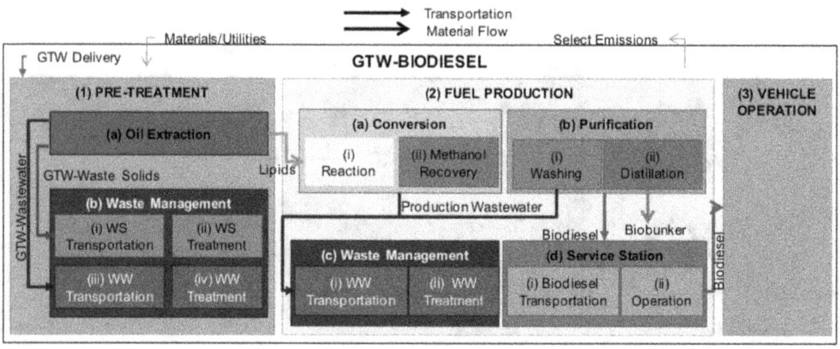

Figure 1.8 System boundary for the grease trap waste (GTW)-biodiesel process. Each of the three main stages include the material and energy inputs and emission outputs for (1) pretreatment, (2) fuel production and (3) vehicle operation. Reproduced from Reference 108 with permission from American Chemical Society.

1.6 Conclusions

Biofuels are beneficial alternatives to the petroleum-derived fuels. These fuels exhibit enhanced sustainability and reduction in the level of greenhouse gas emissions. In addition, the existing engines or vehicles do not require any special modification for utilizing biofuels. Overcoming the challenges associated with productivity and cost in the near future will ensure high demand of bio-derived fuels owing to their beneficial features.

References

1. Bringe, N. A., (2005) Soybean oil composition for biodiesel. In: *The Biodiesel Handbook*, Knothe, G. and Van Gerpen (eds.), AOCS Publishing, USA.

2. Pahl, G. (2008) *Biodiesel: Growing a New Energy Economy*, 2nd edition, Chelsea Green Publishing, USA.
3. Van Gerpen, J. (2005) Biodiesel processing and production. *Fuel Processing Technology*, **86**, 1097-1107.
4. Ma, F., Clements, L. D., and Hanna, M. A. (1998) Biodiesel fuel from animal fat. Ancillary studies on transesterification of beef tallow. *Industrial & Engineering Chemistry Research*, **37**, 3768-3771.
5. Haas, M. J., McAloon, A. J., Yee, W. C., and Foglia, T. A. (2006) A process model to estimate biodiesel production costs. *Bioresource Technology*, **97**, 671-678.
6. Bender, M. (1999) Economic feasibility review for community-scale farmer cooperatives for biodiesel. *Bioresource Technology*, **70**, 81-87.
7. Korbitz, W. (1999) Biodiesel production in Europe and North America, an encouraging prospect. *Renewable Energy*, **16**, 1078-1083.
8. Bozbas, K. (2008) Biodiesel as an alternative motor fuel: production and policies in the European Union. *Renewable and Sustainable Energy Reviews*, **12**, 542-552.
9. Dorado, M., Cruz, F., Palomar, J., and Lopez, F. (2006) An approach to the economics of two vegetable oil-based biofuels in Spain. *Renewable Energy*, **31**, 1231-1237.
10. Tyson, K. S., and McCormick, R. (2001) Biodiesel Handling and Use Guide. *National Renewable Energy Laboratory*, USA. Online: http://biodiesel.org/docs/using-hotline/nrel-handling-and-use.pdf?sfvrsn=4 (assessed 12th June 2018).
11. Hoekman, S. K. (2009) Biofuels in the US - challenges and opportunities. *Renewable Energy*, **34**, 14-22.
12. Larson, E. D. (2008) Biofuel Production Technologies: Status, Prospects and Implications for Trade and Development. *United Nations Conference on Trade and Development*, United Nations. Online: http://unctad.org/en/Docs/ditcted200710_en.pdf (assessed 11th June 2018).
13. Patil, V., Tran, K.-Q., and Giselrod, H. R. (2008) Towards sustainable production of biofuels from microalgae. *International Journal of Molecular Sciences*, **9**, 1188-1195.
14. Gibbons, W. R., and Westby, C. A. (1989) Cofermentation of sweet sorghum juice and grain for production of fuel ethanol and distillers' wet grain. *Biomass*, **18**, 43-57.
15. Suresh, K., and Rao, L. V. (1999) Utilization of damaged sorghum and rice grains for ethanol production by simultaneous saccharification and fermentation. *Bioresource Technology*, **68**, 301-304.
16. Turhollow, A. F., and Heady, E. O. (1986) Large-scale ethanol production from corn and grain sorghum and improving conversion technology. *Energy in Agriculture*, **5**, 309-316.
17. Zhao, R., Bean, S., Wang, D., Park, S., Schober, T., and Wilson, J.

(2009) Small-scale mashing procedure for predicting ethanol yield of sorghum grain. *Journal of Cereal Science*, **49**, 230-238.

18. Love, G., Gough, S., Brady, D., Barron, N., Nigam, P., Singh, D., Marchant, R., and McHale, A. (1998) Continuous ethanol fermentation at 45 °C using Kluyveromyces marxianus IMB3 immobilized in Calcium alginate and kissiris. *Bioprocess Engineering*, **18**, 187-189.
19. Nigam, P., Banat, I., Singh, D., McHale, A., and Marchant, R. (1997) Continuous ethanol production by thermotolerant Kluyveromyces marxianus IMB3 immobilized on mineral Kissiris at 45 C. *World Journal of Microbiology and Biotechnology*, **13**, 283-288.
20. Brady, D., Nigam, P., Marchant, R., and McHale, A. (1997) Ethanol production at 45 C by alginate-immobilized Kluyveromyces marxianus IMB3 during growth on lactose-containing media. *Bioprocess Engineering*, **16**, 101-104.
21. Brady, D., Nigam, P., Marchant, R., Singh, D., and McHale, A. (1997) The effect of Mn^{2+} on ethanol production from lactose using Kluyveromyces marxianus IMB3 immobilized in magnetically responsive matrices. *Bioprocess and Biosystems Engineering*, **17**, 31-34.
22. Riordan, C., Love, G., Barron, N., Nigam, P., Marchant, R., McHale, L., and McHale, A. (1996) Production of ethanol from sucrose at 45 °C by alginate-immoblized preparations of the thermotolerant yeast strain Kluyveromyces marxianus IMB3. *Bioresource Technology*, **55**, 171-173.
23. Love, G., Nigam, P., Barron, N., Singh, D., Marchant, R., and McHale, A. (1996) Ethanol production at 45 °C using preparations of Kluyveromyces marxianus IMB3 immobilized in calcium alginate and kissiris. *Bioprocess and Biosystems Engineering*, **15**, 275-277.
24. Banat, I. M., Nigam, P., and Marchant, R. (1992) Isolation of thermotolerant, fermentative yeasts growing at 52 C and producing ethanol at 45 °C and 50 °C. *World Journal of Microbiology and Biotechnology*, **8**, 259-263.
25. Thegarid, N., Fogassy, G., Schuurman, Y., Mirodatos, C., Stefanidis, S., Iliopoulou, E., Kalogiannis, K., and Lappas, A. (2014) Second-generation biofuels by co-processing catalytic pyrolysis oil in FCC units. *Applied Catalysis B: Environmental*, **145**, 161-166.
26. Schenk, P. M., Thomas-Hall, S. R., Stephens, E., Marx, U. C., Mussgnug, J. H., Posten, C., Kruse, O., and Hankamer, B. (2008) Second generation biofuels: high-efficiency microalgae for biodiesel production. *Bioenergy Research*, **1**, 20-43.
27. Dias, M. O., Junqueira, T. L., Cavalett, O., Cunha, M. P., Jesus, C. D., Rossell, C. E., Maciel Filho, R., and Bonomi, A. (2012) Integrated versus stand-alone second generation ethanol production from sugarcane bagasse and trash. *Bioresource Technology*, **103**, 152-161.
28. Brennan, L., and Owende, P. (2010) Biofuels from microalgae - a review of technologies for production, processing, and extractions of

biofuels and co-products. *Renewable and Sustainable Energy Reviews*, **14**, 557-577.
29. Carere, C. R., Sparling, R., Cicek, N., and Levin, D. B. (2008) Third generation biofuels via direct cellulose fermentation. *International Journal of Molecular Sciences*, **9**, 1342-1360.
30. Saidane-Bchir, F., El Falleh, A., Ghabbarou, E., and Hamdi, M. (2016) 3rd generation bioethanol production from microalgae isolated from slaughterhouse wastewater. *Waste and Biomass Valorization*, **7**, 1041-1046.
31. Zhu, L., Zong, M., and Wu, H. (2008) Efficient lipid production with Trichosporonfermentans and its use for biodiesel preparation. *Bioresource Technology*, **99**, 7881-7885.
32. Samori, C., Torri, C., Samorì, G., Fabbri, D., Galletti, P., Guerrini, F., Pistocchi, R., and Tagliavini, E. (2010) Extraction of hydrocarbons from microalga Botryococcus braunii with switchable solvents. *Bioresource Technology* **101**, 3274-3279.
33. El-Shimi, H., Attia, N. K., El-Sheltawy, S., and El-Diwani, G. (2013) Biodiesel production from Spirulina-platensis microalgae by in-situ transesterification process. *Journal of Sustainable Bioenergy Systems*, **3**, 224-233.
34. Chng, L. M., Chan, D. J., and Lee, K. T. (2016) Sustainable production of bioethanol using lipid-extracted biomass from Scenedesmus dimorphus. *Journal of Cleaner Production*, **130**, 68-73.
35. Abreu, A. P., Fernandes, B., Vicente, A. A., Teixeira, J., and Dragone, G. (2012) Mixotrophic cultivation of Chlorella vulgaris using industrial dairy waste as organic carbon source. *Bioresource Technology*, **118**, 61-66.
36. Srinophakun, P., Thanapimmetha, A., Rattanaphanyapan, K., Sahaya, T., and Saisriyoot, M. (2017) Feedstock production for third generation biofuels through cultivation of Arthrobacter AK19 under stress conditions. *Journal of Cleaner Production*, **142**, 1259-1266.
37. Wang, Y., Wu, H., and Zong, M. (2008) Improvement of biodiesel production by lipozyme TL IM-catalyzed methanolysis using response surface methodology and acyl migration enhancer. *Bioresource Technology*, **99**, 7232-7237.
38. Vicente, G., Martınez, M., and Aracil, J. (2004) Integrated biodiesel production: a comparison of different homogeneous catalysts systems. *Bioresource Technology*, **92**, 297-305.
39. Antolın, G., Tinaut, F., Briceno, Y., Castano, V., Perez, C., and Ramırez, A. (2002) Optimisation of biodiesel production by sunflower oil transesterification. *Bioresource Technology*, **83**, 111-114.
40. Lang, X., Dalai, A. K., Bakhshi, N. N., Reaney, M. J., and Hertz, P. (2001) Preparation and characterization of bio-diesels from various biooils. *Bioresource Technology*, **80**, 53-62.
41. Pramanik, K. (2003) Properties and use of Jatropha curcas oil and

diesel fuel blends in compression ignition engine. *Renewable Energy*, **28**, 239-248.

42. Meher, L. C., Kulkarni, M. G., Dalai, A. K., and Naik, S. N. (2006) Transesterification of karanja (Pongamia pinnata) oil by solid basic catalysts. *European Journal of Lipid Science and Technology*, **108**, 389-397.
43. Abbaszaadeh, A., Ghobadian, B., Omidkhah, M. R., and Najafi, G. (2012) Current biodiesel production technologies: a comparative review. *Energy Conversion and Management*, **63**, 138-148.
44. Bengisu, G. (2014) Alternatif yakıt kaynagı olarak biyoetanol/Bioethanol as an alternative fuel source. *Alınteri Zirai Bilimler Dergisi*, **27**, 43-52.
45. The State of Food and Agriculture, Biofuels: Prospects, Risks and Opportunities (2008). *Food and Agriculture Organization of the United Nations*, Italy. Online: http://www.fao.org/3/a-i0100e.pdf (assessed 9th June 2018).
46. Sustainable Biofuels: Prospects and Challenges (2008). *The Royal Society*, UK. Online: https://royalsociety.org/~/media/royal_society_content/policy/publications/2008/7980.pdf (assessed 1st June 2018).
47. Verma, G., Nigam, P., Singh, D., and Chaudhary, K. (2000) Bioconversion of starch to ethanol in a single-step process by coculture of amylolytic yeasts and Saccharomyces cerevisiae 21. *Bioresource Technology*, **72**, 261-266.
48. Singh, D., Dahiya, J. S., and Nigam, P. (1995) Simultaneous raw starch hydrolysis and ethanol fermentation by glucoamylase from Rhizoctonia solani and Saccharomyces cerevisiae. *Journal of Basic Microbiology*, **35**, 117-121.
49. Sindhu, R., Kuttiraja, M., Binod, P., Sukumaran, R.K., and Pandey, A. (2014) Bioethanol production from dilute acid pretreated Indian bamboo variety (Dendrocalamus sp.) by separate hydrolysis and fermentation. *Industrial Crops and Products*, **52**, 169-176.
50. Trivedi, N., Gupta, V., Reddy, C., and Jha, B. (2013) Enzymatic hydrolysis and production of bioethanol from common macrophytic green alga Ulva fasciata Delile. *Bioresource Technology*, **150**, 106-112.
51. Kwon, E. E., Yi, H., and Jeon, Y. J. (2013) Sequential co-production of biodiesel and bioethanol with spent coffee grounds. *Bioresource Technology*, **136**, 475-480.
52. Enquist-Newman, M., Faust, A. M. E., Bravo, D. D., Santos, C. N. S., Raisner, R. M., Hanel, A., Sarvabhowman, P., Le, C., Regitsky, D. D., and Cooper, S. R. (2014) Efficient ethanol production from brown macroalgae sugars by a synthetic yeast platform. *Nature*, **505**, 239-243.
53. Kumar, S., Gupta, R., Kumar, G., Sahoo, D., and Kuhad, R. C. (2013) Bioethanol production from Gracilaria verrucosa, a red alga, in a bi-

orefinery approach. *Bioresource Technology*, **135**, 150-156.

54. Borines, M. G., de Leon, R. L., and Cuello, J. L. (2013) Bioethanol production from the macroalgae Sargassum spp. *Bioresource Technology*, **138**, 22-29.
55. Shukla, R., Kumar, M., Chakraborty, S., Gupta, R., Kumar, S., Sahoo, D., and Kuhad, R.C. (2016) Process development for the production of bioethanol from waste algal biomass of Gracilaria verrucosa. *Bioresource Technology*, **220**, 584-589.
56. Raghavi, S., Sindhu, R., Binod, P., Gnansounou, E., and Pandey, A. (2016) Development of a novel sequential pretreatment strategy for the production of bioethanol from sugarcane trash. *Bioresource Technology*, **199**, 202-210.
57. Klein, M., Griess, O., Pulidindi, I. N., Perkas, N., and Gedanken, A. (2016) Bioethanol production from Ficus religiosa leaves using microwave irradiation. *Journal of Environmental Management*, **177**, 20-25.
58. Ramey, D. (2004) Butanol. *The Light Party*, USA. Online: http://www.lightparty.com/Energy/Butanol.html (assessed 2nd June 2018).
59. Biobutanol, European Biofuels Technology Platform (2009). Online: http://biofuelstp.eu/newsletters/newsletter_archive.html# (assessed 1st July 2018).
60. Wu, M., Wang, M., Liu, J., and Huo, H. (2007) Life-cycle Assessment of Corn-based Butanol as a Potential Transportation Fuel. *Argonne National Laboratory*, USA. Online: https://anl.box.com/s/s11ha0koxgy4rme4ykgkxw5yrz1omofs (assessed 5th July 2018).
61. Qureshi, N., Saha, B. C., Dien, B., Hector, R. E., and Cotta, M. A. (2010) Production of butanol (a biofuel) from agricultural residues: Part I - Use of barley straw hydrolysate. *Biomass and Bioenergy*, **34**, 559-565.
62. Jiang, W., Zhao, J., Wang, Z., and Yang, S.-T. (2014) Stable high-titer n-butanol production from sucrose and sugarcane juice by Clostridium acetobutylicum JB200 in repeated batch fermentations. *Bioresource Technology*, **163**, 172-179.
63. Moradi, F., Amiri, H., Soleimanian-Zad, S., Ehsani, M. R., and Karimi, K. (2013) Improvement of acetone, butanol and ethanol production from rice straw by acid and alkaline pretreatments. *Fuel*, **112**, 8-13.
64. Du, T.-f., He, A.-y., Wu, H., Chen, J.-n., Kong, X.-p., Liu, J.-l., Jiang, M., and Ouyang, P.-k. (2013) Butanol production from acid hydrolyzed corn fiber with Clostridium beijerinckii mutant. *Bioresource Technology*, **135**, 254-261.
65. Xin, F., Basu, A., Yang, K.-L., and He, J. (2016) Strategies for production of butanol and butyl-butyrate through lipase-catalyzed esterification. *Bioresource Technology*, **202**, 214-219.
66. Kong, X., Xu, H., Wu, H., Wang, C., He, A., Ma, J., Ren, X., Jia, H., Wei, C.,

and Jiang, M. (2016) Biobutanol production from sugarcane bagasse hydrolysate generated with the assistance of gamma-valerolactone. *Process Biochemistry*, **51**, 1538-1543.
67. Houweling-Tan, B., Sperber, B. L., van der Wal, H., Bakker, R. R., and Lopez-Contreras, A. M. (2016) Barley dried distillers grains and solubles (DDGS) as feedstock for production of acetone, butanol and ethanol. *BAOJ Microbio*, **2**, 013.
68. Sharma, Y., Singh, B., and Upadhyay, S. (2008) Advancements in development and characterization of biodiesel: A review. *Fuel*, **87**, 2355-2373.
69. Schuchardt, U., Sercheli, R., and Vargas, R. M. (1998) Transesterification of vegetable oils: a review. *Journal of the Brazilian Chemical Society*, **9**, 199-210.
70. Wang, L., and Yang, J. (2007) Transesterification of soybean oil with nano-MgO or not in supercritical and subcritical methanol. *Fuel*, **86**, 328-333.
71. Kouzu, M., and Hidaka, J.-s. (2012) Transesterification of vegetable oil into biodiesel catalyzed by CaO: A review. *Fuel*, **93**, 1-12.
72. Dalai, A. K., Kulkarni, M. G., and Meher, L. C. (2006) Biodiesel Productions from Vegetable Oils Using Heterogeneous Catalysts and Their Applications as Lubricity Additives. 2006 IEEE EIC Climate Change Conference, Canada, pp. 1-8 (doi: 10.1109/EICCCC.2006.277228.
73. Jitputti, J., Kitiyanan, B., Rangsunvigit, P., Bunyakiat, K., Attanatho, L., and Jenvanitpanjakul, P. (2006) Transesterification of crude palm kernel oil and crude coconut oil by different solid catalysts. *Chemical Engineering Journal*, **116**, 61-66.
74. Patil, P. D., and Deng, S. (2009) Optimization of biodiesel production from edible and non-edible vegetable oils. *Fuel*, **88**, 1302-1306.
75. Singh, A., He, B., Thompson, J., and Van Gerpen, J. (2006) Process optimization of biodiesel production using alkaline catalysts. *Applied Engineering in Agriculture*, **22**, 597-600.
76. Karmakar, A., Karmakar, S., and Mukherjee, S. (2010) Properties of various plants and animals feedstocks for biodiesel production. *Bioresource technology*, **101**, 7201-7210.
77. Canakci, M., and Van Gerpen, J. (1999) Biodiesel production via acid catalysis. *Transactions of the ASAE-American Society of Agricultural Engineers*, **42**, 1203-1210.
78. Vasudevan, P. T., and Briggs, M. (2008) Biodiesel production - current state of the art and challenges. *Journal of Industrial Microbiology & Biotechnology*, **35**, 421.
79. Tan, T., Lu, J., Nie, K., Deng, L., and Wang, F. (2010) Biodiesel production with immobilized lipase: a review. *Biotechnology Advances*, **28**, 628-634.
80. Tan, K. T., Lee, K. T., and Mohamed, A. R. (2009) Production of FAME

by palm oil transesterification via supercritical methanol technology. *Biomass and Bioenergy*, **33**, 1096-1099.
81. Encinar, J.M., González, J.F., Sabio, E., and Ramiro, M.J. (1999) Preparation and properties of biodiesel from Cynara C ardunculus L. oil. *Industrial & Engineering Chemistry Research*, **38**, 2927-2931.
82. van Gerpen, J., Shanks, B., Pruszko, R., Clements, D., and Knothe, G. (2004) Biodiesel analytical methods. *National Renewable Energy Laboratory*, USA. Online: http://www.graham-laming.com/bd/testing biodiesel.pdf (assessed 10th July 2018).
83. Zheng, S., Kates, M., Dube, M., and McLean, D. (2006) Acid-catalyzed production of biodiesel from waste frying oil. *Biomass and Bioenergy*, **30**, 267-272.
84. Demirbas, A., and Demirbas, M. F. (2011) Importance of algae oil as a source of biodiesel. *Energy Conversion and Management*, **52**, 163-170.
85. Leung, D., and Guo, Y. (2006) Transesterification of neat and used frying oil: optimization for biodiesel production. *Fuel Processing Technology*, **87**, 883-890.
86. Rashid, U., and Anwar, F. (2008) Production of biodiesel through optimized alkaline-catalyzed transesterification of rapeseed oil. *Fuel*, **87**, 265-273.
87. Dube, M., Tremblay, A., and Liu, J. (2007) Biodiesel production using a membrane reactor. *Bioresource Technology*, **98**, 639-647.
88. Berchmans, H. J., and Hirata, S. (2008) Biodiesel production from crude Jatropha curcas L. seed oil with a high content of free fatty acids. *Bioresource Technology*, **99**, 1716-1721.
89. Kouzu, M., Kasuno, T., Tajika, M., Sugimoto, Y., Yamanaka, S., and Hidaka, J. (2008) Calcium oxide as a solid base catalyst for transesterification of soybean oil and its application to biodiesel production. *Fuel*, **87**, 2798-2806.
90. Liu, X., Piao, X., Wang, Y., Zhu, S., and He, H. (2008) Calcium methoxide as a solid base catalyst for the transesterification of soybean oil to biodiesel with methanol. *Fuel*, **87**, 1076-1082.
91. Miao, X., and Wu, Q. (2006) Biodiesel production from heterotrophic microalgal oil. *Bioresource Technology*, **97**, 841-846.
92. Wei, N., Oh, E. J., Million, G., Cate, J. H., and Jin, Y.-S. (2015) Simultaneous utilization of cellobiose, xylose, and acetic acid from lignocellulosic biomass for biofuel production by an engineered yeast platform. *ACS Synthetic Biology*, **4**, 707-713.
93. Sarin, R., Sharma, M., Sinharay, S., and Malhotra, R. K. (2007) Jatropha–palm biodiesel blends: an optimum mix for Asia. *Fuel*, **86**, 1365-1371.
94. Pal, N., Cho, E.-B., Kim, D., and Jaroniec, M. (2014) Mn-doped ordered mesoporous ceria-silica composites and their catalytic properties toward biofuel production. *The Journal of Physical Chemistry*

C, **118**, 15892-15901.

95. Sahoo, P., and Das, L. (2009) Process optimization for biodiesel production from Jatropha, Karanja and Polanga oils. *Fuel*, **88**, 1588-1594.
96. Neumann, O., Neumann, A. D., Tian, S., Thibodeaux, C., Shubhankar, S., Müller, J., Silva, E., Alabastri, A., Bishnoi, S. W., and Nordlander, P. (2016) Combining solar steam processing and solar distillation for fully off-grid production of cellulosic bioethanol. *ACS Energy Letters*, **2**, 8-13.
97. Mu, Y., Xiu, Z.-L., and Zhang, D.-J. (2008) A combined bioprocess of biodiesel production by lipase with microbial production of 1, 3-propanediol by Klebsiella pneumoniae. *Biochemical Engineering Journal*, **40**, 537-541.
98. Yuan, X., Liu, J., Zeng, G., Shi, J., Tong, J., and Huang, G. (2008) Optimization of conversion of waste rapeseed oil with high FFA to biodiesel using response surface methodology. *Renewable Energy*, **33**, 1678-1684.
99. Dorado, M. P., Ballesteros, E., López, F. J., and Mittelbach, M. (2004) Optimization of alkali-catalyzed transesterification of Brassica Carinata oil for biodiesel production. *Energy & Fuels*, **18**, 77-83.
100. Johnson, M. B., and Wen, Z. (2009) Production of biodiesel fuel from the microalga Schizochytrium limacinum by direct transesterification of algal biomass. *Energy & Fuels*, **23**, 5179-5183.
101. Studentschnig, A. F., Schober, S., and Mittelbach, M. (2015) Partial hydrogenation of fish oil methyl esters for the production of biofuels. *Energy & Fuels*, **29**, 3776-3779.
102. Phelan, R. M., Sekurova, O. N., Keasling, J. D., and Zotchev, S. B. (2014) Engineering terpene biosynthesis in Streptomyces for production of the advanced biofuel precursor bisabolene. *ACS Synthetic Biology*, **4**, 393-399.
103. Morshed, M., Ferdous, K., Khan, M. R., Mazumder, M., Islam, M., and Uddin, M. T. (2011) Rubber seed oil as a potential source for biodiesel production in Bangladesh. *Fuel*, **90**, 2981-2986.
104. Yang, R., Su, M., Zhang, J., Jin, F., Zha, C., Li, M., and Hao, X. (2011) Biodiesel production from rubber seed oil using poly (sodium acrylate) supporting NaOH as a water-resistant catalyst. *Bioresource Technology*, **102**, 2665-2671.
105. Ghadge, S. V., and Raheman, H. (2006) Process optimization for biodiesel production from mahua (Madhuca indica) oil using response surface methodology. *Bioresource Technology*, **97**, 379-384.
106. Ge, S., Champagne, P., Wang, H., Jessop, P. G., and Cunningham, M. F. (2016) Microalgae recovery from water for biofuel production using CO_2-switchable crystalline nanocellulose. *Environmental Science & Technology*, **50**, 7896-7903.
107. Xie, Y., Ben-David, Y., Shimon, L. J., and Milstein, D. (2016) Highly eff-

icient process for production of biofuel from ethanol catalyzed by ruthenium pincer complexes. *Journal of the American Chemical Society*, **138**, 9077-9080.

108. Hums, M. E., Cairncross, R. A., and Spatari, S. (2016) Life-cycle assessment of biodiesel produced from grease trap waste. *Environmental Science & Technology*, **50**, 2718-2726.

2

Nanocellulose as Reinforcement in Polymer Nanocomposites

2.1 Introduction

Cellulose is regarded as the most abundant organic compound generated from biomass [1]. This material has been utilized for around 150 years in numerous applications. Progressing knowledge about the reactivity and structural features of cellulose has led to the gradual development of novel cellulose-based materials. Independent of its source, cellulose has a white fiber-like structure with a specific gravity of approx. 1.5. The chemical and physical aspects of cellulose have been comprehensively studied since its discovery by Anselme Payen in 1838. It is a linear homopolysaccharide of β-1,4-linked-anhydro-D-glucose units [2,3] having a degree of polymerization (DP) of almost 20,000, but shorter cellulose chains are also observed.

The fundamental chemical framework of cellulose is displayed in Figure 2.1, where a dimer known as cellobiose shows up as a repeated fragment [2].

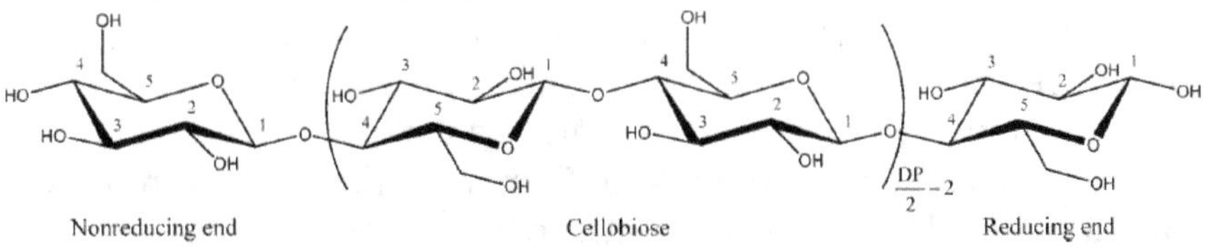

Figure 2.1 The schematic of the chemical framework of cellulose. Reproduced from Reference 2 with permission from American Chemical Society.

Cellulose has four distinct polymorphs namely cellulose I, II, III and IV. Cellulose I exists naturally and possesses two allomorphs, I_α and I_β. Cellulose II, which is the regenerated cellulose, develops subsequent to re-crystallization [4]. The primary difference between

Haleema Saleem and Vikas Mittal
Khalifa University of Science and Technology, Abu Dhabi, UAE

cellulose I and II lies in the arrangement of their atoms: cellulose I chains run in a side-by-side direction, whereas cellulose II has anti-parallel packing. The cellulose III_{II} and III_{I} forms are acquired by cellulose II and I respectively after ammonia treatment, and cellulose IV is subsequently generated by the modification of cellulose III.

The chemical modification of cellulosic materials on an industrial scale has led to the development of many advanced products, such as membranes, coatings, building materials, films, foodstuff and drilling fluids. Presently, the development of nanocellulose in the form of nanocrystals, nano-fibrils, nano-whiskers and nanofibers has gained significant research attention. Novel techniques for the generation of these nanomaterials range from top-down strategies, including physical/synthetic/enzymatic approaches for detachment from agricultural/forest residues and wood, to the bottom-up generation of cellulosic nano-fibrils from glucose by bacteria [5,6]. These materials display unique properties owing to their nanostructure such as hydrophilicity, ease of chemical modification and the generation of adaptable semi-crystalline fiber structures, along with extensive surface area.

Depending on preparation strategies, function as well as dimensions, nanocellulose can be categorized into three primary subcategories, namely micro-fibrillated cellulose (MFC), nanocrystalline cellulose (NCC) and bacterial nanocellulose (BNC) [6]. As indicated by Habibi *et al.* [2], around 36 distinctive cellulose molecules are united by biomass to form units termed as micro-fibrils or elementary fibrils, which are further arranged in bigger units to form micro-fibrillated cellulose [2]. The elementary fibril's diameter is almost 5 nm, while MFC, also known as nano-fibrillated cellulose (NFC), has diameter between 20 to 50 nm. A micro-fibril can be regarded as an adaptable hair strand having crystals of cellulose connected across the axis of the micro-fibril by disorganized amorphous areas [7]. The organized areas are the cellulose chain bundles which are stabilized using a powerful as well as complicated framework of hydrogen bonding [2]. NCC, otherwise called whiskers, comprises of rod-like crystals of cellulose having lengths from 100 nm to a few micrometers and widths in the range 5-70 nm. NCC is prepared by the amorphous phase elimination of a decontaminated cellulose material by acid hydrolysis. Even though comparable in size to MFC, NCC possesses exceptionally restricted flexibility and displays prolonged crystalline rod-like shapes. BNC, additionally known as bio-cellulose or microbial cellulose, is generated by the aerobic bacteria, like acetic acid bacteria of type *Gluconacetobacter* [6,8]. In contrast with the NCC and MFC materials which are segregated from the cellulose sources, BNC is generated as a polymer as well as nanomaterial by biotechnological congregation procedures from the lower molecular weight carbon sources (for example, D-glucose).

In this chapter, substantial advancements and prospects of nanocellulose-based materials are presented. Different preparation and characterization methods of the three types of nanocellulose, namely MFC, NCC and BNC, are also discussed. Finally, the chapter focuses on the potential application of nanocellulose as reinforcing agent for the generation of polymer nanocomposites.

2.2 Types of Nanocellulose

2.2.1 Micro-fibrillated Cellulose

MFC, also known as NFC, cellulose micro-fibrils or micro-fibrillar cellulose, was assessed quite recently, especially with respect to the nanocomposite applications [6,9]. The initial cellulosic material utilized to prepare MFC was wood, as detailed by Herrick *et al.* [10]. The wood pulp was broken down using a high-pressure homogenizer with an aim of acquiring a viscous as well as shear thinning aqueous gel at a lower concentration. Currently, MFC is generated from several cellulosic sources.

Synthesis of MFC

As mentioned earlier, MFC is typically generated from wood by high pressure homogenization as per the techniques advanced at ITT Rayonnier [10]. The bleached kraft pulp is generally employed as a starting material for the generation of MFC. The pulp is generated by the chemical treatment of wood by utilizing a blend of sodium sulfide and sodium hydroxide. Pulping with sulfurous acid salts gives rise to sulfite pulp. Delamination procedure has been observed to be expedited by the incorporation of hydrophilic polymers, like guar gums, carboxymethyl cellulose, poly(acrylic acid), methyl cellulose, carrageenin and hydroxyl-propyl cellulose (HPC). These polymers diminish the clogging trend and allow greater pulp consistencies at the time of homogenization. The sulfite pulps are easier to delaminate relative to the kraft pulps, and higher charge density or/and greater hemi-cellulose content assist the delamination. However, it was noted earlier that almost 27,000 kWh per ton was required for generating a gel-like MFC from a suspension of sulfite-pulp possessing a greater content of hemi-cellulose. The addition of the charged groups within the pulp fibers was observed to improve the fiber wall delamination, and by the incorporation of carboxy-methyl groups, a completely delaminated carboxy-methylated MFC might be generated [11]. The addition of the charged groups has been suggested as a genuinely effective approach to diminish the energy utilization. The use of oxidative transition metal ions to oxidize the fibers has been the subject of many patents, and, in addition, persulfate-oxidized cellulose has been reported as an appropriate substrate to prepare MFC [12]. An innovative method of attaining charged carboxylate groups in the cellulose structure together with MFC generation is the 2,2,6,6-tetramethyl-piperidine-1-oxyl (TEMPO) mediated oxidation of the 1° hydroxy groups at the cellulose C6 location, a useful technique fundamentally advanced by Saito *et al.* [13,14].

A wide range of delamination techniques have been developed including the utilization of micro-fluidizers [15], consolidation of beating, rubbing as well as homogenization [16], super-grinding treatments [17], cryogenic crushing in different configurations [18,19] and greater shear refining. Delamination using ball mills and ultrasonication have additionally been utilized to generate nanofibers [20]. MFC re-dispersion subsequent to drying is a challenge, as the irrevocable fibril conglomeration (a procedure termed as the hornification) brings about a material having ivory-like characteristics which make the material non-dispersible for composite applications. The principle methodology to avoid hornification is the generation of a steric hindrance or presentation of electrostatic groups for obstructing interdependent hydrogen bonding among cellulose chains. Beneficial additives used for this purpose are polyhydroxy-functionalized admixtures, especially carbohydrates or carbohydrate associated compounds, for example, starches, carbohydrate gums, seaweed extracts, glycosides, cellulose derivatives, oligosaccharides and glycol compounds [21]. Another functional strategy is the generation of other functional groups on the cellulose structure, for the purpose of blocking the co-operative hydrogen bonding. In this respect, the carboxy-methylation process is considered to be an extremely viable approach for effectively restricting the cellulose fiber hornification process.

Properties of MFC

MFC displays high surface area, crystallinity and flexibility, along with possessing a large number of hydroxyl groups. The % crystallinity, DP and strength are the important physical characteristics driving the applications of MFC. Also, the degree of crystallinity is dependent on the MFC source. For instance, a greater degree of crystallinity of almost 70% and 78% was observed for soy hull and wheat straw MFC respectively [22]. Iwamoto *et al.* [23] examined the % crystallinity of MFC subsequent to number of passes via a grinder [23]. An enhancement in the number of passes (1 to 30) was observed to diminish the degree of crystallinity, which could be illustrated by CNF hornification under a greater shear rate. The other essential characteristic of MFC is its high specific area, normally examined using modeling tools. Siqueira *et al.* [24] reported the specific area of sisal MFC as 50 m^2g^{-1}, which is almost ten times higher relative to the fibers. It is often difficult to quantify this property because of the strong aggregation of MFC after the drying process. MFC also exhibits particular rheological behavior, characterized by shear thinning as well as pseudo-plasticity [10,25,26].

The thermal degradation characteristics of the MFC films have been examined to a lesser extent. In addition to the preparation procedure, the thermal performance additionally relies on the drying process (which influences the extent of hydrogen bonds). It was observed that the TEMPO-oxidized cellulose exhibited numerous degradation events [27]. The highest extent of degradation occurred during the temperature range 225-231 °C. The weight reduction enhanced remarkably with an enhancement in the sonication time as well as temperature. Quiévy *et al.* [28] also examined the impact of the drying procedure on the thermal resistance of MFC attained by homogenization. The authors analyzed 3 drying procedures: freeze-drying of frozen-cellulose suspension for 48 hours at 40 °C, oven drying for 72 hours at 80 °C and atomization (comprising of spraying of the suspension by progressing it via a nozzle and subsequently drying the wet particles using air at 200-205 °C).

NFC-based hybrids are exceedingly assuring for coatings, lightweight packaging and films [29]. Further, superior performance could be acquired by modifying the NFC hybrid frameworks. In a study by Liimatainen *et al.* [30], talc/NFC hybrid films having a lamellar brick-and-mortar framework were generated. A simple vacuum-filtration method, followed by vacuum drying, was utilized to manufacture the hybrid films from individual talc platelets and dicarboxylic acid cellulose nanofibers (DCC). The DCC nanofiber's lateral dimensions extended from 3-5 nm, as demonstrated by the atomic force micrograph (AFM) in Figure 2.2A, whereas the length was in the micrometer scale. As indicated by the field emission scanning electron microscopy (FESEM), the talc specimen consisted of distinctive platelets with no aggregates (Figure 2.2B). The talc/NFC hybrids, with talc contents from 1 to 50 wt%, were observed to be flexible with Young's modulus and tensile strength values of 12 ± 1 GPa and 211 ± 3 MPa respectively. The acquired enhancement in the mechanical properties was attributed to the effective disintegration process employed for breaking down the talc platelet aggregates.

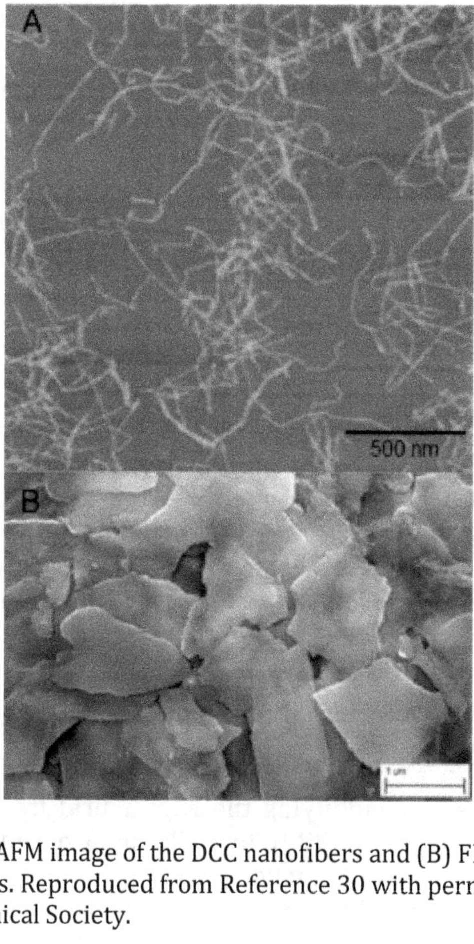

Figure 2.2 (A) AFM image of the DCC nanofibers and (B) FESEM image of the talc platelets. Reproduced from Reference 30 with permission from American Chemical Society.

2.2.2 Nano-crystalline Cellulose

Compared to the cellulose fibers, NCC has several benefits like greater specific strength, nanoscale dimensions, higher modulus and surface area, remarkable optical properties, etc. NCC consists of stiff rod-like crystals few hundred nanometers in length and 10-20 nm in diameter. NCC generated from native cellulose by acid hydrolysis has distinctive morphologies depending upon the hydrolysis conditions. The geometrical aspect ratio and relative degree of crystallinity are essential parameters governing the characteristics of NCC-based materials [31].

Synthesis of NCC

Acid hydrolysis is regarded as the principle procedure utilized to generate NCC. The native cellulose comprises of crystalline as well as amorphous regions, and the crystalline regions have greater density as compared to the amorphous regions. Thus, as the cellulose fibers are subjected to severe acid treatment, the amorphous sections separate, thereby, discharging the individual crystallites. The NCC properties rely upon different factors, such as temperature, reaction time, cellulose sources and types of acid utilized for the hydrolysis process.

NCC materials have been prepared using different sources, for example, cotton [32,33], wood [34], tunicate [35], sisal [36], bacterial cellulose [37], micro-crystalline cellulose [38], ramie [39] and *Valonia* cellulose [40]. NCC obtained from tunicate as well as bacterial cellulose are normally larger in dimensions as compared to cotton and wood [35,41]. This is due to the fact that the bacterial cellulose and tunicate are profoundly crystalline, thus, leading to the generation of bigger nanocrystals. Beck-Candanedo *et al.* [34] compared the characteristics of NCC from hardwood and softwood generated using similar acid-to-pulp proportions, reaction time and temperature [34]. It was noted that the NCC suspensions displayed comparative surface charge and critical concentrations (C_c). It was additionally observed that the longer reaction time and higher acid-to-pulp proportion led to smaller nanocrystals with limited polydispersity index (PDI). Due to the fact that the splitting of the cellulose chains happened arbitrarily during acid hydrolysis, the NCC dimensions were non-uniform. In another study, Bai *et al.* [42] reported a strategy for developing NCC with restricted size distribution by means of differential centrifugation procedure [42]. Bondeson *et al.* [43] improved the reaction environment for the sulfuric acid hydrolysis of micro-crystalline cellulose from Norway spruce, utilizing the response surface approach [43]. The concentration of sulfuric acid and micro-crystalline cellulose, temperature, hydrolysis time and ultrasonic treatment duration were varied. It was noted that the temperature, concentration of acid and reaction time were the most important aspects for NCC generation.

For NCC generation, hydrochloric acid (HCl) and sulfuric acid (H_2SO_4) are broadly utilized. Due to the plenitude of charged sulfate groups existing on the surface, NCC acquired from the H_2SO_4 hydrolysis distributed promptly in H_2O as compared to the material generated from HCl hydrolysis. Also, dissimilarities in the rheological behavior and thermal stability were noted between the NCC generated from HCl and H_2SO_4 [44]. Additionally, NCC has been prepared from the recycled pulp utilizing the microwave supported enzymatic hydrolysis. Filson *et al.* [45] reported a technique to prepare NCC utilizing endoglucanase enzyme, a component of cellulases. The authors observed that the microwave heating resulted in higher NCC yield as compared to normal heating.

Properties of NCC

NCC obtained from the acid hydrolysis of various forest products has the ability to disperse in water because of negatively charged surfaces. At lower concentration, the NCC particles are randomly oriented as an isotropic phase in the aqueous suspensions, and as the concentration attains a critical value, NCC generates a chiral nematic ordering [46]. With an increase in the concentration further, a shear birefringence process is demonstrated by the aqueous suspensions of NCC. The C_c of sulfated NCC generally extends from 1 to 10% (weight/weight), which is considered to be a function of NCC aspect ratio, osmolarity and charge density.

NCC phase behavior is responsive to the presence of electrolytes as well as their counter particles. Impact of the incorporated electrolyte on the NCC phase segregation was analyzed by Dong *et al.* [47]. The authors confirmed that the incorporation of the electrolytes, for example, KCl, HCl and NaCl, remarkably diminished the anisotropic phase volume fraction. In another study, Dong *et al.* [48] also analyzed the impact of the counter ions on the stability and phase segregation of the NCC suspensions. For this purpose, inorganic and poorly basic organic counter ions, along with immensely basic organic tetraalkyl-ammonium salts, were used. The authors also noted that the counter ion type had a remarkable impact on the phase segregation of the suspensions. Similar to the electrolytes, the phase segregation of the suspensions was firmly influenced by the inclusion of macromolecules.

Gray *et al.* [49,50] carried out a comprehensive analysis of the impact of ionic dyes and dextran on the phase equilibrium of the NCC suspensions. Heux *et al.* [41] first stated the ability of the surfactant coating to disperse the NCC whiskers in the non-polar solvents. For this purpose, the NCC whiskers from tunicate as well as cotton were blended using Beycostat NA (BNA) surfactant. Figure 2.3 shows the micrographs of cotton and tunicate suspensions in

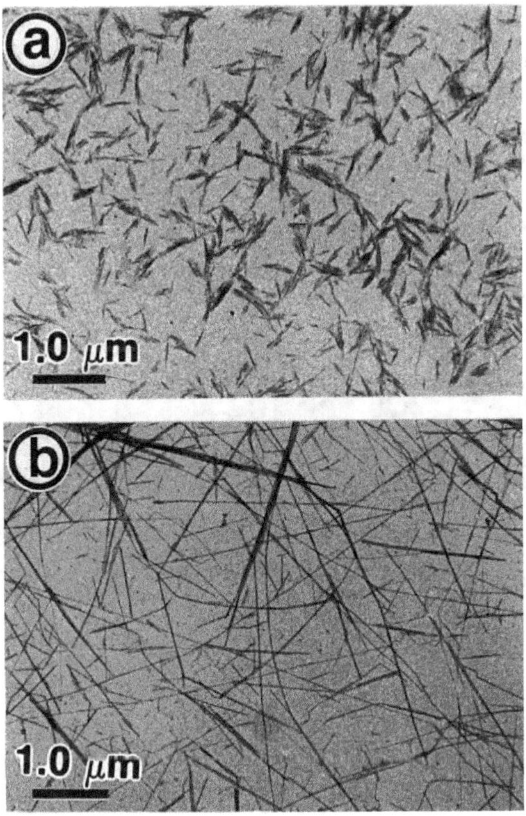

Figure 2.3 TEM micrographs of (a) cotton microcrystals and (b) tunicate microcrystals in toluene. Reproduced from Reference 41 with permission from American Chemical Society.

toluene. For the cotton suspensions (Figure 2.3a), microcrystals 8 nm in width and 200-300 nm in length were obtained. Certain aggregates of a few microcrystals, as bundles, were additionally present. In the case of tunicate suspension (Figure 2.3b) too, a uniform dispersion was observed. Cellulose microcrystals were effectively individualized, with a lateral size of approx. 15 nm and a substantial polydispersity in length, ranging from several hundreds of nanometers to a few micrometers. The authors also illustrated that the chiral nematic phases were generated despite the surfactant layer around the NCC whiskers.

In recent past, thorough analysis of the chiral nematic framework (Figure 2.4) was reported by Elazzouzi-Hafraoui *et al.* [51]. A macroscopic phase segregation with extremely clear phase boundaries as well as sharpened interfaces were noticed. Zhou *et al.* [52] also stated that the NCC material with adsorbed xyloglucan oligosaccharide-poly(ethylene glycol)-polystyrene (XGO-PEG-PS) triblock copolymer demonstrated superior dispersibility in the non-polar solvents. The quantity of the triblock copolymer adsorbed on the cellulose nanocrystals was attained from the decomposition behavior in the thermogravimetric analysis (TGA) (Figure 2.5). The thermal deterioration of primary CNC continued through two independent pyrolysis processes in the ranges of 200-290 °C and 300-400 °C, identical to spheroidal CNC having sulfate groups. Further, the inclusion of the surfactants increased the NCC dispersion in polystyrene (PS)

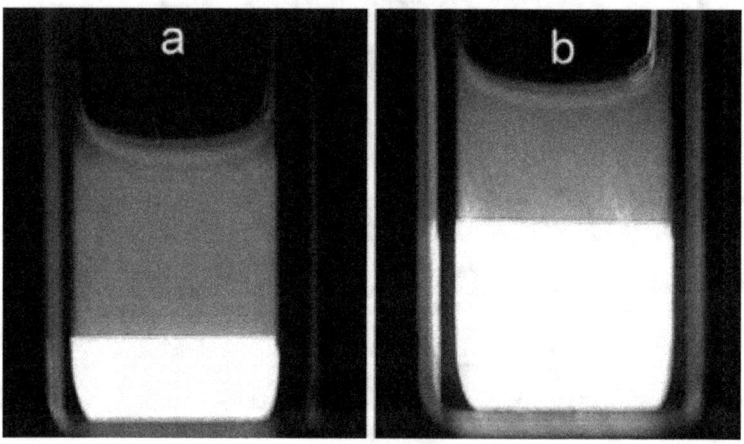

Figure 2.4 Phase separation seen among the cross-polars for varying NCC suspension concentrations of (a) 19.8 wt% and (b) 25.0 wt%. Reproduced from Reference 51 with permission from American Chemical Society.

[53]. In another study, the performance of NCC under the extrinsic field, e.g. AC electric field and magnetic field, was analyzed [54]. The impact of AC electric field on the arrangement as well as orientation of NCC was also examined by Habibi *et al.* [55]. The authors noticed that the application of the AC electric field to the NCC suspension, deposited in the middle of two metallic electrodes, led to the homogeneous arrangement of the NCC molecules. Additionally, the arrangement was significantly affected by the strength of the applied electric field as well as its frequency.

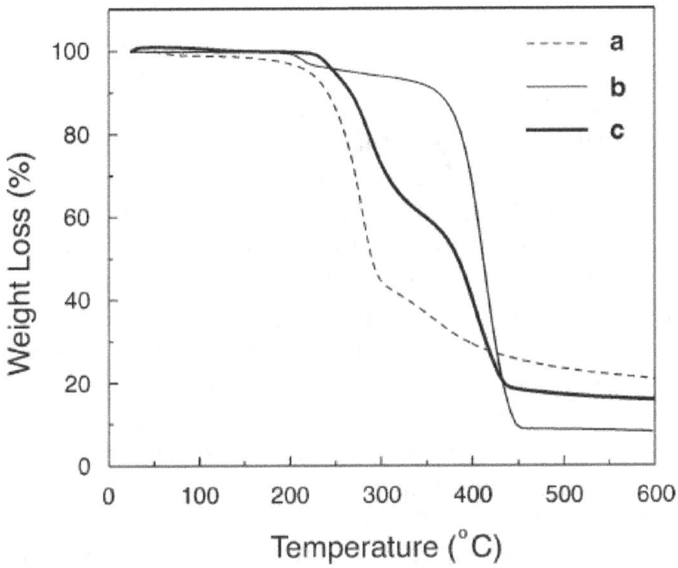

Figure 2.5 TGA curves of (a) pristine CNC, (b) XGO-PEG-PS triblock copolymer and (c) CNC adsorbed with the XGO-PEG-PS. Reproduced from Reference 52 with permission from American Chemical Society.

2.2.3 Bacterial Nanocellulose

BNC is a profoundly crystalline linear glucose polymer mostly obtained from the bacterium *Gluconacetobacter xylinus*, however, other microorganisms also display the capacity to generate BNC, e.g. other types of *Gluconacetobacter*, *Rhizobium* spp., *Agrobacterium tumefaciens* and gram-positive *Sarcina ventriculli* [56,57]. BNC consists of thin nanofibers with diameters ranging from 20-100 nm. It has high surface area and hydrophilic nature, which result in its excellent water absorption ability, enhanced moisture content and good adhesion [58-61]. In addition, BNC exhibits superior tensile strength, durability, biodegradability, insolubility, elasticity and non-allergenic as well as non-toxic features.

Synthesis of BNC

Acetic acid bacteria of the family *Gluconacetobacter* can generate acids as well as cellulose in high yield. The *Acetobacteraceae* are aerobic, gram-negative and rod-like microorganisms of remarkable acid tolerance, high ubiquity and dynamic motility. Remarkably, *Gluconacetobacter xylinus* cultures generate the biofilms of pristine cellulose on non-toxic surfaces [62,63]. The support material and its surface framework, *Gluconacetobacter* strain type, temperature as well as the elements of the culture medium (inclusive of additives) determine the efficiency of cellulose generation.

The bacterial cellulose preparation from the lower molecular weight sugars or alternative carbon sources by means of uridine-diphosphate glucose has been illustrated in the literature [64,65]. The generated cellulose chains discharge as fibers (diameter in nanometer range) into the aqueous culture medium [66]. The cellulose fibrils discharged by bacteria consolidate to form ribbons and ultimately three dimensional (3-D) nanofiber structures.

BNC has identical molecular formula as the plant cellulose, however, it is principally dissimilar due to its nanofiber structure. Subsequent to elementary purification, BNC contains no contaminants and functional groups, apart from the hydroxyl groups. For commercial utilization, cost-proficient procedures for the large scale production of BNC are needed. During the past several years, a wide range of advanced ideas have been examined. Generally, these are based on the consolidation of the agitated as well as static cultivation [67-72].

Properties of BNC

Bacterial cellulose is a biomaterial with extraordinary properties, e.g. mechanical strength, chemical purity, absorbency, flexibility, size and shape compatibility, non-toxicity, inertness and bio-functionality [73]. The undried BNC has remarkable mechanical properties with stress-strain characteristics resembling with delicate tissue. The tensile strength of BNC pellicle has been reported to be 2 MPa in the undried state, which is considered to be high considering 99% H_2O content. In the dry-state, the stress at break of a single BNC fiber is equivalent to steel [74]. Thus, BNC is appropriate for use as a reinforcing material for polymer composites as well as paper [75,76].

2.3 Use of Nanocellulose as a Reinforcement in Polymer Nanocomposites

The initial utilization of nanocellulose as a reinforcing agent for different polymers including high density polyethylene (HDPE), PS and polypropylene (PP) was stated by Boldizar *et al.* [77]. Subsequently, Favier *et al.* [78,79] illustrated the reinforcing effect of a small fraction of nanocellulose on polymer properties. The authors utilized nanocellulose whiskers, obtained from tunicate, for reinforcing butyl acrylate and styrene copolymer latex at a loading of 6 vol%. The nanocomposites displayed superior mechanical properties as compared to the pristine polymer, which was attributed to the generation of an inflexible cellulose whisker percolation framework. Nanocellulose percolation threshold has been assessed to be in the range of 1 to 6 vol%, depending on the source of cellulose [3]. Later, Dufresne and Vignon [80] illustrated the reinforcing effect of potato pulp-extracted NFC on composite properties.

2.3.1 MFC as Reinforcement in Polymer Nanocomposites

MFC is generally utilized as a reinforcing agent for polar/hydrophilic polymers [81]. The preparation of MFC-reinforced nanocomposites with non-polar/hydrophobic polymers is restricted due to the hydrophilic characteristics of the cellulose fibers [81]. Dufresne *et al.* [82] initially illustrated the ability of MFC to enhance the mechanical properties of potato-starch [82]. Subsequently, Nakagaito *et al.* [83] demonstrated that MFC enhanced the tensile properties of a phenolic resin. In another study, Zimmermann *et al.* [84] reported an enhancement in the properties of hydroxy-propylated cellulose (HPC) on MFC incorporation. Other polymer matrices used for the development of MFC reinforced nanocomposites include polyvinyl alcohol (PVA), regenerated cellulose, poly(L-lactic acid) (PLLA) and enzymatically modified rye arabinoxylan [85-90]. Zimmermann *et al.* [84] additionally stated that the mechanical characteristics of PVA were significantly enhanced by the MFC incorporation. The Young's modulus was observed to achieve the highest value at 10 wt% fiber fraction. Lu *et al.* [86] also utilized MFC with diameter ranging from 10-100 nm and web-like framework as the reinforcement for PVA [86]. For this purpose, the authors blended a PVA aqueous solution with MFC suspensions at various proportions (1-15 wt%) and generated the composite samples using film casting. PVA crystallization, melting temperature (T_m) as well as glass transition temperature (T_g) were not affected by MFC. The authors observed 40% and 76% enhancement in modulus and strength respectively at MFC loading of 10 wt%, relative to the pristine PVA specimen. The modulus and strength did not change further on enhancing the MFC content to 15 wt%. This might be due to the accumulation of the nanosized fibers above 10% loading.

Bulota *et al.* [91] generated composite films by blending a MFC suspension with an aqueous solution of PVA using various concentrations as well as blending durations to demonstrate the impact on properties. It was observed that the mechanical properties of the composite films were dependent on the PVA-MFC mixture concentration. Further, longer blending duration negatively influenced the mechanical characteristics of the composites. Besides, the effect of the testing conditions (e.g. relative humidity) was noted to be critical because of the moisture sensitivity of cellulose fibers; higher relative humidity led to lesser strength as well as modulus.

As mentioned earlier, the utilization of hydrophobic polymers to generate MFC-filled nanocomposites is hindered by the absence of compatibility between the two phases [81]. To overcome this limitation, the fibers are required to be surface treated prior to incorporation in the polymer matrix. Suryanegara *et al.* [92] and Iwatake *et al.* [93] utilized comparable preparation strategies to generate MFC-PLA nanocomposites. Iwatake *et al.* [93] observed that the Young's modulus and tensile strength were enhanced by 40% and 25% respectively at 10 wt% MFC content, relative to the pristine polymer. Various PLA grades were utilized by Suryanegara *et al.* [92], and the authors noted small increment in the mechanical properties. Nevertheless, at the similar MFC content, Young's modulus and tensile strength enhanced by almost 35% and 10% respectively.

Lu *et al.* [81] generated hydrophobic epoxy-MFC nanocomposites, and the thermal characteristics were determined utilizing dynamic mechanical thermal analysis (DMTA). Seydibeyoğlu *et al.* [94] employed polyurethane (PU) as hydrophilic polymer matrix and determined the thermal properties utilizing DMTA. Compared to the pristine polymer [81,94], the MFC incorporation led to an enhancement in the storage modulus, E′, at temperatures below or above T_g. This confirmed that the motility of the polymer chains was diminished by the inflexible MFC network and the development of interfacial bonding. Nevertheless, no variation in the T_g was observed on MFC incorporation [94]. Shimazaki *et al.* [95] also reported the development of transparent cellulose nanofiber/epoxy resin nanocomposites (Figure 2.6).

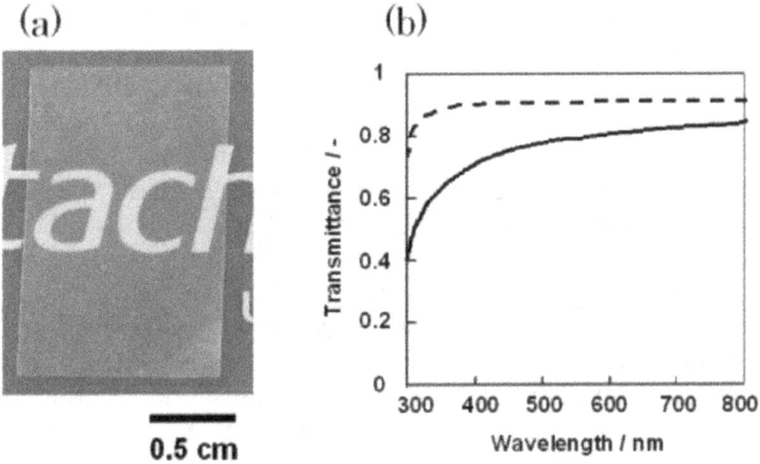

Figure 2.6 (a) Image of cellulose nanofiber/epoxy resin nanocomposite and (b) transmission spectra of nanofiber/epoxy resin nanocomposite (solid line) and epoxy resin without nanofibers (dashed line). Reproduced from Reference 95 with permission from American Chemical Society.

2.3.2 NCC as Reinforcement in Polymer Nanocomposites

NCC has a strong potential for developing lightweight, strong and inexpensive nanocomposites [2,34]. NCC reinforces the polymer matrices by the development of a percolation framework associated with the uniformly dispersed NCC phase [78,96-100].

Khan *et al.* [97] reported biodegradable films of NCC filled chitosan generated using the solution casting technique. The content of NCC varied from 1 to 10 wt%. It was noted that the nanocomposite film with 5 wt% NCC content exhibited 26% improvement in tensile strength, relative to the pure chitosan film. Further, the film demonstrated remarkably improved the barrier properties. Also, the water vapor penetrability (WVP) of the 5 wt% NCC based chitosan film diminished by 27%. In addition, the swelling analysis demonstrated a reduction in the water adsorption on the NCC filled chitosan films. The thermal properties of the polymer were not impacted by the NCC incorporation. The analysis of the surface morphology of the films revealed that the NCC phase was homogenously dispersed in the chitosan matrix.

Huq *et al.* [98] generated NCC filled alginate-based nanocomposites using the solution casting technique. The content of NCC in the nanocomposites varied from 1 to 8 wt%. At 5 wt% NCC content, the nanocomposite film demonstrated 37% higher ultimate tensile strength, relative to the control sample. The inclusion of 5 wt% NCC also reduced the nanocomposite's WVP by 31%. The FTIR analysis confirmed the molecular interaction between the NCC and alginate phases. Also, it was observed that the thermal stability of alginate-based nanocomposite films was improved subsequent to the NCC inclusion.

Li *et al.* [101] generated rod-shaped NCC from micro-crystalline cellulose (MCC) utilizing a physical strategy of high-intensity ultrasonication. The reinforcing ability of the generated NCC was examined by incorporating it in PVA by using the solution casting technique. The diffraction analysis demonstrated that the NCC material possessed the cellulose I crystal framework. Further, it was noted that the NCC crystallinity diminished with an enhancement in the ultrasonication time. The ultrasonic impact was considered to be non-selective, which implies that it could separate the crystalline as well as amorphous cellulose. The storage modulus of NCC/PVA nanocomposites was observed to be remarkably enhanced, relative to pristine PVA. For instance, the modulus of the composite with 8 wt% NCC content was noted to be 2.4 times higher than the pristine PVA. In another study, Lalia *et al.* [102] prepared polyvinylidenefluoride-co-hexafluoropropylene (PVDF-HFP) membranes consisting of 0-4 wt% NCC using the electrospinning method. The impact of NCC on the pore size distribution, morphology, mechanical strength and liquid entry pressure (LEP) of the fibrous membranes was explored. The PVDF-HFP membrane consisting of 2 wt% NCC exhibited Young's modulus and tensile strength values of 105 MPa and 17.2 MPa respectively. The membrane demonstrated a limited pore size distribution with average pore size of 0.2 μm. The inclusion of 2 wt% NCC in the membrane also prompted a significant increment in its LEP from 19 psi to 27 psi.

2.3.3 BNC as Reinforcement in Polymer Nanocomposites

The utilization of BNC as reinforcement in nanocomposites was initially illustrated by Gindl and Keckes [103]. Cellulose acetate butyrate (CAB) matrix was reinforced using different BNC loadings. At 32 vol% BNC loading, the tensile characteristics of the CAB nanocomposite enhanced by almost 5 times, relative to the pristine polymer. It was noted that the BNC filled nanocomposite had tensile strength and modulus as high as 320 MPa and 21 GPa respectively, at BNC loading of 60 vol%. Nakagaito *et al.* [104] also reported the tensile strength of 420 MPa with 84 vol% BNC content in phenolic resin.

Various research studies have been performed to hydrophobize the BNC surface for enhancing the compatibility between the hydrophobic polymer and BNC. The hydrophobization of BNC was reported by employing dodecanoic acid in an esterification process [105]. The contact angles of the PLLA droplets on the surface of the modified BNC demonstrated enhanced wettability. The PLLA nanocomposites with modified BNC (5 wt% content) additionally indicated 12% and 50% enhancement in tensile strength and modulus respectively, relative to PLLA. Nevertheless, compared to PLLA nanocomposites with pristine BNC (5 wt% content), the improvements were marginal (only 5% improvement in tensile modulus, however, no enhancement in tensile strength). Besides, the chemical modification of BNC is considered to be extremely complex [106]. Thus, the alternative strategy of enhancing the BNC loading above 30 vol% may be more beneficial.

Stevanic *et al.* [107] illustrated the feasibility to generate xylan films with enhanced strength by the incorporation of BNC. The blending of BNC with xylan using a high-pressure micro-fluidizer resulted in optically clear nanocomposite films. Generally, the nanocomposite films were stiffer and stronger as compared to the pure polymer. Inclusion of BNC had a minor impact on the T_g of the polymer. The composites exhibited observable reduction in the moisture sensitivity and enhanced stiffness. In another study, Phomrak *et al.* [108] reinforced natural rubber (NR) with BNC and developed the composite films by means of a latex aqueous micro-dispersion technique. The BNC nanofibers were observed to be uniformly distributed and sufficiently dispersed in the NR matrix. It was observed that the crystallinity, opacity and hydrophilicity of the BNC/NR composite films enhanced with an increment in the BNC content. The composite with 80 wt% BNC, a hard as well as solid bio-composite film, demonstrated the Young's modulus and tensile strength of 4,128 MPa and 75 MPa respectively. Further, the 20 wt% BNC filled NR film showed an elastic elongation of 14.6% with relative enhancements in thermal stability, tensile strength and Young's modulus, relative to pristine NR.

2.4 Summary

In this chapter, the synthesis and properties of different nanocellulose types have been briefly reviewed. Subsequently, various studies reporting the use of these nanocellulose types as reinforcement to generate polymer nanocomposites have been summarized.

References

1. Lavoine, N., Desloges, I., Dufresne, A., and Bras, J. (2012) Microfibrillated cellulose–Its barrier properties and applications in cellulosic materials: A review. *Carbohydrate Polymers*, **90**(2), 735-764.
2. Habibi, Y., Lucia, L. A., and Rojas, O. J. (2010) Cellulose nanocrystals: chemistry, self-assembly, and applications. *Chemical Reviews*, **110**(6), 3479-3500.
3. Siqueira, G., Bras, J., and Dufresne, A. (2010) Cellulosic bionanocomposites: a review of preparation, properties and applications. *Polymers*, **2**(4), 728-765.
4. Aulin, C., Ahola, S., Josefsson, P., Nishino, T., Hirose, Y., Österberg, M., and Wågberg, L. (2009) Nanoscale cellulose films with different crystallinities and mesostructures: Their surface properties and interaction with water. *Langmuir*, **25**(13), 7675-7685.
5. Moon, R. J., Martini, A., Nairn, J., Simonsen, J., and Youngblood, J. (2011) Cellulose nanomaterials review: Structure, properties and nano-composites. *Chemical Society Reviews*, **40**(7), 3941-3994.
6. Klemm, D., Kramer, F., Moritz, S., Lindström, T., Ankerfors, M., Gray, D., and Dorris, A. (2011) Nano-celluloses: A new family of nature-based materials. *Angewandte Chemie International Edition*, **50**(24), 5438-5466.
7. Azizi Samir, M. A. S., Alloin, F., and Dufresne, A. (2005) Review of recent research into cellulosic whiskers, their properties and their application in nanocomposite field. *Biomacromolecules*, **6**(2), 612-626.
8. Klemm, D., Schumann, D., Kramer, F., Heßler, N., Koth, D., and Sultanova, B. (2009) Nanocellulose materials - Different cellulose, different functionality. *Macromolecular Symposia*, **280**(1), 60-71.
9. Siró, I., and Plackett, D. (2010) Microfibrillated cellulose and new

nanocomposite materials: a review. *Cellulose*, **17**(3), 459-494.

10. Herrick, F.W. Casebier, R. L. Hamilton, J. K. Sandberg, K. R. (1983) Microfibrillated cellulose: Morphology and accessibility. *Journal of Applied Polymer Science, Applied Polymer Sympoisum*, **37**, 797-813.
11. Wågberg, L., Winter, L., Ödberg, L., and Lindström, T. (1987) On the charge stoichiometry upon adsorption of a cationic polyelectrolyte on cellulosic materials. *Colloids and Surfaces*, **27**(1-3), 163-173.
12. Banker, G. S., and Kumar, V. (1995) Microfibrillated Oxycellulose, patent US5405953.
13. Saito, T., Nishiyama, Y., Putaux, J. L., Vignon, M., and Isogai, A. (2006) Homogeneous suspensions of individualized microfibrils from TEMPO-catalyzed oxidation of native cellulose. *Biomacromolecules*, **7**(6), 1687-1691.
14. Saito, T., and Isogai, A. (2007) Preparation of cellulose single microfibrils from native celluloses by TEMPO-mediated oxidation. *Cellulose Communications*, **14**(2), 62-66.
15. Henriksson, M., Henriksson, G., Berglund, L. A., and Lindström, T. (2007) An environmentally friendly method for enzyme-assisted preparation of microfibrillated cellulose (MFC) nanofibers. *European Polymer Journal*, **43**(8), 3434-3441.
16. Matsuda, Y. (2000) Properties and use of microfibrillated cellulose as papermaking addition. *Sen-i Gakkaishi*, **56**(7), 192-196.
17. Iwamoto, S., Nakagaito, A. N., Yano, H., and Nogi, M. (2005) Optically transparent composites reinforced with plant fiber-based nanofibers. *Applied Physics A: Materials Science and Processing*, **81**(6), 1109-1112.
18. Janardhnan, S., and Sain, M. M. (2007) Isolation of cellulose microfibrils–an enzymatic approach. *Bioresources*, **1**(2), 176-188.
19. Wang, B., and Sain, M. (2007) Dispersion of soybean stock-based nanofiber in a plastic matrix. *Polymer International*, **56**(4), 538-546.
20. Zhao, H. P., Feng, X. Q., and Gao, H. (2007) Ultrasonic technique for extracting nanofibers from nature materials. *Applied Physics Letters*, **90**(7), 073112.
21. Lowys, M. P., Desbrieres, J., and Rinaudo, M. (2001) Rheological characterization of cellulosic microfibril suspensions. Role of polymeric additives. *Food Hydrocolloids*, **15**(1), 25-32.
22. Alemdar, A., and Sain, M. (2008) Isolation and characterization of nanofibers from agricultural residues – Wheat straw and soy hulls. *Bioresource Technology*, **99**(6), 1664-1671.
23. Iwamoto, S., Nakagaito, A. N., and Yano, H. (2007) Nano-fibrillation of pulp fibers for the processing of transparent nano-composites. *Applied Physics A: Materials Science and Processing*, **89**(2), 461-466.
24. Siqueira, G., Bras, J., and Dufresne, A. (2009) New process of chem-

ical grafting of cellulose nanoparticles with a long chain isocyanate. *Langmuir*, **26**(1), 402-411.

25. Pääkkö, M., Ankerfors, M., Kosonen, H., Nykänen, A., Ahola, S., Österberg, M., Ruokolainen, J., Laine, J., Larsson, P.T., Ikkala, O. and Lindström, T. (2007) Enzymatic hydrolysis combined with mechanical shearing and high-pressure homogenization for nanoscale cellulose fibrils and strong gels. *Biomacromolecules*, **8**(6), 1934-1941.
26. Goussé, C., Chanzy, H., Cerrada, M. L., and Fleury, E. (2004) Surface silylation of cellulose microfibrils: preparation and rheological properties. *Polymer*, **45**(5), 1569-1575.
27. Johnson, R. K., Zink-Sharp, A., Renneckar, S. H., and Glasser, W. G. (2009) A new bio-based nanocomposite: fibrillated TEMPO-oxidized celluloses in hydroxypropylcellulose matrix. *Cellulose*, **16**(2), 227-238.
28. Quiévy, N., Jacquet, N., Sclavons, M., Deroanne, C., Paquot, M., and Devaux, J. (2010) Influence of homogenization and drying on the thermal stability of microfibrillated cellulose. *Polymer Degradation and Stability*, **95**(3), 306-314.
29. Ridgway, C. J., and Gane, P. A. (2012) Constructing NFC-pigment composite surface treatment for enhanced paper stiffness and surface properties. *Cellulose*, **19**(2), 547-560.
30. Liimatainen, H., Ezekiel, N., Sliz, R., Ohenoja, K., Sirviö, J. A., Berglund, L., Hormi, O and Niinimäki, J. (2013) High-strength nanocellulose–talc hybrid barrier films. *ACS Applied Materials & Interfaces*, **5**(24), 13412-13418.
31. Peng, B. L., Dhar, N., Liu, H. L., and Tam, K. C. (2011) Chemistry and applications of nanocrystalline cellulose and its derivatives: a nanotechnology perspective. *The Canadian Journal of Chemical Engineering*, **89**(5), 1191-1206.
32. Araki, J., Wada, M., and Kuga, S. (2001) Steric stabilization of a cellulose microcrystal suspension by poly (ethylene glycol) grafting. *Langmuir*, **17**(1), 21-27.
33. Miller, A. F., and Donald, A. M. (2003) Imaging of anisotropic cellulose suspensions using environmental scanning electron microscopy. *Biomacromolecules*, **4**(3), 510-517.
34. Beck-Candanedo, S., Roman, M., and Gray, D. G. (2005) Effect of reaction conditions on the properties and behavior of wood cellulose nanocrystal suspensions. *Biomacromolecules*, **6**(2), 1048-1054.
35. de Souza Lima, M. M., Wong, J. T., Paillet, M., Borsali, R., and Pecora, R. (2003) Translational and rotational dynamics of rodlike cellulose whiskers. *Langmuir*, **19**(1), 24-29.
36. de Rodriguez, N. L. G., Thielemans, W., and Dufresne, A. (2006) Sisal cellulose whiskers reinforced polyvinyl acetate nanocomposites. *Cellulose*, **13**(3), 261-270.

37. Roman, M., and Winter, W. T. (2004) Effect of sulfate groups from sulfuric acid hydrolysis on the thermal degradation behavior of bacterial cellulose. *Biomacromolecules*, **5**(5), 1671-1677.
38. Pranger, L., and Tannenbaum, R. (2008) Biobased nanocomposites prepared by in situ polymerization of furfuryl alcohol with cellulose whiskers or montmorillonite clay. *Macromolecules*, **41**(22), 8682-8687.
39. de Menezes, A. J., Siqueira, G., Curvelo, A. A., and Dufresne, A. (2009) Extrusion and characterization of functionalized cellulose whiskers reinforced polyethylene nano-composites. *Polymer*, **50**(19), 4552-4563.
40. Revol, J. F. (1982) On the cross-sectional shape of cellulose crystallites in Valonia ventricosa. *Carbohydrate Polymers*, **2**(2), 123-134.
41. Heux, L., Chauve, G., and Bonini, C. (2000) Nonflocculating and chiral-nematic self-ordering of cellulose microcrystals suspensions in nonpolar solvents. *Langmuir*, **16**(21), 8210-8212.
42. Bai, W., Holbery, J., and Li, K. (2009) A technique for production of nanocrystalline cellulose with a narrow size distribution. *Cellulose*, **16**(3), 455-465.
43. Bondeson, D., Mathew, A., and Oksman, K. (2006) Optimization of the isolation of nanocrystals from microcrystalline cellulose by acid hydrolysis. *Cellulose*, **13**(2), 171-180.
44. Araki, J., Wada, M., Kuga, S., and Okano, T. (1998) Flow properties of microcrystalline cellulose suspension prepared by acid treatment of native cellulose. *Colloids and Surfaces A: Physicochemical and Engineering Aspects*, **142**(1), 75-82.
45. Filson, P. B., Dawson-Andoh, B. E., and Schwegler-Berry, D. (2009) Enzymatic-mediated production of cellulose nanocrystals from recycled pulp. *Green Chemistry*, **11**(11), 1808-1814.
46. Revol, J. F., Bradford, H., Giasson, J., Marchessault, R. H., and Gray, D. G. (1992) Helicoidal self-ordering of cellulose microfibrils in aqueous suspension. *International Journal Of Biological Macromolecules*, **14**(3), 170-172.
47. Dong, X. M., Kimura, T., Revol, J. F., and Gray, D. G. (1996) Effects of ionic strength on the isotropic– chiral nematic phase transition of suspensions of cellulose crystallites. *Langmuir*, **12**(8), 2076-2082.
48. Dong, X. M., and Gray, D. G. (1997) Effect of counterions on ordered phase formation in suspensions of charged rod-like cellulose crystallites. *Langmuir*, **13**(8), 2404-2409.
49. Edgar, C. D., and Gray, D. G. (2002) Influence of dextran on the phase behavior of suspensions of cellulose nanocrystals. *Macromolecules*, **35**(19), 7400-7406.
50. Beck-Candanedo, S., Viet, D., and Gray, D. G. (2006) Induced phase separation in cellulose nanocrystal suspensions containing ionic dye species. *Cellulose*, **13**(6), 629-635.

51. Elazzouzi-Hafraoui, S., Putaux, J. L., and Heux, L. (2009) Self-assembling and chiral nematic properties of organophilic cellulose nanocrystals. *The Journal of Physical Chemistry B*, **113**(32), 11069-11075.
52. Zhou, Q., Brumer, H., and Teeri, T. T. (2009) Self-organization of cellulose nanocrystals adsorbed with xyloglucan oligosaccharide-poly(ethylene glycol)-polystyrene triblock copolymer. *Macromolecules*, **42**(15), 5430-5432.
53. Kim, J., Montero, G., Habibi, Y., Hinestroza, J. P., Genzer, J., Argyropoulos, D. S., and Rojas, O. J. (2009) Dispersion of cellulose crystallites by nonionic surfactants in a hydrophobic polymer matrix. *Polymer Engineering and Science*, **49**(10), 2054-2061.
54. Sugiyama, J., Chanzy, H., and Maret, G. (1992) Orientation of cellulose microcrystals by strong magnetic fields. *Macromolecules*, **25**(16), 4232-4234.
55. Habibi, Y., Heim, T., and Douillard, R. (2008) AC electric field-assisted assembly and alignment of cellulose nanocrystals. *Journal of Polymer Science, Part B: Polymer Physics*, **46**(14), 1430-1436.
56. Mohammadkazemi, F., Azin, M., and Ashori, A. (2015) Production of bacterial cellulose using different carbon sources and culture media. *Carbohydrate Polymers*, **117**, 518-523.
57. Tanskul, S., Amornthatree, K., and Jaturonlak, N. (2013) A new cellulose-producing bacterium, Rhodococcus sp. MI 2: Screening and optimization of culture conditions. *Carbohydrate polymers*, **92**(1), 421-428.
58. Fu, L., Zhang, J., and Yang, G. (2013) Present status and applications of bacterial cellulose-based materials for skin tissue repair. *Carbohydrate Polymers*, **92**(2), 1432-1442.
59. Numata, Y., Sakata, T., Furukawa, H., and Tajima, K. (2015) Bacterial cellulose gels with high mechanical strength. *Materials Science and Engineering C*, **47**, 57-62.
60. Keshk, S. M. (2014) Bacterial cellulose production and its industrial applications. *Journal of Bioprocessing and Biotechniques*, **4**(2), doi:10.4172/2155-9821.1000150
61. Kurosumi, A., Sasaki, C., Yamashita, Y., and Nakamura, Y. (2009) Utilization of various fruit juices as carbon source for production of bacterial cellulose by Acetobacter xylinum NBRC 13693. *Carbohydrate Polymers*, **76**(2), 333-335.
62. Klemm, D., Heublein, B., Fink, H. P., and Bohn, A. (2005) Cellulose: faszinierendes Biopolymer und nachhaltiger Rohstoff. *Angewandte Chemie*, **117**(22), 3422-3458.
63. Brown, A. J. (1886) XIX. The chemical action of pure cultivations of bacterium aceti. *Journal of the Chemical Society, Transactions*, **49**, 172-187.

64. Yamanaka, S., Watanabe, K., Kitamura, N., Iguchi, M., Mitsuhashi, S., Nishi, Y., and Uryu, M. (1989) The structure and mechanical properties of sheets prepared from bacterial cellulose. *Journal of Materials Science*, **24**(9), 3141-3145.
65. Salmon, S., and Hudson, S. M. (1997) Crystal morphology, biosynthesis, and physical assembly of cellulose, chitin, and chitosan. *Journal of Macromolecular Science, Part C: Polymer Reviews*, **37**(2), 199-276.
66. Jonas, R., and Farah, L. F. (1998) Production and application of microbial cellulose. *Polymer Degradation and Stability*, **59**(1-3), 101-106.
67. Chao, Y. P., Sugano, Y., Kouda, T., Yoshinaga, F., and Shoda, M. (1997) Production of bacterial cellulose by Acetobacter xylinumwith an air-lift reactor. *Biotechnology Techniques*, **11**(11), 829-832.
68. Sakairi, N., Asano, H., Ogawa, M., Nishi, N., and Tokura, S. (1998) A method for direct harvest of bacterial cellulose filaments during continuous cultivation of Acetobacter xylinum. *Carbohydrate Polymers*, **35**(3-4), 233-237.
69. Bungay, H. R., and Serafica, G. C. (2000) Production of Microbial Cellulose, patent US6071727.
70. Chawla, P. R., Bajaj, I. B., Survase, S. A., and Singhal, R. S. (2009) Microbial cellulose: fermentative production and applications. *Food Technology and Biotechnology*, **47**(2), 107-124.
71. Dahman, Y. (2009) Nanostructured biomaterials and biocomposites from bacterial cellulose nanofibers. *Journal of Nanoscience and Nanotechnology*, **9**(9), 5105-5122.
72. Gardner, D. J., Oporto, G. S., Mills, R., and Samir, M. A. S. A. (2008) Adhesion and surface issues in cellulose and nanocellulose. *Journal of Adhesion Science and Technology*, **22**(5-6), 545-567.
73. Dinand, E., Chanzy, H., and Vignon, M. R. (1996) Parenchymal cell cellulose from sugar beet pulp: preparation and properties. *Cellulose*, **3**(1), 183-188.
74. Yano, H., Sugiyama, J., Nakagaito, A. N., Nogi, M., Matsuura, T., Hikita, M., and Handa, K. (2005) Optically transparent composites reinforced with networks of bacterial nanofibers. *Advanced Materials*, **17**(2), 153-155.
75. Yano, H., and Nakahara, S. (2004) Bio-composites produced from plant microfiber bundles with a nanometer unit web-like network. *Journal of Materials Science*, **39**(5), 1635-1638.
76. O'sullivan, A. C. (1997) Cellulose: the structure slowly unravels. *Cellulose*, **4**(3), 173-207.
77. Boldizar, A., Klason, C., Kubat, J., Näslund, P., and Saha, P. (1987) Prehydrolyzed cellulose as reinforcing filler for thermoplastics. *International Journal of Polymeric Materials*, **11**(4), 229-262.

78. Favier, V., Chanzy, H., and Cavaille, J. Y. (1995) Polymer nanocomposites reinforced by cellulose whiskers. *Macromolecules,* **28**(18), 6365-6367.
79. Favier, V., Canova, G. R., Cavaillé, J. Y., Chanzy, H., Dufresne, A., and Gauthier, C. (1995) Nanocomposite materials from latex and cellulose whiskers. *Polymers for Advanced Technologies,* **6**(5), 351-355.
80. Dufresne, A., and Vignon, M. R. (1998) Improvement of starch film performances using cellulose microfibrils. *Macromolecules,* **31**(8), 2693-2696.
81. Lu, J., Askeland, P., and Drzal, L. T. (2008) Surface modification of microfibrillated cellulose for epoxy composite applications. *Polymer,* **49**(5), 1285-1296.
82. Dufresne, A., Dupeyre, D., and Vignon, M. R. (2000) Cellulose microfibrils from potato tuber cells: processing and characterization of starch–cellulose microfibril composites. *Journal of Applied Polymer Science,* **76**(14), 2080-2092.
83. Nakagaito, A. N., and Yano, H. (2004) The effect of morphological changes from pulp fiber towards nano-scale fibrillated cellulose on the mechanical properties of high-strength plant fiber based composites. *Applied Physics A,* **78**(4), 547-552.
84. Zimmermann, T., Pöhler, E., and Geiger, T. (2004) Cellulose fibrils for polymer reinforcement. *Advanced Engineering Materials,* **6**(9), 754-761.
85. Hornung, M., Ludwig, M., and Schmauder, H. P. (2007) Optimizing the production of bacterial cellulose in surface culture: A novel aerosol bioreactor working on a fed batch principle (part 3). *Engineering in Life Sciences,* **7**(1), 35-41.
86. Lu, J., Wang, T., and Drzal, L. T. (2008) Preparation and properties of microfibrillated cellulose polyvinyl alcohol composite materials. *Composites Part A: Applied Science and Manufacturing,* **39**(5), 738-746.
87. Pullawan, T., Wilkinson, A. N., and Eichhorn, S. J. (2010) Discrimination of matrix–fibre interactions in all-cellulose nanocomposites. *Composites Science and Technology,* **70**(16), 2325-2330.
88. Jonoobi, M., Harun, J., Mathew, A. P., and Oksman, K. (2010) Mechanical properties of cellulose nanofiber (CNF) reinforced polylactic acid (PLA) prepared by twin screw extrusion. *Composites Science and Technology,* **70**(12), 1742-1747.
89. Mikkonen, K. S., Pitkänen, L., Liljeström, V., Bergström, E. M., Serimaa, R., Salmén, L., and Tenkanen, M. (2012) Arabinoxylan structure affects the reinforcement of films by microfibrillated cellulose. *Cellulose,* **19**(2), 467-480.
90. Kroon-Batenburg, L. M. J., Kroon, J., and Northolt, M. G. (1986) Chain modulus and intramolecular hydrogen bonding in native

and regenerated cellulose fibers. *Polymer Communications*, **27**(10), 290-292.
91. Bulota, M., Jääskeläinen, A. S., Paltakari, J., and Hughes, M. (2011) Properties of biocomposites: influence of preparation method, testing environment and a comparison with theoretical models. *Journal of Materials Science*, **46**(10), 3387-3398.
92. Suryanegara, L., Nakagaito, A. N., and Yano, H. (2009) The effect of crystallization of PLA on the thermal and mechanical properties of microfibrillated cellulose-reinforced PLA composites. *Composites Science and Technology*, **69**(7), 1187-1192.
93. Iwatake, A., Nogi, M., and Yano, H. (2008) Cellulose nanofiber-reinforced polylactic acid. *Composites Science and Technology*, **68**(9), 2103-2106.
94. Seydibeyoğlu, M. Ö., and Oksman, K. (2008) Novel nano-composites based on polyurethane and micro fibrillated cellulose. *Composites Science and Technology*, **68**(3), 908-914.
95. Shimazaki, Y., Miyazaki, Y., Takezawa, Y., Nogi, M., Abe, K., Ifuku, S., and Yano, H. (2007) Excellent thermal conductivity of transparent cellulose nanofiber/epoxy resin nano-composites. *Biomacromolecules*, **8**(9), 2976-2978.
96. Cao, X., Chen, Y., Chang, P. R., Muir, A. D., and Falk, G. (2008) Starch-based nano-composites reinforced with flax cellulose nanocrystals. *Express Polymer Letters*, **2**(7), 502-510.
97. Khan, A., Khan, R. A., Salmieri, S., Le Tien, C., Riedl, B., Bouchard, J., Chauve, G., Tan, V., Kamal, M.R. and Lacroix, M. (2012) Mechanical and barrier properties of nanocrystalline cellulose reinforced chitosan based nanocomposite films. *Carbohydrate Polymers*, **90**(4), 1601-1608.
98. Huq, T., Salmieri, S., Khan, A., Khan, R. A., Le Tien, C., Riedl, B., Fraschini, C., Bouchard, J., Uribe-Calderon, J., Kamal, M. R. and Lacroix, M. (2012) Nanocrystalline cellulose (NCC) reinforced alginate based biodegradable nanocomposite film. *Carbohydrate Polymers*, **90**(4), 1757-1763.
99. Lange, J., and Wyser, Y. (2003) Recent innovations in barrier technologies for plastic packaging - A review. *Packaging Technology and Science*, **16**(4), 149-158.
100. Wan, W. K., Campbell, G., Zhang, Z. F., Hui, A. J., and Boughner, D. R. (2002) Optimizing the tensile properties of polyvinyl alcohol hydrogel for the construction of a bioprosthetic heart valve stent. *Journal of Biomedical Materials Research Part A*, **63**(6), 854-861.
101. Li, W., Yue, J., and Liu, S. (2012) Preparation of nanocrystalline cellulose via ultrasound and its reinforcement capability for poly (vinyl alcohol) composites. *Ultrasonics Sonochemistry*, **19**(3), 479-485.
102. Lalia, B. S., Guillen, E., Arafat, H. A., and Hashaikeh, R. (2014) Nano-

crystalline cellulose reinforced PVDF-HFP membranes for membrane distillation application. *Desalination*, **332**(1), 134-141.

103. Gindl, W., and Keckes, J. (2004) Tensile properties of cellulose acetate butyrate composites reinforced with bacterial cellulose. *Composites Science and Technology*, **64**(15), 2407-2413.
104. Nakagaito, A. N., Iwamoto, S., and Yano, H. (2005) Bacterial cellulose: the ultimate nano-scalar cellulose morphology for the production of high-strength composites. *Applied Physics A*, **80**(1), 93-97.
105. Lee, K. Y., Blaker, J. J., and Bismarck, A. (2009) Surface functionalisation of bacterial cellulose as the route to produce green polylactide nano-composites with improved properties. *Composites Science and Technology*, **69**(15), 2724-2733.
106. Lee, K. Y., and Bismarck, A. (2012) Susceptibility of never-dried and freeze-dried bacterial cellulose towards esterification with organic acid. *Cellulose*, **19**(3), 891-900.
107. Stevanic, J. S., Joly, C., Mikkonen, K. S., Pirkkalainen, K., Serimaa, R., Rémond, C., Toriz, G., Gatenholm, P., Tenkanen, M., and Salmén, L. (2011) Bacterial nanocellulose-reinforced arabinoxylan films. *Journal of Applied Polymer Science*, **122**(2), 1030-1039.
108. Phomrak, S., and Phisalaphong, M. (2017) Reinforcement of natural rubber with bacterial cellulose via a latex aqueous microdispersion process. *Journal of Nanomaterials*, **2017**, Article ID 4739793.

3

Bio-Inspired Nanomaterials

3.1 Introduction

As a result of recent scientific developments in nanoscience and nanotechnology, research has been focused towards the generation of nanomaterials through green synthesis routes with the aid of biological precursors. The resulting high-performance advanced materials are termed as bio-inspired nanomaterials and exhibit prominent environmental sustainability. Compared to other conventional procedures, nanomaterial fabrication *via* bio-inspiration offers many advantageous characteristics by replacing eco-destructive hazardous materials with eco-friendly biomolecules. These beneficial characteristics of bio-inspired nanomaterials include hierarchical architecture, excellent chemical resistance, accessibility, inexpensiveness, diminished environmental impact, substantial energy effectiveness, low density, balanced stiffness and toughness, etc. [1-8].

As a result, bio-inspired nanomaterials are suitable for diverse biomedical, energy, environmental, antimicrobial, sensing and electronics applications. Among these, biomedical applications like tissue engineering, drug delivery, orthopaedic grafts, protein cages, nervous system, optical implements, cancer rehabilitation, biocatalysis, bio-sensing, bio-imaging, etc., have specifically utilized the bio-generated nanomaterials as bio-templates or bio-mimics owing to their considerable biocompatibility with the human body [7,9-20].

Generally, a vast range of nanomaterials can be designed using either top-down or bottom-up strategies. The top-down strategy includes the size reduction with the application of suitable physicochemical methods, whereas the bottom-up strategy includes the development of nano-framework congregation of atoms, molecules or aggregates. For developing versatile bio-inspired nanomaterials, it is generally recommended to use bottom-up strategies which involve the formation of oriented nano-architectures upon biomolecule

Anish M. Varghese and Vikas Mittal
Khalifa University of Science and Technology, Abu Dhabi, UAE

intervened reduction reactions. Also, the ultimate characteristics such as molecular size, shape, orientation and arrangement of the resultant bio-inspired nanomaterials can be modulated using bottom-up methods [21-24]. Based on the technique followed for the production, the bottom-up strategies are categorized as either direct or indirect production approaches. The direct production category involves various methods, namely precipitation, solvothermal, hydrothermal and ionothermal. The techniques used for the indirect production may involve inclusion, enveloping and modification (Figure 3.1) [18,25]. For designing superior nanomaterials through bio-inspiration, various biomolecules such as microorganisms, viruses, proteins, peptides, deoxyribonucleic acid (DNA), ribonucleic acid (RNA), enzymes, plant extracts and other biomass based materials have displayed great potential [18-20,26-29]. Herein, recent progresses in different strategies developed for the effective generation of various bio-inspired nanomaterials have been reviewed, along with focusing on properties and applications of the developed materials.

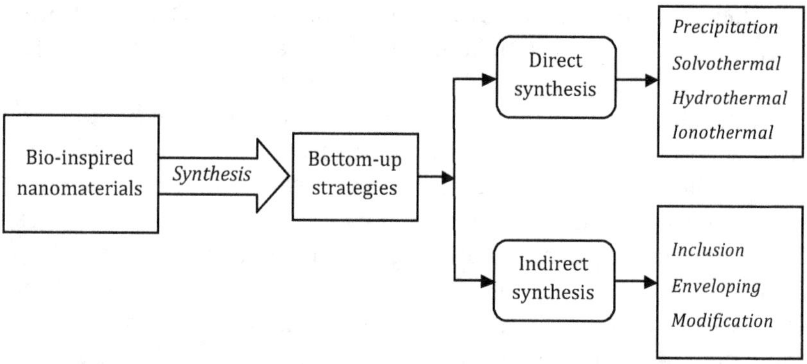

Figure 3.1 Various bottom-up methods for generating bio-inspired nanomaterials.

3.2 Bio-Inspired Gold Nanomaterials

Gold (Au) nanomaterials is a promising class of nanomaterials, which find suitability for vast range of applications. In order to develop environmentally-friendly synthesis approaches for these nanomaterials, bio-inspiration with the aid of various biological precursors has received immense research interest. Also, the bio-inspired synthesis of Au nanomaterials is beneficial to further enrich their biomedical applications. For instance, Dauthal and Mukhopadhyay [30] used the extract of *Delonix regia* plant leaves for the bio-engineered fabrication of Au nanoparticles. The bio-inspired Au nanoparticles demonstrated significant photocatalytic efficiency during the degradation of o-nitroaniline contaminant. Bhau *et al.* [31] reported the application of aqueous *Nepenthes khasiana* plant leaf extract for the green synthesis of Au nanoparticles possessing effective microbial growth resistance. For exploring drug delivery application, Babu *et al.* [32] developed bio-inspired Au nanoparticles with the assistance of ethanol containing leaf extract of *Andrographis paniculata* plant. In another study, Mollick *et al.* [33] employed leaf extract of *Abelmoschus esculents* plant to successfully generate bio-inspired Au nanomaterials for effective cancer treatment.

Also, the leaf extract of various other plants such as *Piper betle* [34], *Butea monosperma* [35], *Chloroxylon swietenia* [36], *Vitis vinifera* [37], *Syzygium aromaticum* [38], *Mentha arvensis* [39], *Vetiveria zizanioides* [40], *Pterocarpus marsupium* [41], *Panax ginseng* [42], *Taxus baccata* [43] and *Bacopa monnieri* [44] have demonstrated substantial usefulness for the eco-friendly bio-inspired synthesis of Au nanoparticles.

In another study, Sett *et al.* [45] attained non-cytotoxic bio-inspired Au nanoparticles through the exploitation of fruit extract of *Dillenia indica* plant. Babu *et al.* [46] detailed the integration of aqueous fruit concentrate of *Solanum indicum* plant during the eco-friendly synthesis of nanoscale Au particles for biomedical applications. Gao *et al.* [47] also reported the potential of kiwifruit concentrate for the beneficial bio-generation of Au nanoparticles for diverse biomedical applications. For enabling superior dermatologic applications, Perde-Schrepler *et al.* [48] developed Au nanoparticles by biological means with the assistance of aqueous fruit extract of *Cornus mas* plant. In another study, Sharma *et al.* [49] employed aqueous fruit extract of *Sapindus mukorossi* plant for the efficient green evolution of Au nanoparticles. Basavegowda *et al.* [50] also reported the favorable reducing capability of *Punica granatum* plant fruit extract during the environmentally-friendly synthesis of Au nanoparticles.

Balamurugan *et al.* [51] reported the use of *Peltophorum pterocarpum* plant aqueous flower extract for the bio-inspired synthesis of globular Au nanoparticles. Geetha *et al.* [52] detailed usefulness of

Couroupita guianensis plant fruit extract for the bio-synthesis of Au nanoparticles with substantial anticancer potential. Noruzi *et al.* [53] also reported the application of flower extract of *Rosa hybrid* plant for the bio-inspired generation of Au nanoparticles. Nirmala *et al.* [54] developed Au nanoparticles through green route for cancer treatment with the assistance of grains extract of *Vitis vinifera* plant. In another work, Patil *et al.* [55] reported the exploitation of *Rhus chinensis* plant seed extract for the green production of Au nanoparticles for effective cancer therapy.

Jun *et al.* [56] applied aqueous root extract of *Polygala tenuifolia* plant for enabling reduction and stabilization of Au nanoparticles during fabrication. The bio-inspired Au nanoparticles displayed potential for anti-coagulant medication owing to blood clotting inhibition capability. Singh *et al.* [57] discussed the eco-friendly synthesis of antibacterial Au nanoparticles with the use of *Red ginseng* plant root extract. In another study, the authors disclosed the practicability of aqueous *Panax ginseng* plant root extract for bio-fabrication of Au nanoparticles with superior microbial germination resistance suitable for vast range of biomedical applications. Kou *et al.* [59] also revealed the use of beet extract for the biological preparation of Au nanoparticles.

Sinha and Ahmaruzzaman [60] reported the bio-generation of Au nanoclusters using aqueous extract of onion peels. The resultant bio-inspired Au nanoclusters exhibited catalytic properties for photo-accelerated degradation of various hazardous dyes. In addition, superior bactericidal features for the Au nanoclusters were also reported. Patil *et al.* [61] also reported the bio-inspired generation of Au nanoparticles using onion peels for the purpose of modifying the characteristics of *Bombyx mori L.* silk and, thereby, to achieve effective biomedical applications. In another study, Castro *et al.* [62] reported orange peel extract assisted bio-fabrication of Au nanoparticles. Babu *et al.* [63] reported the application of *Cocus nucifera* water for the generation of non-toxic bio-inspired nanoscale Au particles. Ganapuram *et al.* [64] utilized *Salmalia malabarica* plant secreted wax during the biological production of nanoscale Au particles with remarkable catalytic performance. Das *et al.* [65] also reported plant sourced quercitin as an effective reducing as well as stabilizing agent during green synthesis of Au nanoparticles.

El-Kassas and El-Sheekh [66] developed Au nanoparticles with the assistance of *Corallina officinalis* red algae extract. The bio-derived Au nanomaterial displayed potential for cancer therapy. In another study, Singh *et al.* [67] used *Padina gymnospora* seawater algae extract for the bio-fabrication of Au nanoparticles for drug delivery system for controlled cancer therapy.

Gao *et al.* [68] reported the surface functionalization of Au nanoparticles by biological means using *Escherichia coli* bacterial cuticles (Figure 3.2). Outstanding stability and activity of the developed material under biological conditions offer high potential for its use in biomedical applications as an effective bactericide. Singh *et al.* [69] described the applicability of *Bhargavaea indica* bacteria for the facile environmental friendly production of Au nanoparticles having flower like morphology. For developing efficient cancer medication delivery system, Kumar *et al.* [70] explored the use of *Delftia* bacteria mediated generation of biogenic nanoscale Au particles. Gopinath and Arumugam [71] reported the green synthesis of Au nanoparticles using *Fusarium solani* fungal extract. Significant reduction potential and stabilization capability were observed for the developed system. Sarkar *et al.* [72] also discussed the green synthesis of Au nanoparticles through the integration of secreted proteins of *Alternaria alternata* fungi.

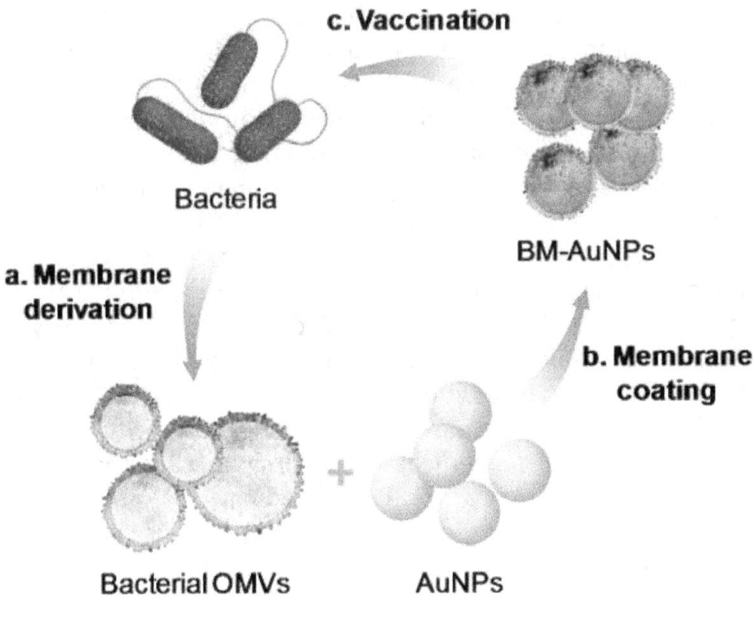

Figure 3.2 Schematic of the synthesis of bacterial membrane capped Au nanoparticles and their antibacterial activity. Reproduced from Reference 68 with permission from American Chemical Society.

For exploring biomedical applications, Liu *et al.* [73] used mussel derived nanostructured polydopamine coated gold nanoparticles with hierarchical shell/core morphology. The resultant nanocomposite structure displayed biocompatibility, low cytotoxicity, appreciable delivery capability and improved *in vivo* stability. In another study, Perez-Mitta *et al.* [74] fabricated nano-fluidic diodes based on self-polymerized dopamine (PDOPA) and Au through bio-synthesis. Black *et al.* [75] reported the surface modification of Au nano-rods by biological means using polydopamine, followed by the introduction of antibody, anti-EGFR coating for generating effective cancer cell detection and subsequent treatment potential. Significant stability was observed for the bio-inspired nanostructure at human body conditions.

With an aim to overcome the clustering of Au nanoparticles and, thus, to achieve superior *in vivo* biomedical applications, Nowinski *et al.* [76] reported the development of stealth peptide encapsulated Au nanoparticles through bio-route. The bio-inspired nanocomposite structure demonstrated safe and non-toxic operating conditions inside the human body. In order to modulate the Au nanoparticles based on demand, Briggs *et al.* [77] reported the application of peptide covering *via* green synthesis (Figure 3.3). Liu *et al.* [78] discussed the modification of Au nanoparticles using biotinylated peptide mediated biological synthesis for sensing of botulinum neurotoxin. Tan *et al.* [79] reported the fabrication of anisotropic DNA-functionalized gold nanoparticles *via* green route. The authors also reported the generation of regioselective analogs of these anisotropic nanostructures for enhancing practicability.

Liang *et al.* [80] reported the Au nanoparticles exhibiting substantial catalytic cyclability, generated using procyanidin affixed egg shell proteins. Wongkrongsak *et al.* [81] described the green synthesis of Au nano-colloid with the assistance of nanoscale fragments of silk fibroin. In order to modulate the interactions of Au nanoparticles with proteins during bio-synthesis and, thereby, to widen the potential for diverse biomedical applications as carriers as well as self-curatives, Moyano *et al.* [82] reported the use of sulfobetaine ligands. The introduction of ligand structure was beneficial to achieve bio-inspired nanoparticles by means of corona-free nanostructure formation.

Table 3.1 presents the comparison of some of the reported biological precursors for the generation of bio-inspired Au based nanoparticles.

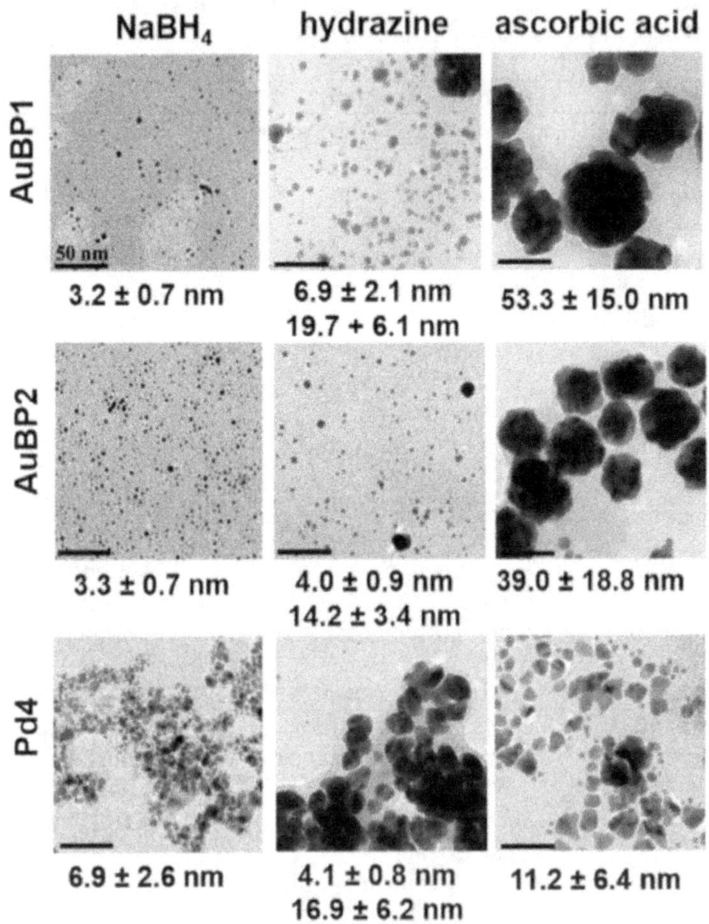

Figure 3.3 Transmission electron micrographs (TEM) of the Au nanoparticles prepared as a function of peptide passivant and reductant (all scale bars are 50 nm). Reproduced from Reference 77 with permission from American Chemical Society.

Ibrahim *et al.* [83] made use of marine bacterial membranes for biological development of nanoparticles based on Au and Au/TiO_2 as well as Au/ZnO biphasic hybrids. The use of marine bacterial membranes in the formulations was beneficial for reduction as well as stabilization of nanoparticles. The resultant nanoparticles demonstrated significant resistance against bacterial growth and UV degradation. Zhou *et al.* [84] also designed Au nanoparticle/metal organic framework biphasic nanocomposite by employing the green

Table 3.1 Comparison of various biological precursors used for the generation of bio-inspired Au nanoparticles

Authors	Biological precursor	Attributes
Dauthal *et al.* [30]	*Delonix regia* leaves extract	Photocatalytic
Bhau *et al.* [31]	*Nepenthes khasiana* leaves extract	Antimicrobial
Mollick *et al.* [33]	*Abelmoschus esculents* leaves extract	Anticancer and antimicrobial
Sett *et al.* [45]	*Dillenia indica* fruit extract	Antibacterial
Geetha *et al.* [52]	*Couroupita guianensis* flower extract	Anticancer
Patil *et al.* [55]	*Rhus chinensis* seed extract	Anticancer
Jun *et al.* [56]	*Polygala tenuifolia* root extract	Anticoagulant
Singh *et al.* [57]	*Red ginseng* plant root extract	Antibacterial
Singh *et al.* [58]	*Panax ginseng* root extract	Antimicrobial
Sinha *et al.* [60]	Onion peels extract	Photocatalytic
Babu *et al.* [63]	*Cocus nucifera* water	Anticancer
Ganapuram *et al.* [64]	*Salmalia malabarica* gum	Photocatalytic
El-Kassas *et al.* [66]	*Corallina officinalis* red algae extract	Anticancer
Singh *et al.* [67]	*Padina gymnospora* marine algae extract	Cancer therapy
Gao *et al.* [68]	*Escherichia coli* bacterial cuticles	Antibacterial
Liu *et al.* [73]	Polydopamine	Biocompatibility and *in vivo* stability
Black *et al.* [75]	Polydopamine and anti-EGFR antibody	*In vivo* stability
Nowinski *et al.* [76]	Stealth peptide	Biocompatibility and non-toxicity
Wongkrongsak *et al.* [81]	Silk fibroin nano-fragments	Oxidation resistance

route involving polydopamine sourced from mussels. The presence of polydopamine in the core/shell morphological nanostructure was advantageous to further enrich the performance in various areas such as biomedical, sensing and catalysis.

Dauthal *et al.* [85] used extract of *Delonix regia* plant leaves in the bio-synthetic formulation of Au/Pd biphasic hybrid nanoparticles as an effective reductant. Appreciable photocatalytic potential and re-usability were acquired for the resultant hybrid nanostructure. Huang *et al.* [86] reported the use of cysteine betaine for the biological functionalization of Au nanoparticles and Au/Ag biphasic nano-shells. Improvement in photo-corrosion resistance as well as anti-microbial properties for the bio-synthesized cysteine betaine enveloped nanostructures were accomplished. For the purpose of effective water treatment, Liu *et al.* [87] fabricated trilayer membrane comprising of nano-Au, nano-TiO_2 and anodized Al_2O_3 through bio-approach using solar energy. The strategy was observed to be preferable for treating contaminated water *via* photocatalysis and subsequent evaporation routes.

3.3 Bio-Inspired Silver Nanomaterials

Similar to Au based nanomaterials, biologically inspired silver (Ag) nanomaterials also offer great potential for use in biomedical applications. For instance, Elangovan *et al.* [88] reported the application of aqueous extract of *Andrographis echioides* plant leaves for the bio-fabrication of Ag nanoparticles exhibiting superior bactericidal behavior. Furthermore, the aqueous leaf extracts of various plants such as *Sesbania grandiflora* [89], *Ocimum sanctum* [90], *Ocimum canum* [91], *Musa paradisiaca* [92], *Vinca rosea* [93], *Eclipta prostrata* [94], *Euphorbia hirta* [95], *Euphorbia prostrate* [96], *Eriobotrya japonica* [97], *Nerium oleander* [98], *Murraya koenigii* [99], *Mimosa pudica* [100], *Sida acuta* [101], *Skimmia laureola* [102], *Chomelia asiatica* [103], Calotropis gigantea [104], *Heliotrpium indicum* [105], *Feronia elephantum* [106], *Nelumbo nucifera* [107], *Dendrophthoe falcata* [108], *Panax ginseng* [42], *Tephrosia purpurea* [109], *Tinospora cordifolia* [110] and *Lowsonia inermis* [111] have demonstrated suitability for the biological generation of Ag nanoparticles with superior antimicrobial and biolarvicidal properties.

Meng *et al.* [112] detailed the green synthesis of nanoscale Ag particles with the assistance of *Rosa chinensis* plant flower extract. Resistance towards gram positive and negative bacterial germination was observed for the bio-derived Ag nanoparticles. In another study, Manikandan *et al.* [113] reported the use of *Rosa indica* plant flower extract for the bio-inspired synthesis of Ag nanoparticles exhibiting bacterial and inflammation resistance as well as attenuated cytotoxicity. Gogoi *et al.* [114] prepared Ag nanoparticles by biological route with the help of alcoholic flower extract of *Nyctanthes arbortristis* plant. Considerable bactericidal features along with non-toxicity for the biogenic Ag nanoparticles were observed.

Ramar *et al.* [115] discussed the usefulness of *Solanum trilobatum* herb aqueous fruit extract for the biological preparation of nanoscale Ag particles with superior cytotoxicity against cancer cells and bactericidal efficacy. The observed features of the bio-inspired Au nanomaterial confirmed suitability for use in cancer therapy. Singh *et al.* [116] reported outstanding reduction and stabilization capability of aqueous fruit extract of *Dillenia indica* plant during the biomimetic generation of Au nanoparticles. Bose *et al.* [117] made use of seed cake extract of *Jatropha curcas* plant for the green evolution of Au nanoparticles with consequential bactericidal effects. Jun *et al.* [118] also observed favorable reduction as well as stabilization of *Polygala tenuifolia* plant root aqueous extarct for the green synthesis of Ag nanoparticles displaying significant bactericidal capability. In another study, Hazarika *et al.* [119] also reported strong resistance towards bacterial as well as fungal growth for bio-inspired Ag nanoparticles based on aqueous root extract of *Thalictrum foliolosum* plant.

Murugan *et al.* [120] utilized the aqueous extract of *Acacia leucophloea* plant for developing bactericidal Ag nanoparticles *via* biological means. Also, the bark extract of *Hovenia dulcis* tree presented feasibility for the bio-inspired generation of Ag nanoparticles for developing high-quality antibiotic with substantial bactericidal proficiency [121]. Santhoshkumar *et al.* [122] acquired superior anti-parasitic effectiveness for nanoscale Ag particles generated with the assistance of *Cissus quadrangularis* plant aqueous stem extract. In another study, Kothai and Jayanthi [123] also synthesized Ag nanoparticles by biological means with the incorporation of aqueous extract of *Camellia sinensis* shrub containing lemon and honey which offered both reduction and stabilization. The authors observed superior homogeneity for the bio-generated Ag nanoparticles using this approach.

For achieving eco-friendly generation of Ag nanoparticles, Castro *et al.* [124] reported the use of orange peel extract. Pronounced bacterial resistance was observed for the bio-inspired system, thus, confirming the feasibility for drug carrying applications. Abubakar *et al.* [125] detailed the use of black carrot root extact for the facile green production of nanoscale Ag particles. In order to realize the effective wound healing applications, Anjum *et al.* [126] reported bio-inspired hydrogel system based on nano-Ag, curcumin and *Aloe vera* plant as a superior dressing material with effective microbial inhibition capability, therapeutic features and non-toxicity.

Aziz *et al.* [127] detailed the synthesis of algae (*Chlorella pyrenoidosa*) inspired Ag nanoparticles with outstanding purity, biocompatible nature and environmental safety (Figure 3.4). These features along with excellent biofouling resistance and photocatalytic capability of the bio-derived nanoparticles make them suitable for extensive waste water management. The resultant bio-inspired Ag nanoparticles also demonstrated outstanding antibacterial performance against various bacterial species. Ebrahiminezhad *et al.* [128]

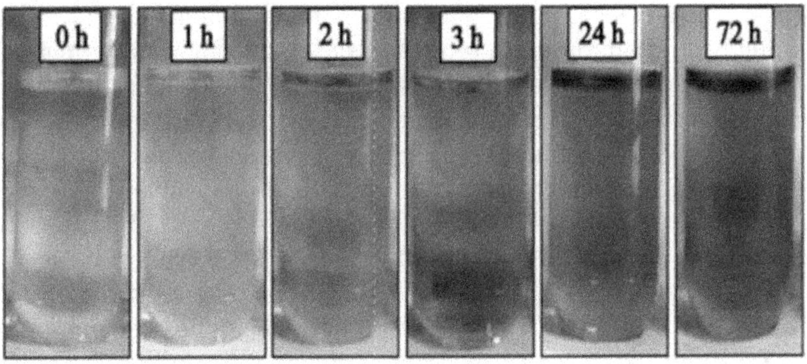

Figure 3.4 Optical images of Ag nanoparticles from start of biosynthesis to the final stage. Reproduced from Reference 127 with permission from American Chemical Society.

reported the use of *Chlorella vulgaris* algae for the biological preparation of nanoscale Ag particles in order to achieve efficient resistance towards bacteria and cancer cell growth. In another study, Aziz *et al.* [129] also described the adequacy of aqueous *Scenedesmus abundans* green algae extract for the biological development of Ag nanoparticles. Significant bioaccumulation along with substantial bactericidal capability were observed for the system. Selvi *et al.* [130] reported cytotoxic behavior towards cancer cells for biocidal Ag nanoparticles engineered with the assistance of *Padina tetrastromatica* algae.

Railean-Plugaru *et al.* [131] deployed acidophilic *Actinobacteria* for the effective green synthesis of Ag nanoparticles with notable bactericidal performance. Anasane *et al.* [132] reported the ability of acidophilic *Actinobacteria* inspired Au nanoparticles for exterior fungal infection treatment owing to antimycotic nature. In another study, Buszewski *et al.* [133] generated antibacterial Ag nanoparticles by the use of *Streptacidiphilus durhamensis* with reduction as well as stabilization capability. In another recent strategy, Golinska *et al.* [134] carried out bio-derivation of Ag nanoparticles with the help of *Pilimelia columellifera* subsp. *pallida* for enabling substantial bactericidal efficiency. Rathod *et al.* [135] revealed the practicability of *Nacardiposis valliformis* based Au nanoparticles for cancer therapy due to their cytotoxic performance against cancer cells. Also, the resultant Ag nanoparticles demonstrated effective bacterial growth resistance.

Gou *et al.* [136] made use of *Lysinibacillus sphaericus* bacterial extract for the facile biological generation of Ag nanoparticles [136]. In another study, Bhatia *et al.* [137] observed antimicrobial efficiency for bio-fabricated Ag nanoparticles with the support of *Lysinibacillus varians* bacterial extract. Kalabegishvili *et al.* [138] reported the preparation of nanoscale Ag particles by green route through the integration of *Sulfolobus islandicus* microbial concentrate. Wang *et al.* [139] also detailed the facile biocidal synthesis of Ag nanoparticles with the assistance of *Microbacterium resistens* bacterial extract. In another recent study, Wang *et al.* [140] observed significant microbial growth resistance for Ag nanoparticles generated by green route through *Bacillus methylotrophicus* bacteria. Jo *et al.* [141] implemented *Pseudomonas deceptionensis* bacteria in order to biologically develop Ag nanoscale particles with high microbial resistance. Also, the bacterial species such as *Weissella oryzae* [142] and *Ralstonia sp.* [143] have presented high potential as effective biological media for the generation of Ag nanocrystals.

Govindappa *et al.* [144] reported the use of *Penicillium* fungal extract for the generation of Ag nanoparticles through biological means. The resultant bio-fabricated Ag nanoparticles exhibited inflammation reduction and superior microbial and oxidation resistance. Rahi *et al.* [145] also discussed the use of *Aloe vera* herb root sourced *Penicillium* fungal extract for the biological preparation of Ag nanoparticles. In another strategy, Shelar and Chavan [146] developed Ag nanoparticles through the inclusion of *Trichoderma harzianum* fungal extract. The bio-inspired Ag nanoparticles displayed potential for agricultural applications for enriching the seed vitality. Fayaz *et al.* [147] reported the synthesis of Ag nanoparticles with noticeable antibacterial effectiveness through the use of

Trichoderma viride fungi. Abdel-Hafez *et al.* [148] reported the usefulness of *Alternaria solani* fungal aqueous extract for the favorable reduction and stabilization of Ag nanoparticles during bio-synthesis. Aziz *et al.* [149] also developed Ag nanoparticles by biological means with the assistance of *Mucor hiemalis* fungal extract. The outstanding antimicrobial performance of the developed bio-inspired Ag nanoparticles confirmed the suitability for extensive biomedical applications. Krishna *et al.* [150] reported the bio-inspired synthesis of Ag nanoparticles using *Trametes sp.* extract and observed significant *Salmonella sp.* bacterial germination resistance for the developed nanoparticles.

Kumar *et al.* [151] reported the inclusion of *Bombyx mori* silk fibroin extract for the biological synthesis of Ag nanoparticles with globular morphology *via* gamma treatment. In this system, the presence of silk fibroin extract was beneficial to successfully accomplish both reduction and stabilization of Ag nanoparticles. In another study, Shivananda *et al.* [152] detailed silk fibroin extract mediated bio-generation of Ag nanoparticles through in-situ approach. Yue *et al.* [153] generated bio-inspired Ag nanoparticles using sericin for enabling high performance fabric finishing applications with superior bactericidal properties. In another recent study, Aramwit *et al.* [154] also reported the practicability of sericin for the bio-generation of Ag nanoparticles with outstanding antimicrobial behavior.

Wu *et al.* [155] reported the use of polydopamine as an effective reductant as well as stabilizer for the environmentally friendly synthesis of Ag/polydopamine hybrid nanocrystals having substantial bactericidal efficacy. In another study, Sinha *et al.* [156] utilized mixture of saturated fatty acid and tryptophan for the green synthesis of Ag nanoparticles. In this system, saturated fatty acid provided stabilization capability, whereas tryptophan offered reduction capability. The bio-derived Ag nanoparticles displayed notable photocatalytic behavior during the reduction reaction of methylene violet pigment. Kim *et al.* [157] developed Ag nanoparticles by biological means through the support of tannic acid. Resistance towards oxidation and bacterial growth was accomplished for the bio-inspired Au nanoparticles.

Francesko *et al.* [158] reported biological modification of Ag nanoparticles using chitosan and aminocellulose for the generation of bactericidal coatings over cork exterior. Aryabadie *et al.* [159] developed antibacterial coating system based on bio-designed Ag nanoparticles through the incorporation of branched poly(amido amine) functionalized chitosan. Wang *et al.* [160] also detailed the green synthesis of ternary nanocomposite comprising of chitosan, gelatin and nano-Ag for diverse biomedical applications as implants as well as dermis and muscles replications owing to its bactericidal features and conductive nature.

Anuradha *et al.* [161] reported the use of arabinogalactan for the green synthesis of Ag nanoparticles possessing outstanding antimicrobial capability and photocatalytic behavior. Huang *et al.* [162] detailed the bio-synthesis of Ag nanoparticles with bacterial germination resistance by the application of chicken egg shell. Sinha and Ahmaruzzaman [163] employed mallard egg shell for the bio-inspired synthesis of Ag and Au/Ag hybrid nanoparticles. Both types of nanoparticles exhibited suitability for water treatment owing to their photocatalytic proficiency during the reduction reactions of various contaminant pigments. Table 3.2 also presents the comparison of some of the biological precursors used for the generation of bio-inspired Ag nanoparticles.

Lu *et al.* [164] used aqueous extract of *Cacumen platycladi* plant for the design of Ag/Pd hybrid nanoparticles. Catalytic performance during the hydrogenation reaction along with outstanding yield and reliability were accomplished for the resultant bio-inspired hybrid nanostructure. Vivekanandhan *et al.* [165] fabricated bio-inspired Ag/ZrO_2 shell/core biphasic hybrid nanostructure using leaf extract of *Anacardium occidentale* plant. Montazer and Kahali [166] also prepared ternary hydrogel system composed of nano-Ag, poly(vinyl alcohol) and tragacanth by biological means. Superior microbial growth resistance was reported for the developed bio-fabricated hydrogel system.

3.4 Bio-Inspired Palladium Nanomaterials

Bio-inspired palladium (Pd) nanomaterials have been developed through the exploitation of various biological precursors. For instance, Sheny *et al.* [167] reported the usage of shrivelled leaves of *Anacardium occidentale* tree for the biological generation of Pd nanoparticles [167]. In another study, Mallikarjuna *et al.* [168] reported the preparation of Pd nanoparticles by using biological route with the aid of aqueous leaf extract of *Piper betle* plant. The authors observed excellent resistance of the developed material against fungal germination.

Table 3.2 Comparison of various biological precursors used for the bio-inspired synthesis of Ag nanoparticles

Authors	Biological precursor	Attributes
Elangovan *et al.* [88]	*Andrographis echioides* leaves extract	Antibacterial
Manikandan *et al.* [113]	*Rosa indica* flower extract	Antibacterial, anti-inflammatory and anticancer
Gogoi *et al.* [114]	*Nyctanthes arbortristis* flower extract	Antibacterial, anticancer
Ramar *et al.* [115]	*Solanum trilobatum* fruit extract	Anticancer, antibacterial
Bose *et al.* [117]	*Jatropha curcas* grain cake extract	Antibacterial
Hazarika *et al.* [119]	*Thalictrum foliolosum* root extract	Antimicrobial
Salunke *et al.* [121]	*Hovenia dulcis* cortex extract	Antibacterial
Castro *et al.* [124]	Orange peel extract	Antibacterial
Anjum *et al.* [126]	Curcumin and *Aloe vera* plant gel	Antimicrobial
Aziz *et al.* [127]	*Chlorella pyrenoidosa* algae	Antibiofouling, photocatalytic
Ebrahiminezhad *et al.* [128]	*Chlorella vulgaris* algae	Antibacterial, anticancer
Aziz *et al.* [129]	*Scenedesmus abundans* algae	Bioaccumulation, antibacterial
Railean-Plugaru *et al.* [131]	*Actinobacteria*	Antibacterial
Rathod *et al.* [135]	*Nacardiposis valliformis* bacteria	Antibacterial
Bhatia *et al.* [137]	*Lysinibacillus varians* bacteria	Antimicrobial
Jo *et al.* [141]	*Pseudomonas deceptionensis* bacteria	Antimicrobial
Govindappa *et al.* [144]	*Penicillium* fungi	Anti-inflammatory, antimicrobial, oxidation resistance
Aziz *et al.* [149]	*Mucor hiemalis* fungi	Antimicrobial
Armwit *et al.* [154]	Sericin	Antimicrobial
Wu *et al.* [155]	Polydopamine	Antibacterial
Kim *et al.* [157]	Tannic acid	Oxidation resistance, antibacterial

Wang *et al.* [160]	Chitosan and gelatin	Antibacterial, conducting nature
Anuradha *et al.* [161]	Arabinogalactan	Antimicrobial, photocatalytic
Huang *et al.* [162]	Chicken egg shell	Antibacterial
Sinha *et al.* [163]	Mallard egg shell	Photocatalytic

Khan *et al.* [169] detailed the environmentally convivial generation of Pd nanoparticles exhibiting catalytic proficiency through the exploitation of aqueous extract of *Pulicaria glutinosa* plant. Lebaschi *et al.* [170] reported *Camellia sinensis* shrub leaf extract supported biological synthesis of Pd nanoparticles for photocatalytic applications. Shaik *et al.* [171] developed Pd nanoparticles using an eco-benign approach with the help of *Origanum vulgare* plant aqueous extract. Also, the eco-friendly preparation of Pd nanoparticles with the help of bio-reductants such as leaf extract of *Withania coagulans* plant [172] and fruit extract of *Berberis* plant [173] has been reported to develop high performance green nanocomposite structures. Ganaie *et al.* [174] demonstrated reduction, capping, and stabilization capability of *Antigonon leptopus* plant leaf, stem and root extract for the green fabrication of Pd nanoparticles. Bedford *et al.* [175] described the design of nanostructured catalyst based on biologically inspired Pd nanoparticles using peptide (Figure 3.5). Modulating catalytic features were observed for the developed bio-derived nanostructure.

Nasrollahzadeh *et al.* [176] reported the environmentally friendly bio-fabrication of Pd/TiO_2 biphasic hybrid nanostructure with catalytic efficacy by using *Myrtus communis* plant leaf extract. In another study, the biological generation of Pd nanocomposite with CuO was carried out using seed extract of *Theobroma cacao* plant [177]. Khodadadi *et al.* [178] employed *Saliva hydrangea* plant extract as an effective bio-reductant for the nature-friendly generation of Pd/apricot seed husk nanoscale composite with substantial catalytic stability. Bordbar and Mortazavimanesh [179] also reported the use of aqueous leaf extract of *Equisetum arvense* plant for the bio-generation of nanocomposite composed of Pd nanoparticles decorated walnut husk.

3.5 Bio-Inspired Iron Oxide Nanomaterials

Bio-inspired iron oxide nanomaterials have been developed using various biological precursors, e.g., plant extracts, algal species, fungi,

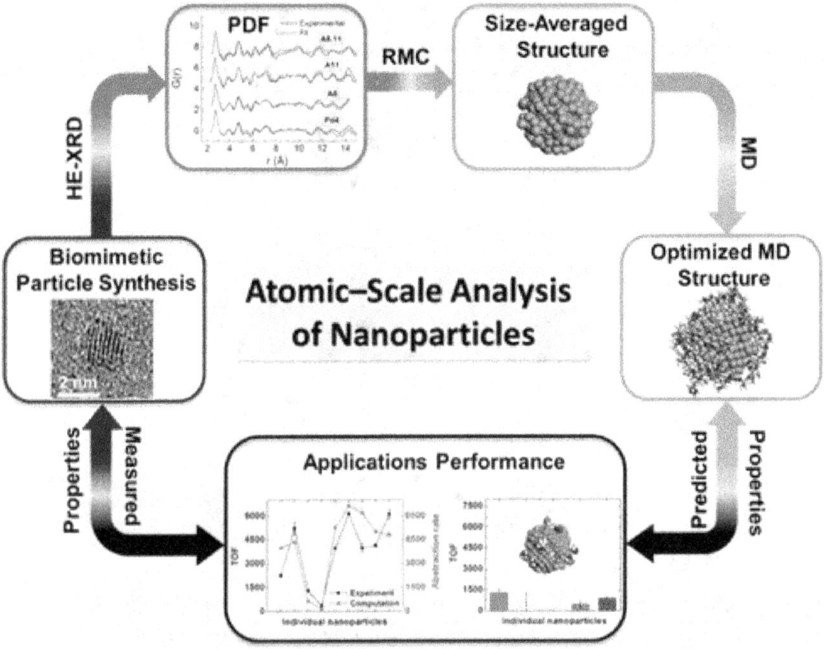

Figure 3.5 Representation of the nanostructure design. Reproduced from Reference 175 with permission from American Chemical Society.

proteins, polydopamine, ascorbic acid, etc. Makarov *et al.* [180] utilized aqueous extracts of plants such as *Hordeum vulgare* and *Rumex acetosa* for the biological production of iron oxide nanoparticles. The nanoparticles demonstrated almost same net protein amount and oxidation resistance in both cases. However, *Rumex acetosa* inspired iron oxide nanoparticles displayed higher stability and congregation resistance as compared to *Hordeum vulgare*. It was related to the acidic content in the resultant iron oxide nanoparticles as the particles with greater acidic concentration exhibited greater stability. Figure 3.6 demonstrates the morphology of iron oxide nanocrystals obtained through the use of two plants extracts. Balamurughan *et al.* [181] discussed the development of Fe_2O_3 nanoparticles by biological means by the use of aqueous leaf extract of *Ocimum sanctum* plant. Superior reduction and stabilization capability was observed for the developed system. In another study, Balamurugan *et al.* [182] also reported the preparation of bio-engineered β-Fe_2O_3 nanoparticles through the application of leaf extract of *Eucalyptus globules* plant.

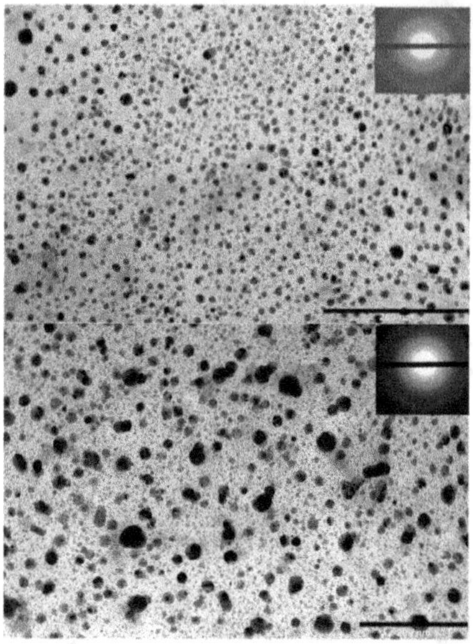

Figure 3.6 TEM micrographs of nanoparticles synthesized in (top) *Hordeum vulgare* extract and (bottom) *Rumex acetosa* extract. Reproduced from Reference 180 with permission from American Chemical Society.

Kumar *et al.* [183] used aqueous extract of *Andean blackberry* plant leaves for the bio-fabrication of Fe_3O_4 nanoparticles. Thermogravimetric analysis of the bio-engineered nanostructure revealed about 95% metallic content and 5% biomolecular coating. Exceptional catalytic efficiency during the light induced reduction of various organic pigments was accomplished. Also, bio-inspired Fe_3O_4 nanoparticles demonstrated appreciable resistance towards oxidation reactions. Kombaiah *et al.* [184] detailed the environmentally friendly design of $ZnFe_2O_3$ through the exploitation of aqueous leaf extract of *Hibiscus rosa-sinensis* plant.

Yew *et al.* [185] reported the biological evolution of high performance Fe_2O_3 nanoparticles with the support of *Kappaphycus alvarezii* red algae extract. Globular shaped morphology was observed for the bio-inspired Fe_2O_3 nanoparticles. In another study, El-Kassas *et al.* [186] revealed outstanding reduction as well as stabilization capability for brown algae such as *Padina pavonica* and *Sargassam acinarium* during the biological fabrication of Fe_2O_3 nanoparticles. Also, the resultant bio-inspired nanoparticles displayed superior Pb removal efficiency. Mohamed *et al.* [187] also reported the biological fabrication of iron nanoparticles displaying superior antimicrobial performance through the use of *Alternaria alternata* fungal extract.

Kashyap *et al.* [188] carried out the biological synthesis of iron oxide nanoparticles through nucleation of micelle-bound iron species with the aid of bacterial protein named Mms6 (Figure 3.7). In another study, Zhang *et al.* [189] also discussed the development of eco-friendly phosphatidylcholin capped Fe_3O_4 nanoparticles for developing effective adsorption and absorption materials.

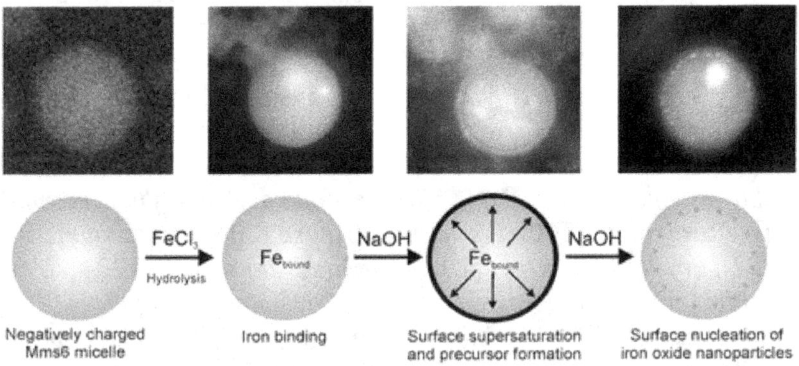

Figure 3.7 Illustration of bacterial protein (Mms-6) assisted iron oxide nanoparticle nucleation. Reproduced from Reference 188 with permission from American Chemical Society.

In another study, Wang *et al.* [190] utilized polydopamine as a stabilizer for the development of bio-inspired Fe_3O_4 nanoparticles as sorbent material for water treatment. Si *et al.* [191] obtained biocompatibility for polydopamine functionalized Fe_3O_4 nano-spheres, which confirmed their usefulness for catalytic applications. Zeng *et al.* [192] reported the biological stabilization of Fe_3O_4 nanoparticles with the assistance of polydopamine to generate uniform surface distribution of nanoscale Au particles for maintaining appreciable catalytic activity during degradation of various dyes. Zhu and Pan [193] also carried out immobilization of Fe_2O_3 nanoparticles over foam surface with the help of mussels. The biological means of immobilization of Fe_2O_3 nanoparticles was favorable to achieve superior hydrophobicity, which is helpful for the segregation of oil from its mixture with water. Sood *et al.* [194] developed biphasic hybrid nanostructure by depositing Au nanoparticles on the surfaces of Fe_2O_3 nanoparticles through bio-approach with the support of ascorbic acid for generating effective biomedical applications. Combined magnetic and plasmonic resonance behaviors were acquired for the bio-inspired hybrid nano-system owing to the synergistic effect of the components.

3.6 Bio-Inspired Zinc Oxide Nanomaterials

For designing environmentally-friendly zinc oxide (ZnO) nanomaterials, green synthesis approaches have been reported using biological precursors like plant extracts, bacterial serum, fungi, chitosan, polyamine, etc. Shah *et al.* [195] described the eco-friendly development of ZnO nanoparticles through the integration of *Camellia sinensis* leaf extract. The resultant bio-inspired ZnO nanomaterial exhibited antibacterial as well as antifungal properties. Also, the leaf extracts of various other plants such as *Anisochilus carnosus* [196], *Lobelia leshenaultiana* [197], *Mimosa pudica* [198], *Vitex negundo* [199], *Vitex trifolia* [200], *Corymbia citridora* [201], *Pongamia pinnata* [202], *Phyllanthus niruri* [203], *Corriandrum sativum* [204], *Limonia acidissima* [205], *Heritiera fomes* [206], *Sonneratia apetala* [206], *Lantana aculeata* [207], *Azadirachta indica* [208], *Agathosma betulina* [209] and *Aleo vera* [210] have displayed feasibility as effective biological agents with appreciable reduction as well as stabilization proficiency for the environment friendly preparation of ZnO nanoparticles by biological means.

Jamdagni *et al.* [211] reported the use of flower extract of *Nyctanthes arbor-tristis* plant for the bio-fabrication of ZnO nanoparticles possessing resistance against fungal proliferation. In another study, Diallo *et al.* [212] prepared ZnO nanoparticles *via* nature-friendly approach with the support of *Aspalathus linearis* plant flower extract as effective reducing and stabilization agent. Dobrucka and Dlugaszewska [213] used flower extract of *Trifolium pratense* plant as an effective medium for the biological fabrication of ZnO nanoparticles. Sharma *et al.* [214] disclosed the feasibility of flower extract of *Jacaranda mimosifolia* plant for the generation of ZnO nanoparticles by biological approach. Azizi *et al.* [215] also used flower extract of *Anchusa italica* plant as an efficient agent for the bio-fabrication of ZnO nanoparticles.

Jafarirad *et al.* [216] reported the use of *Rosa canina* plant fruit extract as an effective reducing medium for the environmentally friendly production of bactericidal ZnO nanoparticles. Nethravathi *et al.* [217] also developed bio-inspired nanoscale ZnO particles by the application of fruit extract of *Garcinia xanthochymus* plant. Also, the fruit extracts of *Carisa edulis* plant [218] and *Terminalia chebula* plant [219] have presented suitability for the green synthesis of ZnO nanocrystals as superior biological precursors.

Karnan and Selvakumar [220] employed aqueous fruit peel extract of *Nephelium lappaceum* plant for the bio-engineered generation of ZnO nanoparticles with photocatalytic performance. In another study, Yuvakkumar *et al.* [221] reported the eco-friendly synthesis of ZnO nano-chains through the exploitation of rambutan peels. Supraja *et al.* [222] observed antibacterial properties of bio-inspired ZnO nanoparticles developed with the help of stem bark extract of *Bosewellia ovalifoliolata* plant. Stan *et al.* [223] observed the effectiveness of extracts of various plants like onion, garlic and parsley for the eco-amiable generation of high performance photocatalytic ZnO nanoparticles. In another work, Sharma [224] revealed the formation of bactericidal flower-shaped ZnO nanomaterials through the exploitation of mucilage of *Carica papaya* fruit. Krishna *et al.* [225] detailed lemon extract mediated biological generation of ZnO nanoparticles. The resultant bio-inspired nanostructure displayed feasibility for cancer therapy owing to its good biocompatibility, absence of side effects and anti-tubercular behavior. In another study, Colak and Karakose [226] reported the application of lemon skin extract for developing ultra-thin ZnO nano-film through biomimetic way.

Buazar *et al.* [227] utilized potato extract to develop ZnO nanoparticles in an environmental friendly manner for enabling diverse biological applications. In another study, Sutradhar and Saha [228] described tomato extract supported synthesis of ZnO nanoparticles for designing high performance ternary nanocomposite comprising GO and nano-TiO_2 and to accomplish diverse photovoltaic applications. Also, the suitability of biological reducing agents such as aqueous coffee powder extract [229], black tea residue extract [230] and green tea leaf extract [231] for the eco-friendly evolution of ZnO nanoparticles has been reported.

Pandimurugan and Thambidurai [232] reported the use of *Padina tetrastromatica* brown algal species as an effective agent for the generation of nanoscale ZnO particles. Ghayempour and Montazer [233] reported superior photocatalytic and bactericidal features for ZnO nanoparticles bio-designed with the aid of *tragacanth gum*. In another study, the authors also used *tragacanth gum* for the green synthesis of hierarchical ZnO nano-rods possessing photocatalytic efficacy and resistance towards various bacteria and fungi.

Kundu *et al.* [235] reported the biomimetic generation of ZnO nanoparticles by using *actinobacteria*. The resultant bio-inspired ZnO nanoparticles displayed potential for biomedical and engineering applications due to the resistance against UV radiations and bacterial proliferation. In another study, Shamsuzzaman *et al.* [236] described *Candida albicans* fungi supported green production of ZnO nanoparticles for catalytic applications. Dhillon *et al.* [237] observed microbial as well as biofouling resistance for chitosan mediated biodesigned ZnO nanoparticles. Abiraman *et al.* [238] also reported the generation of ZnO nano-rods by biological route with the application of chitosan. Effective biofouling resistance for the ZnO nano-rod based coating systems was observed. Manna *et al.* [239] observed photocatalytic features and biofouling resistance for nanoarchitecture based on polyamine inspired ZnO. Kirthi *et al.* [240] also used starch solution for bio-generation of ZnO nanoparticles. Resistance against parasite germination was observed for the starch inspired nanoparticles.

Zhao *et al.* [241] reported the use of leaf extract of *Tribulus terrestris* plant for the environmentally-friendly preparation of ZnO/Ag core/shell biphasic nanocomposite with gas detection capability. Ansari *et al.* [242] described the generation of biologically inspired Ag/ZnO biphasic nanostructures with the aid of electrochemically active biofilm. The biphasic nanocomposite exhibited suitability for use in photoelectric cell design as electrode structure and light activated reactions as catalyst. Figure 3.8 illustrates the schematic of photocatalytic mechanism of the Ag/ZnO biphasic nanostructure during the degradation of dye molecules. In another study, Sivakumar *et al.* [243] also detailed the generation of ZnO/Ag biphasic hybrid nano-wires by biological means using chitosan. Superior antimicrobial properties were observed for the developed hybrid nano-wires.

3.7 Bio-Inspired Tin Dioxide Nanomaterials

With an aim to reduce the usage of noxious reagents during the production of tin dioxide (SnO_2) based nanomaterials, diverse green synthesis strategies with the aid of biological agents have been introduced. For instance, Diallo *et al.* [244] reported the environmentally friendly preparation of SnO_2 nanoparticles using leaves essence of *Aspalathus linearis* plant. The resultant bio-inspired SnO_2 nanoparticles demonstrated suitability for the light induced reduction of

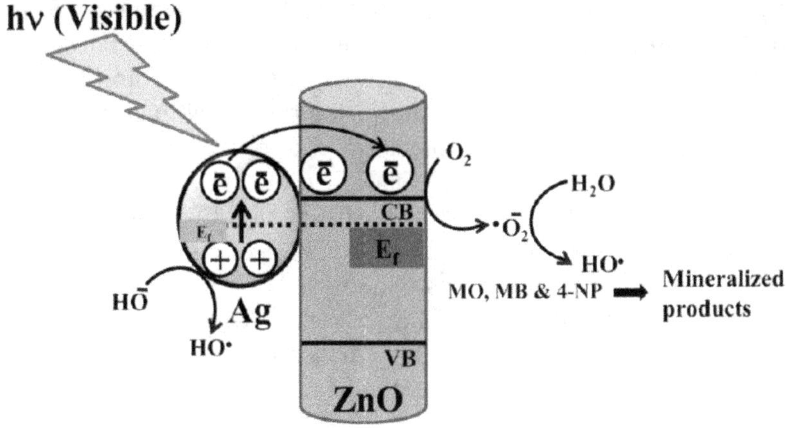

Figure 3.8 Schematic photocatalytic mechanism of biofilm inspired Ag/ZnO biphasic nanostructures during the degradation of various dye molecules such as methyl orange (MO), methylene blue (MB) and p-nitrophenol (4-NP). Reproduced from Reference 242 with permission from American Chemical Society.

various pigment molecules. In a recent study, Osuntokun *et al.* [245] revealed the use of aqueous extract of cauliflower for the biological generation of SnO_2 nanoparticles. Catalytic capability for the SnO_2 nanoparticles for the photo-degradation of methylene blue was reported. In another study, Bhattacharjee and Ahmaruzzaman [246] employed sugarcane extract to generate 3D SnO_2 quantum dots for achieving effective catalytic capability for photo-degradation of organic compounds and to create effective wastewater management system.

Srivastava and Mukhopadhyay [247] reported the development of bio-engineered SnO_2 nanoparticles with the aid of *Erwinia herbicola* bacterial serum for photocatalytic purposes. The presence of microbes in the formulation was beneficial in terms of reduction as well as stabilization during the evolution of bio-inspired nanoparticles. Wang *et al.* [248] detailed the design of lithium ion battery anode structure using bio-inspired SnO_2 nano-rods. *Escherichia coli* outer membranes mediated biological synthesis of SnO_2 nano-rods was carried out. Bhattacharjee *et al.* [249] reported the generation of SnO_2 nanoparticles with the aid of glycerol and water for photo-degradation of methyl violet 6B and methylene blue. In another study, the authors detailed the incorporation of combination of glycerol and alcohol for the bio-fabrication of SnO_2 nanoparticles with globular morphology and photocatalytic effectiveness [250]. In another strategy, the authors also designed bio-inspired SnO_2 nanoparticles with the aid of urea along with cationic or anionic surfactants [251]. Photocatalytic capability for the developed system for light induced reduction of various organic dyes was observed.

Tammina and Mandal [252] made use of tyrosine for biological development of SnO_2 nanoparticles with an aim to achieve superior photocatalytic performance. About 100% reduction efficiency for the developed bio-inspired system towards violet 4 BSN colorant was observed. Begum *et al.* [253] detailed the synthesis of bio-inspired SnO_2 nanoparticles with 3D morphology by means of L-lysine monohydrate incorporation. The SnO_2 nanoparticles displayed catalytic performance during light-induced reduction of various organic pigments. Li *et al.* [254] developed highly oriented butterfly wings inspired carbon/SnO_2 biphasic nanocomposite, as demonstrated in Figure 3.9, for lithium-ion battery applications as a superior anode structure. Considerable reversible storage capacity along with excellent cyclability and rate capability were observed for the resultant anode structure.

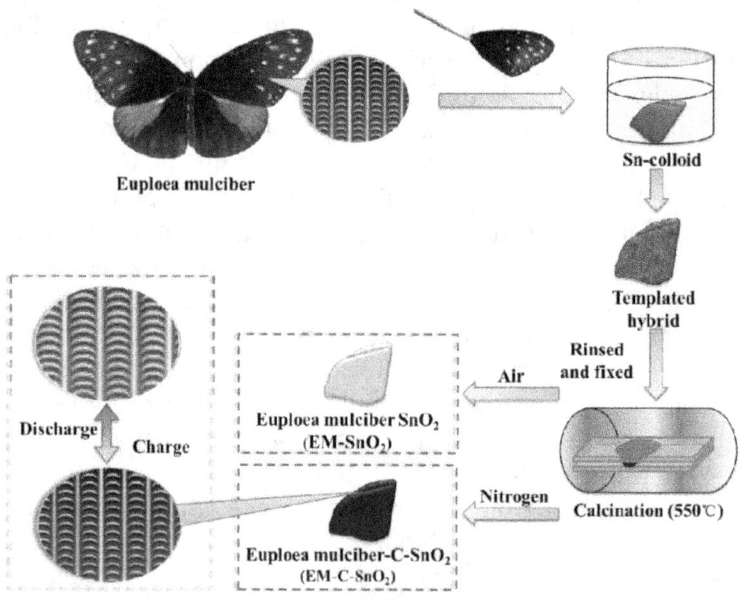

Figure 3.9 Schematic of *Euploea mulciber* butterfly wings assisted green synthesis of C/SnO_2 nanocomposite Reproduced from Reference 254 with permission from American Chemical Society.

3.8 Bio-Inspired Titanium Dioxide Nanomaterials

Rajakumar *et al.* [255] reported the environmental friendly synthesis of TiO_2 nanoparticles through the use of *Solanum trilobatum* plant aqueous leaf extract. By the application of the simple synthetic approach, high performance biomimetic TiO_2 nanoparticles with parasitic proliferation resistance were obtained. In another study, the authors revealed the reduction as well as stabilization potential of *Mangifera indica* plant aqueous leaf extract during the bio-fabrication of TiO_2 nanoparticles [256]. Anti-parasitic performance of the bio-derived TiO_2 nanoparticles against blood consuming parasites was observed. Also, the leaf extract of *Eclipta prostrata* plant was observed to be beneficial for the biological generation of TiO_2 nanoparticles with globular morphology [257]. Velayutham *et al.* [258] reported leaf extract sourced from *Catharanthus roseus* plant as an effective reducing agent for the green fabrication of TiO_2 nanoparticles. Effective anti-parasitic behavior for the bio-generated TiO_2 nanoparticles was observed.

Sankar *et al.* [259] reported superior photocatalytic characteristics against hazardous dye molecules for bio-engineered TiO_2 nanoparticles generated using *Azadirachta indica* plant leaf extract. Santhoshkumar *et al.* [260] used *Psidium guajava* plant leaf extract for bio-generation of antibacterial TiO_2 nanoparticles in a facile manner. In another work, Sundrarajan and Gowri [261] applied aqueous leaf extract derived from *Nyctanthes arbor-tristis* plant for the generation of TiO_2 nanoparticles. Hudlikar *et al.* [262] also revealed the usefulness of *Jatropha curcas* plant extract for the effective bio-generation of TiO_2 nanoparticles.

Suman *et al.* [263] detailed the generation of anti-insect TiO_2 nanoparticles using root extract of *Morinda citrifolia* plant. Roopan *et al.* [264] made use of aqueous fruit skin extract of *Annona squamosa* plant for the facile generation of biomimetic TiO_2 nanoparticles. Marimuthu *et al.* [265] also reported superior acaricidal characteristics for bio-inspired TiO_2 nanoparticles engineered with the assistance of *Calotropis gigantea* plant fruit extract. Kirthi *et al.* [266] reported rapid biological fabrication of TiO_2 nanoparticles in the presence of *Bacillus subtilis* bacteria. In another study, Jayaseelan *et al.* [267] used *Aeromonas hydrophila* bacterial extract for the environmental friendly production of TiO_2 nanoparticles. The obtained bio-inspired TiO_2 nanomaterial demonstrated effective antibacterial performance. For exploring solar cell applications, Ordenes-

Aenishanslins *et al.* [268] demonstrated the suitability of bio-inspired TiO_2 nanoparticles generated with the assistance of *Bacillus mycoides* bacteria. Rajakumar *et al.* [269] also discussed the application of *Aspergillus flavus* fungi for the development of bio-engineered bactericidal nanoscale TiO_2 particles. In a recent study, Kandiah *et al.* [270] demonstrated the green generation of chondroitin-4-sulfate infused TiO_2 nanostructure for biomedical applications as bone tissue engineering scaffold material.

3.9 Bio-Inspired Copper Nanomaterials

Prasad *et al.* [271] used biological precursor based on *Broccoli* plant for the environmentally friendly bio-generation of Cu nanoparticles possessing photocatalytic proficiency. In another study, the use of *Gundelia tournefortii* plant aqueous extract was reported as an effective support for the bio-production of Cu nanoparticles with substantial catalytic activity. Gopinath *et al.* [273] described the usefulness of leaf extract of *Nerium oleander* plant for the fabrication of Cu nanocrystals. Superior antibacterial ability was obtained for the biomimetic Cu nanoparticles. Kaur *et al.* [274] discussed the biomimetic development of Cu nanoparticles possessing substantial microbial resistance using *Punica granatum* plant peel extract as a reductant and stabilizer. Gerits *et al.* [275] reported the application of agro-waste of grape and coffee for the biological generation of high performance Cu nanoparticles.

Nazrollahzadeh *et al.* [276] reported the eco-benign development of CuO nanoparticles with the aid of precursor based on leaf extract of *Thymus vulgaris* plant. The bio-inspired CuO nanoparticles demonstrated superior photocatalytic characteristics. Karimi and Mohsenzadeh [277] employed flower extract of *Aloe vera* plant as a biological precursor for the facile environmentally-benign generation of CuO nanoscale crystals. Abiraman *et al.* [278] reported chitosan mediated bio-generation of CuO nano-leaves with superior yield. Effective biofouling resistance for the nano-leaves based coating systems was observed.

3.10 Bio-Inspired Nanomaterials Based on Other Metal Oxides

The eco-friendly green synthesis of some other metal oxide nanomaterials has also been reported with the aid of plant extract based supports. These systems include *Aspalathus linearis* plant leaf extract inspired RuO_2 [279], Co_3O_4 [280] and Eu_2O_3 nanoparticles [281]; *Acalypha indica* plant leaf extract inspired RuO_2 nanoparticles [282]; *Callistemon viminalis* plant flower extract inspired NiO nanoparticles [283]; *Agathosma betulina* plant leaf extract inspired NiO nanoparticles [284]; *Nephelium lappaceum* plant fruit skin extract inspired NiO nanoparticles [285] and mesoporous NiO nanomaterial [286]; *Hibiscus sabdariffa* plant flower extract inspired CdO nanoparticles [287] and CeO_2 nanoparticles [288]; *Leucas aspera* plant leaf extract inspired CeO_2 nanoparticles [289]; *Agathosma betulina* plant leaf extract inspired CdO nanoparticles [290]; *Achillea wilhelmsii* plant flower extract inspired CdO nanoparticles [291]; *Callistemon viminalis* plant flower extract inspired Sm_2O_3 nanoparticles [292]; *Callistemon viminalis* plant flower extract inspired Cr_2O_3 nanoparticles [293]; and *Aloe vera* plant extract inspired $MgO:Fe^{3+}$ nanocrystals [294].

3.11 Bio-Inspired Silica Based Nanomaterials

For the green synthesis of silica, various biological precursors like polydopamine, phospholipids, chitosan, alginate and collagen have demonstrated significant usefulness. Xia *et al.* [295] reported the use of polydopamine as biological stabilizer for generating eco-friendly silica nanoparticles for protein recognition and separation. In another study, Chang *et al.* [296] also designed polydopamine capped silica nanoparticles for the fabrication of targeted cancer drug carriers. For achieving superior drug carrying applications, Zheng *et al.* [297] reported the design of biocompatible nano-silica exhibiting meso-porosity by biological functionalization with the aid of polydopamine. Lai *et al.* [298] utilized polydopamine as a biological precursor for the stabilization of nanoscale silica spheres to achieve surface distribution of Au nanocrystals and, thus, superior bio-sensing properties.

Tripathi *et al.* [299] reported the bio-synthesis of Si nanoparticles with the aid of mussel sourced polydopamine for designing membranes in combination with polyacrylonitrile. The resultant nanocomposite based membrane structure displayed suitability for separation and water treatment applications. Tian *et al.* [300] and Heng *et al.* [301] also reported the stabilization of silica nanoparticles using polydopamine in order to design high performance polymer nanocomposite structures. Gupta *et al.* [302] reported the development of hierarchical nanocomposite comprised of phospholipids and nano-silica by biological means. The nanocomposite displayed considerable potential for applications ranging from biomedical to energy storage due to its modulating structural features and reliability under both air and aqueous conditions. To explore drug delivery applications, Feng *et al.* [303] reported the use of mesoporous nanostructure composed of alginate/chitosan bilayer encapsulated silica nanoparticles.

Xavier *et al.* [304] fabricated biologically inspired hybrid hydrogels composed of 2D nano-silicates and collagen. The resultant multifunctional hydrogels are promising candidates for bone tissue engineering applications due to their properties such as compatibility with living systems, mechanical properties, porosity, evolution of the mineralized structure, etc. Guo *et al.* [305] observed photoluminescence features for nano-hybrid system based on chitosan mediated bio-inspired alumino-silicate (MMT) and Cd/Te bimetallic quantum dots. In another study, Altinisik *et al.* [306] also detailed the bio-fabrication of composite system comprised of MMT, silver ions and chitosan for generating fabric coating with bactericidal efficacy.

3.12 Bio-Inspired Graphene Based Nanomaterials

To widen the biomedical applications of graphene based materials, diverse green synthesis approaches have been developed. For instance, Khenfouch *et al.* [307] reported the use of aqueous leaf extract of *Artemisia herba-alba* shrub for the facile biological reduction of GO. Graphene with multilayered morphology and effective electrical properties was achieved. Sadhukhan *et al.* [308] fabricated graphene by means of facile *Mangifera indica* plant aqueous leaf extract assisted deoxygenation of GO. Gurunathan *et al.* [309] also reported the facile green generation of reduced GO by using aqueous leaf extract of *Ginkgo biloba* tree. The produced bio-graphene presented viability for cancer treatment due to its biocompatibility. In another study, Ghosh *et al.* [310] reported eco-friendly graphene using effective bio-reductant based on *Paederia foetide* plant leaf extract.

Hou *et al.* [311] used aqueous fruit extract of *Lycium barbarum* plant as an effective reductant and stabilizer for the bio-friendly graphene production for biomedical applications. Kuila *et al.* [312] disclosed the usefulness of *Daucus carota* plant root extract as a reductant for the deoxygenation of GO, which resulted in the synthesis of graphene with layered morphology. In another study, Feng *et al.* [313] used aqueous extract of *Costus igneus* (insulin) and *Chinese wolfberry* plants for the eco-benign deoxygenation of GO. Firdhouse and Lalitha [314] fabricated graphene using a green route with the support of *Eichhornia crassipes* plant extract as a biological precursor. Chu *et al.* [315] reported black *Glycine max* plant extract assisted eco-friendly generation of graphene for energy storage application as a high-performance electrode structure with substantial porosity. In another study, the aqueous extract of *Hibiscus sabdarriffa* plant demonstrated effectiveness as a green precursor for the synthesis of bio-inspired graphene.

For achieving environmentally friendly production of graphene, Thakur and Karak [317] used three biological reduction precursors such as aqueous leaf extracts of *Colocasia esculenta* plant and *Mesua ferrea* plant as well as aqueous fruit skin extract of orange. The reductants improved the storage capacity and conducting nature of the resultant graphene. Yaragalla *et al.* [318] detailed the nature-friendly fabrication of graphene with the exploitation of aqueous grape seed extract as an effective bio-reductant. The resultant biologically derived graphene demonstrated resistance against cancer cell growth and bacterial proliferation. Plant derived resveratrol also presented superior reduction and stabilization capability during the bio-engineered evolution of graphene possessing effectiveness for targeted cancer therapy applications [319]. In another study, Zaid *et al.* [320] reported the utilization of bio-sourced β-carotene for the bio-synthesis of graphene sheets.

Saikia *et al.* [321] reported the biologically inspired synthesis of Au coated graphene nanostructure with the aid of *Piper pedicellatum* plant leaf extract. The resultant bio-inspired nanocomposite demonstrated photocatalytic effectiveness for the reduction of noxious dye molecules. Simsikova *et al.* [322] also detailed the biological development of Au/graphene biphasic nanoparticles with the aid of *Camellia sinensis* plant leaf extract for high performance photocatalytic applications. In another study, Gurunathan *et al.* [323] used aqueous extract of *Tilia amurensis* plant for reduction as well as stabilization during eco-friendly synthesis of Ag decorated graphene nanocomposite with substantial resistance against cancer cell proliferation. Khan *et al.* [324] also reported the facile and biologically inspired Ag infused graphene nanoscale composite using reducing agent based on *Pulicaria glutinosa* plant aqueous extract. Superior anticancer performance was observed for the biocompatible graphene nanocomposite. In another recent study, Gurunathan *et al.* [325] discussed the bio-inspired synthesis of graphene using microbial deoxygenation of GO with the support of *Escherichia coli* bacteria. Akhavan and Ghaderi [326] also followed *Escherichia coli* bacterial serum induced eco-friendly synthesis of graphene with significant antibacterial properties. In another strategy, Gurunathan *et al.* [327] reported the employment of *Pseudomonas aeruginosa* bacterial extract for the biological deoxygenation of GO to fabricate graphene for biomedical applications. The authors also used *Bacillus marisflavi* bacterial serum as a bio-reductant for the development of biocompatible graphene with considerable cytotoxicity in human breast cancer cells [328]. Biological reductant based on *Escherichia fergusoni* bacterial serum also presented high efficiency for deoxygenation of GO during graphene generation [329].

Khanra *et al.* [330] reported the facile *Saccharomyces cerevisiae* fungi assisted biological fabrication of graphene exhibiting substantial water dispersibility. The exploitation of fungal based precursor was favorable with respect to reduction of GO and stabilization of developed graphene. Gurunathan *et al.* [331] also reported anticancer attributes for *Ganoderma* mushroom extract assisted graphene. Zhang and Gurunathan [332] noted the practicability of nicotinamide (vitamin B3) for the synthesis of bio-inspired graphene. The bio-derived graphene presented substantial usefulness for biomedical applications as high quality tissue engineering scaffold material owing to its biocompatibility and gene expression characteristic. Gurunathan *et al.* [333] also developed eco-friendly graphene by biological means with the implementation of humanin polypeptide. In another study, SuryaBhaskaram *et al.* [334] made use of glucose as reducing agent for the synthesis of graphene in an environmentally affable manner.

Feng *et al.* [335] carried out eco-friendly GO reduction reaction by biological means using various starch containing substances like solvable starch, corn dust, potatoes and saccharine potatoes. The materials displayed effective reduction as well as stabilization proficiency during the biological reduction of GO. Ma *et al.* [336] deployed carboxymethyl starch polysaccharide as an efficient reductant and stabilizer for environmentally friendly synthesis of graphene. Zhang *et al.* [337] reported the generation of graphene possessing superior photocatalytic capability using biological reduction precursor glycylglycine. Shin *et al.* [338] developed bio-inspired hydrogel based on GO and gelatin methacrylate as a high-performance scaffold structure for tissue engineering application owing to the modulating mechanical and electrical features.

Tran *et al.* [339] reported the potential of α-amino acids such as L-aspartic acid and valine for GO reduction. Bo *et al.* [340] used caffeic acid for the biological development of graphene from GO for energy depository devices and sensors. Choi *et al.* [341] also detailed the fabrication of graphene using uric acid assisted effective deoxygenation of GO. The synthesized graphene demonstrated suitability for cancer therapy due to its cytotoxicity against human ovarian cancer cells. In another study, Ortega-Amaya *et al.* [342] also reported the implementation of citric acid for reduction as well as stabilization during the eco-friendly synthesis of graphene quantum dots.

For exploring biomedical applications of graphene quantum dots, Nurunnabi *et al.* [343] carried out surface modifications with the aid of mussel derived polydopamine. Polymerization time was observed to have significant influence on the particle size of the resultant polydopamine stabilized graphene quantum dots (Figure 3.10). Owing

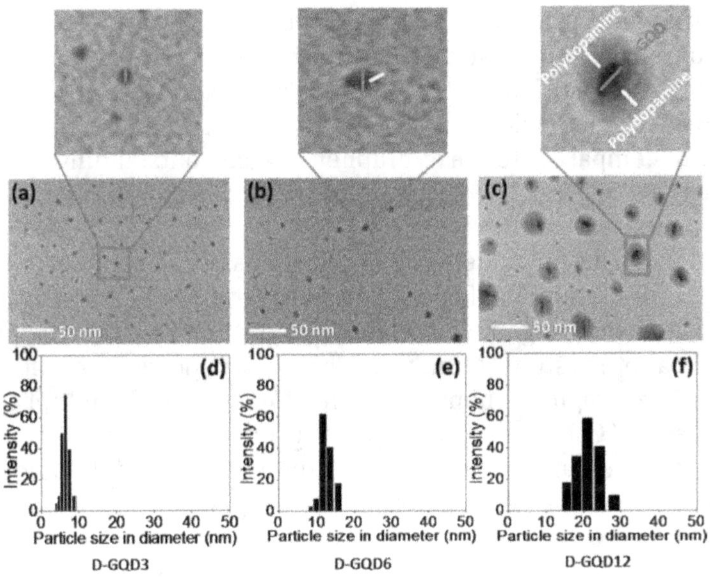

Figure 3.10 Variation of particle size with polymerization time of polydopamine modified graphene quantum dots. TEM images of sample after a) 3, b) 6, and c) 12 h of polymerization and particle size distributions of sample after d) 3, e) 6, and f) 12 h of polymerization. Reproduced from Reference 343 with permission from American Chemical Society.

to the attributes such as reliability in water, photoluminescence intensity and non-toxicity, the bio-engineered nanomaterial is suitable for use in medical imaging as well as drug delivery applications. Jung *et al.* [344] described the use of biological precursor polydopamine for the eco-friendly stabilization of graphene foam to achieve surface dispersion of nanoscale Pt particles. The resultant bio-inspired composite structure displayed substantial hydrogen depository properties due to its high capacity. Ning *et al.* [345] reported dielectric performance like negligible dielectric loss and enriched breakdown stability for composite structure composed of bio-derived polydopamine covered GO and carboxylated acrylonitrile-butadiene rubber. The observed features of the bio-inspired composite confirm its use in electrical and electronics applications as a superior dielectric material.

Cheng *et al.* [346] reported the generation of bio-inspired graphene-based artificial nacre with stabilized mechanical and electrical performance suitable for extensive engineering as well as biomedical applications. The authors attributed the impressive performance of the artificial nacre to the bonding effect at the interface. In another study, Gong *et al.* [347] developed bio-inspired ternary nanocomposite comprised of natural nacre, graphene and double walled carbon nanotubes (DWCNTs). The resultant nanocomposite displayed about 3-fold improvement in both toughness and tensile strength compared to neat graphene. Also, outstanding thermal conductivity of >390 S/cm was reported for the nanocomposite system.

For exploring drug delivery applications, Ordikhani *et al.* [348] reported the use of biologically derived graphene/chitosan nanocomposite exhibiting biocompatibility and superior bactericidal features. In another study, Mazaheri *et al.* [349] reported the suitability of chitosan/graphene nanocomposite for stem cell multiplication mold design. Chowdhuri *et al.* [350] used chitosan mediated biofabricated graphene to develop nanocomposite with ZnO nanoparticles for biomedical and environmental applications. The resultant nanocomposite displayed bactericidal nature together with the formation of chemically sensitive oxygen based fragments. Xie *et al.* [351] also fabricated nanocomposite structure based on chitosan assisted bio-inspired graphene and calcium phosphate for bone tissue engineering application as an effective scaffold material due to biocompatibility, antimicrobial performance, non-toxicity, tissue reclamation and growth rate (Figure 3.11).

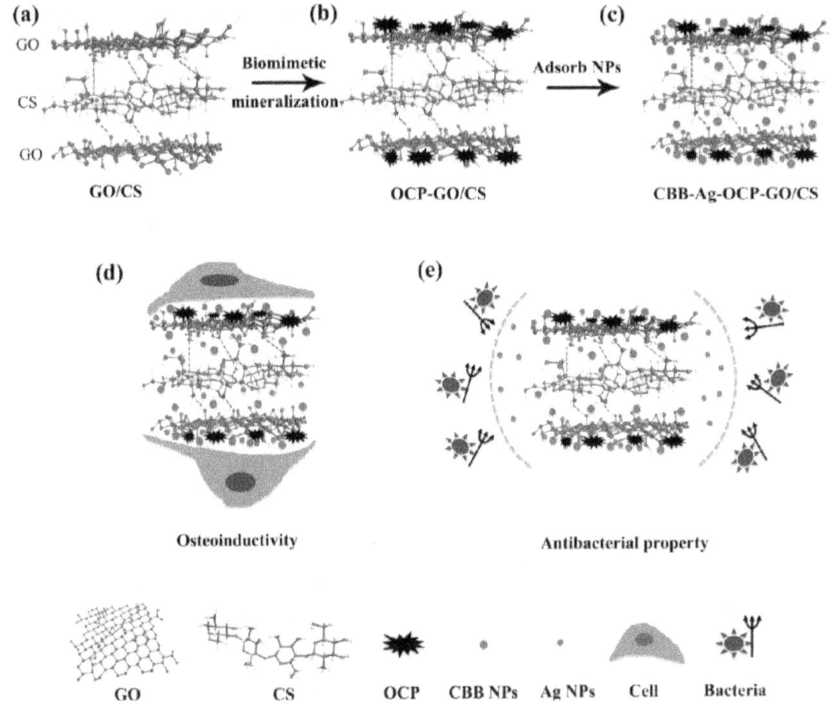

Figure 3.11 Schematic of the synthesis of biomimetic mineralized hierarchical graphene oxide/chitosan scaffolds (OCP: octacalcuim phosphate; CS: chitosan; CBB NPs: CS-BMP-BSA nanoparticles). Reproduced from Reference 351 with permission from American Chemical Society.

Reddy *et al.* [352] also observed superior photocatalytic characteristics for graphene/ZnO nanocomposite designed with the aid of polyamine. Moreover, the resultant composite structure exhibited appreciable photo-corrosion resistance.

3.13 Other Bio-Inspired Carbonaceous Nanomaterials

Bhunia *et al.* [353] revealed the formation of carbon nanoparticles with the aid of vitamin B1 exhibiting blue and green fluorescence (Figure 3.12). For exploring cancer therapy application, Singh *et al.* [354] developed biomimetic multi-walled carbon nanotubes (MWCNTs) using vitamin E succinate as stabilizer. Fujie *et al.* [355] synthesized high performance nano-membranes for biological implements with the support of C2C12 skeletal myoblasts inspired

MWCNTs. The nano-membrane structure presented pliability, modulating structural features, mechanical properties, binding potential and cellular arrangement capability.

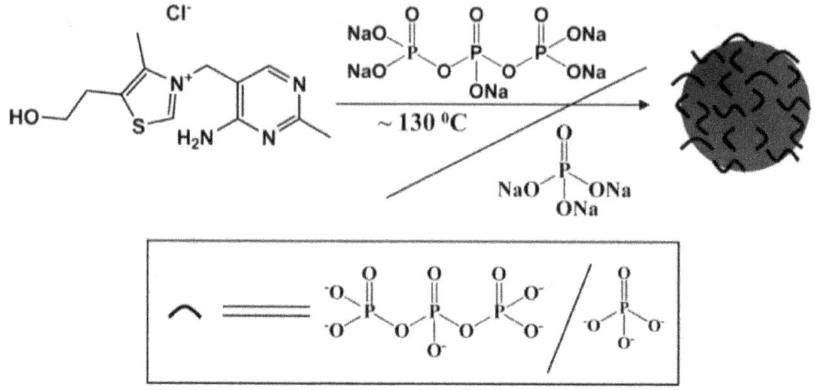

Figure 3.12 Synthetic strategy for the preparation of fluorescent carbon nanoparticles. Reproduced from Reference 353 with permission from American Chemical Society.

Zhang *et al.* [356] reported substantial biocompatibility for polydopamine mediated biological functionalization of MWCNTs [356]. In another study, Yan *et al.* [357] described the biomimetic generation of polydopamine capped carbon nanotubes (CNTs) for biomedical applications due to biocompatible nature and capability for tissue strengthening through mineral deposition. Zhang *et al.* [358] also reported the use of polydopamine for the effective surface stabilization of MWCNTs in order to develop polymer nanocomposite structures.

3.14 Bio-Inspired Polymer Nanomaterials

A large number of bio-inspired polymer nanomaterials have been reported in the literature. This section presents only a brief overview of some recent studies.

Xu *et al.* [359] fabricated bio-inspired nanostructure comprised of glutaraldehyde crosslinked polydopamine covered polystyrene nanofibers for extraction of medicinal substances from fish muscles. The superior extraction capability resulted due to the unique features such as sensitivity and stability together with biofouling resistance. Ritz *et al.* [360] synthesized apolipoprotein H capped polystyrene nanoparticles for drug delivery applications. Improved cellular ingestion was noticed for the developed system. Li *et al.* [361] carried out environmentally amiable synthesis of particulate nanocomposite structure comprised of polystyrene and nano-silica using polydopamine as a functional precursor.

Tamang *et al.* [362] detailed the design of nano-generator using DNA inspired poly(vinylidene fluoride) (PVDF) sheet (Figure 3.13). The nano-generator structure displayed outstanding energy translation capability which confirmed its suitability for use in movable self-energized implements. Merindol *et al.* [363] reported the biological generation of nanocomposite structure composed of nanocellulose fibers and poly(vinyl amine). The layered transparent nanocomposite structure exhibited self-healing, humidity-sensitive shape memory behavior and modulating mechanical properties feasible for wide range of applications (Figure 3.14). For instance, the dry films were strong and brittle, whereas the wet films were soft and ductile. The observed changes were attributed to the polycation mobility in the polymer caused by plasticization effect of water.

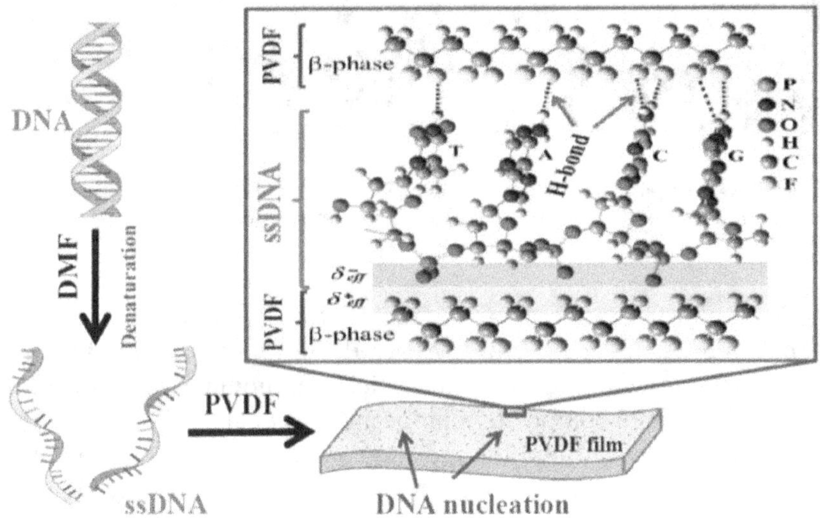

Figure 3.13 Schematic of DNA denaturation and subsequent formation of DNA-PVDF film. Reproduced from Reference 362 with permission from American Chemical Society.

Fang *et al.* [364] developed bio-inspired poly(lactic-co-glycolic acid) (PLGA) nanoparticles with a layer of membrane coating derived from cancer cells, as presented in Figure 3.14. The cancer cell inspired PLGA nanoparticles offer suitability for cancer treatment purposes as either anticancer vaccine or drug carrier due to outstanding antigen carrying as well as targeting capability. Steinbach *et al.* [365] used cell penetrating peptides for the eco-friendly generation of stabilized PLGA nanoparticles exhibiting superior potential for targeted drug delivery application. In another recent study, Zhu *et al.* [366] also reported the biomimetic functionalization of vitamin E succinate/poly(lactic acid) biphasic nanoparticles using polydopamine for effective application in cancer medication delivery.

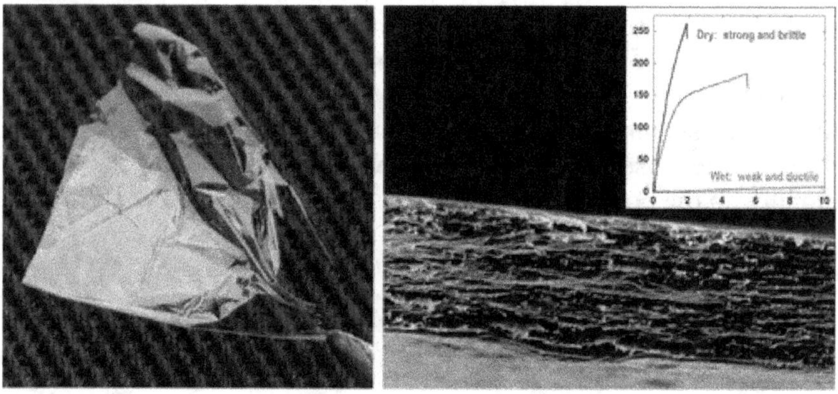

Figure 3.14 Representation of the nanocomposite structure composed of nanocellulose fibers and poly(vinyl amine). Reproduced from Reference 363 with permission from American Chemical Society.

To explore antimicrobial characteristics of polysulfone based membranes, Tang *et al.* [367] reported the functionalization using polydopamine actuated Ag nanoparticles. Biofouling resistance for the resultant nanocomposite membrane was observed owing to the synergistic impact of binding resistance and antimicrobial features offered by polydopamine and Ag nanoparticles respectively. Jiang *et al.* [368] reported the eco-benign stabilization of polycaprolactone nanofibers using polydopamine as biological precursor for achieving use as drug carriers. Rush *et al.* [369] detailed the generation of nucleic acid coupled carboxylated norbornyl polymer nanoparticles for controlling mRNA count in the human body as well as to analyze the genetic appearance. In a recent study, Tay *et al.* [370] detailed the fabrication of nano-sensor based on nano-snail mediated DNA for perceiving the mRNA structure in the biological systems. The resultant nano-sensor demonstrated superior performance owing to the synergistic effect of sensitivity, accuracy and stability. For exploring effective cancer drug delivery, Gujrati *et al.* [371] fabricated nanostructure based on *Escherichia coli* cuticles and short interfering RNA. The drug conveyance system displayed stability during animal trials without any harmful effects.

3.15 Conclusions

Bio-inspired synthesis of nanomaterials with an objective to substitute noxious chemical agents by biologically originated precursors acting as reducing agents, capping agents and stabilizers has received widespread scientific attention. In this chapter, a detailed overview of the recent developments related to the green synthesis/bio-inspired synthesis of nanomaterials has been presented. Specifically, the review scrutinized bio-inspired nanomaterials based on gold, silver, palladium, iron oxide, zinc oxide, tin oxide, titanium dioxide, copper, graphene and polymers. For the generation of different bio-inspired nanomaterials, various biological precursors demonstrating substantial performance include extracts of leaves, flower, fruit, fruit peel, seed, seed peel, root, stem and bark of diverse plants; serum of various bacteria, fungi and viruses; algae; polydopamine; proteins; peptides; amino acids; fibroin and sericin of silk; egg shell proteins; chitosan; alginate; etc.

References

1. Gao, S., Li, Z., and Zhang, H. (2010) Bioinspired green synthesis of nanomaterials and their applications. *Current Nanoscience,* **6**, 452-468.
2. Wang, J., Lin, L., Cheng, Q., and Jiang, L. (2012) A strong bio-inspired layered PNIPAM-clay nanocomposite hydrogel. *Angewandte Chemie International Edition,* **51**, 4676-4680.
3. Ke, P. C., and Qiao, R. (2007) Carbon nanomaterials in biological systems. *Journal of Physics: Condensed Matter,* **19**, 373101.
4. Nel, A. E., Madler, L., Velegol, D., Xia, T., Hoek, E. M., Somasundaran, P., Klaessig, F., Castranova, V., and Thompson, M. (2009) Understanding biophysicochemical interactions at the nano-bio interface. *Nature Materials,* **8**, 543-557.
5. Sanchez, C., Arribart, H., and Guille, M. M. G. (2005) Biomimetism

and bioinspiration as tools for the design of innovative materials and systems. *Nature Materials,* **4**, 277-288.

6. Xia, F., and Jiang, L. (2008) Bio-inspired, smart, multiscale interfacial materials. *Advanced Materials,* **20**, 2842-2858.
7. Bhattacharya, P., Du, D., and Lin, Y. (2014) Bioinspired nanoscale materials for biomedical and energy applications. *Journal of The Royal Society Interface,* **11**, 20131067.
8. Machado, R., and Sencadas, V. (2016) Advanced techniques for characterizing bioinspired materials. In: *Bioinspired Materials for Medical Applications,* Rodrigues, L., and Mota, M. (eds), Woodhead Publishing, UK, Chapter 7.
9. Wong, K., and Liu, X. (2012) Nanomedicine: a primer for surgeons. *Pediatric Surgery International,* **28**, 943-951.
10. Yang, L. (2015) *Nanotechnology-Enhanced Orthopedic Materials: Fabrications, Applications and Future Trends,* Woodhead Publishing, UK.
11. Tran, N., and Webster, T. J. (2009) Nanotechnology for bone materials. *Wiley Interdisciplinary Reviews: Nanomedicine and Nanobiotechnology,* **1**, 336-351.
12. Haidar, Z. S. (2010) Bio-inspired/-functional colloidal core-shell polymeric-based nanosystems: technology promise in tissue engineering, bioimaging and nanomedicine. *Polymers,* **2**, 323-352.
13. Patwardhan, S. V. (2011) Biomimetic and bioinspired silica: recent developments and applications. *Chemical Communications,* **47**, 7567-7582.
14. Das, S., Carnicer-Lombarte, A., Fawcett, J. W., and Bora, U. (2016) Bio-inspired nano tools for neuroscience. *Progress in Neurobiology,* **142**, 1-22.
15. Hinman, S. S., and Cheng, Q. (2016) Bioinspired assemblies and plasmonic interfaces for electrochemical biosensing. *Journal of Electroanalytical Chemistry,* **781**, 136-146.
16. Yao, X., Song, Y., and Jiang, L. (2011) Applications of bio-inspired special wettable surfaces. *Advanced Materials,* **23**, 719-734.
17. Darmanin, T., and Guittard, F. (2014) Recent advances in the potential applications of bioinspired superhydrophobic materials. *Journal of Materials Chemistry A,* **2**, 16319-16359.
18. Das, S. K., and Marsili, E. (2011) Bioinspired metal nanoparticle: Synthesis, properties and application. In: *Materials,* Rahman, M. M. (ed.), Intech, UK, DOI: 10.5772/25305.
19. Huang, J., Lin, L., Sun, D., Chen, H., Yang, D., and Li, Q. (2015) Bio-inspired synthesis of metal nanomaterials and applications. *Chemical Society Reviews,* **44**, 6330-6374.
20. Leiro, V., Moreno, P., Sarmento, B., Durao, J., Gales, L., Pego, A., and Barrias, C. (2016) Design and preparation of biomimetic and bio-inspired materials. In: *Bioinspired Materials for Medical Applica-*

tions, Rodrigues, L., and Mota, M. (eds), Woodhead Publishing, UK, Chapter 1.

21. Zinjarde, S. S. (2012) Bio-inspired nanomaterials and their applications as antimicrobial agents. *Chronicles of Young Scientists,* **3**(1), 74-81.
22. Bedford, N., Munro, C., and Knecht, M. (2016) Peptide binding for bio-based nanomaterials. *Methods in Enzymology,* **580**, 581-598.
23. Cai, Y., and Yao, J. (2010) Effect of proteins on the synthesis and assembly of calcium phosphate nanomaterials. *Nanoscale,* **2**, 1842-1848.
24. Xu, A.-W., Ma, Y., and Colfen, H. (2007) Biomimetic mineralization. *Journal of Materials Chemistry,* **17**, 415-449.
25. Kumar, P., and Kim, K.-H. (2016) Recent progress and innovation in carbon capture and storage using bioinspired materials. *Applied Energy,* **172**, 383-397.
26. Wang, M., Odoom-Wubah, T., Chen, H., Jing, X., Kong, T., Sun, D., Huang, J., and Li, Q. (2013) Microorganism-mediated synthesis of chemically difficult-to-synthesize Au nanohorns with excellent optical properties in the presence of hexadecyltrimethylammonium chloride. *Nanoscale,* **5**, 6599-6606.
27. Shi, J., Jiang, Y., Zhang, S., Yang, D., and Jiang, Z. (2016) Biomimetic/bioinspired design of enzyme@capsule nano/microsystems. *Methods in Enzymology,* **571**, 87-112.
28. Kumar, C. V. (2016) *Rational Design of Enzyme-Nanomaterials,* Academic Press, USA.
29. Aizenberg, J., and Fratzl, P. (2009) Biological and biomimetic materials. *Advanced Materials,* **21**, 387-388.
30. Dauthal, P., and Mukhopadhyay, M. (2016) Phyto-synthesis and structural characterization of catalytically active gold nanoparticles biosynthesized using Delonix regia. *3 Biotech,* **6**, 1-9.
31. Bhau, B., Ghosh, S., Puri, S., Borah, B., Sarmah, D., and Khan, R. (2015) Green synthesis of gold nanoparticles from the leaf extract of Nepenthes khasiana and antimicrobial assay. *Advanced Materials Letters,* **6**(1), 55-58.
32. Babu, P. J., Saranya, S., Sharma, P., Tamuli, R., and Bora, U. (2012) Gold nanoparticles: sonocatalytic synthesis using ethanolic extract of Andrographis paniculata and functionalization with polycaprolactone-gelatin composites. *Frontiers of Materials Science,* **6**, 236-249.
33. Mollick, M. M. R., Bhowmick, B., Mondal, D., Maity, D., Rana, D., Dash, S. K., Chattopadhyay, S., Roy, S., Sarkar, J., and Acharya, K., (2014) Anticancer (in vitro) and antimicrobial effect of gold nanoparticles synthesized using Abelmoschus esculentus (L.) pulp extract via a green route. *RSC Advances,* **4**, 37838-37848.
34. Punuri, J. B., Sharma, P., Sibyala, S., Tamuli, R., and Bora, U. (2012)

Piper betle-mediated green synthesis of biocompatible gold nanoparticles. *International Nano Letters,* **2**, 18.

35. Patra, S., Mukherjee, S., Barui, A. K., Ganguly, A., Sreedhar, B., and Patra, C. R. (2015) Green synthesis, characterization of gold and silver nanoparticles and their potential application for cancer therapeutics. *Materials Science and Engineering: C,* **53**, 298-309.
36. Balasubramani, G., Ramkumar, R., Krishnaveni, N., Sowmiya, R., Deepak, P., Arul, D., and Perumal, P. (2015) GC-MS analysis of bioactive components and synthesis of gold nanoparticle using Chloroxylon swietenia DC leaf extract and its larvicidal activity. *Journal of Photochemistry and Photobiology B: Biology,* **148**, 1-8.
37. Ismail, E. H., Khalil, M. M., Al Seif, F. A., and El-Magdoub, F. (2014) Biosynthesis of gold nanoparticles using extract of grape (vitis vinifera) leaves and seeds. *Progress in Nanotechnology and Nanomaterials,* **3**, 1-12.
38. Raghunandan, D., Ravishankar, B., Sharanbasava, G., Mahesh, D. B., Harsoor, V., Yalagatti, M. S., Bhagawanraju, M., and Venkataraman, A. (2011) Anti-cancer studies of noble metal nanoparticles synthesized using different plant extracts. *Cancer Nanotechnology,* **2**, 57-65.
39. Babu, P. J., Sharma, P., Borthakur, B. B., Das, R. K., Nahar, P., and Bora, U. (2010) Synthesis of gold nanoparticles using Mentha arvensis leaf extract. *International Journal of Green Nanotechnology: Physics and Chemistry,* **2**, P62-P68.
40. Swain, S., Barik, S., Behera, T., Nayak, S., Sahoo, S., Mishra, S., and Swain, P. (2016) Green synthesis of gold nanoparticles using root and leaf extracts of Vetiveria zizanioides and Cannabis sativa and its antifungal activities. *BioNanoScience,* **6**, 205-213.
41. Dhamecha, D., Jalalpure, S., Jadhav, K., and Sajjan, D. (2016) Green synthesis of gold nanoparticles using Pterocarpus marsupium: Characterization and biocompatibility studies. *Particulate Science and Technology,* **34**, 156-164.
42. Singh, P., Kim, Y. J., and Yang, D. C. (2016) A strategic approach for rapid synthesis of gold and silver nanoparticles by Panax ginseng leaves. *Artificial Cells, Nanomedicine, and Biotechnology,* **44**, 1949-1957.
43. Kajani, A. A., Bordbar, A.-K., Esfahani, S. H. Z., and Razmjou, A. (2016) Gold nanoparticles as potent anticancer agent: green synthesis, characterization, and in vitro study. *RSC Advances,* **6**, 63973-63983.
44. Babu, P. J., Sharma, P., Saranya, S., and Bora, U. (2013) Synthesis of gold nanoparticles using ethonolic leaf extract of Bacopa monnieri and UV irradiation. *Materials Letters,* **93**, 431-434.
45. Sett, A., Gadewar, M., Sharma, P., Deka, M., and Bora, U. (2016) Green synthesis of gold nanoparticles using aqueous extract of

Dillenia indica. *Advances in Natural Sciences: Nanoscience and Nanotechnology,* **7**, 025005.

46. Babu, P. J., Sharma, P., Saranya, S., Tamuli, R., and Bora, U. (2013) Green synthesis and characterization of biocompatible gold nanoparticles using Solanum indicum fruits. *Nanomaterials and Nanotechnology,* **3**, doi:10.5772/56608.
47. Gao, Y., Hu, Y., Li, C., Zhao, X., Huang, X., and Liu, R. (2014) Kiwifruit as reducing reagent for green synthesis of gold nanoparticles at room temperature. *Nanoscience and Nanotechnology Letters,* **6**, 118-123.
48. Perde-Schrepler, M., David, L., Olenic, L., Potara, M., Fischer-Fodor, E., Virag, P., Imre-Lucaci, F., Brie, I., and Florea, A. (2016) Gold nanoparticles synthesized with a polyphenols-rich extract from Cornelian cherry (Cornus mas) fruits: Effects on human skin cells. *Journal of Nanomaterials,* 2016, 6986370.
49. Sharma, P., Babu, P., and Bora, U. (2012) Sapindus mukorossi aqueous fruit extract as reducing, capping and dispersing agents in synthesis of gold nanoparticles. *IET Micro & Nano Letters,* **7**, 1296-1299.
50. Basavegowda, N., Sobczak-Kupiec, A., Fenn, R. I., and Dinakar, S. (2013) Bioreduction of chloroaurate ions using fruit extract Punica granatum (Pomegranate) for synthesis of highly stable gold nanoparticles and assessment of its antibacterial activity. *IET Micro & Nano Letters,* **8**, 400-404.
51. Balamurugan, M., Kaushik, S., and Saravanan, S. (2016) Green synthesis of gold nanoparticles by using Peltophorum pterocarpum flower extracts. *Nano Biomedical Engineering,* **8**, 213-218.
52. Geetha, R., Ashokkumar, T., Tamilselvan, S., Govindaraju, K., Sadiq, M., and Singaravelu, G. (2013) Green synthesis of gold nanoparticles and their anticancer activity. *Cancer Nanotechnology,* **4**(4-5), 91-98.
53. Noruzi, M., Zare, D., Khoshnevisan, K., and Davoodi, D. (2011) Rapid green synthesis of gold nanoparticles using Rosa hybrida petal extract at room temperature. *Spectrochimica Acta Part A: Molecular and Biomolecular Spectroscopy,* **79**, 1461-1465.
54. Nirmala, J. G., Akila, S., Nadar, M. M., Narendhirakannan, R., and Chatterjee, S. (2016) Biosynthesized Vitis vinifera seed gold nanoparticles induce apoptotic cell death in A431 skin cancer cells. *RSC Advances,* **6**, 82205-82218.
55. Patil, M. P., Ngabire, D., Thi, H. H. P., Kim, M.-D., and Kim, G.-D. (2016) Eco-friendly synthesis of gold nanoparticles and evaluation of their cytotoxic activity on cancer cells. *Journal of Cluster Science,* **28**(1), 119-132.
56. Jun, S. H., Kim, H.-S., Koo, Y. K., Park, Y., Kim, J., Cho, S., and Park, Y. (2014) Root extracts of Polygala tenuifolia for the green synthesis

of gold nanoparticles. *Journal of Nanoscience and Nanotechnology,* **14**, 6202-6208.

57. Singh, P., Kim, Y. J., Wang, C., Mathiyalagan, R., El-Agamy Farh, M., and Yang, D. C. (2016) Biogenic silver and gold nanoparticles synthesized using red ginseng root extract, and their applications. *Artificial Cells, Nanomedicine, and Biotechnology,* **44**, 811-816.
58. Singh, P., Kim, Y. J., Wang, C., Mathiyalagan, R., and Yang, D. C. (2016) The development of a green approach for the biosynthesis of silver and gold nanoparticles by using Panax ginseng root extract, and their biological applications. *Artificial Cells, Nanomedicine, and Biotechnology,* **44**, 1150-1157.
59. Kou, J., and Varma, R. S. (2012) Beet juice utilization: Expeditious green synthesis of noble metal nanoparticles (Ag, Au, Pt, and Pd) using microwaves. *RSC Advances,* **2**, 10283-10290.
60. Sinha, T., and Ahmaruzzaman, M. (2016) A new and facile strategy for the one-pot fabrication of luminescent gold nanoclusters and their prospective application. *RSC Advances,* **6**, 44-56.
61. Patil, R., Naika, H. R., Rayar, S., Balashanmugam, N., Uppar, V., and Bhattacharyya, A. (2016) Green synthesis of gold nanoparticles: Its effect on cocoon and silk traits of mulberry silkworm (Bombyx mori L.). *Particulate Science and Technology,* **35**(3), 291-297.
62. Castro, L., Blázquez, M. L., González, F., Munoz, J. A., and Ballester, A. (2013) Gold, silver and platinum nanoparticles biosynthesized using orange peel extract. *Advanced Materials Research,* **825**, 556-559.
63. Babu, P. J., Das, R. K., Kumar, A., and Bora, U. (2011) Microwave-mediated synthesis of gold nanoparticles using coconut water. *International Journal of Green Nanotechnology,* **3**, 13-21.
64. Ganapuram, B. R., Alle, M., Dadigala, R., Dasari, A., Maragoni, V., and Guttena, V. (2015) Catalytic reduction of methylene blue and Congo red dyes using green synthesized gold nanoparticles capped by salmalia malabarica gum. *International Nano Letters,* **5**, 215-222.
65. Das, D. K., Chakraborty, A., Bhattacharjee, S., and Dey, S. (2013) Biosynthesis of stabilised gold nanoparticle using an aglycone flavonoid, quercetin. *Journal of Experimental Nanoscience,* **8**, 649-655.
66. El-Kassas, H. Y., and El-Sheekh, M. M. (2014) Cytotoxic activity of biosynthesized gold nanoparticles with an extract of the red seaweed Corallina officinalis on the MCF-7 human breast cancer cell line. *Asian Pacific Journal of Cancer Prevention,* **15**, 4311-4317.
67. Singh, M., Kumar, M., Manikandan, S., Chandrasekaran, N., Mukherjee, A., and Kumaraguru, A., (2014) Drug delivery system for controlled cncer therapy using physico-chemically stabilized bioconjugated gold nanoparticles synthesized from marine macroalgae, Padina gymnospora. *Journal of Nanomedicine & Nanotechnology,* doi:10.4172/2157-7439.S5-009.

68. Gao, W., Fang, R. H., Thamphiwatana, S., Luk, B. T., Li, J., Angsantikul, P., Zhang, Q., Hu, C.-M. J., and Zhang, L. (2015) Modulating antibacterial immunity via bacterial membrane-coated nanoparticles. *Nano Letters,* **15**, 1403-1409.
69. Singh, P., Kim, Y. J., Wang, C., Mathiyalagan, R., and Yang, D. C. (2016) Microbial synthesis of flower-shaped gold nanoparticles. *Artificial Cells, Nanomedicine, and Biotechnology,* **44**, 1469-1474.
70. Kumar, C. G., Poornachandra, Y., and Mamidyala, S. K. (2014) Green synthesis of bacterial gold nanoparticles conjugated to resveratrol as delivery vehicles. *Colloids and Surfaces B: Biointerfaces,* **123**, 311-317.
71. Gopinath, K., and Arumugam, A. (2014) Extracellular mycosynthesis of gold nanoparticles using Fusarium solani. *Applied Nanoscience,* **4**, 657-662.
72. Sarkar, J., Ray, S., Chattopadhyay, D., Laskar, A., and Acharya, K. (2012) Mycogenesis of gold nanoparticles using a phytopathogen Alternaria alternata. *Bioprocess and Biosystems Engineering,* **35**, 637-643.
73. Liu, X., Cao, J., Li, H., Li, J., Jin, Q., Ren, K., and Ji, J. (2013) Mussel-inspired polydopamine: a biocompatible and ultrastable coating for nanoparticles in vivo. *ACS Nano,* **7**, 9384-9395.
74. Perez-Mitta, G., Tuninetti, J. S., Knoll, W., Trautmann, C., Toimil-Molares, M. E., and Azzaroni, O. (2015) Polydopamine meets solid-state nanopores: A bioinspired integrative surface chemistry approach to tailor the functional properties of nanofluidic diodes. *Journal of the American Chemical Society,* **137**, 6011-6017.
75. Black, K. C., Yi, J., Rivera, J. G., Zelasko-Leon, D. C., and Messersmith, P. B. (2013) Polydopamine-enabled surface functionalization of gold nanorods for cancer cell-targeted imaging and photothermal therapy. *Nanomedicine,* **8**, 17-28.
76. Nowinski, A. K., White, A. D., Keefe, A. J., and Jiang, S. (2014) Biologically inspired stealth peptide-capped gold nanoparticles. *Langmuir,* **30**, 1864-1870.
77. Briggs, B. D., Li, Y., Swihart, M. T., and Knecht, M. R. (2015) Reductant and sequence effects on the morphology and catalytic activity of peptide-capped Au nanoparticles. *ACS Applied Materials & Interfaces,* **7**, 8843-8851.
78. Liu, X., Wang, Y., Chen, P., Wang, Y., Zhang, J., Aili, D., and Liedberg, B. (2014) Biofunctionalized gold nanoparticles for colorimetric sensing of botulinum neurotoxin a light chain. *Analytical Chemistry,* **86**, 2345-2352.
79. Tan, L. H., Xing, H., Chen, H., and Lu, Y. (2013) Facile and efficient preparation of anisotropic DNA-functionalized gold nanoparticles and their regioselective assembly. *Journal of the American Chemical Society,* **135**, 17675-17678.

80. Liang, M., Su, R., Huang, R., Qi, W., Yu, Y., Wang, L., and He, Z. (2014) Facile in situ synthesis of silver nanoparticles on procyanidin-grafted eggshell membrane and their catalytic properties. *ACS Applied Materials & Interfaces,* **6**, 4638-4649.
81. Wongkrongsak, S., Tangthong, T., and Pasanphan, W. (2016) Electron beam induced water-soluble silk fibroin nanoparticles as a natural antioxidant and reducing agent for a green synthesis of gold nanocolloid. *Radiation Physics and Chemistry,* **118**, 27-34.
82. Moyano, D. F., Saha, K., Prakash, G., Yan, B., Kong, H., Yazdani, M., and Rotello, V. M. (2014) Fabrication of corona-free nanoparticles with tunable hydrophobicity. *ACS Nano,* **8**, 6748-6755.
83. Ibrahim, N. A., Eid, B. M., and Abdel-Aziz, M. S. (2016) Green synthesis of AuNPs for eco-friendly functionalization of cellulosic substrates. *Applied Surface Science,* **389**, 118-125.
84. Zhou, J., Wang, P., Wang, C., Goh, Y. T., Fang, Z., Messersmith, P. B., and Duan, H. (2015) Versatile core-shell nanoparticle@metal-organic framework nanohybrids: Exploiting mussel-inspired polydopamine for tailored structural integration. *ACS Nano,* **9**, 6951-6960.
85. Dauthal, P., and Mukhopadhyay, M. (2016) AuPd bimetallic nanoparticles: single step biofabrication, structural characterization and catalytic activity. *Journal of Industrial and Engineering Chemistry,* **35**, 45-53.
86. Huang, C.-J., Chu, S.-H., Wang, L.-C., Li, C.-H., and Lee, T. R. (2015) Bioinspired zwitterionic surface coatings with robust photostability and fouling resistance. *ACS Applied Materials & Interfaces,* **7**, 23776-23786.
87. Liu, Y., Lou, J., Ni, M., Song, C., Wu, J., Dasgupta, N. P., Tao, P., Shang, W., and Deng, T. (2015) Bioinspired bifunctional membrane for efficient clean water generation. *ACS Applied Materials & Interfaces,* **8**, 772-779.
88. Elangovan, K., Elumalai, D., Anupriya, S., Shenbhagaraman, R., Kaleena, P., and Murugesan, K. (2015) Phyto mediated biogenic synthesis of silver nanoparticles using leaf extract of Andrographis echioides and its bio-efficacy on anticancer and antibacterial activities. *Journal of Photochemistry and Photobiology B: Biology,* **151**, 118-124.
89. Das, J., Das, M. P., and Velusamy, P. (2013) Sesbania grandiflora leaf extract mediated green synthesis of antibacterial silver nanoparticles against selected human pathogens. *Spectrochimica Acta Part A: Molecular and Biomolecular Spectroscopy,* **104**, 265-270.
90. Rao, Y. S., Kotakadi, V. S., Prasad, T., Reddy, A., and Gopal, D. S. (2013) Green synthesis and spectral characterization of silver nanoparticles from Lakshmi tulasi (Ocimum sanctum) leaf extract. *Spectrochimica Acta Part A: Molecular and Biomolecular Spectrosc-*

opy, **103**, 156-159.

91. Jayaseelan, C., and Rahuman, A. A. (2012) Acaricidal efficacy of synthesized silver nanoparticles using aqueous leaf extract of Ocimum canum against Hyalomma anatolicum anatolicum and Hyalomma marginatum isaaci (Acari: Ixodidae). *Parasitology Research,* **111**, 1369-1378.
92. Jayaseelan, C., Rahuman, A. A., Rajakumar, G., Santhoshkumar, T., Kirthi, A. V., Marimuthu, S., Bagavan, A., Kamaraj, C., Zahir, A. A., Elango, G., Velayutham, K., Rao, B. V. K., Karthik, L., and Raveendran, S. (2012) Efficacy of plant-mediated synthesized silver nanoparticles against hematophagous parasites. *Parasitology Research,* **111**, 921-933.
93. Subarani, S., Sabhanayakam, S., and Kamaraj, C. (2013) Studies on the impact of biosynthesized silver nanoparticles (AgNPs) in relation to malaria and filariasis vector control against Anopheles stephensi Liston and Culex quinquefasciatus Say (Diptera: Culicidae). *Parasitology Research,* **112**, 487-499.
94. Rajakumar, G., and Rahuman, A. A. (2011) Larvicidal activity of synthesized silver nanoparticles using Eclipta prostrata leaf extract against filariasis and malaria vectors. *Acta Tropica,* **118**, 196-203.
95. Priyadarshini, K. A., Murugan, K., Panneerselvam, C., Ponarulselvam, S., Hwang, J.-S., and Nicoletti, M. (2012) Biolarvicidal and pupicidal potential of silver nanoparticles synthesized using Euphorbia hirta against Anopheles stephensi Liston (Diptera: Culicidae). *Parasitology Research,* **111**, 997-1006.
96. Zahir, A. A., and Rahuman, A. A. (2012) Evaluation of different extracts and synthesised silver nanoparticles from leaves of Euphorbia prostrata against Haemaphysalis bispinosa and Hippobosca maculata. *Veterinary Parasitology,* **187**, 511-520.
97. Rao, B., and Tang, R.-C. (2017) Green synthesis of silver nanoparticles with antibacterial activities using aqueous Eriobotrya japonica leaf extract. *Advances in Natural Sciences: Nanoscience and Nano-Technology,* **8**, 015014.
98. Roni, M., Murugan, K., Panneerselvam, C., Subramaniam, J., and Hwang, J.-S. (2013) Evaluation of leaf aqueous extract and synthesized silver nanoparticles using Nerium oleander against Anopheles stephensi (Diptera: Culicidae). *Parasitology Research,* **112**, 981-990.
99. Suganya, A., Murugan, K., Kovendan, K., Kumar, P. M., and Hwang, J.-S. (2013) Green synthesis of silver nanoparticles using Murraya koenigii leaf extract against Anopheles stephensi and Aedes aegypti. *Parasitology Research,* **112**, 1385-1397.
100. Ganaie, S., Abbasi, T., and Abbasi, S. (2015) Green synthesis of silver nanoparticles using an otherwise worthless weed Mimosa

(Mimosa pudica): feasibility and process development toward shape/size control. *Particulate Science and Technology,* **33**, 638-644.

101. Veerakumar, K., Govindarajan, M., and Rajeswary, M. (2013) Green synthesis of silver nanoparticles using Sida acuta (Malvaceae) leaf extract against Culex quinquefasciatus, Anopheles stephensi, and Aedes aegypti (Diptera: Culicidae). *Parasitology Research,* **112**, 4073-4085.
102. Ahmed, M. J., Murtaza, G., Mehmood, A., and Bhatti, T. M. (2015) Green synthesis of silver nanoparticles using leaves extract of Skimmia laureola: characterization and antibacterial activity. *Materials Letters,* **153**, 10-13.
103. Muthukumaran, U., Govindarajan, M., and Rajeswary, M. (2015) Mosquito larvicidal potential of silver nanoparticles synthesized using Chomelia asiatica (Rubiaceae) against Anopheles stephensi, Aedes aegypti, and Culex quinquefasciatus (Diptera: Culicidae). *Parasitology Research,* **114**, 989-999.
104. Baskaralingam, V., Sargunar, C. G., Lin, Y. C., and Chen, J. C. (2012) Green synthesis of silver nanoparticles through Calotropis gigantea leaf extracts and evaluation of antibacterial activity against Vibrio alginolyticus. *Nanotechnology Development,* **2**, doi:10.4081/nd.2012.e3.
105. Veerakumar, K., Govindarajan, M., Rajeswary, M., and Muthukumaran, U. (2014) Mosquito larvicidal properties of silver nanoparticles synthesized using Heliotropium indicum (Boraginaceae) against Aedes aegypti, Anopheles stephensi, and Culex quinquefasciatus (Diptera: Culicidae). *Parasitology Research,* **113**, 2363-2373.
106. Veerakumar, K., Govindarajan, M., Rajeswary, M., and Muthukumaran, U. (2014) Low-cost and eco-friendly green synthesis of silver nanoparticles using Feronia elephantum (Rutaceae) against Culex quinquefasciatus, Anopheles stephensi, and Aedes aegypti (Diptera: Culicidae). *Parasitology Research,* **113**, 1775-1785.
107. Santhoshkumar, T., Rahuman, A. A., Rajakumar, G., Marimuthu, S., Bagavan, A., Jayaseelan, C., Zahir, A. A., Elango, G., and Kamaraj, C. (2011) Synthesis of silver nanoparticles using Nelumbo nucifera leaf extract and its larvicidal activity against malaria and filariasis vectors. *Parasitology Research,* **108**, 693-702.
108. Kora, A. J., and Arunachalam, J. (2012) Leaf Extract of Dendrophthoe falcata: A renewable source for the green synthesis of antibacterial silver nanoparticles. *Journal of Biobased Materials and Bioenergy,* **6**, 158-164.
109. Ajitha, B., Reddy, Y. A. K., and Reddy, P. S. (2014) Biogenic nanoscale silver particles by Tephrosia purpurea leaf extract and their inborn antimicrobial activity. *Spectrochimica Acta Part A: Molecular and Biomolecular Spectroscopy,* **121**, 164-172.

110. Jayaseelan, C., Rahuman, A. A., Rajakumar, G., Kirthi, A. V., Santhoshkumar, T., Marimuthu, S., Bagavan, A., Kamaraj, C., Zahir, A. A., and Elango, G. (2011) Synthesis of pediculocidal and larvicidal silver nanoparticles by leaf extract from heartleaf moonseed plant, Tinospora cordifolia Miers. *Parasitology Research,* **109**, 185-194.
111. Marimuthu, S., Rahuman, A. A., Santhoshkumar, T., Jayaseelan, C., Kirthi, A. V., Bagavan, A., Kamaraj, C., Elango, G., Zahir, A. A., and Rajakumar, G. (2012) Lousicidal activity of synthesized silver nanoparticles using Lawsonia inermis leaf aqueous extract against Pediculus humanus capitis and Bovicola ovis. *Parasitology Research,* **111**, 2023-2033.
112. Meng, Y., and Sun, Y. (2016) Development of biogenic silver nanoparticle using Rosa chinensis flower extract and its antibacterial property. *Journal of Nanoscience and Nanotechnology,* **16**(4), 3969-3972.
113. Manikandan, R., Manikandan, B., Raman, T., Arunagirinathan, K., Prabhu, N. M., Basu, M. J., Perumal, M., Palanisamy, S., and Munusamy, A. (2015) Biosynthesis of silver nanoparticles using ethanolic petals extract of Rosa indica and characterization of its antibacterial, anticancer and anti-inflammatory activities. *Spectrochimica Acta Part A: Molecular and Biomolecular Spectroscopy,* **138**, 120-129.
114. Gogoi, N., Babu, P. J., Mahanta, C., and Bora, U. (2015) Green synthesis and characterization of silver nanoparticles using alcoholic flower extract of Nyctanthes arbortristis and in vitro investigation of their antibacterial and cytotoxic activities. *Materials Science and Engineering: C,* **46**, 463-469.
115. Ramar, M., Manikandan, B., Marimuthu, P. N., Raman, T., Mahalingam, A., Subramanian, P., Karthick, S., and Munusamy, A. (2015) Synthesis of silver nanoparticles using Solanum trilobatum fruits extract and its antibacterial, cytotoxic activity against human breast cancer cell line MCF 7. *Spectrochimica Acta Part A: Molecular and Biomolecular Spectroscopy,* **140**, 223-228.
116. Singh, S., Saikia, J. P., and Buragohain, A. K. (2013) A novel 'green'synthesis of colloidal silver nanoparticles (SNP) using Dillenia indica fruit extract. *Colloids and Surfaces B: Biointerfaces,* **102**, 83-85.
117. Bose, A., Keharia, H., and Deshpande, M. (2013) Eco-friendly phyto-synthesis of silver nanoparticles using Jatropha seedcake extract. *Chinese Physics Letters,* **30**, 128103.
118. Jun, S. H., Cha, S.-H., Kim, J., Cho, S., and Park, Y. (2015) Crystalline silver nanoparticles by using Polygala tenuifolia root extract as a green reducing agent. *Journal of Nanoscience and Nanotechnology,* **15**, 1567-1574.
119. Hazarika, S. N., Gupta, K., Shamin, K. N. A. M., Bhardwaj, P., Boruah,

R., Yadav, K. K., Naglot, A., Deb, P., Mandal, M., and Doley, R. (2016) One-pot facile green synthesis of biocidal silver nanoparticles. *Materials Research Express,* **3**, 075401.

120. Murugan, K., Senthilkumar, B., Senbagam, D., and Al-Sohaibani, S. (2014) Biosynthesis of silver nanoparticles using Acacia leucophloea extract and their antibacterial activity. *International Journal of Nanomedicine,* **9**, 2431-2438.
121. Salunke, B. K., Sawant, S. S., and Kim, B. S. (2016) Enhancement of antibacterial effect by biosynthesized silver nanoparticles with antibiotics. *Journal of Nanoscience and Nanotechnology,* **16**, 7191-7194.
122. Santhoshkumar, T., Rahuman, A. A., Bagavan, A., Marimuthu, S., Jayaseelan, C., Kirthi, A. V., Kamaraj, C., Rajakumar, G., Zahir, A. A., and Elango, G. (2012) Evaluation of stem aqueous extract and synthesized silver nanoparticles using Cissus quadrangularis against Hippobosca maculata and Rhipicephalus (Boophilus) microplus. *Experimental Parasitology,* **132**, 156-165.
123. Kothai, S., and Jayanthi, B. (2014) Ultrasound Intensified green synthesis of silver nanoparticles using Camellia sinensis extract fortified with lemon and honey. *International Journal of ChemTech Research,* **6**, 248-253.
124. Castro, L., Blazquez, M. L., González, F., Muñoz, J. A., and Ballester, A. (2015) Biosynthesis of silver and platinum nanoparticles using orange peel extract: characterisation and applications. *IET Nano-Biotechnology,* **9**, 252-258.
125. Abubakar, A. S., Salisu, I. B., Chahal, S., Sahni, G., and Pudake, R. N. (2014) Biosynthesis and characterization of silver nano particles using black carrot root extract. *International Journal of Current Research and Review,* **6**(17), 5-8.
126. Anjum, S., Gupta, A., Sharma, D., Gautam, D., Bhan, S., Sharma, A., Kapil, A., and Gupta, B. (2016) Development of novel wound care systems based on nanosilver nanohydrogels of polymethacrylic acid with Aloe vera and curcumin. *Materials Science and Engineering: C,* **64**, 157-166.
127. Aziz, N., Faraz, M., Pandey, R., Shakir, M., Fatma, T., Varma, A., Barman, I., and Prasad, R. (2015) Facile algae-derived route to biogenic silver nanoparticles: synthesis, antibacterial, and photocatalytic properties. *Langmuir,* **31**, 11605-11612.
128. Ebrahiminezhad, A., Bagheri, M., Taghizadeh, S.-M., Berenjian, A., and Ghasemi, Y. (2016) Biomimetic synthesis of silver nanoparticles using microalgal secretory carbohydrates as a novel anticancer and antimicrobial. *Advances in Natural Sciences: Nanoscience and Nanotechnology,* **7**, 015018.
129. Aziz, N., Fatma, T., Varma, A., and Prasad, R. (2014) Biogenic synthesis of silver nanoparticles using Scenedesmus abundans and

evaluation of their antibacterial activity. *Journal of Nanoparticles,* 2014, Article ID 689419.

130. Selvi, B. C. G., Madhavan, J., and Santhanam, A. (2016) Cytotoxic effect of silver nanoparticles synthesized from Padina tetrastromatica on breast cancer cell line. *Advances in Natural Sciences: Nanoscience and Nanotechnology,* **7**, 035015.
131. Railean-Plugaru, V., Pomastowski, P., Wypij, M., Szultka-Mlynska, M., Rafinska, K., Golinska, P., Dahm, H., and Buszewski, B. (2016) Study of silver nanoparticles synthesized by acidophilic strain of Actinobacteria isolated from the of Picea sitchensis forest soil. *Journal of Applied Microbiology,* **120**(5), 1250-1263.
132. Anasane, N., Golinska, P., Wypij, M., Rathod, D., Dahm, H., and Rai, M. (2015) Acidophilic actinobacteria synthesised silver nanoparticles showed remarkable activity against fungi-causing superficial mycoses in humans. *Mycoses,* **59**(3), 157-166.
133. Buszewski, B., Railean-Plugaru, V., Pomastowski, P., Rafinska, K., Szultka-Mlynska, M., Golinska, P., Wypij, M., Laskowski, D., and Dahm, H. (2016) Antimicrobial activity of biosilver nanoparticles produced by a novel Streptacidiphilus durhamensis strain. *Journal of Microbiology, Immunology and Infection.* **51**(1), 45-54.
134. Golinska, P., Wypij, M., Rathod, D., Tikar, S., Dahm, H., and Rai, M. (2015) Synthesis of silver nanoparticles from two acidophilic strains of Pilimelia columellifera subsp. pallida and their antibacterial activities. *Journal of Basic Microbiology,* **56**(5), 541-556.
135. Rathod, D., Golinska, P., Wypij, M., Dahm, H., and Rai, M. (2016) A new report of Nocardiopsis valliformis strain OT1 from alkaline Lonar crater of India and its use in synthesis of silver nanoparticles with special reference to evaluation of antibacterial activity and cytotoxicity. *Medical Microbiology and Immunology,* **205**, 435-447.
136. Gou, Y., Zhou, R., Ye, X., Gao, S., and Li, X. (2015) Highly efficient in vitro biosynthesis of silver nanoparticles using Lysinibacillus sphaericus MR-1 and their characterization. *Science and Technology of Advanced Materials,* **16**, 015004.
137. Bhatia, D., Mittal, A., and Malik, D. K. (2016) Antimicrobial activity of PVP coated silver nanoparticles synthesized by Lysinibacillus varians. *3 Biotech,* **6**(2), 196.
138. Kalabegishvili, T. L., Murusidze, I. G., Prangishvili, D. A., Kvachadze, L. I., Kirkesali, E. I., Rcheulishvili, A. N., Ginturi, E. N., Janjalia, M. B., Tsertsvadze, G. I., Gabunia, V. M., Vladimirovna, F. M., Inga, Z., and Sergeevich, P. S. (2015) Silver nanoparticles in Sulfolobus islandicus biomass for technological applications. *Advanced Science, Engineering and Medicine,* **7**, 797-804.
139. Wang, C., Singh, P., Kim, Y. J., Mathiyalagan, R., Myagmarjav, D., Wang, D., Jin, C.-G., and Yang, D. C. (2016) Characterization and an-

timicrobial application of biosynthesized gold and silver nanoparticles by using Microbacterium resistens. *Artificial Cells, Nanomedicine, and Biotechnology,* **44**, 1714-1721.

140. Wang, C., Kim, Y. J., Singh, P., Mathiyalagan, R., Jin, Y., and Yang, D. C. (2016) Green synthesis of silver nanoparticles by Bacillus methylotrophicus, and their antimicrobial activity. *Artificial Cells, Nanomedicine, and Biotechnology,* **44**, 1127-1132.
141. Jo, J. H., Singh, P., Kim, Y. J., Wang, C., Mathiyalagan, R., Jin, C.-G., and Yang, D. C. (2016) Pseudomonas deceptionensis DC5-mediated synthesis of extracellular silver nanoparticles. *Artificial Cells, Nanomedicine, and Biotechnology,* **44**, 1576-1581.
142. Singh, P., Kim, Y. J., Wang, C., Mathiyalagan, R., and Yang, D. C. (2016) Weissella oryzae DC6-facilitated green synthesis of silver nanoparticles and their antimicrobial potential. *Artificial Cells, Nanomedicine, and Biotechnology,* **44**, 1569-1575.
143. Ashengroph, M. (2014) Extracellular synthesis of silver nanoparticles by Ralstonia sp. SM8 isolated from the Sarcheshmeh copper mine. *Biological Journal of Microorganism,* **3**(9), 53-64.
144. Govindappa, M., Farheen, H., Chandrappa, C., Rai, R. V., and Raghavendra, V. B. (2016) Mycosynthesis of silver nanoparticles using extract of endophytic fungi, Penicillium species of Glycosmis mauritiana, and its antioxidant, antimicrobial, anti-inflammatory and tyrokinase inhibitory activity. *Advances in Natural Sciences: Nanoscience and Nanotechnology,* **7**, 035014.
145. Rahi, D., and Parmar, A. (2014) Mycosynthesis of silver nanoparticles by an endophytic Penicillium species of Aloe vera root, evaluation of their antibacterial and antibiotic enhancing activity. *International Journal of Nanomaterials Biostructures,* **4**(3), 46-51.
146. Shelar, G. B., and Chavan, A. M. (2015) Myco-synthesis of silver nanoparticles from Trichoderma harzianum and its impact on germination status of oil seed. *Biolife,* **3**, 109-113.
147. Fayaz, A. M., Balaji, K., Girilal, M., Yadav, R., Kalaichelvan, P. T, and Venketesan, R. (2010) Biogenic synthesis of silver nanoparticles and their synergistic effect with antibiotics: a study against gram-positive and gram-negative bacteria. *Nanomedicine: Nanotechnology, Biology and Medicine,* **6**, 103-109.
148. Abdel-Hafez, S. I., Nafady, N. A., Abdel-Rahim, I. R., Shaltout, A. M., Daros, J.-A., and Mohamed, M. A. (2016) Assessment of protein silver nanoparticles toxicity against pathogenic Alternaria solani. *3 Biotech,* **6**, 199.
149. Aziz, N., Pandey, R., Barman, I., and. Prasad, R (2016) Leveraging the attributes of Mucor hiemalis-derived silver nanoparticles for a synergistic broad-spectrum antimicrobial platform. *Frontiers in Microbiology,* **7**, 1984.
150. Krishna, G., Kumar, S., Pranitha, V., Alha, M., and Charaya, S. (2015)

Biogenic synthesis of silver nanoparticles and their synergistic effect with antibiotics: A study against salmonella SP. *International Journal of Pharmacy and Pharmaceutical Sciences,* **7**, 84-88.
151. Kumar, R. M., Rao, B. L., Asha, S., Narayana, B., Byrappa, K., Wang, Y., Yao, D., and Sangappa Y. (2015) Gamma radiation assisted biosynthesis of silver nanoparticles and their characterization. *Advance Materials Letters,* **6**, 1088-1093.
152. Shivananda, C., Madhu Kumar, R., Narayana, B., Byrappa, K., Renu, P., Wang, Y., and Sangappa, Y. (2016) Preparation and characterisation of silk fibroin-silver nanoparticles (SF–AgNPs) composite films. *Materials Research Innovations,* **21**(4), 210-214.
153. Yue, X., Lin, H., Yan, T., Zhang, D., Lin, H., and Chen, Y. (2014) Synthesis of silver nanoparticles with sericin and functional finishing to cotton fabrics. *Fibers and Polymers,* **15**, 716-722.
154. Aramwit, P., Bang, N., Ratanavaraporn, J., and Ekgasit, S. (2014) Green synthesis of silk sericin-capped silver nanoparticles and their potent anti-bacterial activity. *Nanoscale Research Letters,* **9**(1), 79.
155. Wu, C., Zhang, G., Xia, T., Li, Z., Zhao, K., Deng, Z., Guo, D., and Peng, B. (2015) Bioinspired synthesis of polydopamine/Ag nanocomposite particles with antibacterial activities. *Materials Science and Engineering: C,* **55**, 155-165.
156. Sinha, T., Ahmaruzzaman, M., and Bhattacharjee, A. (2014) A simple approach for the synthesis of silver nanoparticles and their application as a catalyst for the photodegradation of methyl violet 6B dye under solar irradiation. *Journal of Environmental Chemical Engineering,* **2**, 2269-2279.
157. Kim, T. Y., Cha, S.-H., Cho, S., and Park, Y. (2016) Tannic acid-mediated green synthesis of antibacterial silver nanoparticles. *Archives of Pharmacal Research,* **39**, 465-473.
158. Francesko, A., Blandon, L., Vazquez, M., Petkova, P., Morato, J., Pfeifer, A., Heinze, T., Mendoza, E., and Tzanov, T. (2015) Enzymatic functionalization of cork surface with antimicrobial hybrid biopolymer/silver nanoparticles. *ACS Applied Materials & Interfaces,* **7**, 9792-9799.
159. Aryabadie, S., Sadeghi-Kiakhani, M., and Arami, M. (2015) Antimicrobial and dyeing studies of treated cotton fabrics by prepared chitosan-PAMAM dendrimer/Ag nano-emulsion. *Fibers and Polymers,* **16**, 2529-2537.
160. Wang, Y., Guo, X., Pan, R., Han, D., Chen, T., Geng, Z., Xiong, Y., and Chen, Y. (2015) Electrodeposition of chitosan/gelatin/nanosilver: A new method for constructing biopolymer/nanoparticle composite films with conductivity and antibacterial activity. *Materials Science and Engineering: C,* **53**, 222-228.
161. Anuradha, K., Bangal, P., and Madhavendra, S. S. (2016) Macromol-

ecular arabinogalactan polysaccharide mediated synthesis of silver nanoparticles, characterization and evaluation. *Macromolecular Research,* **24**, 152-162.

162. Huang, M., Du, L., and Feng, J.-X. (2016) Photochemical synthesis of silver nanoparticles/eggshell membrane composite, its characterization and antibacterial activity. *Science of Advanced Materials,* **8**, 1641-1647.
163. Sinha, T., and Ahmaruzzaman, M. (2015) High-value utilization of egg shell to synthesize silver and gold–silver core shell nanoparticles and their application for the degradation of hazardous dyes from aqueous phase-A green approach. *Journal of Colloid and Interface Science,* **453**, 115-131.
164. Lu, F., Sun, D., Huang, J., Du, M., Yang, F., Chen, H., Hong, Y., and Li, Q. (2014) Plant-mediated synthesis of Ag–Pd alloy nanoparticles and their application as catalyst toward selective hydrogenation. *ACS Sustainable Chemistry & Engineering,* **2**, 1212-1218.
165. Vivekanandhan, S., Venkateswarlu, M., Rawls, H., Misra, M., Mohanty, A., and Satyanarayana, N. (2015) Synthesis and characterization of AgNP:ZrO_2 functional nanomaterials by leaf extract assisted bioreduction process. *Ceramics International,* **41**, 3305-3311.
166. Montazer, M., and Kahali, P. (2016) A novel polyvinyl alcohol-tragacanth/nano silver hydrogel on polyester fabric through in situ synthesis method. *Journal of Industrial Textiles,* **45**, 1635-1651.
167. Sheny, D., Philip, D., and Mathew, J. (2012) Rapid green synthesis of palladium nanoparticles using the dried leaf of Anacardium occidentale. *Spectrochimica Acta Part A: Molecular and Biomolecular Spectroscopy,* **91**, 35-38.
168. Mallikarjuna, K., Sushma, N. J., Reddy, B. S., Narasimha, G., and Raju, B. D. P. (2013) Palladium nanoparticles: single-step plant-mediated green chemical procedure using Piper betle leaves broth and their anti-fungal studies. *International Journal of Chemical and Analytical Science,* **4**, 14-18.
169. Khan, M., Khan, M., Kuniyil, M., Adil, S. F., Al-Warthan, A., Alkhathlan, H. Z., Tremel, W., Tahir, M. N., and Siddiqui, M. R. H. (2014) Biogenic synthesis of palladium nanoparticles using Pulicaria glutinosa extract and their catalytic activity towards the Suzuki coupling reaction. *Dalton Transactions,* **43**, 9026-9031.
170. Lebaschi, S., Hekmati, M., and Veisi, H. (2017) Green synthesis of palladium nanoparticles mediated by black tea leaves (Camellia sinensis) extract: Catalytic activity in the reduction of 4-nitrophenol and Suzuki-Miyaura coupling reaction under ligand-free conditions. *Journal of Colloid and Interface Science,* **485**, 223-231.
171. Shaik, M. R., Ali, Z. J. Q., Khan, M., Kuniyil, M., Assal, M. E., Alkhathl-

an, H. Z., Al-Warthan, A., Siddiqui, M. R. H., Khan, M., and Adil, S. F. (2017) Green synthesis and characterization of palladium nanoparticles using Origanum vulgare L. extract and their catalytic activity. *Molecules,* **22**, doi:10.3390/molecules22010165.

172. Atarod, M., Nasrollahzadeh, M., and Sajadi, S. M. (2016) Green synthesis of Pd/RGO/Fe_3O_4 nanocomposite using Withania coagulans leaf extract and its application as magnetically separable and reusable catalyst for the reduction of 4-nitrophenol. *Journal of Colloid and Interface Science,* **465**, 249-258.
173. Nasrollahzadeh, M., Sajadi, S. M., Rostami-Vartooni, A., Alizadeh, M., and Bagherzadeh, M. (2016) Green synthesis of the Pd nanoparticles supported on reduced graphene oxide using barberry fruit extract and its application as a recyclable and heterogeneous catalyst for the reduction of nitroarenes. *Journal of Colloid and Interface Science,* **466**, 360-368.
174. Ganaie, S., Abbasi, T., and Abbasi, S. (2016) Low-cost, environment-friendly synthesis of palladium nanoparticles by utilizing a terrestrial weed Antigonon leptopus. *Particulate Science and Technology,* **34**, 201-208.
175. Bedford, N. M., Ramezani-Dakhel, H., Slocik, J. M., Briggs, B. D., Ren, Y., Frenkel, A. I., Petkov, V., Heinz, H., Naik, R. R., and Knecht, M. R. (2015) Elucidation of peptide-directed palladium surface structure for biologically tunable nanocatalysts. *ACS Nano,* **9**, 5082-5092.
176. Nasrollahzadeh, M., and Sajadi, S. M. (2016) Green synthesis, characterization and catalytic activity of the Pd/TiO_2 nanoparticles for the ligand-free Suzuki-Miyaura coupling reaction. *Journal of Colloid and Interface Science,* **465**, 121-127.
177. Nasrollahzadeh, M., Sajadi, S. M., Rostami-Vartooni, A., and Bagherzadeh, M. (2015) Green synthesis of Pd/CuO nanoparticles by Theobroma cacao L. seeds extract and their catalytic performance for the reduction of 4-nitrophenol and phosphine-free Heck coupling reaction under aerobic conditions. *Journal of Colloid and Interface Science,* **448**, 106-113.
178. Khodadadi, B., Bordbar, M., and Nasrollahzadeh, M. (2017) Green synthesis of Pd nanoparticles at Apricot kernel shell substrate using Salvia hydrangea extract: Catalytic activity for reduction of organic dyes. *Journal of Colloid and interface Science,* **490**, 1-10.
179. Bordbar, M., and Mortazavimanesh, N. (2016) Green synthesis of Pd/walnut shell nanocomposite using Equisetum arvense L. leaf extract and its application for the reduction of 4-nitrophenol and organic dyes in a very short time. *Environmental Science and Pollution Research,* **24**, 4093-4104.
180. Makarov, V. V., Makarova, S. S., Love, A. J., Sinitsyna, O. V., Dudnik, A. O., Yaminsky, I. V., Taliansky, M. E., and Kalinina, N. O. (2014) Biosynthesis of stable iron oxide nanoparticles in aqueous extracts

of Hordeum vulgare and Rumex acetosa plants. *Langmuir,* **30**, 5982-5988.

181. Balamurughan, M., Mohanraj, S., Kodhaiyolii, S., and Pugalenthi, V. (2014) Ocimum sanctum leaf extract mediated green synthesis of iron oxide nanoparticles: spectroscopic and microscopic studies. *Journal of Chemical and Pharmaceutical Sciences,* **2014**, 201-204.
182. Balamurugan, M., Saravanan, S., and Soga, T. (2014) Synthesis of iron oxide nanoparticles by using Eucalyptus globulus plant extract. *e-Journal of Surface Science and Nanotechnology,* **12**, 363-367.
183. Kumar, B., Smita, K., Cumbal, L., Debut, A., Galeas, S., and Guerrero, V. H. (2016) Phytosynthesis and photocatalytic activity of magnetite (Fe_3O_4) nanoparticles using the Andean blackberry leaf. *Materials Chemistry and Physics,* **179**, 310-315.
184. Kombaiah, K., Vijaya, J. J., Kennedy, L. J., and Bououdina, M. (2016) Studies on the microwave assisted and conventional combustion synthesis of Hibiscus rosa-sinensis plant extract based $ZnFe_2O_4$ nanoparticles and their optical and magnetic properties. *Ceramics International,* **42**, 2741-2749.
185. Yew, Y. P., Shameli, K., Miyake, M., Kuwano, N., Khairudin, N. B. B. A., Mohamad, S. E. B., and Lee, K. X. (2016) Green synthesis of magnetite (Fe_3O_4) nanoparticles using seaweed (Kappaphycus alvarezii) extract. *Nanoscale Research Letters,* **11**, 1-7.
186. El-Kassas, H. Y., Aly-Eldeen, M. A., and Gharib, S. M. (2016) Green synthesis of iron oxide (Fe_3O_4) nanoparticles using two selected brown seaweeds: characterization and application for lead bioremediation. *Acta Oceanologica Sinica,* **35**, 89-98.
187. Mohamed, Y., Azzam, A., Amin, B., and Safwat, N. (2015) Mycosynthesis of iron nanoparticles by Alternaria alternata and its antibacterial activity. *African Journal of Biotechnology,* **14**, 1234-1241.
188. Kashyap, S., Woehl, T. J., Liu, X., Mallapragada, S. K., and Prozorov, T. (2014) Nucleation of iron oxide nanoparticles mediated by Mms6 protein in situ. *ACS Nano,* **8**, 9097-9106.
189. Zhang, S., Niu, H., Zhang, Y., Liu, J., Shi, Y., Zhang, X., and Cai, Y. (2012) Biocompatible phosphatidylcholine bilayer coated on magnetic nanoparticles and their application in the extraction of several polycyclic aromatic hydrocarbons from environmental water and milk samples. *Journal of Chromatography A,* **1238**, 38-45.
190. Wang, Y., Wang, S., Niu, H., Ma, Y., Zeng, T., Cai, Y., and Meng, Z. (2013) Preparation of polydopamine coated Fe_3O_4 nanoparticles and their application for enrichment of polycyclic aromatic hydrocarbons from environmental water samples. *Journal of Chromatography A,* **1283**, 20-26.
191. Si, J., and Yang, H. (2011) Preparation and characterization of biocompatible Fe_3O_4@polydopamine spheres with core/shell nanost-

ructure. *Materials Chemistry and Physics,* **128**, 519-524.

192. Zeng, T., Niu, H.-y., Ma, Y.-r., Li, W.-h., and Cai, Y.-q. (2013) In situ growth of gold nanoparticles onto polydopamine-encapsulated magnetic microspheres for catalytic reduction of nitrobenzene. *Applied Catalysis B: Environmental,* **134**, 26-33.
193. Zhu, Q., and Pan, Q. (2014) Mussel-inspired direct immobilization of nanoparticles and application for oil–water separation. *ACS Nano,* **8**, 1402-1409.
194. Sood, A., Arora, V., Shah, J., Kotnala, R., and Jain, T. K. (2016) Ascorbic acid-mediated synthesis and characterisation of iron oxide/gold core-shell nanoparticles. *Journal of Experimental Nanoscience,* **11**, 370-382.
195. Shah, R. K., Boruah, F., and Parween, N. (2015) Synthesis and characterization of ZnO nanoparticles using leaf extract of Camellia sinesis and evaluation of their antimicrobial efficacy. *International Journal of Current Microbiology and Applied Science,* **4**, 444-450.
196. Anbuvannan, M., Ramesh, M., Viruthagiri, G., Shanmugam, N., and Kannadasan, N. (2015) Anisochilus carnosus leaf extract mediated synthesis of zinc oxide nanoparticles for antibacterial and photocatalytic activities. *Materials Science in Semiconductor Processing,* **39**, 621-628.
197. Banumathi, B., Malaikozhundan, B., and Vaseeharan, B. (2016) Invitro acaricidal activity of ethnoveterinary plants and green synthesis of zinc oxide nanoparticles against Rhipicephalus (Boophilus) microplus. *Veterinary Parasitology,* **216**, 93-100.
198. Fatimah, I., Pradita, R. Y., and Nurfalinda, A. (2016) Plant extract mediated of ZnO nanoparticles by using ethanol extract of Mimosa pudica leaves and coffee powder. *Procedia Engineering,* **148**, 43-48.
199. Ambika, S., and Sundrarajan, M. (2015) Antibacterial behaviour of Vitex negundo extract assisted ZnO nanoparticles against pathogenic bacteria. *Journal of Photochemistry and Photobiology B: Biology,* **146**, 52-57.
200. Elumalai, K., Velmurugan, S., Ravi, S., Kathiravan, V., and Raj, G. A. (2015) Bio-approach: Plant mediated synthesis of ZnO nanoparticles and their catalytic reduction of methylene blue and antimicrobial activity. *Advanced Powder Technology,* **26**, 1639-1651.
201. Zheng, Y., Fu, L., Han, F., Wang, A., Cai, W., Yu, J., Yang, J., and Peng, F. (2015) Green biosynthesis and characterization of zinc oxide nanoparticles using Corymbia citriodora leaf extract and their photocatalytic activity. *Green Chemistry Letters and Reviews,* **8**, 59-63.
202. Sundrarajan, M., Ambika, S., and Bharathi, K. (2015) Plant-extract mediated synthesis of ZnO nanoparticles using Pongamia pinnata and their activity against pathogenic bacteria. *Advanced Powder Technology,* **26**, 1294-1299.

203. Anbuvannan, M., Ramesh, M., Viruthagiri, G., Shanmugam, N., and Kannadasan, N. (2015) Synthesis, characterization and photocatalytic activity of ZnO nanoparticles prepared by biological method. *Spectrochimica Acta Part A: Molecular and Biomolecular Spectroscopy,* **143**, 304-308.
204. Hassan, S. S., El Azab, W. I., Ali, H. R., and Mansour, M. S. (2015) Green synthesis and characterization of ZnO nanoparticles for photocatalytic degradation of anthracene. *Advances in Natural Sciences: Nanoscience and Nanotechnology,* **6**, 045012.
205. Patil, B. N., and Taranath, T. C. (2016) Limonia acidissima L. leaf mediated synthesis of zinc oxide nanoparticles: A potent tool against Mycobacterium tuberculosis. *International Journal of Mycobacteriology,* **5**, 197-204.
206. Thatoi, P., Kerry, R. G., Gouda, S., Das, G., Pramanik, K., Thatoi, H., and Patra, J. K. (2016) Photo-mediated green synthesis of silver and zinc oxide nanoparticles using aqueous extracts of two mangrove plant species, Heritiera fomes and Sonneratia apetala and investigation of their biomedical applications. *Journal of Photochemistry and Photobiology B: Biology,* **163**, 311-318.
207. Narendhran, S., and Sivaraj, R. (2016) Biogenic ZnO nanoparticles synthesized using L. aculeata leaf extract and their antifungal activity against plant fungal pathogens. *Bulletin of Materials Science,* **39**, 1-5.
208. Madan, H., Sharma, S., Suresh, D., Vidya, Y., Nagabhushana, H., Rajanaik, H., Anantharaju, K., Prashantha, S., and Maiya, P. S. (2016) Facile green fabrication of nanostructure ZnO plates, bullets, flower, prismatic tip, closed pine cone: Their antibacterial, antioxidant, photoluminescent and photocatalytic properties. *Spectrochimica Acta Part A: Molecular and Biomolecular Spectroscopy,* **152**, 404-416.
209. Thema, F., Manikandan, E., Dhlamini, M., and Maaza, M. (2015) Green synthesis of ZnO nanoparticles via Agathosma betulina natural extract. *Materials Letters,* **161**, 124-127.
210. Ali, K., Dwivedi, S., Azam, A., Saquib, Q., Al-Said, M. S., Alkhedhairy, A. A., and Musarrat, J. (2016) Aloe vera extract functionalized zinc oxide nanoparticles as nanoantibiotics against multi-drug resistant clinical bacterial isolates. *Journal of Colloid and Interface Science,* **472**, 145-156.
211. Jamdagni, P., Khatri, P., and Rana, J. (2016) Green synthesis of zinc oxide nanoparticles using flower extract of Nyctanthes arbortristis and their antifungal activity. *Journal of King Saud University-Science,* **30**(2), 168-175.
212. Diallo, A., Ngom, B., Park, E., and Maaza, M. (2015) Green synthesis of ZnO nanoparticles by Aspalathus linearis: Structural & optical properties. *Journal of Alloys and Compounds,* **646**, 425-430.

213. Dobrucka, R., and Długaszewska, J. (2016) Biosynthesis and antibacterial activity of ZnO nanoparticles using Trifolium pratense flower extract. *Saudi Journal of Biological Sciences,* **23**, 517-523.
214. Sharma, D., Sabela, M. I., Kanchi, S., Mdluli, P. S., Singh, G., Stenstrom, T. A., and Bisetty, K. (2016) Biosynthesis of ZnO nanoparticles using Jacaranda mimosifolia flowers extract: Synergistic antibacterial activity and molecular simulated facet specific adsorption studies. *Journal of Photochemistry and Photobiology B: Biology,* **162**, 199-207.
215. Azizi, S., Mohamad, R., Bahadoran, A., Bayat, S., Rahim, R. A., Ariff, A., and Saad, W. Z. (2016) Effect of annealing temperature on antimicrobial and structural properties of bio-synthesized zinc oxide nanoparticles using flower extract of Anchusa italica. *Journal of Photochemistry and Photobiology B: Biology,* **161**, 441-449.
216. Jafarirad, S., Mehrabi, M., Divband, B., and Kosari-Nasab, M. (2016) Biofabrication of zinc oxide nanoparticles using fruit extract of Rosa canina and their toxic potential against bacteria: A mechanistic approach. *Materials Science and Engineering: C,* **59**, 296-302.
217. Nethravathi, P., Shruthi, G., Suresh, D., Nagabhushana, H., and Sharma, S. (2015) Garcinia xanthochymus mediated green synthesis of ZnO nanoparticles: Photoluminescence, photocatalytic and antioxidant activity studies. *Ceramics International,* **41**, 8680-8687.
218. Fowsiya, J., Madhumitha, G., Al-Dhabi, N. A., and Arasu, M. V. (2016) Photocatalytic degradation of Congo red using Carissa edulis extract capped zinc oxide nanoparticles. *Journal of Photochemistry and Photobiology B: Biology,* **162**, 395-401.
219. Rana, N., Chand, S., and Gathania, A. K. (2016) Green synthesis of zinc oxide nano-sized spherical particles using Terminalia chebula fruits extract for their photocatalytic applications. *International Nano Letters,* **6**, 91-98.
220. Karnan, T., and Selvakumar, S. A. S. (2016) Biosynthesis of ZnO nanoparticles using rambutan (Nephelium lappaceumL.) peel extract and their photocatalytic activity on methyl orange dye. *Journal of Molecular Structure,* **1125**, 358-365.
221. Yuvakkumar, R., Suresh, J., Saravanakumar, B., Nathanael, A. J., Hong, S. I., and Rajendran, V. (2015) Rambutan peels promoted biomimetic synthesis of bioinspired zinc oxide nanochains for biomedical applications. *Spectrochimica Acta Part A: Molecular and Biomolecular Spectroscopy,* **137**, 250-258.
222. Supraja, N., Prasad, T., Krishna, T. G., and David, E. (2016) Synthesis, characterization, and evaluation of the antimicrobial efficacy of Boswellia ovalifoliolata stem bark-extract-mediated zinc oxide nanoparticles. *Applied Nanoscience,* **6**, 581-590.
223. Stan, M., Popa, A., Toloman, D., Dehelean, A., Lung, I., and Katona, G.

(2015) Enhanced photocatalytic degradation properties of zinc oxide nanoparticles synthesized by using plant extracts. *Materials Science in Semiconductor Processing,* **39**, 23-29.

224. Sharma, S. (2016) ZnO nano-flowers from Carica papaya milk: Degradation of Alizarin Red-S dye and antibacterial activity against Pseudomonas aeruginosa and Staphylococcus aureus. *Optik-International Journal for Light and Electron Optics,* **127**, 6498-6512.
225. Krishna, P. G., Ananthaswamy, P. P., Trivedi, P., Chaturvedi, V., Mutta, N. B., Sannaiah, A., Erra, A., and Yadavalli, T. (2017) Antitubercular activity of ZnO nanoparticles prepared by solution combustion synthesis using lemon juice as bio-fuel. *Materials Science and Engineering: C,* **75**, 1026-1033.
226. Colak, H. and Karakose, E. (2017) Green synthesis and characterization of nanostructured ZnO thin films using Citrus aurantifolia (lemon) peel extract by spin-coating method. *Journal of Alloys and Compounds,* **690**, 658-662.
227. Buazar, F., Bavi, M., Kroushawi, F., Halvani, M., Khaledi-Nasab, A., and Hossieni, S. (2016) Potato extract as reducing agent and stabiliser in a facile green one-step synthesis of ZnO nanoparticles. *Journal of Experimental Nanoscience,* **11**, 175-184.
228. Sutradhar, P., and Saha, M. (2016) Green synthesis of zinc oxide nanoparticles using tomato (Lycopersicon esculentum) extract and its photovoltaic application. *Journal of Experimental Nanoscience,* **11**, 314-327.
229. Sutradhar, P., Debbarma, M., and Saha, M. (2016) Microwave synthesis of zinc oxide nanoparticles using coffee powder extract and its application for solar cell. *Synthesis and Reactivity in Inorganic, Metal-Organic, and Nano-Metal Chemistry,* **46**, 1622-1627.
230. Hassan, S. S., Abdel-Shafy, H. I., and Mansour, M. S. (2016) Removal of pharmaceutical compounds from urine via chemical coagulation by green synthesized ZnO-nanoparticles followed by microfiltration for safe reuse. *Arabian Journal of Chemistry,* doi: 10.1016/j.arabjc.2016.04.009.
231. Sutradhar, P., and Saha, M. (2015) Synthesis of zinc oxide nanoparticles using tea leaf extract and its application for solar cell. *Bulletin of Materials Science,* **38**, 653-657.
232. Pandimurugan, R., and Thambidurai, S. (2016) Novel seaweed capped ZnO nanoparticles for effective dye photodegradation and antibacterial activity. *Advanced Powder Technology,* **27**, 1062-1072.
233. Ghayempour, S., and Montazer, M. (2017) Ultrasound irradiation based in-situ synthesis of star-like Tragacanth gum/zinc oxide nanoparticles on cotton fabric. *Ultrasonics Sonochemistry,* **34**, 458-465.

234. Ghayempour, S., Montazer, M., and Rad, M. M. (2016) Tragacanth gum biopolymer as reducing and stabilizing agent in biosonosynthesis of urchin-like ZnO nanorod arrays: a low cytotoxic photocatalyst with antibacterial and antifungal properties. *Carbohydrate Polymers,* **136**, 232-241.
235. Kundu, D., Hazra, C., Chatterjee, A., Chaudhari, A., and Mishra, S. (2014) Extracellular biosynthesis of zinc oxide nanoparticles using Rhodococcus pyridinivorans NT2: Multifunctional textile finishing, biosafety evaluation and in vitro drug delivery in colon carcinoma. *Journal of Photochemistry and Photobiology B: Biology,* **140**, 194-204.
236. Shamsuzzaman, Mashrai, A., Khanam, H., and Aljawfi, R. N. (2013) Biological synthesis of ZnO nanoparticles using C. albicans and studying their catalytic performance in the synthesis of steroidal pyrazolines. *Arabian Journal of Chemistry,* **10**, S1530-S1536.
237. Dhillon, G. S., Kaur, S., and Brar, S. K. (2014) Facile fabrication and characterization of chitosan-based zinc oxide nanoparticles and evaluation of their antimicrobial and antibiofilm activity. *International Nano Letters,* **4**, 1-11.
238. Abiraman, T., Kavitha, G., Rengasamy, R., and Balasubramanian, S. (2016) Antifouling behavior of chitosan adorned zinc oxide nanorods. *RSC Advances,* **6**, 69206-69217.
239. Manna, J., Goswami, S., Shilpa, N., Sahu, N., and Rana, R. K. (2015) Biomimetic method to assemble nanostructured Ag@ZnO on cotton fabrics: application as self-cleaning flexible materials with visible-light photocatalysis and antibacterial activities. *ACS Applied Materials & Interfaces,* **7**, 8076-8082.
240. Kirthi, A. V., Rahuman, A. A., Rajakumar, G., Marimuthu, S., Santhoshkumar, T., Jayaseelan, C., and Velayutham, K. (2011) Acaricidal, pediculocidal and larvicidal activity of synthesized ZnO nanoparticles using wet chemical route against blood feeding parasites. *Parasitology Research,* **109**, 461-472.
241. Zhao, Z.-Y., Wang, M.-H., and Liu, T.-T. (2015) Tribulus terrestris leaf extract assisted green synthesis and gas sensing properties of Ag-coated ZnO nanoparticles. *Materials Letters,* **158**, 274-277.
242. Ansari, S. A., Khan, M. M., Ansari, M. O., Lee, J., and Cho, M. H. (2013) Biogenic synthesis, photocatalytic, and photoelectrochemical performance of Ag–ZnO nanocomposite. *The Journal of Physical Chemistry C,* **117**, 27023-27030.
243. Sivakumar, P., Muralidharan, N., Santhosh, V., and Rajendran, P. (2014) Synthesis of Cs-ZnO-Ag nanocomposite from crustacean shells and its antibacterial activity. *Journal of Chemical and Pharmaceutical Sciences,* **2014**, 155-156.
244. Diallo, A., Manikandan, E., Rajendran, V., and Maaza, M. (2016) Physical & enhanced photocatalytic properties of green synthe-

sized SnO_2 nanoparticles via Aspalathus linearis. *Journal of Alloys and Compounds,* **681**, 561-570.

245. Osuntokun, J., Onwudiwe, D. C., and Ebenso, E. E. (2017) Biosynthesis and photocatalytic properties of SnO_2 nanoparticles prepared using aqueous extract of cauliflower. *Journal of Cluster Science,* **28**(4), 1883-1896.
246. Bhattacharjee, A., and Ahmaruzzaman, M. (2015) Photocatalytic-degradation and reduction of organic compounds using SnO_2 quantum dots (via a green route) under direct sunlight. *RSC Advances,* **5**, 66122-66133.
247. Srivastava, N., and Mukhopadhyay, M. (2014) Biosynthesis of SnO_2 nanoparticles using bacterium Erwinia herbicola and their photocatalytic activity for degradation of dyes. *Industrial & Engineering Chemistry Research,* **53**, 13971-13979.
248. Wang, M., Ping, H., Xie, H., Chen, B., Yan, M., Fang, W., and Fu, Z. (2016) Confined-space synthesis of hierarchical SnO_2 nanorods assembled by ultrasmall nanocrystals for energy storage. *RSC Advances,* **6**, 81809-81813.
249. Bhattacharjee, A., Ahmaruzzaman, M., Devi, T. B., and Nath, J. (2016) Photodegradation of methyl violet 6B and methylene blue using tin-oxide nanoparticles (synthesized via a green route). *Journal of Photochemistry and Photobiology A: Chemistry,* **325**, 116-124.
250. Bhattacharjee, A., and Ahmaruzzaman, M. (2015) A green approach for the synthesis of SnO_2 nanoparticles and its application in the reduction of p-nitrophenol. *Materials Letters,* **157**, 260-264.
251. Bhattacharjee, A., Ahmaruzzaman, M., and Sinha, T. (2014) Surfactant effects on the synthesis of durable tin-oxide nanoparticles and its exploitation as a recyclable catalyst for the elimination of toxic dye: a green and efficient approach for wastewater treatment. *RSC Advances,* **4**, 51418-51429.
252. Tammina, S. K., and Mandal, B. K. (2016) Tyrosine mediated synthesis of SnO_2 nanoparticles and their photocatalytic activity towards Violet 4 BSN dye. *Journal of Molecular Liquids,* **221**, 415-421.
253. Begum, S., Devi, T. B., and Ahmaruzzaman, M. (2016) l-lysine monohydrate mediated facile and environment friendly synthesis of SnO_2 nanoparticles and their prospective applications as a catalyst for the reduction and photodegradation of aromatic compounds. *Journal of Environmental Chemical Engineering,* **4**, 2976-2989.
254. Li, Y., Meng, Q., Ma, J., Zhu, C., Cui, J., Chen, Z., Guo, Z., Zhang, T., Zhu, S., and Zhang, D. (2015) Bioinspired carbon/SnO_2 composite anodes prepared from a photonic hierarchical structure for lithium batteries. *ACS Applied Materials & Interfaces,* **7**, 11146-11154.
255. Rajakumar, G., Rahuman, A. A., Jayaseelan, C., Santhoshkumar, T.,

Marimuthu, S., Kamaraj, C., Bagavan, A., Zahir, A. A., Kirthi, A. V., Elango, G., Arora, P., Karthikeyan, R., Manikandan, S., and Jose, S. (2014) Solanum trilobatum extract-mediated synthesis of titanium dioxide nanoparticles to control Pediculus humanus capitis, Hyalomma anatolicum anatolicum and Anopheles subpictus. *Parasitology Research,* **113**, 469-479.

256. Rajakumar, G., Rahuman, A. A., Roopan, S. M., Chung, I.-M., Anbarasan, K., and Karthikeyan, V. (2015) Efficacy of larvicidal activity of green synthesized titanium dioxide nanoparticles using Mangifera indica extract against blood-feeding parasites. *Parasitology Research,* **114**, 571-581.
257. Rajakumar, G., Rahuman, A. A., Priyamvada, B., Khanna, V. G., Kumar, D. K., and Sujin, P. (2012) Eclipta prostrata leaf aqueous extract mediated synthesis of titanium dioxide nanoparticles. *Materials Letters,* **68**, 115-117.
258. Velayutham, K., Rahuman, A. A., Rajakumar, G., Santhoshkumar, T., Marimuthu, S., Jayaseelan, C., Bagavan, A., Kirthi, A. V., Kamaraj, C., Zahir, A. A., and Elango, G. (2012) Evaluation of Catharanthus roseus leaf extract-mediated biosynthesis of titanium dioxide nanoparticles against Hippobosca maculata and Bovicola ovis. *Parasitology Research,* **111**, 2329-2337.
259. Sankar, R., Rizwana, K., Shivashangari, K. S., and Ravikumar, V. (2015) Ultra-rapid photocatalytic activity of Azadirachtaindica engineered colloidal titanium dioxide nanoparticles. *Applied Nanoscience,* **5**, 731-736.
260. Santhoshkumar, T., Rahuman, A. A., Jayaseelan, C., Rajakumar, G., Marimuthu, S., Kirthi, A. V., Velayutham, K., Thomas, J., Venkatesan, J., and Kim, S.-K. (2014) Green synthesis of titanium dioxide nanoparticles using Psidium guajava extract and its antibacterial and antioxidant properties. *Asian Pacific Journal of Tropical Medicine,* **7**, 968-976.
261. Sundrarajan, M., and Gowri, S. (2011) Green synthesis of titanium dioxide nanoparticles by Nyctanthes arbor-tristis leaves extract. *Chalcogenide Letters,* **8**, 447-451.
262. Hudlikar, M., Joglekar, S., Dhaygude, M., and Kodam, K. (2012) Green synthesis of TiO_2 nanoparticles by using aqueous extract of Jatropha curcas L. latex. *Materials Letters,* **75**, 196-199.
263. Suman, T. Y., Ravindranath, R. R. S., Elumalai, D., Kaleena, P. K., Ramkumar, R., Perumal, P., Aranganathan, L., and Chitrarasu, P. S. (2015) Larvicidal activity of titanium dioxide nanoparticles synthesized using Morinda citrifolia root extract against Anopheles stephensi, Aedes aegypti and Culex quinquefasciatus and its other effect on non-target fish. *Asian Pacific Journal of Tropical Disease,* **5**, 224-230.
264. Roopan, S. M., Bharathi, A., Prabhakarn, A., Rahuman, A. A., Velayu-

tham, K., Rajakumar, G., Padmaja, R., Lekshmi, M., and Madhumitha, G. (2012) Efficient phyto-synthesis and structural characterization of rutile TiO_2 nanoparticles using Annona squamosa peel extract. *Spectrochimica Acta Part A: Molecular and Biomolecular Spectroscopy,* **98**, 86-90.

265. Marimuthu, S., Rahuman, A. A., Jayaseelan, C., Kirthi, A. V., Santhoshkumar, T., Velayutham, K., Bagavan, A., Kamaraj, C., Elango, G., Iyappan, M., Sivaa, C., Karthika, L., and Rao, K. V. B. (2013) Acaricidal activity of synthesized titanium dioxide nanoparticles using Calotropis gigantea against Rhipicephalus microplus and Haemaphysalis bispinosa. *Asian Pacific Journal of Tropical Medicine,* **6**, 682-688.
266. Kirthi, A. V., Rahuman, A. A., Rajakumar, G., Marimuthu, S., Santhoshkumar, T., Jayaseelan, C., Elango, G., Zahir, A. A., Kamaraj, C., and Bagavan, A. (2011) Biosynthesis of titanium dioxide nanoparticles using bacterium Bacillus subtilis. *Materials Letters,* **65**, 2745-2747.
267. Jayaseelan, C., Rahuman, A. A., Roopan, S. M., Kirthi, A. V., Venkatesan, J., Kim, S.-K., Iyappan, M., and Siva, C. (2013) Biological approach to synthesize TiO_2 nanoparticles using Aeromonas hydrophila and its antibacterial activity. *Spectrochimica Acta Part A: Molecular and Biomolecular Spectroscopy,* **107**, 82-89.
268. Ordenes-Aenishanslins, N. A., Saona, L. A., Duran-Toro, V. M., Monras, J. P., Bravo, D. M., and Perez-Donoso, J. M. (2014) Use of titanium dioxide nanoparticles biosynthesized by Bacillus mycoides in quantum dot sensitized solar cells. *Microbial Cell Factories,* **13**, 90.
269. Rajakumar, G., Rahuman, A. A., Roopan, S. M., Khanna, V. G., Elango, G., Kamaraj, C., Zahir, A. A., and Velayutham, K. (2012) Fungus-mediated biosynthesis and characterization of TiO_2 nanoparticles and their activity against pathogenic bacteria. *Spectrochimica Acta Part A: Molecular and Biomolecular Spectroscopy,* **91**, 23-29.
270. Kandiah, K., Venkatachalam, R., Wang, C., Valiyaveettil, S., and Ganesan, K. (2015) In vitro and preliminary in vivo toxicity screening of high-surface-area TiO_2–chondroitin-4-sulfate nanocomposites for bone regeneration application. *Colloids and Surfaces B: Biointerfaces,* **128**, 347-356.
271. Prasad, P. R., Kanchi, S., and Naidoo, E. (2016) In-vitro evaluation of copper nanoparticles cytotoxicity on prostate cancer cell lines and their antioxidant, sensing and catalytic activity: One-pot green approach, *Journal of Photochemistry and Photobiology B: Biology,* **161**, 375-382.
272. Nasrollahzadeh, M., Maham, M., and Sajadi, S. M. (2015) Green synthesis of CuO nanoparticles by aqueous extract of Gundelia tournefortii and evaluation of their catalytic activity for the synthesis of N-monosubstituted ureas and reduction of 4-nitrophenol. *Journal*

of Colloid and Interface Science, **455**, 245-253.

273. Gopinath, M., Subbaiya. R., Selvam, M. M., and Suresh, D. (2014) Synthesis of copper nanoparticles from Nerium oleander leaf aqueous extract and its antibacterial activity. *International Journal of Current Microbiology and Applied Science,* **3**, 814-818.
274. Kaur, P., Thakur, R., and Chaudhury, A. (2016) Biogenesis of copper nanoparticles using peel extract of Punica granatum and their antimicrobial activity against opportunistic pathogens. *Green Chemistry Letters and Reviews,* **9**, 33-38.
275. Gerits, N. (2016) Synthesis of copper nanoparticles using grape stalk and spent coffee extract. Online: https://dugi-doc.udg.edu/handle/10256/13232 (assessed 13th May 2018).
276. Nasrollahzadeh, M., Sajadi, S. M., Rostami-Vartooni, A., and Hussin, S. M. (2016) Green synthesis of CuO nanoparticles using aqueous extract of Thymus vulgaris L. leaves and their catalytic performance for N-arylation of indoles and amines. *Journal of Colloid and Interface Science,* **466**, 113-119.
277. Karimi, J., and Mohsenzadeh, S. (2015) Rapid, green, and eco-friendly biosynthesis of copper nanoparticles using flower extract of Aloe vera. *Synthesis and Reactivity in Inorganic, Metal-Organic, and Nano-Metal Chemistry,* **45**, 895-898.
278. Abiraman, T., Ramanathan, E., Kavitha, G., Rengasamy, R., and Balasubramanian, S. (2017) Synthesis of chitosan capped copper oxide nanoleaves using high intensity (30kHz) ultrasound sonication and their application in antifouling coatings. *Ultrasonics Sonochemistry,* **34**, 781-791.
279. Ismail, E., Diallo, A., Khenfouch, M., Dhlamini, S., and Maaza, M. (2016) RuO_2 nanoparticles by a novel green process via Aspalathus linearis natural extract & their water splitting response. *Journal of Alloys and Compounds,* **662**, 283-289.
280. Diallo, A., Beye, A., Doyle, T., Park, E., and Maaza, M. (2015) Green synthesis of Co_3O_4 nanoparticles via Aspalathus linearis: physical properties. *Green Chemistry Letters and Reviews,* **8**, 30-36.
281. Diallo, A., Mothudi, B. M., Manikandan, E., and Maaza, M. (2016) Luminescent Eu_2O_3 nanocrystals by Aspalathus linearis' extract: structural and optical properties. *Journal of Nanophotonics,* **10**, 026010.
282. Kannan, S., and Sundrarajan, M. (2015) Green synthesis of ruthenium oxide nanoparticles: Characterization and its antibacterial activity. *Advanced Powder Technology,* **26**, 1505-1511.
283. Sone, B., Fuku, X., and Maaza, M. (2016) Physical & Electrochemical Properties of Green Synthesized Bunsenite NiO Nanoparticles via Callistemon Viminalis Extracts. *International Journal of Electrochemical Science,* **11**, 8204-8220.
284. Thema, F., Manikandan, E., Gurib-Fakim, A., and Maaza, M. (2016)

Single phase Bunsenite NiO nanoparticles green synthesis by Agathosma betulina natural extract. *Journal of Alloys and Compounds,* **657**, 655-661.

285. Yuvakkumar, R., Suresh, J., Nathanael, A. J., Sundrarajan, M., and Hong, S. (2014) Rambutan (Nephelium lappaceum L.) peel extract assisted biomimetic synthesis of nickel oxide nanocrystals. *Materials Letters,* **128**, 170-174.
286. Yuvakkumar, R., Lee. S., Lee. G. W., and Hong, S. I. (2015) An environment benign biomimetic synthesis of mesoporous NiO concentric stacked doughnuts architecture. *Microporous and Mesoporous Materials,* **207**, 185-194.
287. Thovhogi, N., Park, E., Manikandan, E., Maaza, M., and Gurib-Fakim, A. (2016) Physical properties of CdO nanoparticles synthesized by green chemistry via Hibiscus Sabdariffa flower extract. *Journal of Alloys and Compounds,* **655**, 314-320.
288. Thovhogi, N., Diallo, A., Gurib-Fakim, A., and Maaza, M. (2015) Nanoparticles green synthesis by Hibiscus Sabdariffa flower extract: main physical properties. *Journal of Alloys and Compounds,* **647**, 392-396.
289. Malleshappa, J., Nagabhushana, H., Sharma, S., Vidya, Y., Anantharaju, K., Prashantha, S., Prasad, B. D., Naika, H. R., Lingaraju, K., and Surendra, B. (2015) Leucas aspera mediated multifunctional CeO_2 nanoparticles: structural, photoluminescent, photocatalytic and antibacterial properties. *Spectrochimica Acta Part A: Molecular and Biomolecular Spectroscopy,* **149**(5), 452-462.
290. Thema, F., Beukes, P., Gurib-Fakim, A., and Maaza, M. (2015) Green synthesis of Monteponite CdO nanoparticles by Agathosma betulina natural extract. *Journal of Alloys and Compounds,* **646**, 1043-1048.
291. Karimi Andeani, J., and Mohsenzadeh, S. (2012) Phytosynthesis of cadmium oxide nanoparticles from Achillea wilhelmsii flowers. *Journal of Chemistry,* **2013**, Article ID 147613.
292. Sone, B., Manikandan, E., Gurib-Fakim, A., and Maaza, M. (2015) Sm_2O_3 nanoparticles green synthesis via Callistemon viminalis' extract. *Journal of Alloys and Compounds,* **650**, 357-362.
293. Sone, B., Manikandan, E., Gurib-Fakim, A., and Maaza, M. (2016) Single-phase α-Cr_2O_3 nanoparticles green synthesis using Callistemon viminalis red flower extract. *Green Chemistry Letters and Reviews,* **9**, 85-90.
294. Anilkumar, M., Nagaswarupa, H., Nagabhushana, H., Sharma, S., Vidya, Y., Anantharaju, K., Prashantha, S., Shivakuamra, C., and Gurushantha, K. (2015) Bio-inspired route for the synthesis of spherical shaped $MgO:Fe^{3+}$ nanoparticles: Structural, photoluminescence and photocatalytic investigation. *Spectrochimica Acta Part A: Molecular and Biomolecular Spectroscopy,* **149**, 703-713.

295. Xia, Z., Lin, Z., Xiao, Y., Wang, L., Zheng, J., Yang, H., and Chen, G. (2013) Facile synthesis of polydopamine-coated molecularly imprinted silica nanoparticles for protein recognition and separation. *Biosensors and Bioelectronics,* **47**, 120-126.
296. Chang, D., Gao, Y., Wang, L., Liu, G., Chen, Y., Wang, T., Tao, W., Mei, L., Huang, L., and Zeng, X. (2016) Polydopamine-based surface modification of mesoporous silica nanoparticles as pH-sensitive drug delivery vehicles for cancer therapy. *Journal of Colloid and Interface Science,* **463**, 279-287.
297. Zheng, Q., Lin, T., Wu, H., Guo, L., Ye, P., Hao, Y., Guo, Q., Jiang, J., Fu, F., and Chen, G. (2014) Mussel-inspired polydopamine coated mesoporous silica nanoparticles as pH-sensitive nanocarriers for controlled release. *International Journal of Pharmaceutics,* **463**, 22-26.
298. Lai, G., Zhang, H., Yong, J., and Yu, A. (2013) In situ deposition of gold nanoparticles on polydopamine functionalized silica nanosphere for ultrasensitive nonenzymatic electrochemical immunoassay. *Biosensors and Bioelectronics,* **47**, 178-183.
299. Tripathi, B. P., Dubey, N. C., Subair, R., Choudhury, S., and Stamm, M. (2016) Enhanced hydrophilic and antifouling polyacrylonitrile membrane with polydopamine modified silica nanoparticles. *RSC Advances,* **6**, 4448-4457.
300. Tian, J., Zhang, H., Liu, M., Deng, F., Huang, H., Wan, Q., Li, Z., Wang, K., He, X., and Zhang, X. (2015) A bioinspired strategy for surface modification of silica nanoparticles. *Applied Surface Science,* **357**, 1996-2003.
301. Heng, C., Liu, M., Wang, K., Deng, F., Huang, H., Wan, Q., Hui, J., Zhang, X., and Wei, Y. (2015) Biomimic preparation of highly dispersible silica nanoparticles based polymer nanocomposites. *Ceramics International,* **41**, 15075-15082.
302. Gupta, G., Iyer, S., Leasure, K., Virdone, N., Dattelbaum, A. M., Atanassov, P. B., and Lopez, G. P. (2013) Stable and fluid multilayer phospholipid–silica thin films: mimicking active multi-lamellar biological assemblies. *ACS Nano,* **7**, 5300-5307.
303. Feng, W., Nie, W., He, C., Zhou, X., Chen, L., Qiu, K., Wang, W., and Yin, Z. (2014) Effect of pH-responsive alginate/chitosan multilayers coating on delivery efficiency, cellular uptake and biodistribution of mesoporous silica nanoparticles based nanocarriers. *ACS Applied Materials & Interfaces,* **6**, 8447-8460.
304. Xavier, J. R., Thakur, T., Desai, P., Jaiswal, M. K., Sears, N., Cosgriff-Hernandez, E., Kaunas, R., and Gaharwar, A. K. (2015) Bioactive nanoengineered hydrogels for bone tissue engineering: a growth-factor-free approach. *ACS Nano,* **9**, 3109-3118.
305. Guo, Y., Ge, X., Guan, J., Wu, L., Zhao, F., Li, H., Mu, X., Jiang, Y., and Chen, A. (2016) A novel method for fabricating hybrid biobased

nanocomposites film with stable fluorescence containing CdTe quantum dots and montmorillonite-chitosan nanosheets. *Carbohydrate Polymers,* **145**, 13-19.

306. Altinisik, A., Bozaci, E., Akar, E., Seki, Y., Yurdakoc, K., Demir, A., and Ozdogan, E. (2013) Development of antimicrobial cotton fabric using bionanocomposites. *Cellulose,* **20**, 3111-3121.
307. Khenfouch, M., Minnis Ndimba, R., Diallo, A., Khamlich, S., Hamzah, M., Dhlamini, M., Mothudi, B., Baitoul, M., Srinivasu, V., and Maaza, M. (2016) Artemisia herba-alba Asso eco-friendly reduced few-layered graphene oxide nanosheets: structural investigations and physical properties. *Green Chemistry Letters and Reviews,* **9**, 122-131.
308. Sadhukhan, S., Ghosh, T. K., Rana, D., Roy, I., Bhattacharyya, A., Sarkar, G., Chakraborty, M., and Chattopadhyay, D. (2016) Studies on synthesis of reduced graphene oxide (RGO) via green route and its electrical property. *Materials Research Bulletin,* **79**, 41-51.
309. Gurunathan, S., Han, J. W., Park, J. H., Eppakayala, V., and Kim, J.-H. (2014) Ginkgo biloba: a natural reducing agent for the synthesis of cytocompatible graphene. *International Journal of Nanomedicine,* **9**, 363-377.
310. Ghosh, T. K., Gope, S., Rana, D., Roy, I., Sarkar, G., Sadhukhan, S., Bhattacharya, A., Pramanik, K., Chattopadhyay, S., Chakraborty, M., and Chattopadhyay D. (2016) Physical and electrical characterization of reduced graphene oxide synthesized adopting green route. *Bulletin of Materials Science,* **39**, 543-550.
311. Hou, D., Liu, Q., Cheng, H., Zhang, H., and Wang, S. (2017) Green reduction of graphene oxide via Lycium barbarum extract. *Journal of Solid State Chemistry,* **246**, 351-356.
312. Kuila, T., Bose, S., Khanra, P., Mishra, A. K., Kim, N. H., and Lee, J. H. (2012) A green approach for the reduction of graphene oxide by wild carrot root. *Carbon,* **50**, 914-921.
313. Feng, B., Xie, J., Dong, C., Zhang, S., Cao, G., and Zhao, X., (2014) From graphite oxide to nitrogen and sulfur co-doped few-layered graphene by a green reduction route via Chinese medicinal herbs. *RSC Advances,* **4**, 17902-17907.
314. Firdhouse, M. J., and Lalitha, P. (2014) Phyto-reduction of graphene oxide using the aqueous extract of Eichhornia crassipes (Mart.) Solms. *International Nano Letters,* **4**, 103-108.
315. Chu, H.-J., Lee, C.-Y., and Tai, N.-H. (2016) Green preparation using black soybeans extract for graphene-based porous electrodes and their applications in supercapacitors. *Journal of Power Sources,* 322, 31-39.
316. Alshamsi, H. A., and Ali, S. K. (2016) Kinetics of Green Reduction of Graphene Oxide via Hibiscus Sabdarriffa L Aqueous Solution. *International Journal of ChemTech Research,* **9**, 321-331.

317. Thakur, S., and Karak, N. (2012) Green reduction of graphene oxide by aqueous phytoextracts. *Carbon,* **50**, 5331-5339.
318. Yaragalla, S., Rajendran, R., Jose, J., AlMaadeed, M. A., Kalarikkal, N., and Thomas, S. (2016) Preparation and characterization of green graphene using grape seed extract for bioapplications. *Materials Science and Engineering: C,* **65**, 345-353.
319. Gurunathan, S., Han, J. W., Kim, E. S., Park, J. H., and Kim, J.-H. (2015) Reduction of graphene oxide by resveratrol: a novel and simple biological method for the synthesis of an effective anticancer nanotherapeutic molecule. *International Journal of Nanomedicine,* **10**, 2951-2969.
320. Zaid, R. M., Chong, F. C., Teo, E. Y. L., Ng, E.-P., and Chong, K. F. (2015) Reduction of graphene oxide nanosheets by natural beta carotene and its potential use as supercapacitor electrode. *Arabian Journal of Chemistry,* **8**, 560-569.
321. Saikia, I., Hazarika, M., and Tamuly, C. (2016) Green synthesis of Au-rGO nanocomposite and its catalytic activity in nitro-reduction and degradation of dyes. *Materials Research Express,* **3**, 105007.
322. Simsikova, M., Bartos, M., Kesa, P., and Sikola, T. (2016) Green approach for preparation of reduced graphene oxide decorated with gold nanoparticles and its optical and catalytic properties. *Materials Chemistry and Physics,* **177**, 339-345.
323. Gurunathan, S., Han, J. W., Park, J. H., Kim, E., Yun-Jung, Y.-J., Kwon, D.-N., and Kim, J.-H. (2015) Reduced graphene oxide-silver nanoparticle nanocomposite: a potential anticancer nanotherapy *International Journal of Nanomedicine,* **10**, 6257-6276.
324. Khan, M., Khan, M., Al-Marri, A. H., Al-Warthan, A., Alkhathlan, H. Z., Siddiqui, M. R. H., Nayak, V. L., Kamal, A., and Adil, S. F. (2016) Apoptosis inducing ability of silver decorated highly reduced graphene oxide nanocomposites in A549 lung cancer. *International Journal of Nanomedicine,* **11**, 873-883.
325. Gurunathan, S., Han, J. W., Eppakayala, V., and Kim, J.-H. (2013) Microbial reduction of graphene oxide by Escherichia coli: a green chemistry approach. *Colloids and Surfaces B: Biointerfaces,* **102**, 772-777.
326. Akhavan, O., and Ghaderi, E. (2012) Escherichia coli bacteria reduce graphene oxide to bactericidal graphene in a self-limiting manner. *Carbon,* **50**, 1853-1860.
327. Gurunathan, S., Han, J. W., Eppakayala, V., and Kim, J.-H. (2013) Biocompatibility of microbially reduced graphene oxide in primary mouse embryonic fibroblast cells. *Colloids and Surfaces B: Biointerfaces,* **105**, 58-66.
328. Gurunathan, S., Han, J. W., Eppakayala, V., and Kim, J.-H. (2013) Green synthesis of graphene and its cytotoxic effects in human breast cancer cells. *International Journal of Nanomedicine,* **8**, 1015-

1027.

329. Gurunathan, S., Han, J. W., Eppakayala, V., Jeyaraj, M., and. Kim, J.-H (2013) An environmentally friendly approach to the reduction of graphene oxide by Escherichia fergusoni. *Journal of Nanoscience and Nanotechnology,* **13**, 2091-2098.
330. Khanra, P., Kuila, T., Kim, N. H., Bae, S. H., Yu, D.-s., and Lee, J. H. (2012) Simultaneous bio-functionalization and reduction of graphene oxide by baker's yeast. *Chemical Engineering Journal,* **183**, 526-533.
331. Gurunathan, S., Han, J.W., Park, J. H., and Kim J.H. (2014) An in vitro evaluation of graphene oxide reduced by Ganoderma spp. in human breast cancer cells (MDA-MB-231). *International Journal of Nanomedicine,* **9**, 1783-1797.
332. Zhang, X.-F., and Gurunathan, S. (2016) Biofabrication of a novel biomolecule-assisted reduced graphene oxide: an excellent biocompatible nanomaterial. *International Journal of Nanomedicine,* **11**, 6635-6649.
333. Gurunathan, S., Han, J., and Kim, J. H. (2013) Humanin: a novel functional molecule for the green synthesis of graphene. *Colloids and Surfaces B: Biointerfaces,* **111**, 376-383.
334. SuryaBhaskaram, D., Cheruku, R., and Govindaraj, G. (2014) Sugar assisted graphene: A green synthesis approach. *International Journal of ChemTech Research,* **6**(6), 3291-3293.
335. Feng, Y., Feng, N., and Du, G. (2013) A green reduction of graphene oxide via starch-based materials. *RSC Advances,* **3**, 21466-21474.
336. Ma, J., Wang, X., Liu, Y., Wu, T., Liu, Y., Guo, Y., Li, R., Sun, X., Wu, F., and Li, C. (2013) Reduction of graphene oxide with l-lysine to prepare reduced graphene oxide stabilized with polysaccharide polyelectrolyte. *Journal of Materials Chemistry A,* **1**, 2192-2201.
337. Zhang, C., Chen, M., Xu, X., Zhang, L., Zhang, L., Xia, F., Li, X., Liu Y, Hu, W., and Gao, J. (2014) Graphene oxide reduced and modified by environmentally friendly glycylglycine and its excellent catalytic performance. *Nanotechnology,* **25**, 135707.
338. Shin, S. R., Aghaei-Ghareh-Bolagh, B., Dang, T. T., Topkaya, S. N., Gao, X., Yang, S. Y., Jung, S. M., Oh. J. H., Dokmeci, M. R., Tang, X. S., and Khademhosseini, A. (2013) Cell-laden microengineered and mechanically tunable hybrid hydrogels of gelatin and graphene oxide. *Advanced Materials,* **25**, 6385-6391.
339. Tran, D. N., Kabiri, S., and Losic, D. (2014) A green approach for the reduction of graphene oxide nanosheets using non-aromatic amino acids. *Carbon,* **76**, 193-202.
340. Bo, Z., Shuai, X., Mao, S., Yang, H., Qian, J., Chen, J., Yan, J., and Cen, K. (2014) Green preparation of reduced graphene oxide for sensing and energy storage applications. *Scientific Reports,* **4**, 4684.
341. Choi, Y.-J., Kim, E., Han, J., Kim, J.-H., and Gurunathan, S. (2016) A

novel biomolecule-mediated reduction of graphene oxide: A multifunctional anti-cancer agent. *Molecules,* **21**(3), 375.

342. Ortega-Amaya, R., Matsumoto, Y., Flores-Conde, A., Perez-Guzman, M., and Ortega-Lopez, M. (2016) In situ formation of rGO quantum dots during GO reduction via interaction with citric acid in aqueous medium. *Materials Research Express,* **3**, 105601.
343. Nurunnabi, M., Khatun, Z., Nafiujjaman, M., Lee, D.-g., and Lee, Y.-k. (2013) Surface coating of graphene quantum dots using mussel-inspired polydopamine for biomedical optical imaging. *ACS Applied Materials & Interfaces,* **5**, 8246-8253.
344. Jung, H., Park, K. T., Gueye, M. N., So, S. H., and Park, C. R. (2016) Bio-inspired graphene foam decorated with Pt nanoparticles for hydrogen storage at room temperature. *International Journal of Hydrogen Energy,* **41**, 5019-5027.
345. Ning, N., Ma, Q., Liu, S., Tian, M., Zhang, L., and Nishi, T. (2015) Tailoring dielectric and actuated properties of elastomer composites by bioinspired poly (dopamine) encapsulated graphene oxide. *ACS Applied Materials & Interfaces,* **7**, 10755-10762.
346. Cheng, Q., Duan, J., Zhang, Q., and Jiang, L. (2015) Learning from nature: constructing integrated graphene-based artificial nacre. *ACS Nano,* **9**(3), 2231-2234.
347. Gong, S., Cui, W., Zhang, Q., Cao, A., Jiang, L., and Cheng, Q. (2015) Integrated ternary bioinspired nanocomposites via synergistic toughening of reduced graphene oxide and double-walled carbon nanotubes. *ACS Nano,* **9**, 11568-11573.
348. Ordikhani, F., Farani, M. R., Dehghani, M., Tamjid, E., and Simchi, A. (2015) Physicochemical and biological properties of electrodeposited graphene oxide/chitosan films with drug-eluting capacity. *Carbon,* **84**, 91-102.
349. Mazaheri, M., Akhavan, O., and Simchi, A. (2014) Flexible bactericidal graphene oxide–chitosan layers for stem cell proliferation. *Applied Surface Science,* **301**, 456-462.
350. Chowdhuri, A. R., Tripathy, S., Chandra, S., Roy, S., and Sahu, S. K, (2015) A ZnO decorated chitosan–graphene oxide nanocomposite shows significantly enhanced antimicrobial activity with ROS generation. *RSC Advances,* **5**, 49420-49428.
351. Xie, C., Lu, X., Han, L., Xu, J., Wang, Z., Jiang, L., Wang, K., Zhang, H., Ren, F., and Tang, Y. (2016) Biomimetic mineralized hierarchical graphene oxide/chitosan scaffolds with adsorbability for immobilization of nanoparticles for biomedical applications. *ACS Applied Materials & Interfaces,* **8**, 1707-1717.
352. Reddy, T. N., Manna, J., and Rana, R. K. (2015) Polyamine-mediated interfacial assembly of rGO-ZnO nanostructures: a bio-inspired approach and enhanced photocatalytic properties. *ACS Applied Materials & Interfaces,* **7**, 19684-19690.

353. Bhunia, S. K., Pradhan, N., and Jana, N. R. (2014) Vitamin B1 derived blue and green fluorescent carbon nanoparticles for cell-imaging application. *ACS Applied Materials & Interfaces,* **6**, 7672-7679.
354. Singh, R. P., Sharma, G., Singh, S., Kumar, M., Pandey, B. L., Koch, B., and Muthu, M. S. (2016) Vitamin E TPGS conjugated carbon nanotubes improved efficacy of docetaxel with safety for lung cancer treatment. *Colloids and Surfaces B: Biointerfaces,* **141**, 429-442.
355. Fujie, T., Ahadian, S., Liu, H., Chang, H., Ostrovidov, S., Wu, H., Bae, H., Nakajima, K., Kaji, H., and Khademhosseini. A. (2013) Engineered nanomembranes for directing cellular organization toward flexible biodevices. *Nano Letters,* **13**, 3185-3192.
356. Zhang, X., Zeng, G., Tian, J., Wan, Q., Huang, Q., Wang, K., Zhang, Q., Liu, M., Deng, F., and Wei, Y. (2015) PEGylation of carbon nanotubes via mussel inspired chemistry: preparation, characterization and biocompatibility evaluation. *Applied Surface Science,* **351**, 425-432.
357. Yan, P., Wang, J., Wang, L., Liu, B., Lei, Z., and Yang, S. (2011) The in vitro biomineralization and cytocompatibility of polydopamine coated carbon nanotubes. *Applied Surface Science,* **257**, 4849-4855.
358. Zhang, X., Huang, Q., Liu, M., Tian, J., Zeng, G., Li, Z., Wang, K., Zhang, Q., Wan, Q., Deng, F., and Wei, Y. (2015) Preparation of amine functionalized carbon nanotubes via a bioinspired strategy and their application in Cu^{2+} removal. *Applied Surface Science,* **343**, 19-27.
359. Xu, J., Huang, S., Wu, R., Jiang, R., Zhu, F., Wang, J., and Ouyang, G. (2015) Bioinspired polydopamine sheathed nanofibers for high-efficient in vivo solid-phase microextraction of pharmaceuticals in fish muscle. *Analytical Chemistry,* **87**, 3453-3459.
360. Ritz, S., Schottler, S., Kotman, N., Baier, G., Musyanovych, A., Kuharev, J., Landfester, K., Schild, H., Jahn, O., Tenzer, S., and Mailander, V. (2015) Protein corona of nanoparticles: distinct proteins regulate the cellular uptake. *Biomacromolecules,* **16**, 1311-1321.
361. Li, Z., Wu, C., Zhao, K., Peng, B., and Deng, Z. (2015) Polydopamine-assisted synthesis of raspberry-like nanocomposite particles for superhydrophobic and superoleophilic surfaces. *Colloids and Surfaces A: Physicochemical and Engineering Aspects,* **470**, 80-91.
362. Tamang, A., Ghosh, S. K., Garain, S., Alam, M. M., Haeberle, J. r., Henkel, K., Schmeisser, D., and Mandal, D. (2015) DNA-Assisted β-phase nucleation and alignment of molecular dipoles in PVDF Film: A realization of self-poled bioinspired flexible polymer nanogenerator for portable electronic devices. *ACS Applied Materials & Interfaces,* **7**, 16143-16147.
363. Merindol, R., Diabang, S., Felix, O., Roland, T., Gauthier, C., and Decher, G. (2015) Bio-inspired multiproperty materials: strong,

self-healing, and transparent artificial wood nanostructures. *ACS Nano,* **9**, 1127-1136.

364. Fang, R. H., Hu, C.-M. J., Luk, B. T., Gao, W., Copp, J. A., Tai, Y., O'Connor, D. E., and Zhang, L., (2014) Cancer cell membrane-coated nanoparticles for anticancer vaccination and drug delivery. *Nano Letters,* **14**, 2181-2188.
365. Steinbach, J. M., Seo, Y.-E., and Saltzman, W. M. (2016) Cell penetrating peptide-modified poly (lactic-co-glycolic acid) nanoparticles with enhanced cell internalization. *Acta Biomaterialia,* **30**, 49-61.
366. Zhu, D., Tao, W., Zhang, H., Liu, G., Wang, T., Zhang, L., Zeng, X., and Mei, L. (2016) Docetaxel (DTX)-loaded polydopamine-modified TPGS-PLA nanoparticles as a targeted drug delivery system for the treatment of liver cancer. *Acta Biomaterialia,* 30, 144-154.
367. Tang, L., Livi, K. J., and Chen, K. L. (2015) Polysulfone membranes modified with bioinspired polydopamine and silver nanoparticles formed in situ to mitigate biofouling. *Environmental Science & Technology Letters,* **2**, 59-65.
368. Jiang, J., Xie, J., Ma, B., Bartlett, D. E., Xu, A., and Wang, C.-H. (2014) Mussel-inspired protein-mediated surface functionalization of electrospun nanofibers for pH-responsive drug delivery. *Acta Bio-Materialia,* **10**, 1324-1332.
369. Rush, A. M., Nelles, D. A., Blum, A. P., Barnhill, S. A., Tatro, E. T., Yeo, G. W., and Gianneschi, N. C., (2014) Intracellular mRNA regulation with self-assembled locked nucleic acid polymer nanoparticles. *Journal of the American Chemical Society,* **136**, 7615-7618.
370. Tay, C. Y., Yuan, L., and Leong, D. T. (2015) Nature-Inspired DNA nanosensor for real-time in situ detection of mRNA in living cells. *ACS Nano,* **9**, 5609-5617.
371. Gujrati, V., Kim, S., Kim, S.-H., Min, J. J., Choy, H. E., Kim, S. C., and Jon, S. (2014) Bioengineered bacterial outer membrane vesicles as cell-specific drug-delivery vehicles for cancer therapy. *ACS Nano,* **8**, 1525-1537.

4

Polymers and Polymer Nanocomposites for Tissue Engineering and Artificial Joints

4.1 Introduction

Biomaterials are defined as biologically stable materials with either natural or synthetic origins having superior compatibility and practicality acceptable to cure, ameliorate or restore tissues, joints, etc. [1]. Polymeric biomaterials have received immense research attention for various biomedical applications, as compared to biomaterials based on metals and ceramics. The biomedical performance of polymeric biomaterials is derived from their modulating structural, mechanical, physical, biological, chemical and surface properties, along with cost effectiveness and ease of precessing [1-4]. Due to these unique features in combination with the developing technologies, polymer based biomaterials exhibit high potential for tissue engineering, artificial joints, implants, drug delivery, surgical and wound dressing, biosensors, bio-separation, medical devices, etc. [2,5,6].

4.2 Polymeric Materials for Tissue Engineering

Tissue engineering is a multidisciplinary system to rehabilitate, ameliorate or substitute the performance of living tissues by acquiring the benefits from engineering, bioscience and materials through the application of cells and scaffolds [7-10]. In order to ensure effective tissue engineering practices, the scaffold structure should possess tissue compatibility, biodegradation without the evolution of noxious products, tissue manageability, porosity and satisfactory mechanical properties [11-13]. Due to their substantial suitability to emulate the structure of extracellular matrix and, thereby, to control the germination of new tissues, polymer nanomaterials based tissue engineering scaffold materials have drawn significant research interest. Furthermore, in order to enable effective development of neo-tissues, the features such as structural arrangements, composi-

Anish M. Varghese and Vikas Mittal
Khalifa University of Science and Technology, Abu Dhabi, UAE

tion and extent of crosslinking of polymeric scaffold materials can be modulated [8,14-16]. Tissue engineering scaffolds based on natural polymers such as alginate, cellulose, chitosan, hyaluronic acid, starch, agarose, gelatin, collagen, elastin, fibrin, etc., and their derivatives and composites as well as synthetic polymers such as poly(vinyl alcohol) (PVA), polyacrylamide, poly(acrylic acid) (PAA), poly(ethylene glycol), polydioxanone, poly(glycerol sebacate), polycaprolactone (PCL), poly(lactic acid) (PLA), poly(glycolide), etc., and their copolymers, derivatives and composites in the forms of hydrogels, nanofibers and films have presented appropriateness for a myriad range of applications [8,17-23].

Water swollen 3-D networks of polymeric materials, known as hydrogels, exhibit specific potential for tissue engineering applications as scaffold materials, conveyance medium or space padding substances owing to the presence of multi-functionality [24,25]. These hydrogels exhibit improved structural compatibility with extracellular matrix, transfer of nutrients and gases due to the presence of appreciable amount of water, effective conveyance of cells in combination with bioactive molecular fragments, successful differentiation and consecutive generation of new cell structure through the regulated breakdown of scaffold structure, etc. [4,26,27]. In addition, tissue engineering scaffolds based on polymeric nanofibers demonstrate effective biomimetic practicability due to their distinctive features such as size compatibility with molecules associated with extracellular matrix, high surface area responsible for enhanced extent of cell growth and drug delivery, improved adsorption degree of protein molecules responsible for cell growth and beneficial mechanical features resulting due to ordered molecular arrangement [9,14,28-32]. For the successful generation of polymeric nanofibers, electrospinning technique is promising owing to the benefits like accessibility, adaptability, tunability and faster processing steps, along with applicability of different collection techniques [33-37].

Conducting polymers namely polyaniline, poly(3,4-ethylenedioxythiophene) (PEDOT), polythiophene, polypyrrole, etc., and their derivatives, blends and composites have also been extensively utilized specifically for the design of electrically conducting tissue engineering scaffold structures [11,38-43]. Due to outstanding electrical features along with other properties, conducting polymer based scaffold structures find suitability for the effective electrical signal transfer between electrically active cells including neurons, myoblasts, cardiac muscle cells, bone cells, stem cells, etc., and, thus, maintain beneficial cellular responses under the influence of low electrical potential [43-48]. Also, the conductive polymer composites based on carbonaceous nanomaterials such as graphene and carbon nanotubes as well as metallic nanowires have been used to fabricate scaffold structures possessing enriched cell proliferation and segregation rate due to efficient electroactive cell modulations caused by powerful propagation of impulses [49-54]. In this section, recent studies associated with tissue engineering applications of polymeric nanomaterials have been summarized.

4.2.1 Polymeric Hydrogels

Jo *et al.* [49] described the development of conductive scaffold structure based on composite hydrogel composed of polyacrylamide and graphene. The composite hydrogel demonstrated higher rate of myogenic cell growth and corresponding divergence as compared to hydrogels of polyacrylamide/graphene oxide (GO) composite and neat polyacrylamide. The developed hydrogel system exhibited a superior electrical conductivity of 10^{-4} S/cm and tensile modulus value of 50 kPa, together with about 10 fold decrease in electrochemical impedance value. The implementation of electrical stimulation of myoblasts growing on the hydrogels significantly enhanced the myogenic gene expression. In another study, Li *et al.* [55] studied the practicability of hydrogel scaffold based on polyacrylamide/GO composite. Among various compositions, the one containing 0.4% of GO exhibited most favorable peripheral nerve cell adhesion, growth rate and biofactors release.

To generate superior tissue regeneration, Lam *et al.* [27] studied the influence of various variables such as nature of crosslinking agent, crosslinker-to-macromer carbon-carbon double bond ratio, incubator composition, molecular weight of poly(ethylene glycol) (PEG) and extent of gelatin crosslinking for the design of PEG/gelatin composite hydrogel. In another study, Patel *et al.* [56] detailed the development of blend hydrogel composed of PEG dimethacrylate and methacrylic acid for high performance tissue engineering applications as scaffold structure. Modulating mechanical features together with biocompatibility were observed for the hydrogel system. Multipotent as well as pluripotent undifferentiated cells displayed better viability, adhesion and growth mechanisms on the hydrogel based scaffold structure.

Henke *et al.* [57] reported the usefulness of scaffold structure based on hydrogel foams of oligo(PEG fumarate) for cartilage tissue regeneration. The presence of bovine chondrocytes on the porous scaffold along with outstanding cytocompatibility were observed. In a study reporting the composites of oligo(PEG fumarate), Kinard *et al.* [58] discussed the development of scaffold structure based on oligo(PEG fumarate)/gelatin composite hydrogels for the osteoblast proliferation. Composite composition was observed to effectively control the biological and mechanical performance of hydrogel scaffold structure.

Liu *et al.* [59] reported the development of crosslinked composite hydrogel scaffold system exhibiting surpassing electrical conductivity for nerve tissue engineering (Figure 4.1). For this purpose, a mixture of acrylated GO and PEG acrylated carbon nanotubes (CNTs)

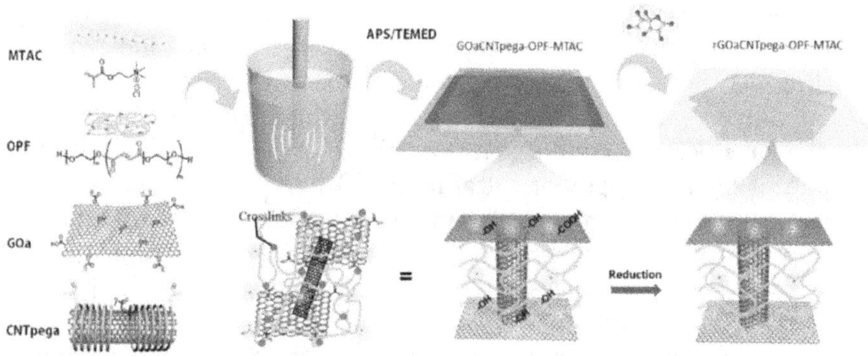

Figure 4.1 Schematic of the composite synthesis. Reproduced from Reference 59 with permission from American Chemical Society.

was introduced into a hybrid system composed of oligo(PEG fumarate) (OPF) and (2-(methacryloyloxy) ethyl)trimethylammonium chloride (MTAC), followed by crosslinking under specific conditions with the aid of crosslinking agents. The resultant hydrogel system displayed high electrical conductivity and excellent cytocompatibility to neural cells. Also, appreciable neural cell growth and differentiation were observed on the positively charged conductive composite hydrogel based scaffold.

Hong *et al.* [60] reported the development of hydrogel based injectable tissue regeneration carrier using PEG-poly(L-alanine) together with mesenchymal stem cells and growth factors. Superior cell expression and differentiation, along with high extent of growth ingestion for the hydrogel system were observed, as compared to hyaluronic acid based hydrogel carrier system. For enabling effective adipogenesis of mesenchymal stem cells, Patel *et al.* [61] discussed the use of 3D scaffold structure based on PEG-poly(L-alanine)/GO composite hydrogel (Figure 4.2). The better performance of the developed scaffold structure was related to the combined effect of superior biological and chemical features of GO and appreciable cytocompatibility of the copolymer component.

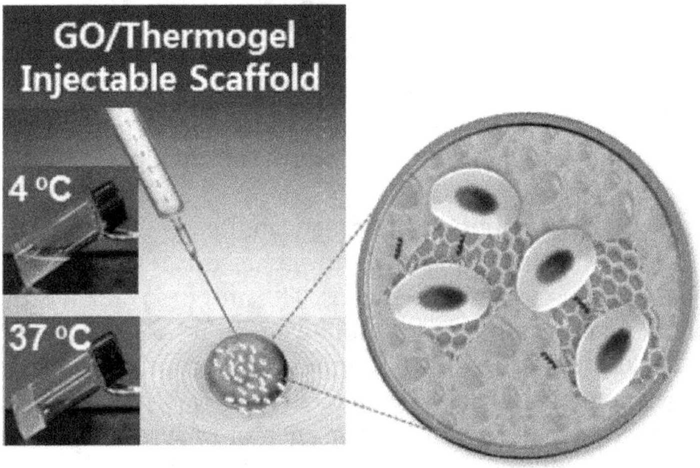

Figure 4.2 Schematic of the culture of tonsil-derived mesenchymal stem cells in the composite injectable system. Reproduced rom Reference 61 with permission from American Chemical Society.

Wang *et al.* [62] discussed the implementation of chitosan hydrogels as effective tissue engineering carriers for the differentiation of cardiac stem cells sourced from brown fat, and, thereby, to restore the functions of impaired cardiac tissues. For this purpose, the authors recommended the application of an injectable tissue regeneration approach. From the results of *in vivo* examinations, the supporting effect of chitosan based hydrogels for the strengthening of cardiac tissue proliferation was confirmed. Deng *et al.* [63] discussed the application of composite hydrogel comprised of alginate and chitosan as a high quality scaffold structure for the treatment of myocardial infarction through the effective regeneration of cardiomyocytes. For injectable strategy, significant recovery of cardiac functions was observed by effective cardiomyocytes regeneration. The composite hydrogel system displayed better performance than the individual components in terms of both tissue regeneration and mechanical features.

Giammanco *et al.* [64] reported the composite hydrogel system based on ferric ion impregnated alginate/acrylamide blend exhibiting photo-sensitive behavior as an effective carrier for the regeneration of cartilage tissues. The resultant carrier system displayed outstanding cytocompatibility and appreciable hydrolytic stability. The presence of ferric ions in the hydrogel based carrier system was beneficial to modulate the architecture as well as biological and mechanical features through photochemical actions. It was useful to tailor the extent of chondrocytes growth and expression on the photo-sensitive platform. Two times higher cell expression on photochemically treated system was observed as compared to dark control due to the enriched porous nature and adequate mechanical properties.

Naseri *et al.* [65] reported the practicability of gelatin/sodium alginate/nanocellulose composite hydrogel as a carrier for the effectual regeneration of cartilage tissues. The authors reported the use of $CaCl_2$ and genipin treatment, followed by freeze-drying, for the generation of dual crosslinked composite hydrogel system. Appreciable mesenchymal stem cells attachment was observed on the scaffold substrate owing to the presence of ordered porous structure, biocompatibility and significant ingestion of buffer solution. In addition, improved mechanical performance was noted for the nanocellulose based composite hydrogel system. Figure 4.3 also depicts the crosslinking reactions between the components. In another study, Kumar *et al.* [66] designed composite hydrogel based carrier material by the addition of nanocellulose into the blend of poly(vinyl alcohol) (PVA) and sodium alginate in order to engineer fibroblast cell proliferation. Influence of crosslinker nature on the ultimate performance of hydrogel scaffold structure was observed. Among various crosslinkers, borax based system exhibited optimal performance including hydrophilicity, porosity as well as thermal, mechanical and biological characteristics. Furthermore, optimal degree of fibroblast cell attachment and growth on borax based composite hydrogel system was achieved.

For the successful regeneration of complex anisotropic tissue structures, Rhee *et al.* [67] reported the applicability of collagen based hydrogels with the aid of 3D printing technology. In another study, Vazquez-Portalatin *et al.* [68] reported the fabrication of scaffold structure using composite hydrogel systems composed of type I

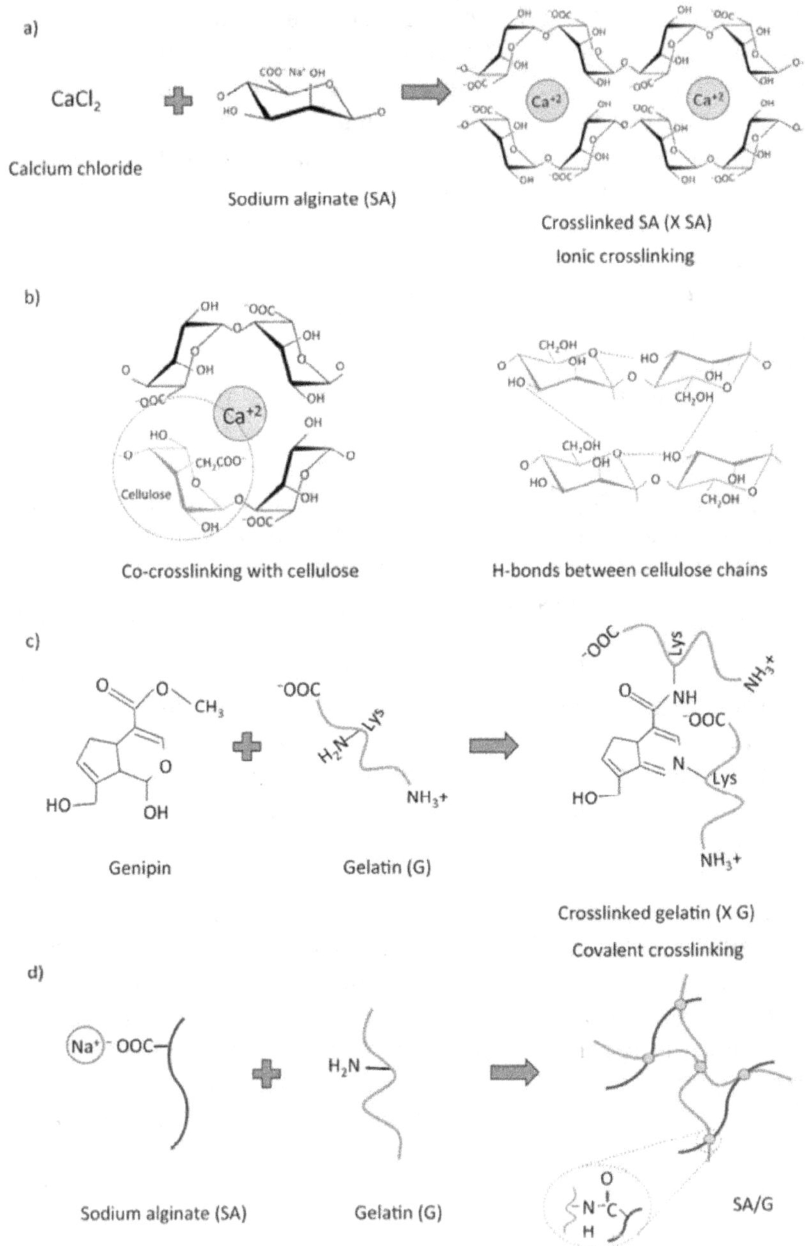

Figure 4.3 Crosslinking reactions of a) sodium alginate/$CaCl_2$, b) sodium alginate/CNC, H-bonds between CNC, c) genipin/gelatin, and d) sodium alginate/gelatin. Reproduced from Reference 65 with permission from American Chemical Society.

and type II collagens for successful regeneration of articular cartilage. In the hydrogel based scaffold system, type I collagen offered good mechanical features, whereas type II collagen offered superior biological potential. For effective cartilage tissue regeneration, hydrogel system with type I collagen to type II collagen ratio of 3:1 demonstrated significant suitability, as compared to other compositions owing to effective mechanical properties together with adequate biological responses. For effectual neural tissue regeneration, Koppes *et al.* [69] reported the development of conductive hydrogel scaffold substrate through the inclusion of single-walled CNTs (SWCNTs) into type I collagen. The addition of SWCNTs to the hydrogel composition was effective to improve the proliferation of neural cells appreciably due to the increased conduction of electrical cues. A significant increase in electrical conductivity (up to 1.7 times) of the hydrogel system was observed with SWCNTs, without considerably altering the Young's modulus. In another study, Kim *et al.* [70] also highlighted the usefulness of hydrogels based on transglutaminase crosslinked resilin protein for durable cardiovascular cell adhesion.

Domingues *et al.* [71] followed injectable strategy for studying the tissue engineering suitability of carrier system based on composite hydrogel of aldehyde functionalized hyaluronic acid, hydrazide functionalized hyaluronic acid and aldehyde functionalized nanocellulose. The presence of nanocellulose in the composite hydrogel enhanced its mechanical performance considerably. Moreover, a decrease in swelling degree and increase in degradation resistance were observed with enhancement in nanocellulose concentration in the composite system. Biological performance of the hydrogel was analyzed using stem cells sourced from human adipose by following *in vitro* examinations. The applied stem cells demonstrated more desirable cell attachment and distribution on the hydrogel, which correspondingly signaled significant proliferation capability. The performance of the composite hydrogel could also be modulated for different tissue engineering applications by varying its composition.

Table 4.1 presents the summary of polymeric hydrogels reported recently for tissue engineering.

4.2.2 Polymer Nanofibers

Nedjari *et al.* [72] reported electrospun three-dimensional PCL nanofibers based scaffold structure for the effective restoration of bone,

Table 4.1 Summary of polymeric hydrogels reported recently for tissue engineering

Reference	Polymeric hydrogel	Applications
Jo *et al.* [49]	Polyacrylamide/graphene composite	Growth and differentiation of myogenic cells
Li *et al.* [55]	Polyacrylamide/GO composite	Growth, and adhesion of peripheral nerve cells
Patel *et al.* [56]	PEG dimethacrylate/ methacrylic acid blend	Viability, adhesion and growth of multipotent and pluripotent undifferentiated cells
Henke *et al.* [57]	Oligo(PEG fumarate)	Juvenile bovine chondrocytes allocation
Kinard *et al.* [58]	Oligo(PEG fumarate)/gelatin composite	Osteoblast proliferation
Liu *et al.* [59]	Acrylated GO/PEG acrylated CNTs/oligo(PEG fumarate)/(2-(methacryloyloxy) ethyl)trimethylammonium chloride	Neural cell growth and differentiation
Hong *et al.* [60]	PEG-poly(L-alanine)	Mesenchymal stem cell expression and differentiation
Patel *et al.* [61]	PEG-poly(L-alanine)/GO composite	Adipogenesis of mesenchymal stem cells
Wang *et al.* [62]	Chitosan	Differentiation of cardiac stem cells
Deng *et al.* [63]	Alginate/chitosan composite	Regeneration of cardiomyocytes
Giammanco *et al.* [64]	Ferric ion impregnated alginate/acrylamide	Regeneration of cartilage tissues
Naseri *et al.* [65]	Sodium alginate/ gelatin/nanocellulose composite	Regeneration of cartilage tissues
Kumar *et al.* [66]	PVA/sodium alginate	Fibroblast cell proliferation
Rhee *et al.* [67]	3D printed collagen	Proliferation of fibrochondrocytes
Vazquez-Portalatin *et al.* [68]	Collagen I/collagen II composite	Regeneration of articular cartilage
Koppes *et al.* [69]	Collagen I/SWCNTs composite	Proliferation of neural cells
Kim *et al.* [70]	Transglutaminase crosslinked resilin protein	Durable cardiovascular cell adhesion

interface and cardiac tissues due to uniform pore dimensions and distribution. Song *et al.* [73] developed nanofiber scaffold structure based on PCL composite with GO *via* electrospinning technique for effective differentiation of mouse marrow mesenchymal stem cells and rat pheochromocytoma cells. The application of the bio-composite scaffold was beneficial to improve the differentiation of cells along with superior adhesion and biocompatibility. Hardy *et al.* [74] also discussed the design of conductive PCL/polypyrrole /polystyrenesulfonate ternary composite nanofiber scaffold structure for bone tissue engineering applications. An improvement in differentiation of mesenchymal stem cells with respect to growth on the conductive composite scaffold was observed through the application of electrical stimulation due to the elevation in the magnitude of calcium ousting and alkaline phosphatase exploitation. To engineer tendon tissues successfully, Domingues *et al.* [75] also reported the exploitation of composite nanofiber based carrier developed by electrospinning of nanocellulose incorporated blend of PCL and chitosan.

Zhang *et al.* [76] prepared scaffold structure based on nanofibers of PLA composite with nanocellulose containing PEG grafting with the aid of electrospinning for the purpose of osteoblast cell proliferation. Improved level of osteoblastic cell interaction and differentiation on the composite scaffold substrate was observed, along with good noncytotoxic behavior and biocompatibility. In another study, the authors carried out the design of scaffold structure based on electrospun composite nanofibers composed of PLA and PEG grafted GO. The application of PEG grafted GO was advantageous to enhance the hydrophilicity, thermal properties and tensile strength of PLA without altering cytocompatible behavior. Appreciable mice fibroblast cell interaction and proliferation on the biocompatible composite nanofiber based scaffold structure were observed. In a recent study, Zhang *et al.* [78] reported the application of GO coating over tissue engineering scaffold based on ordered nanofibers of poly-L-lactic acid (PLLA). After the implementation of GO coating, the orderly arrangement of PLLA nanofibers remained unchanged. The resultant scaffold structure displayed higher hydrophilic behavior, surface irregularity and cell adhesion. As a result, significant improvement in the proliferation of nerve cells as well pheochromocytoma cells on ordered PLLA nanofibers was observed. Also, improved pheochromocytoma cell division and outgrowth occurred on the GO coated PLLA scaffold system.

Rajzer *et al.* [32] detailed the modification of PCL as well as PLA electrospun nanofibers based bone tissue regeneration platforms through osteogenon loading in order to improve cell attachment and, thereby, growth and differentiation. The inclusion of osteogenon was beneficial to upgrade the biological activities through enrichment in hydrophilic nature and mineralization.

4.2.3 Conducting Polymers

Stewart *et al.* [79] discussed the preparation of dodecylbenzenesulfonate doped polypyrrole based neural stem cell scaffold structure. The resultant scaffold system presented superior cell multiplication capability due to the combined effect of biological compatibility and electrical stimulation. Zhang *et al.* [80] also reported the application of dodecylbenzenesulfonic acid doped polypyrrole for enabling effective growth of neural stem cells by means of electrical stimulation and, thus, to heal schizophrenia. In another study, Zhang *et al.* [50] discussed the inclusion of polypyrrole into PCL matrix for the successful generation of conductive composite scaffold structure. The presence of polypyrrole in the PCL matrix was favorable to enrich compressive modulus and hydrophilic nature along with electrical conductivity. Yang *et al.* [81] reported the fabrication of conductive scaffold structure possessing satisfactory mechanical performance, using polypyrrole/hyaluronic acid composite hydrogel. Attributes like porous nature, configurational stability and significant electrical conductivity were observed for the developed system. Compared to other compositions, polypyrrole/hyaluronic acid composite hydrogel containing 50 mM of pyrrole displayed optimal biological performances such as viability, attachment and growth rate. Scaffold structure based on composite hydrogel of polypyrrole and alginate was also reported by Yang *et al.* [48] for the purpose of neural tissue regeneration. Owing to the combined effect of electrical conductivity of polypyrrole and biocompatibility of alginate, the resultant composite hydrogel system exhibited effective neural stem cell regeneration proficiency. From the results of *in vitro* analysis, considerable neural stem cell interaction and proliferation could be observed on the conducting scaffold structure. Owing to the effective conveyance of both electrical and mechanical cues, differentiation of neural stem cells was achieved on the composite scaffold.

Pires *et al.* [42] detailed the development of neural stem cell scaffold exhibiting no cytotoxity, substantial stability and electroactive behavior through the use of 3-glycidoxypropyltrimethoxysilane crosslinked PEDOT:polystyrene sulfonate (PSS). Effective neural stem cell differentiation with the formation of neurites on the transparent conducting polymer based scaffold structure was observed. In another study, Srivastava *et al.* [82] disclosed the practicability of scaffold structure designed by PEDOT:PSS covered styrene ethylene butylene styrene (SEBS) block copolymer film to control the neuronal differentiation and growth of embryonic stem cells. In another recent study, Yang *et al.* [47] detailed the preparation of cardiac tissue regeneration scaffold using poly(thiophene-3-acetic acid)/methacrylic anhydride treated aminated gelatin composite hydrogels possessing significant electrical conductivity and dual network structure. The presence of gelatin in the composite hydrogel composition was advantageous to enrich the biological compatibility of polythiophene. The performance of the composite hydrogels could be tailored by controlling the composition. Adequate cardiac stem cell viability as well as subsequent growth and differentiation proficiencies were observed on the scaffold structure. The application of electrical cues was effective to ameliorate the performance of the developed scaffold.

4.2.4 Other Polymer Composites and Blends

Atak *et al.* [83] reported chitosan nanocomposites with hydroxyapatite as a high quality scaffold material for bone tissue engineering. Adequate biocompatibility and minimal cytotoxic behavior were observed for the bio-composite based scaffold materials, and the scaffold based on amine functionalized nano-hydroxyapatite exhibited supporting behavior during osteoblastic growth mechanism. In another study, Huang *et al.* [84] reported the use of scaffold structure generated using chitosan/tea polyphenol modified GO (TPG) composite for bone tissue engineering. The tensile properties of the composite (such as strength and modulus) were observed to improve due to the incorporation of tea polyphenol modified GO, as compared to unmodified GO. Due to the presence of features like biocompatibility and hydrophilicity, the resultant scaffold structure exhibited appreciable osteoblastic proliferation response. Figure 4.4 also shows the scanning electron images of the fracture surfaces of pure polymer and polymer nanocomposite. In another recent study, Serra *et al.* [85] employed freeze drying technique for the effective fabrication of the bone tissue engineering scaffold material based on

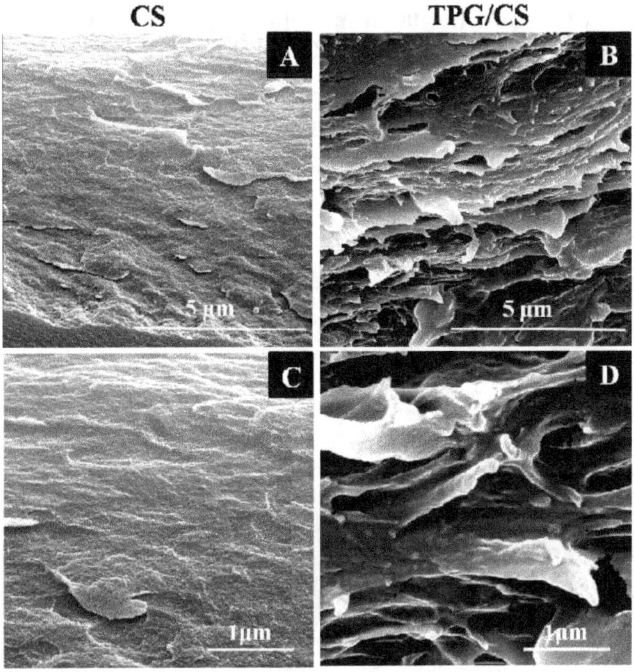

Figure 4.4 Scanning electron images of the fracture surfaces of chitosan (A,C) and TPG/chitosan composites (B,D). Reproduced from Reference 84 with permission from American Chemical Society.

chitosan/gelatin/β tri-calcium phosphate ternary composite. The resultant biocompatible scaffold system exhibited significant osteoblastic cell adherence and corresponding cell growth. Furthermore, the scaffold substrate displayed substantial resistance against biofilm formation and microbial activity, together with favorable accumulation of calcium. With an objective to achieve adipocytes adhesion on biologically degradable Vicryl scaffold structure, Beringer *et al.* [86] reported the suitability of collagen VI/heparin sulfate blend coating. Improved adipocytes adhesion and distribution on collagen VI/heparin sulfate covered scaffold structure was observed, as compared to the uncoated material. The developed system exhibited potential for diverse wound healing applications.

Jin *et al.* [87] detailed the preparation of bio-composite composed of nanoscale bacterial cellulose (BC) fibers and GO films for designing scaffold structure suitable for proliferation of mesenchymal stem cells (Figure 4.5). The presence of nanoscale bacterial cellulose fibers in the composite formulation supported the attachment and growth of mesenchymal stem cells. Improved electrical and mechanical features were observed for the bio-composite scaffold, along with hydrophilic nature.

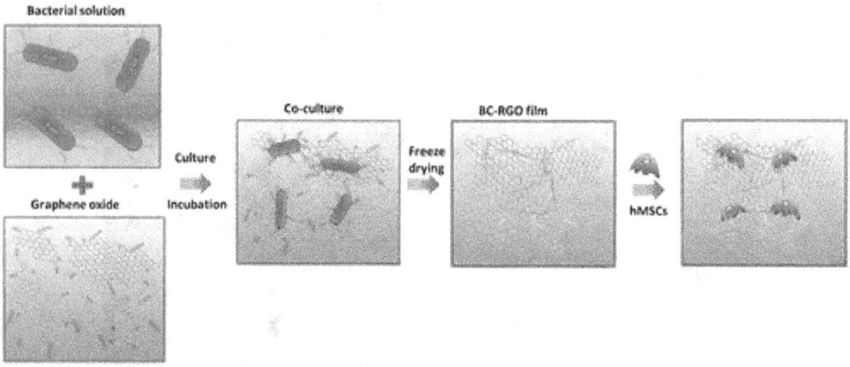

Figure 4.5 Schematic of the synthesis of BC-reduced GO films for tissue engineering scaffolds. Reproduced from Reference 87 with permission from American Chemical Society.

Naskar *et al.* [88] designed tissue engineering scaffold structure based on silk fibroin/carbon nanofiber bio-composite using solvent evaporation approach (Figure 4.6). The resultant nanocomposite exhibited enhanced electrical conductivity (up to 6.4×10^{-6} Mho cm, 3 orders of magnitude higher than the pure polymer) and Young's modulus (up to 1423 MPa, which is 12.5 times than the pure polymer) due to the presence of carbon nanofibers. From the biological examination with fibroblast cells, excellent biocompatibility, attachment and multiplication were observed for the composite system. A study by Aznar-Cervantes *et al.* [89] highlighted the applicability of graphene coating to improve the electrical conductivity of silk fibroin based tissue engineering scaffold structure and, thereby, to enhance the biological responses of cells through electrical stimulation. The biocompatibility of silk fibroin was maintained after coating with graphene, when seeded with fibroblasts.

For the purpose of bone as well as cartilage tissue engineering, Scaffaro *et al.* [90] reported the design of three-layered scaffolds exhibiting a pore size gradient, developed using melt mixing polylactic acid (PLA) and two water-soluble porogen agents: sodium chloride (NaCl) and polyethylene glycol (PEG). Appreciable porous behavior was noted for the scaffold structure, and its degree and size distribution could be modulated as per the need [91]. Also, the double layered version of the developed system demonstrated high potential to engineer interface tissues [92].

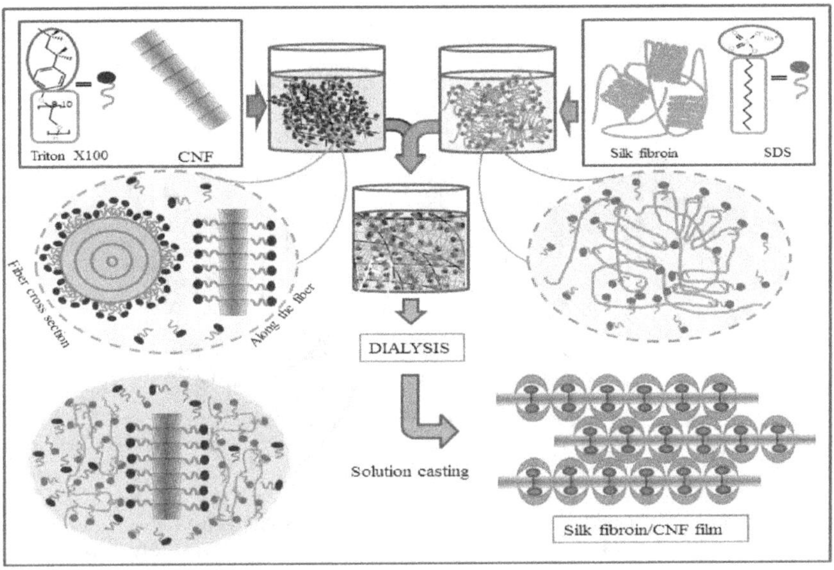

Figure 4.6 Schematic demonstration of the synthesis and molecular arrangements of silk fibroin/carbon nanofiber composite film. Reproduced from Reference 88 with permission from American Chemical Society.

Zhang *et al.* [93] detailed the development of high performance poly(lactic-co-glycolic acid) (PLGA) composite with phospholipid functionalized nanoscale diamond exhibiting outstanding mechanical characteristics for enabling effective regeneration of bone tissues as a supporting platform. The incorporation of phospholipid functionalized nanoscale diamond in the formulation of PLGA composite was beneficial to enhance the hydrophilicity, hardness (550% increase) and tensile modulus (100% increase). Excellent cytocompatibility and improved mineralization competence were observed for the composite scaffold. Also, it was observed that the nanodiamond addition to the PLGA matrix had no diminishing effect on the extent of osteoblast cell viability, growth and differentiation. In another study, Jiang *et al.* [94] revealed the applicability of ternary composite system containing PLGA, nanoscale hydroxyapatite and bamboo fiber as an effective osteogenesis platform. The addition of bamboo fiber in the composite composition was advantageous to improve biocompatibility (Figure 4.7), biological decomposition, porous nature and compressive strength. As a combined effect of these improved features, the resultant tissue engineering platform displayed enhancement in cell adhesion and growth.

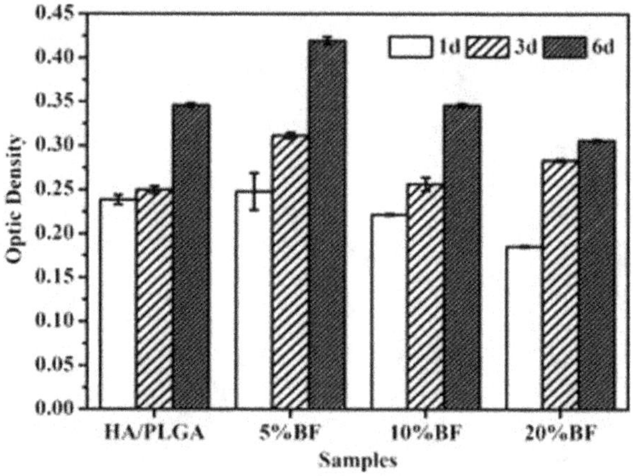

Figure 4.7 MG-63 cell biocompatibility of PLGA/30% nano-hydroxyapatite/bamboo fiber composites with 0, 5, 10, and 20% bamboo fiber studied using MTT assay. Reproduced from Reference 94 with permission from American Chemical Society.

Pan *et al.* [95] studied the influence of polydopamine surface functionalization as well as BMP-2 sourced peptide loading on the performance of aspartic acid incorporated PLGA-PEG copolymer based bone tissue engineering scaffold structure during the proliferation of osteoblast cells. Peptide loading was beneficial to increase the osteoblast cell proliferation on the surface of the copolymer. Also, the polydopamine surface functionalization of copolymer helped to achieve peptide conveyance in a steady way and, thus, effective proliferation of osteoblast cells.

4.3 Artificial Joints

The main role of orthopaedic joints is to provide good pliability to the body at the time of motion. In addition, these are beneficial to diminish the bone-bone friction. Artificial joints are widely recommended to partially or fully replace the joints suffering from osteoarthritis as a consequence of ageing, nutritional insufficiency, overweight or injury [96-98]. For the replacement of orthopaedic joints in hand, fingers, elbow, shoulder, neck, back, hip, knee, ankle, leg, etc., polymer based artificial joint materials have received considerable research interest owing to their unique features such as lightweight, flexibility and improved wear resistance [99]. Polymeric materials can be used in the formulation of artificial joints as surface modifiers of hard materials like metals and ceramics to yield better tribological behavior and, thereby, enriched performance. For various artificial joint applications, feasibility of polymeric materials such as ultra-high molecular weight polyethylene (UHMWPE), poly(aryl-ether-ether-ketone), poly(tetrafluoroethylene), polycarbonate, poly(acrylic acid), poly(glycolic acid), poly(2-methacryloyloxy ethyl phosphorylcholine) and silicones has been studied [98-106]. Compared to other polymeric materials, UHMWPE has received wide acceptance as an artificial joint constituent because of its distinctive features such as adequate biocompatibility; resistance against abrasion, chemicals and moisture; low coefficient of friction; appreciable mechanical features including toughness and stiffness; effective fatigue properties and adequate light transmission capability [107-111].

During the application of long rolling and sliding movements, there is a higher possibility for the failure of polymer layer on the artificial joint surfaces by means of wearing, which further enhances in the presence of surface irregularities and higher loading conditions [112-117]. A number of studies have been reported to enhance the wear resistance, eliminate the generation and deposition of wear debris, strengthen the mechanical performance and reduce the friction of UHMWPE based artificial joint constituents. The ultimate aim of these studies was to upgrade the life expectancy through the improvement in tribological behavior. For this, the strategies include self-strengthening; addition of functional fillers, lubricants, and vitamin E; gamma rays induced crosslinking; surface finishing, etc. [109,116,118-120].

With the growing requirement to design artificial joint structures with low friction, PVA based hydrogels have also been introduced as effective lubricating agents. The application of PVA hydrogel is beneficial to form joint cartilage fluid film on the artificial joint surface [121-124]. Improved wear resistance and lower friction have been observed for artificial joint system based on PVA hydrogels prepared with the aid of combined cast-drying and freezing-thawing technique [123-126].

In this section, recent studies associated with the use of polymeric materials for artificial joint applications have been reviewed.

4.3.1 UHMWPE

Huang *et al.* [127] reported the modification of UHMWPE through the formation of self-strengthened composite with low molecular weight polyethylene (LMWPE) for enabling longevity in artificial joint applications even in the presence of unfavorable working conditions. For this, the authors made use of oscillatory shear generated by an injection molding machine to generate the material composed of almost equal amount of crosslinked UHMWPE and LMWPE. Due to the synergistic effect of oscillatory shear movements and pre-additives, interlocking shish-kebab morphology was obtained, which correspondingly resulted in significant improvement in tensile strength (improved from 27.6 MPa for compression-molded UHMWPE to 81.2 MPa), impact strength (increased from 29.6 to 35.2 kJ m^{-2}), fatigue strength and wear resistance. The improved mechanical and tribological characteristics of the material exhibited the potential for long life artificial joint bearing applications. In another study, the authors observed improved mechanical behavior and wear resistance for self-strengthened UHMWPE with about 10 wt% ultra LMWPE, developed by applying intense shear flow during injection molding [110]. For gaining improved mechanical and tribological features acceptable to design artificial joint bearing surfaces, Xu *et al.* [128] also reported the use of the system composed of about 40 wt% UHMWPE and 60 wt% LMWPE. The application of oscillatory shear movements generated by an injection molding machine was beneficial to yield enhancement in mechanical properties (tensile strength increased from 27.6 MPa for UHMWPE to 78.4 MPa for the developed material along the flow direction) as well as wear resistance through the formation of interlocked shish-kebab morphology (Figure 4.8) possessing improved crystallinity. In another study, Xu *et al.* [129] also detailed the preparation of self-strengthened system based on 30 wt% UHMWPE and 70% high density polyethylene (HDPE), and considerable enhancement in properties such as tensile strength (increased from 26.4 MPa for HDPE to 68.5 MPa for the developed material), impact strength (increased from 6.3 to 34.1 kJ m^{-2}) and wear resistance (wear rate reduced from 22.1 to 7.6 mg MC^{-1}) was achieved due to the presence of aligned crystal structure.

It is necessary to upgrade the mechanical attributes and anti-wear characteristics of UHMWPE concurrently for generating artificial joint bearing surfaces with prolonged life span. Taking this into

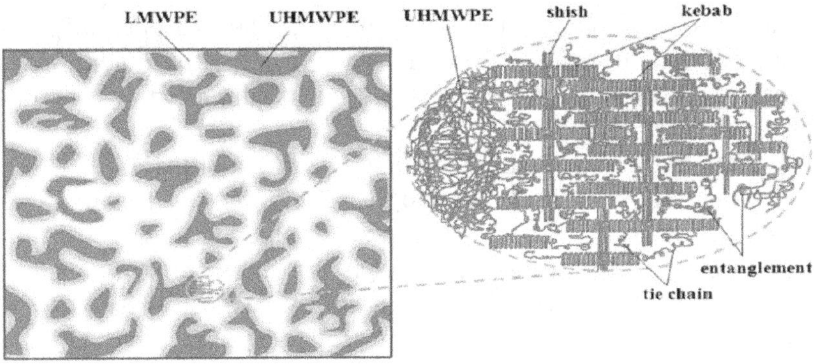

Figure 4.8 Schematic demonstration for interlocking shish-kebab morphology of UHMWPE/LMWPE. Reproduced from Reference 128 with permission from American Chemical Society.

account, Huang *et al.* [130] reported the radiation induced crosslinking of UHMWPE and subsequent melt mixing and processing in extruder and oscillatory shear injection molding machine. At the time of melt mixing, about 5 wt% of ultra LMWPE was incorporated in the UHMWPE matrix as a filler to aid in processing. As a result of oscillatory shear movements, the formation of aligned lamellar like crystal structure was obtained, which correspondingly helped to reduce the molecular defects. Due to the comprehensive effect of the obtained features, enhancement in anti-wear, antioxidation and mechanical properties was obtained. Xiong and Xiong [131] also investigated the impact of irradiation dosage on the performance of UHMWPE.

Takahashi *et al.* [132] discussed the modification of UHMWPE through the addition of vitamin E and subsequent radiation crosslinking. Vitamin E reinforcement acted as an effective antioxidant, which enriched the oxidation resistance of crosslinked UHMWPE without any diminishing effect on the mechanical features. Vitamin E incorporation in crosslinked UHMWPE was observed to eliminate the strain softening induced wearing [133]. In another study, the authors observed enhanced impingement resistance for vitamin E incorporated crosslinked UHMWPE due to vitamin E influenced elimination of toughness reduction during crosslinking reaction [134]. In a similar study, Chen *et al.* [135] highlighted the capability of vitamin E reinforcement to strengthen the anti-wear characteristic of crosslinked UHMWPE. Fu *et al.* [136] also highlighted the usefulness of naturally sourced polyphenols as effective multifunctional antioxidants in UHMWPE matrix. The presence of polyphenols such as dodecyl gallate and gallic acid in the polymer did not negatively impact the UHMWPE crosslinking reaction in comparison with vitamin E.

With an aim to enrich the tribological features of UHMWPE powders for effective artificial joint applications, Deng *et al.* [137] reported the development of composite structures by the introduction of acrylic acid grafts using ultraviolet light. Due to the influence of PAA grafts, UHMWPE exhibited reduction in friction as well as enrichment in wear resistance in the presence of water and artificial serum based lubricants, and one containing 3.5% PAA graft ratio displayed optimal results. Also, significant enhancement in the degree of wetting for the PAA reinforced UHMWPE system was observed. In another study, the authors investigated the effect of acrylamide grafting on the tribological behavior of UHMWPE powder [138]. The incorporation of acrylamide grafting was advantageous to enhance the wettability of UHMWPE due to its hydrophilic nature. Compared to neat UHMWPE, acrylamide grafted system demonstrated superior wear resistance in lubricants based on water and synthetic serum. Moreover, the effect of 3-dimethyl-(3-(N-methacrylamido)propyl) ammonium propane sulfonate (DMAPA) grafting on the mechanical and morphological characteristics of UHMWPE powder was investigated in another study [139]. Chen *et al.* [140] also discussed the introduction of PVA grafts on the UHMWPE surface *via* chemical route to generate effective joint prosthesis applications. In another study, Uehara *et al.* [141] reported 3 folds enhancement in wear resistance for crosslinked UHMWPE through the introduction of phospholipid polymer grafts. Panjwani *et al.* [142] also reported that the introduction of perfluoropolyether coating on UHMWPE offered much enriched lubrication and thermal features.

Gu *et al.* [143] examined the impact of strontium doped calcium polyphosphate addition on the properties of UHMWPE composites. The resultant composite exhibited enhanced hardness and anti-wear characteristics. The developed composite system had the capability to arrest aseptic loosening.

Samad and Sinha [144] examined the tribological and mechanical performance of SWCNTs reinforced UHMWPE. The friction coefficient and wear rate of UHMWPE were minimized by the incorporation of 0.1 wt% SWCNTs. Also, the composite system exhibited improvement in mechanical performance. The introduction of perfluoropolyether covering on this nanocomposite was beneficial to further minimize the friction coefficient. For developing artificial joint materials possessing minimal wear rate, Suner *et al.* [145] reported the use of multi-walled CNTs (MWCNTs) filled UHMWPE. An appreciable improvement in wear resistance was acquired for UHMWPE by the addition of MWCNTs. Low cytotoxicity and osteolytic potential were noted for wear debris generated from the UHMWPE/MWCNTs composite during abrasion in comparison with neat UHMWPE. In another study, Maksimkin *et al.* [146] explored the effect of bulk orientation of fluorinated MWCNTs on the tribological response of UHMWPE composites. The wear resistance of the oriented nanocomposites was remarkably improved due to the net effect of the nanofibrillar structural generation in the composites and the reinforcing effect of MWCNTs. Kumar *et al.* [147] also compared UHMWPE composites with CNTs of different aspect ratio. Higher mechanical properties, crystallinity and thermal stability were observed for UHMWPE composite with higher aspect ratio CNTs, as compared to the composite with lower aspect ratio CNTs.

Kolanthai *et al.* [148] studied the synergistic effect of nanoscale carbonaceous materials and subsequent hot rolling treatment on the properties of UHMWPE. The monoclinic and orthorhombic crystallites was observed in the composites. The carbonaceous nanomaterials diminished the crystallinity of UHMWPE through the size reduction of crystallites, and the composite containing graphene displayed higher tendency than MWCNTs. The application of hot rolling was favorable to enhance the surface texture of neat UHMWPE, but the presence of graphene and MWCNTs had an impairing effect on the surface texture after hot rolling, which was much higher in the case of graphene.

Lahiri *et al.* [149] reported the preparation of graphene nanoplatelets reinforced UHMWPE for achieving improved joint prosthesis performance. Graphene concentration and dispersion as well as graphene-UHMWPE binding significantly influenced the final properties of the nanocomposites. UHMWPE composite with 0.1 wt% of graphene exhibited enhancement in tensile strength (71%) and fracture toughness (54%), whereas further increase in graphene content (1 wt%) displayed diminishing effect. The tensile modulus and yield strength of the composites increased with corresponding increase in graphene concentration. Moreover, the osteoblast biocompatibility was higher for the composite with 0.1 wt% graphene than 1 wt% composite. Depletion in relative survivability was about

16% for the composite with 0.1 wt% graphene after 120 h incubation, whereas it was about 86% for 1 wt% composite system. Figure 4.9 also shows the stress-strain plots of the pure polymer and graphene composites. In another study, the authors discussed the tribological response of the composite system [150]. The friction coefficient of the UHMWPE composites was minimized with increase in graphene content because of its shearing capability resulting from weak van der Waals interactions. In addition, about four folds improvement in wear resistance was attained, owing to the synergistic influence of toughness and lubricating capability.

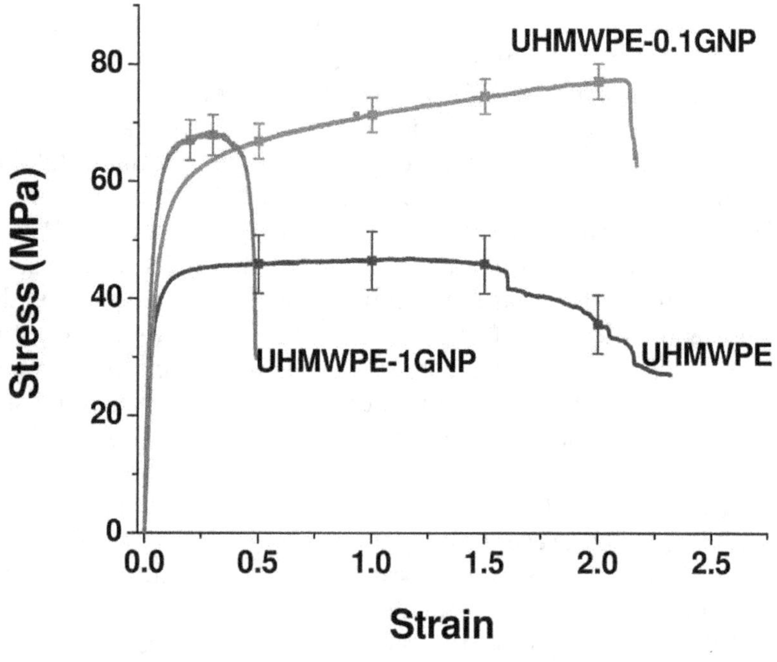

Figure 4.9 Stress-strain plots of the pure polymer and graphene nanocomposites. Reproduced from Reference 149 with permission from American Chemical Society.

Xu *et al.* [151] reported the development of fluorinated graphene filled UHMWPE nanocomposites through the combinatorial ultrasonication and thermoforming technique. Lowering of friction coefficient and enhancement in hardness were observed with increase in graphene concentration from 0.1 to 1 wt%. Moreover, a notable decrease in wear debris deposition amount was observed. The presence of graphene had no adverse effect on the biocompatibility of

UHMWPE. In another study, Suner and Emami [152] reported the potential of ball milled UHMWPE/GO nanocomposite for joint applications. Ball milling yielded uniform dispersion of GO in UHMWPE matrix. Enhanced thermal stability for UHMWPE/GO composite was observed. Also, tensile properties, fracture toughness, microhardness, antioxidation capability and hydrophilicity of UHMWPE were enhanced through the incorporation of GO up to 0.5 wt% [153]. In another study, Tai *et al.* [154] reported notable increase in anti-wear properties and hardness for solvent mixed UHMWPE composite with 1 wt% GO. Increase in anti-wear performance of the composite was related to the change of tribological wear response from fatigue to abrasive, which correspondingly formed lubricating surface layer. Chen *et al.* [155] also reported enhanced mechanical properties such as yield strength and hardness as well as adequate cytocompatibility for UHMWPE/GO composites generated *via* combinatorial ultra-sonication and compression molding.

Mirsalehi *et al.* [156] detailed the development of nano-hydroxyapatite filled UHMWPE composites for artificial joint applications. The presence of 50 wt% nano-hydroxyapatite in the UHMWPE composite formulation was advantageous not only to considerably enhance the tensile modulus and hardness but also to significantly minimize the friction. In another study, Shi *et al.* [157] reported a combination of sol-gel strategy and compression molding to develop high performance UHMWPE/hydroxyapatite nanocomposites suitable for artificial hip joint applications (Figure 4.10). Minimized friction coefficient and wear rate were observed for the composites due to the heat induced planar alignment of the UHMWPE crystal unit present at (110) plane and the resultant higher surface crystallinity. Maksimkin *et al.* [158] made use of mechano-activation method to achieve hydroxyapatite dispersion in the UHMWPE matrix and, thereby, enhanced tribological and mechanical properties. In another study, Kang *et al.* [159] reported the use of both nano- and micro-hydroxyapatite to improve the tribological and mechanical characteristics of UHMWPE. In the case of mechanical properties, the composite with nano-hydroxyapatite displayed higher extent of enhancement as compared to micro-filler. The optimal performance was noted for UHMWPE composites containing 10 wt% of nano-filler and 15 wt% of micro-filler respectively. In another recent study, Panin *et al.* [160] also observed enhancement in wear resistance of UHMWPE composites based on nano- as well as micro-hydroxyapatite particulates irradiated by nitrogen ion beams.

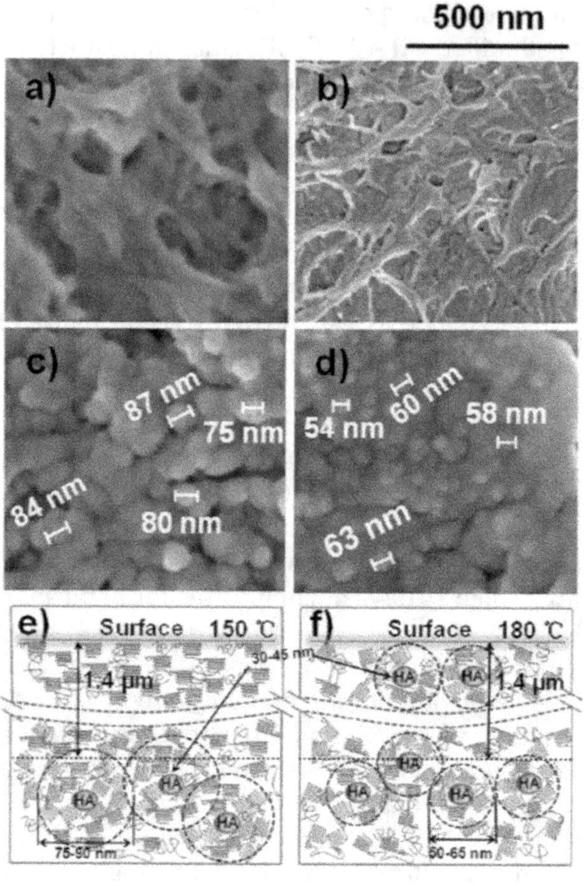

Figure 4.10 Scanning electron images of pristine UHMWPE molded at 150 °C (a) and 180 °C (b) and UHMWPE/HA composites with 23.5 vol % HA content molded at 150 °C (c) and 180 °C (d). Models (e) and (f) are schematic designs for UHMWPE/HA composites molded at 150 and 180 °C respectively. Reproduced from Reference 157 with permission from American Chemical Society.

To generate effective articular cartilage replacement materials, Senatov *et al.* [161] detailed the fabrication of 3 wt% nano-Al_2O_3 reinforced UHMWPE composites with the aid of a universal mill. Mechanical activation was observed to impact the compression properties and anti-wear potential of the nanocomposites by means of defects reduction and, thus, molecular strengthening through self-diffusion. Good cytocompatibility was noted for mechanically activated UHMWPE/Al_2O_3 nanocomposite. In another study, the authors compared the effect of Al_2O_3 nanoparticles and microspheres on the properties of UHMWPE [162]. Both forms of Al_2O_3 were favorable to enhance the polymer performance and had almost similar effect on the mechanical properties. UHMWPE composite containing nano-Al_2O_3 exhibited higher wear resistance in comparison to microspheres.

Wang *et al.* [163] detailed the modification of UHMWPE by the incorporation of nanocellulose for generating effective and durable artificial joint materials. Ball milling was used for the generation of UHMWPE-nanocellulose composites. The wear rate of hydrophobic UHMWPE was considerably minimized after the addition of hydrophilic nanocellulose. Upon abrasion, nanocellulose in the UHMWPE matrix was unfastened, which subsequently acted as an effective lubricant and shielded the UHMWPE matrix from damage. Furthermore, the UHMWPE/nanocellulose composites displayed improved cytocompatibility with macrophage as well as osteoblast cells.

Chang *et al.* [164] reported the fabrication of UHMWPE nanocomposites with silane modified ZnO by combinatorial ball milling and hot pressing technique. The addition of ZnO nanoparticles up to 10 wt% was favorable to enhance the wear resistance and compressive strength of UHMWPE. Lowering in friction coefficient was observed with increase in nano-ZnO concentration in the composite formulation. The mechanical properties such as hardness, Young's modulus and compression modulus were also observed to improve with increase in nano-ZnO concentration. Moreover, the composites displayed better bacterial resistance and it was optimal for the composite containing 20 wt% of nano-ZnO.

Wang *et al.* [165] reported the generation of high performance UHMWPE/crosslinked UHMWPE powder composite for durable artificial joint applications. Significant enhancement in Young's modulus, anti-wear activity and ductility of UHMWPE was attained with the incorporation of 25 wt% crosslinked UHMWPE powder. In another study, Gong *et al.* [166] used mechanical activation to enhance the performance of UHMWPE composites with alendronate sodium. Wang *et al.* [102] also revealed superior lubrication capability for UHMWPE based layered composite system composed of UHMWPE film, TiO_2 porous layer and 3-dimethyl-(3-(N-methacrylamido)propyl) ammonium propane sulfonate polymer brushes. The resultant composite system was observed to display improved wettability, along with substantial reduction in friction and wear rate.

4.3.2 PVA Hydrogels Based Materials

To generate joint cartilage prosthesis with lubrication capability, Yarimitsu *et al.* [124] reported the use of PVA hydrogels generated through combinatorial freeze-thawing and cast-drying technique. PVA hydrogel prepared under the drying conditions of 8 °C and 50% RH displayed a low friction coefficient (0.013), which gradually increased with increase of sliding distance. The application of extra drying conditions of 20 °C and 40% RH in addition to an initial PVA hydrogel drying at 8 °C and 50% RH was found to be beneficial for achieving low friction coefficient and, thus, superior wear resistance. Murakami *et al.* [167] studied the effect of three PVA hydrogel synthesis methods such as freeze-thawing, cast-drying and combination of the two on the friction coefficient. In saline water based lubricant system, hydrogel synthesized *via* combination of freeze-thawing and cast-drying technique displayed lower friction coefficient suitable for artificial joint cartilage applications. Further insights were also reported by the authors in other studies [168,169]. Osaheni *et al.* [170] detailed the generation of high performance PVA based composite hydrogels by the addition of poly(2-(methacryloyloxy)ethyl) dimethyl-(3-sulfopropyl) ammonium hydroxide (PMDAH). Due to the boundary lubricating efficiency of PMDAH, the hydrogel displayed considerable decline of about 80% in friction. The inclusion of PMDAH had no adverse effect on the mechanical as well as physical properties of PVA. Further, good cytocompatibility was noted for the developed hydrogel system. Shi *et al.* [171] reported the hydrogels composed of PVA and polyvinylpyrrolidone for generating artificial joint cartilage with the aid of cyclic freezing/thawing technique. Owing to the porous network structure, the PVA/polyvinylpyrrolidone hydrogel exhibited analogous biphasic lubrication process as seen in human articular cartilage.

To enable effective articular cartilage replacement, Li *et al.* [172] reported high performance hydroxyapatite reinforced PVA composite hydrogel. The resultant composite hydrogel system exhibited higher compressive strength and modulus in comparison with PVA hydrogel due to effective hydroxyapatite distribution in the pores and pore walls of PVA hydrogel network. Chen *et al.* [173] reported composite hydrogel composed of PVA and hydroxyapatite for synthetic joint cartilage. The application of rising swing angle as well as contact load on the composite hydrogel had a significant effect on the reaction force and contact stress, whereas a diminishing effect on the fluid load support. Furthermore, substantial fluid load support behavior was noticed for the composite hydrogel under the influence of swing friction due to the lubrication *via* fluid release. At the same time, the mechanical features of the composite were observed to rehabilitate after the application of swing friction as a result of flow back tendency of released fluid. In another study, the authors reported almost similar behavior for hydrogels based on PVA-hydroxyapatite-silk composite [174]. Also, the ternary composite based hydrogel exhibited inverse relation of fluid load support with start-up friction resistance.

Jing *et al.* [175] reported the generation of PVA based interpenetrating hydrogels through the incorporation hydroxyl functionalized boron nitride nanosheets (Figure 4.11). The incorporation a small amount of filler (0.12 wt%) was favorable to significantly increase compressive, tensile strengths and Young's modulus by 45, 43, and 63% respectively, due to the combined impact of hydrogen bonding in the nanocomposite structure and reinforcing effect of hydroxyl functionalized boron nitride nanosheets. Also, the nanocomposite hydrogels displayed enhancement in the thermal properties such as conductivity (5%) and diffusivity (15%), as compared to PVA hydrogel. The nanocomposite hydrogels also exhibited cytocompatible behavior, thus, confirming their potential for large number of applications.

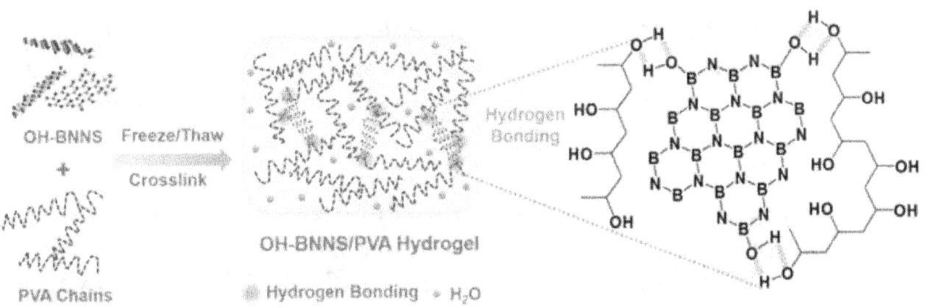

Figure 4.11 Schematic demonstration for the synthesis of PVA/hydroxylated boron nitride nanosheet (OH-BNNS) interpenetrating network hydrogel. Reproduced from Reference 175 with permission from American Chemical Society.

For generating superior synthetic articular cartilage, Chen *et al.* [176] reported the fabrication of β-cyclodextrin aldehyde crosslinked GO filled PVA interpenetrating hydrogel. For this, the authors made use of freeze-thawing technique. Compression modulus of the β-GO/PVA hydrogels were increased by 5.3 fold, as compared to pure PVA hydrogels at 3 $mg.ml^{-1}$ β-GO. An increase of 2.5 folds was observed in breaking elongation at the same β-GO concentration. In addition, β-cyclodextrin aldehyde crosslinked GO had no unfavorable effect on the cytocompatibility of PVA hydrogel.

4.4 Conclusions

Over the last few years, the research efforts for the design of polymeric tissue engineering and artificial joint materials have displayed increasing trend due to excellent properties and biomedical suitability of these materials. Tissue engineering scaffolds based on various natural and synthetic polymeric materials and composites in the form of hydrogels, nanofibers, films, etc., have displayed suitability for a variety of tissue regeneration applications. For efficient artificial joints, UHMWPE based materials have drawn great research interest, along with PVA lubricating hydrogels. Overall, the multitude of polymers and polymer composites developed for tissue engineering and artificial joint applications is envisaged to grow further with the ongoing development in synthetic methodologies.

References

1. Klee, D., and Hocker, H. (1999) Polymers for biomedical applications: improvement of the interface compatibility. *Biomedical Applications Polymer Blends, Advances in Polymer Science*, Eastmond G. C., Hocker H., and Klee D. (eds.), Volume 149, Springer, Germany, pp. 1-57.
2. Kulshrestha, A. S., and Mahapatro, A. (2008) Polymers for biomedical applications. In: *Polymers for Biomedical Applications*, Mahapatro, A., and Kulshrestha, A. S. (eds.), ACS Publications, USA, pp. 1-7.
3. Majumdar, J., and Edmond, L. (2006) Polymers in biomedical applications. *Australasian Biotechnology*, **16**, 14-15.
4. Gaharwar, A. K., Peppas, N. A., and Khademhosseini, A. (2014) Nanocomposite hydrogels for biomedical applications. *Biotechnology and Bioengineering*, **111**, 441-453.
5. Simanek, E. (2010) Polymers for biomedical applications. *Molecular Pharmaceutics*, **7**(4), 921.
6. Tonzani, S. (2013) Polymers for biomedical applications. *Journal of Applied Polymer Science*, **129**, 527.
7. Ravichandran, R., Sundarrajan, S., Venugopal, J. R., Mukherjee, S.,

and Ramakrishna, S. (2012) Advances in polymeric systems for tissue engineering and biomedical applications. *Macromolecular Bioscience*, **12**, 286-311.

8. Barnes, C. P., Sell, S. A., Boland, E. D., Simpson, D. G., and Bowlin, G. L. (2007) Nanofiber technology: designing the next generation of tissue engineering scaffolds. *Advanced Drug Delivery Reviews*, **59**, 1413-1433.
9. Pham, Q. P., Sharma, U., and Mikos, A. G. (2006) Electrospinning of polymeric nanofibers for tissue engineering applications: a review. *Tissue Engineering*, **12**, 1197-1211.
10. Wolfe, P. S., Sell, S. A., and Bowlin, G. L. (2011) Natural and synthetic scaffolds. In: *Tissue Engineering*, Pallua N., and Suscheck C. (eds.), Springer, Germany, pp. 41-67.
11. Guo, B., Glavas, L., and Albertsson, A.-C. (2013) Biodegradable and electrically conducting polymers for biomedical applications. *Progress in Polymer Science*, **38**, 1263-1286.
12. Van Vlierberghe, S., Dubruel, P., and Schacht, E. (2011) Biopolymer-based hydrogels as scaffolds for tissue engineering applications: a review. *Biomacromolecules*, **12**, 1387-1408.
13. Chen, G., Ushida, T., and Tateishi, T. (2002) Scaffold design for tissue engineering. *Macromolecular Bioscience*, **2**, 67-77.
14. Dahlin, R. L., Kasper, F. K., and Mikos, A. G. (2011) Polymeric nanofibers in tissue engineering. *Tissue Engineering, Part B: Reviews*, **17**, 349-364.
15. Spiller, K. L., Maher, S. A., and Lowman, A. M. (2011) Hydrogels for the repair of articular cartilage defects. *Tissue Engineering, Part B: Reviews*, **17**, 281-299.
16. Lee, K. Y., and Mooney, D. J. (2001) Hydrogels for tissue engineering. *Chemical Reviews*, **101**, 1869-1880.
17. Carrow, J. K., and Gaharwar, A. K. (2015) Bioinspired polymeric nanocomposites for regenerative medicine. *Macromolecular Chemistry and Physics*, **216**, 248-264.
18. Yang, J., Zhang, Y. S., Yue, K., and Khademhosseini, A. (2017) Cell-laden hydrogels for osteochondral and cartilage tissue engineering. *Acta biomaterialia*, **57**, 1-25.
19. Wan, C., and Chen, B. (2011) Poly (ε-caprolactone)/graphene oxide biocomposites: mechanical properties and bioactivity. *Biomedical Materials*, **6**, 055010.
20. Shen, J., Yan, B., Li, T., Long, Y., Li, N., and Ye, M. (2012) Study on graphene-oxide-based polyacrylamide composite hydrogels. *Composites, Part A: Applied Science and Manufacturing*, **43**, 1476-1481.
21. Shin, H., Ruhe, P. Q., Mikos, A.G., and Jansen, J.A. (2003) In vivo bone and soft tissue response to injectable, biodegradable oligo (poly(ethylene glycol) fumarate) hydrogels. *Biomaterials*, **24**, 3201-3211.

22. Lee, K., Kim, H., Khil, M., Ra, Y., and Lee, D. (2003) Characterization of nano-structured poly (ε-caprolactone) nonwoven mats via electrospinning. *Polymer*, **44**, 1287-1294.
23. Cross, L. M., Thakur, A., Jalili, N. A., Detamore, M., and Gaharwar, A. K. (2016) Nanoengineered biomaterials for repair and regeneration of orthopedic tissue interfaces. *Acta Biomaterialia*, **42**, 2-17.
24. Calo, E., and Khutoryanskiy, V.V. (2015) Biomedical applications of hydrogels: A review of patents and commercial products. *European Polymer Journal*, **65**, 252-267.
25. Hoffman, A. S. (2012) Hydrogels for biomedical applications. *Advanced Drug Delivery Reviews*, **64**, 18-23.
26. Shin, H., Temenoff, J. S., and Mikos, A. G. (2003) In vitro cytotoxicity of unsaturated oligo[poly(ethylene glycol) fumarate] macromers and their cross-linked hydrogels. *Biomacromolecules*, **4**, 552-560.
27. Lam, J., Kim, K., Lu, S., Tabata, Y., Scott, D. W., Mikos, A. G., and Kurtis Kasper, F. (2014) A factorial analysis of the combined effects of hydrogel fabrication parameters on the in vitro swelling and degradation of oligo(poly(ethylene glycol) fumarate) hydrogels. *Journal of Biomedical Materials Research, Part A*, **102**, 3477-3487.
28. Inozemtseva, O., Salkovskiy, Y. E., Severyukhina, A., Vidyasheva, I., Petrova, N., Metwally, H., Stetciura, I., and Gorin, D. A. (2015) Electrospinning of functional materials for biomedicine and tissue engineering. *Russian Chemical Reviews*, **84**, 251.
29. Saw, S. H., Wang, K., Yong, T., and Ramakrishna, S. (2007) Polymeric nanofibers in tissue engineering. In: *Nanotechnologies for the Life Sciences*, Kumar, C. S. (ed.), Wiley-VCH, Germany, doi:10.1002/9783527610419.ntls0096.
30. Yoo, H. S., Kim, T. G., and Park, T. G. (2009) Surface-functionalized electrospun nanofibers for tissue engineering and drug delivery. *Advanced Drug Delivery Reviews*, **61**, 1033-1042.
31. Kwon, I. K., Kidoaki, S., and Matsuda, T. (2005) Electrospun nano- to microfiber fabrics made of biodegradable copolyesters: structural characteristics, mechanical properties and cell adhesion potential. *Biomaterials*, **26**, 3929-3939.
32. Rajzer, I., Menaszek, E., and Castano, O. (2017) Electrospun polymer scaffolds modified with drugs for tissue engineering. *Materials Science and Engineering C*, **77**, 493-499.
33. Stefani, I., and Cooper-White, J. (2016) Development of an in-process UV-crosslinked, electrospun PCL/aPLA-co-TMC composite polymer for tubular tissue engineering applications. *Acta Biomaterialia*, **36**, 231-240.
34. Hasan, A., Memic, A., Annabi, N., Hossain, M., Paul, A., Dokmeci, M. R., Dehghani, F., and Khademhosseini, A. (2014) Electrospun scaffolds for tissue engineering of vascular grafts. *Acta Biomaterialia*, **10**, 11-25.

35. Bhardwaj, N., and Kundu, S. C. (2010) Electrospinning: a fascinating fiber fabrication technique. *Biotechnology Advances*, **28**, 325-347.
36. Subbiah, T., Bhat, G., Tock, R., Parameswaran, S., and Ramkumar, S. (2005) Electrospinning of nanofibers. *Journal of Applied Polymer Science*, **96**, 557-569.
37. Wendorff, J. H., Agarwal, S., and Greiner, A. (2012) *Electrospinning: Materials, Processing, and Applications*, John Wiley & Sons, USA.
38. Balint, R., Cassidy, N. J., and Cartmell, S. H. (2014) Conductive polymers: towards a smart biomaterial for tissue engineering. *Acta Biomaterialia*, **10**, 2341-2353.
39. Mawad, D., Stewart, E., Officer, D. L., Romeo, T., Wagner, P., Wagner, K., and Wallace, G. G. (2012) A single component conducting polymer hydrogel as a scaffold for tissue engineering. *Advanced Functional Materials*, **22**, 2692-2699.
40. Kim, D. H., Richardson-Burns, S. M., Hendricks, J. L., Sequera, C., and Martin, D. C. (2007) Effect of immobilized nerve growth factor on conductive polymers: electrical properties and cellular response. *Advanced Functional Materials*, **17**, 79-86.
41. Asplund, M., Thaning, E., Lundberg, J., Sandberg-Nordqvist, A., Kostyszyn, B., Inganas, O., and von Holst, H. (2009) Toxicity evaluation of PEDOT/biomolecular composites intended for neural communication electrodes. *Biomedical Materials*, **4**, 045009.
42. Pires, F., Ferreira, Q., Rodrigues, C. A., Morgado, J., and Ferreira, F. C. (2015) Neural stem cell differentiation by electrical stimulation using a cross-linked PEDOT substrate: expanding the use of biocompatible conjugated conductive polymers for neural tissue engineering. *Biochimica et Biophysica Acta (BBA) - General Subjects*, **1850**, 1158-1168.
43. Green, R., and Abidian, M. R. (2015) Conducting polymers for neural prosthetic and neural interface applications. *Advanced Materials*, **27**, 7620-7637.
44. Ghasemi-Mobarakeh, L., Prabhakaran, M. P., Morshed, M., Nasr-Esfahani, M. H., Baharvand, H., Kiani, S., Al-Deyab, S. S., and Ramakrishna, S. (2011) Application of conductive polymers, scaffolds and electrical stimulation for nerve tissue engineering. *Journal of Tissue Engineering and Regenerative Medicine*, **5**, e17-e35.
45. Hardy, J. G., Lee, J. Y., and Schmidt, C. E. (2013) Biomimetic conducting polymer-based tissue scaffolds. *Current Opinion in Biotechnology*, **24**, 047-054.
46. Bendrea, A.-D., Cianga, L., and Cianga, I. (2011) Review paper: Progress in the field of conducting polymers for tissue engineering applications. *Journal of Biomaterials Applications*, **26**, 3-84.
47. Yang, B., Yao, F., Hao, T., Fang, W., Ye, L., Zhang, Y., Wang, Y., Li, J., and Wang, C. (2016) Development of electrically conductive dou-

ble-network hydrogels via one-step facile strategy for cardiac tissue engineering. *Advanced Healthcare Materials*, **5**, 474-488.

48. Yang, S., Jang, L., Kim, S., Yang, J., Yang, K., Cho, S. W., and Lee, J. Y. (2016) Polypyrrole/alginate hybrid hydrogels: electrically conductive and soft biomaterials for human mesenchymal stem cell culture and potential neural tissue engineering applications. *Macromolecular Bioscience*, **16**, 1653-1661.
49. Jo, H., Sim, M., Kim, S., Yang, S., Yoo, Y., Park, J.-H., Yoon, T. H., Kim, M.-G., and Lee, J. Y. (2017) Electrically conductive graphene/polyacrylamide hydrogels produced by mild chemical reduction for enhanced myoblast growth and differentiation. *Acta Biomaterialia*, **48**, 100-109.
50. Zhang, J., Li, M., Kang, E.-T., and Neoh, K. G. (2016) Electrical stimulation of adipose-derived mesenchymal stem cells in conductive scaffolds and the roles of voltage-gated ion channels. *Acta Biomaterialia*, **32**, 46-56.
51. Ramazani, S., and Karimi, M. (2015) Aligned poly (ε-caprolactone)/graphene oxide and reduced graphene oxide nanocomposite nanofibers: Morphological, mechanical and structural properties. *Materials Science and Engineering C*, **56**, 325-334.
52. Dvir, T., Timko, B. P., Brigham, M. D., Naik, S. R., Karajanagi, S. S., Levy, O., Jin, H., Parker, K. K., Langer, R., and Kohane, D. S. (2011) Nanowired three-dimensional cardiac patches. *Nature Nanotechnology*, **6**, 720-725.
53. Shin, S. R., Bae, H., Cha, J. M., Mun, J. Y., Chen, Y.-C., Tekin, H., Shin, H., Farshchi, S., Dokmeci, M. R., and Tang, S. (2011) Carbon nanotube reinforced hybrid microgels as scaffold materials for cell encapsulation. *ACS Nano*, **6**, 362-372.
54. Jin, G., and Li, K. (2014) The electrically conductive scaffold as the skeleton of stem cell niche in regenerative medicine. *Materials Science and Engineering C*, **45**, 671-681.
55. Li, G., Zhao, Y., Zhang, L., Gao, M., Kong, Y., and Yang, Y. (2016) Preparation of graphene oxide/polyacrylamide composite hydrogel and its effect on Schwann cells attachment and proliferation. *Colloids and Surfaces B: Biointerfaces*, **143**, 547-556.
56. Patel, N. R., Whitehead, A. K., Newman, J. J., and Caldorera-Moore, M. E. (2016) Poly(ethylene glycol) hydrogels with tailorable surface and mechanical properties for tissue engineering applications. *ACS Biomaterials Science & Engineering*, **3**(8), 1494-1498.
57. Henke, M., Baumer, J., Blunk, T., and Tessmar, J. (2014) Foamed oligo(poly(ethylene glycol) fumarate) hydrogels as versatile prefabricated scaffolds for tissue engineering. *Journal of Tissue Engineering and Regenerative Medicine*, **8**, 248-252.
58. Kinard, L. A., Chu, C.-Y., Tabata, Y., Kasper, F. K., and Mikos, A. G. (2013) Bone morphogenetic protein-2 release from composite hy-

drogels of oligo(poly(ethylene glycol) fumarate) and gelatin. *Pharmaceutical Research*, **30**, 2332-2343.

59. Liu, X., Miller, A. L., Park, S., Waletzki, B. E., Zhou, Z., Terzic, A., and Lu, L. (2017) Functionalized carbon nanotube and graphene oxide embedded electrically conductive hydrogel synergistically stimulates nerve cell differentiation. *ACS Applied Materials & Interfaces*, **9**, 14677-14690.
60. Hong, J. H., Lee, H. J., and Jeong, B. (2017) Injectable polypeptide thermogel as a tissue engineering system for hepatogenic differentiation of tonsil-derived mesenchymal stem cells. *ACS Applied Materials & Interfaces*, **9**, 11568-11576.
61. Patel, M., Moon, H. J., Ko, D. Y., and Jeong, B. (2016) Composite system of graphene oxide and polypeptide thermogel as an injectable 3D scaffold for adipogenic differentiation of tonsil-derived mesenchymal stem cells. *ACS Applied Materials & Interfaces*, **8**, 5160-5169.
62. Wang, H., Shi, J., Wang, Y., Yin, Y., Wang, L., Liu, J., Liu, Z., Duan, C., Zhu, P., and Wang, C. (2014) Promotion of cardiac differentiation of brown adipose derived stem cells by chitosan hydrogel for repair after myocardial infarction. *Biomaterials*, **35**, 3986-3998.
63. Deng, B., Shen, L., Wu, Y., Shen, Y., Ding, X., Lu, S., Jia, J., Qian, J., and Ge, J. (2015) Delivery of alginate-chitosan hydrogel promotes endogenous repair and preserves cardiac function in rats with myocardial infarction. *Journal of Biomedical Materials Research, Part A*, **103**, 907-918.
64. Giammanco, G. E., Carrion, B., Coleman, R. M., and Ostrowski, A. D. (2016) Photoresponsive polysaccharide-based hydrogels with tunable mechanical properties for cartilage tissue engineering. *ACS Applied Materials & Interfaces*, **8**, 14423-14429.
65. Naseri, N., Deepa, B., Mathew, A. P., Oksman, K., and Girandon, L. (2016) Nanocellulose-based interpenetrating polymer network (IPN) hydrogels for cartilage applications. *Biomacromolecules*, **17**, 3714-3723.
66. Kumar, A., Lee, Y., Kim, D., Rao, K. M., Kim, J., Park, S., Haider, A., and Han, S. S. (2017) Effect of crosslinking functionality on microstructure, mechanical properties, and in vitro cytocompatibility of cellulose nanocrystals reinforced poly(vinyl alcohol)/sodium alginate hybrid scaffolds. *International Journal of Biological Macromolecules*, **95**, 962-973.
67. Rhee, S., Puetzer, J. L., Mason, B. N., Reinhart-King, C. A., and Bonassar, L. J. (2016) 3D bioprinting of spatially heterogeneous collagen constructs for cartilage tissue engineering. *ACS Biomaterials Science & Engineering*, **2**, 1800-1805.
68. Vazquez-Portalatin, N., Kilmer, C. E., Panitch, A., and Liu, J. C. (2016) Characterization of collagen type I and II blended hydro-

gels for articular cartilage tissue engineering. *Biomacromolecules,* **17**, 3145-3152.

69. Koppes, A. N., Keating, K. W., McGregor, A. L., Koppes, R. A., Kearns, K. R., Ziemba, A. M., McKay, C. A., Zuidema, J. M., Rivet, C. J., Gilbert, R. J., and Thompson, D. M. (2016) Robust neurite extension following exogenous electrical stimulation within single walled carbon nanotube-composite hydrogels. *Acta Biomaterialia,* **39**, 34-43.
70. Kim, Y., Gill, E. E., and Liu, J. C. (2016) Enzymatic cross-linking of resilin-based proteins for vascular tissue engineering applications. *Biomacromolecules,* **17**, 2530-2539.
71. Domingues, R. M. A., Silva, M., Gershovich, P., Betta, S., Babo, P., Caridade, S. G., Mano, J. F., Motta, A., Reis, R. L., and Gomes, M. E. (2015) Development of injectable hyaluronic acid/cellulose nanocrystals bionanocomposite hydrogels for tissue engineering applications. *Bioconjugate Chemistry,* **26**, 1571-1581.
72. Nedjari, S., Schlatter, G., and Hebraud, A. (2015) Thick electrospun honeycomb scaffolds with controlled pore size. *Materials Letters,* **142**, 180-183.
73. Song, J., Gao, H., Zhu, G., Cao, X., Shi, X., and Wang, Y. (2015) The preparation and characterization of polycaprolactone/graphene oxide biocomposite nanofiber scaffolds and their application for directing cell behaviors. *Carbon,* **95**, 1039-1050.
74. Hardy, J. G., Villancio-Wolter, M. K., Sukhavasi, R. C., Mouser, D. J., Aguilar, D., Geissler, S. A., Kaplan, D. L., and Schmidt, C. E. (2015) Electrical stimulation of human mesenchymal stem cells on conductive nanofibers enhances their differentiation toward osteogenic outcomes. *Macromolecular Rapid Communications,* **36**, 1884-1890.
75. Domingues, R., Chiera, S., Gershovich, P., Motta, A., Reis, R. L., and Gomes, M. E. (2016) Enhancing the biomechanical performance of anisotropic nanofibrous scaffolds in tendon tissue engineering: reinforcement with cellulose nanocrystals. *Advanced Healthcare Materials,* **5**, 1364-1375.
76. Zhang, C., Salick, M. R., Cordie, T. M., Ellingham, T., Dan, Y., and Turng, L.-S. (2015) Incorporation of poly (ethylene glycol) grafted cellulose nanocrystals in poly (lactic acid) electrospun nanocomposite fibers as potential scaffolds for bone tissue engineering. *Materials Science and Engineering C,* **49**, 463-471.
77. Zhang, C., Wang, L., Zhai, T., Wang, X., Dan, Y., and Turng, L.-S. (2016) The surface grafting of graphene oxide with poly(ethylene glycol) as a reinforcement for poly(lactic acid) nanocomposite scaffolds for potential tissue engineering applications. *Journal of the Mechanical Behavior of Biomedical Materials,* **53**, 403-413.
78. Zhang, K., Zheng, H., Liang, S., and Gao, C. (2016) Aligned PLLA nanofibrous scaffolds coated with graphene oxide for promoting

neural cell growth. *Acta Biomaterialia*, **37**, 131-142.

79. Stewart, E., Kobayashi, N. R., Higgins, M. J., Quigley, A. F., Jamali, S., Moulton, S. E., Kapsa, R. M., Wallace, G. G., and Crook, J. M. (2014) Electrical stimulation using conductive polymer polypyrrole promotes differentiation of human neural stem cells: a biocompatible platform for translational neural tissue engineering. *Tissue Engineering, Part C: Methods*, **21**, 385-393.
80. Zhang, Q., Esrafilzadeh, D., Crook, J. M., Kapsa, R., Stewart, E. M., Tomaskovic-Crook, E., Wallace, G. G., and Huang, X.-F. (2017) Electrical stimulation using conductive polymer polypyrrole counters reduced neurite outgrowth of primary prefrontal cortical neurons from NRG1-KO and DISC1-LI mice. *Scientific Reports*, **7**, 42525.
81. Yang, J., Choe, G., Yang, S., Jo, H., and Lee, J. Y. (2016) Polypyrrole-incorporated conductive hyaluronic acid hydrogels. *Biomaterials Research*, **20**, 31.
82. Srivastava, N., Venugopalan, V., Divya, M., Rasheed, V., James, J., and Narayan, K. (2013) Neuronal differentiation of embryonic stem cell derived neuronal progenitors can be regulated by stretchable conducting polymers. *Tissue Engineering, Part A*, **19**, 1984-1993.
83. Atak, B. H., Buyuk, B., Huysal, M., Isik, S., Senel, M., Metzger, W., and Cetin, G. (2017). Preparation and characterization of amine functional nano-hydroxyapatite/chitosan bionanocomposite for bone tissue engineering applications. *Carbohydrate Polymers*, **164**, 200-213.
84. Huang, Q., Hao, L., Xie, J., Gong, T., Liao, J., and Lin, Y. (2015) Tea polyphenol–functionalized graphene/chitosan as an experimental platform with improved mechanical behavior and bioactivity. *ACS Applied Materials & Interfaces*, **7**, 20893-20901.
85. Serra, I., Fradique, R., Vallejo, M., Correia, T., Miguel, S., and Correia, I. (2015) Production and characterization of chitosan/gelatin/β-TCP scaffolds for improved bone tissue regeneration. *Materials Science and Engineering C*, **55**, 592-604.
86. Beringer, L. T., Li, S., Kallick, E. J., Shields, K. J., Faight, E. M., Cartieri, F., Aballay, A., Edington, H., and Averick, S. (2016) Promoting adipogenesis using a collagen VI–heparin sulfate coating: Applications in tissue engineering for wound healing. *Industrial & Engineering Chemistry Research*, **55**, 12687-12692.
87. Jin, L., Zeng, Z., Kuddannaya, S., Wu, D., Zhang, Y., and Wang, Z. (2015) Biocompatible, free-standing film composed of bacterial cellulose nanofibers–graphene composite. *ACS Applied Materials & Interfaces*, **8**, 1011-1018.
88. Naskar, D., Bhattacharjee, P., Ghosh, A. K., Mandal, M., and Kundu, S. C. (2017) Carbon nanofiber reinforced nonmulberry silk protein fibroin nanobiocomposite for tissue engineering applications. *ACS Applied Materials & Interfaces*, **9**(23), 19356-19370.

89. Aznar-Cervantes, S., Martinez, J. G., Bernabeu-Esclapez, A., Lozano-Perez, A. A., Meseguer-Olmo, L., Otero, T. F., and Cenis, J. L. (2016) Fabrication of electrospun silk fibroin scaffolds coated with graphene oxide and reduced graphene for applications in biomedicine. *Bioelectrochemistry*, **108**, 36-45.
90. Scaffaro, R., Lopresti, F., Botta, L., Rigogliuso, S., and Ghersi, G. (2016) Preparation of three-layered porous PLA/PEG scaffold: relationship between morphology, mechanical behavior and cell permeability. *Journal of the Mechanical Behavior of Biomedical Materials*, **54**, 8-20.
91. Scaffaro, R., Lopresti, F., Botta, L., Rigogliuso, S., and Ghersi, G. (2016) Melt processed PCL/PEG scaffold with discrete pore size gradient for selective cellular infiltration. *Macromolecular Materials and Engineering*, **301**, 182-190.
92. Scaffaro, R., Lopresti, F., Botta, L., Rigogliuso, S., and Ghersi, G. (2016) Integration of PCL and PLA in a monolithic porous scaffold for interface tissue engineering. *Journal of the Mechanical Behavior of Biomedical Materials*, **63**, 303-313.
93. Zhang, F., Song, Q., Huang, X., Li, F., Wang, K., Tang, Y., Hou, C., and Shen, H. (2016) A novel high mechanical property PLGA composite matrix loaded with nanodiamond–phospholipid compound for bone tissue engineering. *ACS Applied Materials & Interfaces*, **8**, 1087-1097.
94. Jiang, L., Li, Y., Xiong, C., Su, S., and Ding, H. (2017) Preparation and properties of bamboo fiber/nano-hydroxyapatite/poly(lactic-co-glycolic) composite scaffold for bone tissue engineering. *ACS Applied Materials & Interfaces*, **9**, 4890-4897.
95. Pan, H., Zheng, Q., Guo, X., Wu, Y., and Wu, B. (2016) Polydopamine-assisted BMP-2-derived peptides immobilization on biomimetic copolymer scaffold for enhanced bone induction in vitro and in vivo. *Colloids and Surfaces B: Biointerfaces*, **142**, 1-9.
96. Zimmerli, W., Trampuz, A., and Ochsner, P. E. (2004) Prosthetic-joint infections. *New England Journal of Medicine*, **351**, 1645-1654.
97. Pope, B. J., Taylor, J. K., Dixon, R. H., Gardinier, C. F., Pope, L. M., Blackburn, D. C., Vail, M. A., and Jensen, K. M. (2002) Prosthetic Joint Having At Least One Superhard Articulation Surface, US patent US6398815B1.
98. Liang, H., Shi, B., Fairchild, A., and Cale, T. (2004) Applications of plasma coatings in artificial joints: an overview. *Vacuum*, **73**, 317-326.
99. Yousef, S. (2016) Polymer nanocomposite artificial joints. In: *Carbon Nanotubes: Current Progress of their Polymer Composites*, Berber, M., and Hafez, I. H. (eds.), InTech, UK, doi: 10.5772/62269.
100. Bistolfi, A., and Bellare, A. (2011) The relative effects of radiation crosslinking and type of counterface on the wear resistance of ul-

trahigh-molecular-weight polyethylene. *Acta Biomaterialia*, **7**, 3398-3403.

101. Hyon, S.-H., and Oka, M. (2001) Ultra High Molecular Weight Polyethylene Molded Article for Artificial Joints and Method of Preparing the Same, US patent US6168626B1.
102. Wang, K., Xiong, D., Deng, Y., and Niu, Y. (2017) Ultra-lubricated surface of Ti_6Al_4V fabricated with combination of porous TiO_2 layer, ultra-high molecular weight polyethylene film and hydrophilic polymer brushes. *Materials & Design*, **114**, 18-24.
103. Kyomoto, M., Iwasaki, Y., Moro, T., Konno, T., Miyaji, F., Kawaguchi, H., Takatori, Y., Nakamura, K., and Ishihara, K. (2007) High lubricious surface of cobalt–chromium–molybdenum alloy prepared by grafting poly(2-methacryloyloxyethyl phosphorylcholine). *Biomaterials*, **28**, 3121-3130.
104. Lee, S.-D., Hsiue, G.-H., Chang, P.C.-T., and Kao, C.-Y. (1996) Plasma-induced grafted polymerization of acrylic acid and subsequent grafting of collagen onto polymer film as biomaterials. *Biomaterials*, **17**, 1599-1608.
105. Kobayashi, M., and Takahara, A. (2010) Tribological properties of hydrophilic polymer brushes under wet conditions. *The Chemical Record*, **10**, 208-216.
106. Ishihara, K. (2015) Highly lubricated polymer interfaces for advanced artificial hip joints through biomimetic design. *Polymer Journal*, **47**, 585-597.
107. Gomez-Barrena, E., Puertolas, J.-A., Munuera, L., and Konttinen, Y. T. (2008) Update on UHMWPE research from the bench to the bedside. *Acta Orthopaedica*, **79**, 832-840.
108. Slouf, M., Eklova, S., Kumstatova, J., Berger, S., Synkova, H., Sosna, A., Pokorny, D., Spundova, M., and Entlicher, G. (2007) Isolation, characterization and quantification of polyethylene wear debris from periprosthetic tissues around total joint replacements. *Wear*, **262**, 1171-1181.
109. Martinez-Morlanes, M., Castell, P., Martinez-Nogues, V., Martinez, M., Alonso, P., and Puertolas, J. (2011) Effects of gamma-irradiation on UHMWPE/MWNT nanocomposites. *Composites Science and Technology*, **71**, 282-288.
110. Huang, Y.-F., Xu, J.-Z., Li, J.-S., He, B.-X., Xu, L., and Li, Z.-M. (2014) Mechanical properties and biocompatibility of melt processed, self-reinforced ultrahigh molecular weight polyethylene. *Biomaterials*, **35**, 6687-6697.
111. Lu, S. H., Liang, G. Z., Zhou, Z. W., and Li, F. (2006) Structure and properties of UHMWPE fiber/carbon fiber hybrid composites. *Journal of Applied Polymer Science*, **101**, 1880-1884.
112. Williams, J. A. (2005) Wear and wear particles - some fundamentals. *Tribology International*, **38**, 863-870.

113. Xue, Y., Wu, W., Jacobs, O., and Schadel, B. (2006) Tribological behaviour of UHMWPE/HDPE blends reinforced with multi-wall carbon nanotubes. *Polymer Testing*, **25**, 221-229.
114. Delport, H., Banks, S., De Schepper, J., and Bellemans, J. (2006) A kinematic comparison of fixed-and mobile-bearing knee replacements. *Bone & Joint Journal*, **88**, 1016-1021.
115. Valenza, A., Visco, A., Torrisi, L., and Campo, N. (2004) Characterization of ultra-high-molecular-weight polyethylene (UHMWPE) modified by ion implantation. *Polymer*, **45**, 1707-1715.
116. Baena, J. C., Wu, J., and Peng, Z. (2015) Wear performance of UHMWPE and reinforced UHMWPE composites in arthroplasty applications: a review. *Lubricants*, **3**, 413-436.
117. Osterle, W., Klaffke, D., Griepentrog, M., Gross, U., Kranz, I., and Knabe, C. (2008) Potential of wear resistant coatings on Ti-6Al-4V for artificial hip joint bearing surfaces. *Wear*, **264**, 505-517.
118. Scholz, M.-S., Blanchfield, J. P., Bloom, L. D., Coburn, B. H., Elkington, M., Fuller, J. D., Gilbert, M. E., Muflahi, S. A., Pernice, M. F., Rae, S. I., Trevarthen, J. A., White, S. C., Weaver, P. M., and Bond, I. P. (2011) The use of composite materials in modern orthopaedic medicine and prosthetic devices: a review. *Composites Science and Technology*, **71**, 1791-1803.
119. Mattei, L., Di Puccio, F., Piccigallo, B., and Ciulli, E. (2011) Lubrication and wear modelling of artificial hip joints: A review. *Tribology International*, **44**, 532-549.
120. Tomita, N., Kitakura, T., Onmori, N., Ikada, Y., and Aoyama, E. (1999) Prevention of fatigue cracks in ultrahigh molecular weight polyethylene joint components by the addition of vitamin E. *Journal of Biomedical Materials Research, Part A*, **48**, 474-478.
121. Suzuki, A., and Sasaki, S. (2015) Swelling and mechanical properties of physically crosslinked poly(vinyl alcohol) hydrogels. *Proceedings of the Institution of Mechanical Engineers, Part H: Journal of Engineering in Medicine*, **229**, 828-844.
122. Yarimitsu, S., Nakashima, K., Sawae, Y., and Murakami, T. (2013) Influence of phospholipid and protein constituents on tribological properties of artificial hydrogel cartilage material. *Journal of Biomechanical Science and Engineering*, **8**, 257-267.
123. Murakami, T., Yarimitsu, S., Nakashima, K., Yamaguchi, T., Sawae, Y., Sakai, N., and Suzuki, A. (2014) Superior lubricity in articular cartilage and artificial hydrogel cartilage. *Proceedings of the Institution of Mechanical Engineers, Part J: Journal of Engineering Tribology*, **228**, 1099-1111.
124. Yarimitsu, S., Yoshida, A., Sasaki, S., Murakami, T., and Suzuki, A. (2016) Evaluation of lubrication property of poly(vinyl alcohol) hybrid gel for artificial articular cartilage. *Tribology Online*, **11**, 360-365.

125. Suzuki, A., Sasaki, S., Sasaki, S., Noh, T., Nakashima, K., Yarimitsu, S., and Murakami, T. (2014) Elution and Wear of PVA Hydrogels by Reciprocating Friction. *5th World Tribology Congress, WTC 2013*, Italy, volume 3, pp. 2283-2286.
126. Noh, T., Bando, Y., Ota, K., Sasaki, S., and Suzuki, A. (2015) Tear force of physically crosslinked poly (vinyl alcohol) gels with different submicrometer-scale network structures. *Journal of Applied Polymer Science*, **132**, doi:10.1002/app.41356.
127. Huang, Y.-F., Xu, J.-Z., Xu, J.-Y., Zhang, Z.-C., Hsiao, B. S., Xu, L., and Li, Z.-M. (2014) Self-reinforced polyethylene blend for artificial joint application. *Journal of Materials Chemistry B*, **2**, 971-980.
128. Xu, L., Chen, C., Zhong, G.-J., Lei, J., Xu, J.-Z., Hsiao, B. S., and Li, Z.-M. (2012) Tuning the superstructure of ultrahigh-molecular-weight polyethylene/low-molecular-weight polyethylene blend for artificial joint application. *ACS Applied Materials & Interfaces*, **4**, 1521-1529.
129. Xu, L., Huang, Y.-F., Xu, J.-Z., Ji, X., and Li, Z.-M. (2014) Improved performance balance of polyethylene by simultaneously forming oriented crystals and blending ultrahigh-molecular-weight polyethylene. *RSC Advances*, **4**, 1512-1520.
130. Huang, Y.-F., Zhang, Z.-C., Xu, J.-Z., Xu, L., Zhong, G.-J., He, B.-X., and Li, Z.-M. (2016) Simultaneously improving wear resistance and mechanical performance of ultrahigh molecular weight polyethylene via cross-linking and structural manipulation. *Polymer*, **90**, 222-231.
131. Xiong, L., and Xiong, D. (2012) The influence of irradiation dose on mechanical properties and wear resistance of molded and extruded ultra high molecular weight polyethylene. *Journal of the Mechanical Behavior of Biomedical Materials*, **9**, 73-82.
132. Takahashi, Y., Yamamoto, K., Shishido, T., Masaoka, T., Tateiwa, T., Puppulin, L., and Pezzotti, G. (2014) Strain-induced microstructural rearrangement in ultra-high molecular weight polyethylene for hip joints: a comparison between conventional and vitamin E-infused highly-crosslinked liners. *Journal of the Mechanical Behavior of Biomedical Materials*, **31**, 31-44.
133. Takahashi, Y., Sugano, N., Puppulin, L., Zhu, W., and Pezzotti, G. (2014) Raman spectroscopic study of remelting and annealing-induced effects on microstructure and compressive deformation behavior of highly crosslinked UHMWPE for total hip arthroplasty. *Journal of Biomedical Materials Research, Part B: Applied Biomaterials*, **102**, 1762-1770.
134. Takahashi, Y., Tateiwa, T., Pezzotti, G., Shishido, T., Masaoka, T., and Yamamoto, K. (2016) Improved resistance to neck-liner impingement in second-generation highly crosslinked polyethylene - the role of vitamin E and crosslinks. *The Journal of Arthroplasty*,

31, 2926-2932.

135. Chen, G., Ni, Z., Qian, S., and Zhao, Y. (2016) Biotribological behaviour of Vitamin E-blended highly cross-linked UHMWPE in a hip joint simulator. *Industrial Lubrication and Tribology*, **68**, 548-553.
136. Fu, J., Shen, J., Gao, G., Xu, Y., Hou, R., Cong, Y., and Cheng, Y. (2013) Natural polyphenol-stabilised highly crosslinked UHMWPE with high mechanical properties and low wear for joint implants. *Journal of Materials Chemistry B*, **1**, 4727-4735.
137. Deng, Y., Xiong, D., and Wang, K. (2013) Biotribological properties of UHMWPE grafted with AA under lubrication as artificial joint. *Journal of Materials Science: Materials in Medicine*, **24**, 2085-2091.
138. Deng, Y., and Xiong, D. (2015) Fabrication and properties of UHMWPE grafted with acrylamide polymer brushes. *Journal of Polymer Research*, **22**, 195.
139. Deng, Y., Xiong, D., and Wang, K. (2014) The mechanical properties of the ultra high molecular weight polyethylene grafted with 3-dimethy (3-(N-methacryamido) propyl) ammonium propane sulfonate. *Journal of the Mechanical Behavior of Biomedical Materials*, **35**, 18-26.
140. Chen, K., Zhang, D., Cui, X., and Wang, Q. (2015) Preparation of ultrahigh-molecular-weight polyethylene grafted with polyvinyl alcohol hydrogel as an artificial joint. *RSC Advances*, **5**, 24215-24223.
141. Uehara, S., Yarimitsu, S., Moro, T., Kyomoto, M., Watanabe, K., Tanaka, S., Ishihara, K., and Murakami, T. (2016) Evaluation of wear suppression for phospholipid polymer-grafted ultra-high molecular weight polyethylene at concentrated contact. *Tribology International*, **101**, 264-272.
142. Panjwani, B., Satyanarayana, N., and Sinha, S. K. (2011) Tribological characterization of a biocompatible thin film of UHMWPE on Ti_6Al_4V and the effects of PFPE as top lubricating layer. *Journal of the Mechanical Behavior of Biomedical Materials*, **4**, 953-960.
143. Gu, Z., Huang, B., Li, Y., Tian, M., Li, L., and Yu, X. (2016) Strontium-doped calcium polyphosphate/ultrahigh molecular weight polyethylene composites: A new class of artificial joint components with enhanced biological efficacy to aseptic loosening. *Materials Science and Engineering C*, **61**, 526-533.
144. Samad, M. A., and Sinha, S. K. (2011) Dry sliding and boundary lubrication performance of a UHMWPE/CNTs nanocomposite coating on steel substrates at elevated temperatures. *Wear*, **270**, 395-402.
145. Suner, S., Bladen, C. L., Gowland, N., Tipper, J. L., and Emami, N. (2014) Investigation of wear and wear particles from a UHMWPE/multi-walled carbon nanotube nanocomposite for total joint replacements. *Wear*, **317**, 163-169.
146. Maksimkin, A. V., Senatov, F. S., Danilov, V. D., Mostovaya, K. S., Kal-

oshkin, S. D., Gorshenkov, M. V., Kharitonov, A. P., and Chukov, D. I. (2016) Transformation of the lamellar structure into nanofibrillar structure in the bulk oriented ultra high molecular weight polyethylene: mechanical and tribological properties. *Mendeleev Communications*, **26**, 350-352.

147. Kumar, R. M., Sharma, S. K., Kumar, B. M., and Lahiri, D. (2015) Effects of carbon nanotube aspect ratio on strengthening and tribological behavior of ultra high molecular weight polyethylene composite. *Composites, Part A: Applied Science and Manufacturing*, **76**, 62-72.
148. Kolanthai, E., Kalsar, R., Bose, S., Suwas, S., and Chatterjee, K. (2014) Combinatorial effect of rolling and carbonaceous nanoparticles on the evolution of crystallographic texture and structural properties of ultra high molecular weight polyethylene. *Physical Chemistry Chemical Physics*, **16**, 23108-23117.
149. Lahiri, D., Dua, R., Zhang, C., de Socarraz-Novoa, I., Bhat, A., Ramaswamy, S., and Agarwal, A. (2012) Graphene nanoplatelet-induced strengthening of ultrahigh molecular weight polyethylene and biocompatibility in vitro. *ACS Applied Materials & Interfaces*, **4**, 2234-2241.
150. Lahiri, D., Hec, F., Thiesse, M., Durygin, A., Zhang, C., and Agarwal, A. (2014) Nanotribological behavior of graphene nanoplatelet reinforced ultra high molecular weight polyethylene composites. *Tribology International*, **70**, 165-169.
151. Xu, L., Zheng, Y., Yan, Z., Zhang, W., Shi, J., Zhou, F., Zhang, X., Wang, J., Zhang, J., and Liu, B. (2016) Preparation, tribological properties and biocompatibility of fluorinated graphene/ultrahigh molecular weight polyethylene composite materials. *Applied Surface Science*, **370**, 201-208.
152. Suner, S., and Emami, N. (2014) Investigation of graphene oxide as reinforcement for orthopaedic applications. *Tribology-Materials, Surfaces & Interfaces*, **8**, 1-6.
153. Suner, S., Joffe, R., Tipper, J., and Emami, N. (2015) Ultra high molecular weight polyethylene/graphene oxide nanocomposites: Thermal, mechanical and wettability characterisation. *Composites, Part B: Engineering*, **78**, 185-191.
154. Tai, Z., Chen, Y., An, Y., Yan, X., and Xue, Q. (2012) Tribological behavior of UHMWPE reinforced with graphene oxide nanosheets. *Tribology Letters*, **46**, 55-63.
155. Chen, Y., Qi, Y., Tai, Z., Yan, X., Zhu, F., and Xue, Q. (2012) Preparation, mechanical properties and biocompatibility of graphene oxide/ultrahigh molecular weight polyethylene composites. *European Polymer Journal*, **48**, 1026-1033.
156. Mirsalehi, S. A., Khavandi, A., Mirdamadi, S., Naimi-Jamal, M. R., and Kalantari, S. M. (2015) Nanomechanical and tribological behavior

of hydroxyapatite reinforced ultrahigh molecular weight polyethylene nanocomposites for biomedical applications. *Journal of Applied Polymer Science*, 132, doi:10.1002/ app.42052.

157. Shi, X., Bin, Y., Hou, D., and Matsuo, M. (2013) Surface characterization for ultrahigh molecular weight polyethylene/hydroxyapatite gradient composites prepared by the gelation/crystallization method. *ACS Applied Materials & Interfaces*, **5**, 1768-1780.
158. Maksimkin, A., Kaloshkin, S., Tcherdyntsev, V., Senatov, F., and Danilov, V. (2012) Structure and properties of ultra-high molecular weight polyethylene filled with disperse hydroxyapatite. *Inorganic Materials: Applied Research*, **3**, 288-295.
159. Kang, X., Zhang, W., and Yang, C. (2016) Mechanical properties study of micro-and nano-hydroxyapatite reinforced ultrahigh molecular weight polyethylene composites. *Journal of Applied Polymer Science*, 133, doi:10.1002/app.42869.
160. Panin, S. V., Kornienko, L., Chaikina, M., Sergeev, V. P., Ivanova, L., and Shilko, S. (2014) Nano-and micro-structured UHMWPE composites filled with hydroxyapatite irradiated by nitrogen ion beams for bio-medical applications. *Russian Physics Journal*, **56**, 1137-1143.
161. Senatov, F., Kopylov, A., Anisimova, N. Y., Kiselevsky, M., and Maksimkin, A. (2015) UHMWPE-based nanocomposite as a material for damaged cartilage replacement. *Materials Science and Engineering C*, **48**, 566-571.
162. Senatov, F., Gorshenkov, M., Kaloshkin, S., Tcherdyntsev, V., Anisimova, N. Y., Kopylov, A., and Kiselevsky, M. (2014) Biocompatible polymer composites based on ultrahigh molecular weight polyethylene perspective for cartilage defects replacement. *Journal of Alloys and Compounds*, **586**, S544-S547.
163. Wang, S., Feng, Q., Sun, J., Gao, F., Fan, W., Zhang, Z., Li, X., and Jiang, X. (2015) Nanocrystalline cellulose improves the biocompatibility and reduces the wear debris of ultrahigh molecular weight polyethylene via weak binding. *ACS Nano*, **10**, 298-306.
164. Chang, B. P., Akil, H. M., Nasir, R. B. M., Bandara, I., and Rajapakse, S. (2014) The effect of ZnO nanoparticles on the mechanical, tribological and antibacterial properties of ultra-high molecular weight polyethylene. *Journal of Reinforced Plastics and Composites*, **33**, 674-686.
165. Wang, H., Xu, L., Zhang, M., Li, R., Xing, Z., Hu, J., Wang, M., and Wu, G. (2017) More wear-resistant and ductile UHMWPE composite prepared by the addition of radiation crosslinked UHMWPE powder. *Journal of Applied Polymer Science*, **134**, doi:10.1002/app.44643.
166. Gong, K., Qu, S., Liu, Y., Wang, J., Zhang, Y., Jiang, C., and Shen, R. (2016) The mechanical and tribological properties of UHMWPE

loaded ALN after mechanical activation for joint replacements. *Journal of the Mechanical Behavior of Biomedical Materials*, **61**, 334-344.

167. Murakami, T., Sakai, N., Yamaguchi, T., Yarimitsu, S., Nakashima, K., Sawae, Y., and Suzuki, A. (2015) Evaluation of a superior lubrication mechanism with biphasic hydrogels for artificial cartilage. *Tribology International*, **89**, 19-26.
168. Yarimitsu, S., Sasaki, S., Murakami, T., and Suzuki, A. (2016) Evaluation of lubrication properties of hydrogel artificial cartilage materials for joint prosthesis. *Biosurface and Biotribology*, **2**, 40-47.
169. Murakami, T., Yarimitsu, S., Sakai, N., Nakashima, K., Yamaguchi, T., Sawae, Y., and Suzuki, A. (2017) Superior lubrication mechanism in poly (vinyl alcohol) hybrid gel as artificial cartilage. *Proceedings of the Institution of Mechanical Engineers, Part J: Journal of Engineering Tribology*, **231**(9), 1160-1170.
170. Osaheni, A. O., Finkelstein, E. B., Mather, P. T., and Blum, M. M. (2016) Synthesis and characterization of a zwitterionic hydrogel blend with low coefficient of friction. *Acta Biomaterialia*, **46**, 245-255.
171. Shi, Y., and Xiong, D. (2013) Microstructure and friction properties of PVA/PVP hydrogels for articular cartilage repair as function of polymerization degree and polymer concentration. *Wear*, **305**, 280-285.
172. Li, W., Wang, D., Yang, W., and Song, Y. (2016) Compressive mechanical properties and microstructure of PVA–HA hydrogels for cartilage repair. *RSCAdvances*, **6**, 20166-20172.
173. Chen, K., Zhang, D., Cui, X., and Wang, Q. (2015) Research on swing friction lubrication mechanisms and the fluid load support characteristics of PVA–HA composite hydrogel. *Tribology International*, **90**, 412-419.
174. Chen, K., Zhang, D., Dai, Z., Wang, S., and Ge, S. (2014) Research on the interstitial fluid load support characteristics and start-up friction mechanisms of PVA-HA-silk composite hydrogel. *Journal of Bionic Engineering*, **11**, 378-388.
175. Jing, L., Li, H., Tay, R. Y., Sun, B., Tsang, S. H., Cometto, O., Lin, J., Teo, E. H. T., and Tok, A. I. Y. (2017) Biocompatible hydroxylated boron nitride nanosheets/poly(vinyl alcohol) interpenetrating hydrogels with enhanced mechanical and thermal responses. *ACS Nano*, **11**, 3742-3751.
176. Chen, J., Shi, X., Ren, L., and Wang, Y. (2017) Graphene oxide/PVA inorganic/organic interpenetrating hydrogels with excellent mechanical properties and biocompatibility. *Carbon*, **111**, 18-27.

5

Recent Advances in Nanocarbon-based Polymer Composite Materials as Membranes in Fuel Cell Applications

5.1 Introduction

The most significant environmental hazards in the world today is the air pollution from the internal combustion engines. The environmental need for clean and efficient energy has paved the way to the development of fuel cells. Fuel cells currently attract significant research attention because of their advantages compared to other energy producing devices such as saving in conventional fuels such as oil or natural gas, higher energy conversion efficiency than diesel or gas engines, no moving parts, lack of sonic pollution compared to internal combustion engines, no atmospheric polluting emissions of NO_x, SO_x, CO, CO_2, etc. Fuel cells generate electrical energy by an electrochemical reaction in which the fuel (e.g. H_2, ethanol, or methanol) combines with oxygen (from air) to produce water, heat and electricity [1].

Fuel cells can be categorized based on different parameters which include the electrolyte type employed in them, fuel cell structure, type of reactants, operating conditions, etc. Fuel cells are of five types, based on the nature of electrolyte used in them; i) alkaline fuel cells (AFC) using alkaline electrolyte, ii) phosphoric acid fuel cells (PAFC) using phosphoric acid electrolyte, iii) proton exchange membrane fuel cells (PEMFCs) using proton exchange membrane electrolyte, iv) molten carbonate fuel cells (MOFC) using molten carbonate salt electrolyte, and v) solid oxide fuel cells (SOFC) using solid oxide electrolyte [2]. Among these, PEMFCs have merits like higher energy efficiency, quiet operation and environmental friendliness for stationary, mobile and portable applications [3]. Direct methanol fuel cells (DMFC) are the most suited energy sources for

Seba Sara Varghese[a,b], Saino Hanna Varghese[c], Sundaram Swaminathan[d], Krishna Kumar Singh[b] and Vikas Mittal[a]
[a]Khalifa University of Science and Technology, Abu Dhabi, UAE; [b]Birla Institute of Technology and Science, Dubai, UAE; [c]Catholicate College, India; [d]DIT University, India

transportation purposes because of the use of inexpensive and portable fuel, simple structural design, higher energy density per unit volume as compared to PEMFCs at low operating temperatures, and simple operation [4-6].

The proton exchange membrane (PEM) is the indispensable part of the PEMFCs and DMFCs, which acts as the electrolyte that transports protons from anode to the cathode and effectively divides the anode and cathode reactants [7]. The materials used as PEMs should have high proton conduction, low fuel or water crossover, low electronic conduction, moderate cost, durability, high thermal and mechanical stability above 120 °C, good mechanical strength, and high hydrolytic and oxidative stability [8].

The currently used polymer electrolyte membranes (PEMs) are perfluorosulfonated (PFSA) membranes, as they possess excellent proton conducting property, poor electrical conduction, good mechanical, physical, chemical and thermal stabilities at moderate temperatures [9]. Among various PFSA polymers, Nafion membrane is the most commonly used PEM for both PEMFCs and DMFCs. The hydrophilic side chains of the Nafion membrane with sulfonic acid functional groups are responsible for its remarkable proton conduction property, as the proton conductive clusters formed by the sulfonic acid groups contribute to the proton transport through the Nafion membrane [10]. For instance, Figure 5.1 demonstrates 2-D SAXS scattering patterns of as-received Nafion, showing a slight

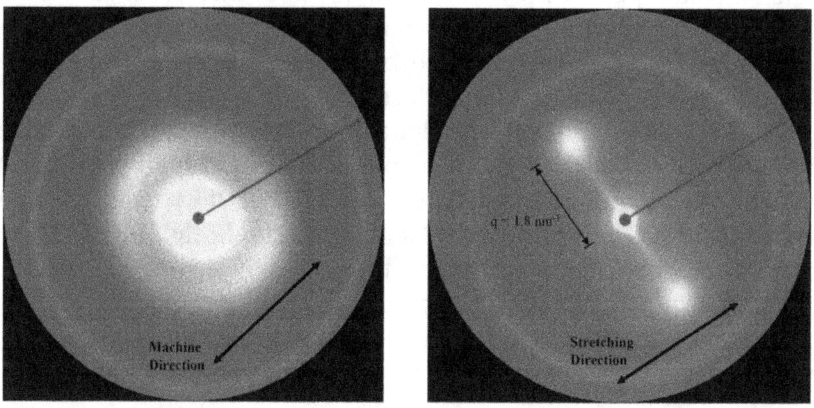

(a) (b)

Figure 5.1 2-D SAXS scattering patterns of (a) as-received Nafion, and (b) uniaxially oriented Nafion (λ_b = 5.4). Reproduced from Reference 10 with permission from American Chemical Society.

morphological anisotropy in the ionic domains from the membrane calendaring process in comparison with uniaxially oriented Nafion [10]. However, high cost, high fuel gas/liquid permeability, water dependence for proton conduction, low proton conductivity above 100 °C and under low humidity, low thermal and mechanical stability above 120 °C, limit the application of Nafion [11]. These problems need to be resolved for enabling practical applications of the fuel cells, as the properties of PEMs determine the performance of the PEMFCs and DMFCs. Alternative low cost membranes such as modified PFSA polymers, sulfonated aromatic hydrocarbon based polymers, acid-functionalized aromatic hydrocarbon polymers, etc., were also proposed as potential PEMs in PEMFCs and DMFCs.

PEMFCs operating at high temperature (>100 °C) have shown significantly enhanced performance as high temperature operation improves the CO tolerance of the electrocatalysts, enhances electrode reaction rates, reduces the need for excess precious Pt catalysts and simplifies water management systems of PEMFCs. So polymer electrolytes having desirable transport properties and good thermo-mechanical stability at elevated temperatures for PEMFCs have received great scientific attention. The limited capability of Nafion membrane at high temperature makes it unsuited for high temperature applications because of the dramatic reduction of water in these membranes at temperatures above 100 °C. Hence research has also centered on the search for polymer membranes suitable for usage at elevated temperature and under reduced humidity.

The high methanol crossover of Nafion membrane from the anode to the cathode is a serious problem in DMFCs, as it causes fuel wastage, catalyst poisoning and reduces the efficiency and the cell performance. Several studies have also been performed to lower the methanol permeability through Nafion while preserving the good proton conductivity. The slow oxidation kinetics of methanol below 100 °C and significant methanol permeation through PEM in DMFCs results in low energy efficiencies of ~20-25% [12-17]. Hence, many research studies have been directed on boosting the energy efficiency of DMFCs for enabling portable power applications.

The development of desirable PEMs for fuel cells has been one of the hottest topics for more than 40 years. The high thermal stability, low fuel crossover, high CO tolerance, zero water-osmotic drag and excellent mechanical properties of polybenzimidazole (PBI)-based membranes make them attractive as promising electrolyte materials for high temperature applications [18]. PBI-based membranes can achieve desired proton conductivities only by the use of dopants such as sulfuric or phosphoric acid (PA). Previous experimental researches have shown that the characteristics of PBI-based PEMs could be improved by suitable modifications such as sulfonation, acid-base blend, and incorporation of fillers or additives [19]. PEMFCs and DMFCs with Nafion-based membranes also need some modifications to lower the methanol crossover, as well as to make these highly proton conductive and mechanically strong. The modification of polymer membranes with nano-fillers has demonstrated to be one of the promising approaches for improving the membrane's properties and for achieving desired fuel cell performance.

To satisfy the needs of good mechanical strength, proton transport property and reduced methanol crossover for PEMs under fuel cell operating conditions, extensive research efforts were carried out to modify Nafion and other polymer membranes with nanocarbon based materials such as carbon nanotubes (CNTs) and graphene oxide (GO). Various polymer composite membranes based on CNTs and GO have been prepared for PEMs in fuel cells. The nanocarbon-modified polymer membranes have exhibited superior features like excellent mechanical properties and resistance to methanol crossover, while preserving the proton conductivity as high as possible, compared to unmodified polymer membranes and, thus, remarkably improved fuel cell performance. This chapter reviews the state-of-the-art research on composite membranes based on polymers and nanocarbon materials such as CNTs, graphene oxide (GO) used as PEMs for fuel cells. Firstly, polymer composite membranes based on CNTs and functionalized CNTs are discussed. The next section focuses on recent reports that describe the modification of widely used polymer membranes with GO and functionalized GO. The chapter concludes with future prospects of nanocarbon-based polymer composite PEMs for applications in fuel cells.

5.2 Types of Polymer/Nanocarbon Composite Membranes

5.2.1 Polymer/CNT Composite Membranes

CNTs, the one-dimensional carbon allotrope with cylindrical nanostructure [20], have demonstrated great utility in applications such as electronics, sensors, hydrogen storage, gas adsorption, biomedicine, spintronics, energy storage, photonics, etc. [21-26], because of advanced characteristics like high aspect ratio, large specific surface area, excellent thermal and electrical conductivity, good mechanical strength and nanometer scale size. CNTs have also attracted particular scientific attention in fuel cell applications as materials for electrode supports and as catalyst supports [27-30]. Figure 5.2 demonstrates the schematic of the use of multi-walled carbon nanotubes as a platinum support for proton exchange membrane fuel cells [27]. Various research studies have revealed that the incorporation of CNTs could significantly improve the mechanical performance and the electrical conductivity of polymer matrices, hence, CNTs have also been used as nanofillers for reinforcement of polymers [31-34]. The use of CNTs as filler in polymer membranes is limited by the short-circuiting problem in fuel cells arising from the high electrical conductivity of CNTs [35, 36].

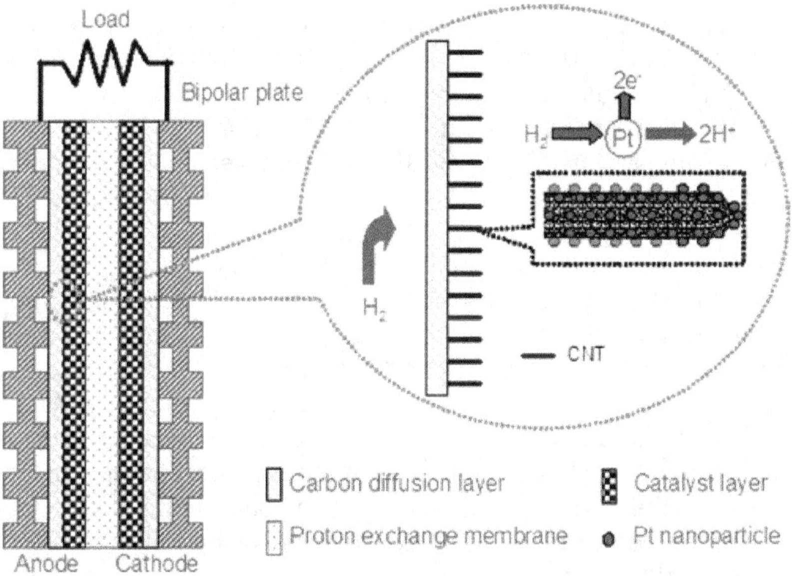

Figure 5.2 Schematic of the use of multi-walled carbon nanotubes as a platinum support for proton exchange membrane fuel cells. Reproduced from Reference 27 with permission from American Chemical Society.

Jana and Bhunia [37] prepared silane based nanostructured composite membranes having different proportion of aqueous orthophosphoric acid, nano-clay (Cloisite® 30B) and multi-walled carbon nanotubes (MWCNTs) by sol–gel method and investigated the dependence of thermal stability and proton conductivity of the composite membrane on the content of orthophosphoric acid, nano-clay and MWCNT in the composites. The C–O–C and Si–O–Si bonds between the silanes resulted in improved thermal stability of the composite membranes up to 200 °C. The thermal stability was observed to decrease with the increase in phosphoric acid content, whereas an increase in thermal stability was observed in the case of increased quantity of nano-clay and nanotubes in the membranes. At 80 °C, 100 °C, 120 °C and 140 °C and at 30% relative humidity (RH), the measured proton conductivity of the membranes increased with increase in weight proportion of nano-clay and MWCNTs in the composite membranes, which could be attributed to the presence of chemisorbed water in the composites. The uniform distribution of the nanoparticles throughout the polymer matrix was evident from the phase morphology study [37].

The use of Nafion/CNT as proton-conducting membrane for FCs has been investigated by several research groups. Liu *et al.* [35] demonstrated that the risk of short-circuit in Nafion/CNTs composites could be avoided by keeping the CNTs content lower than the percolation threshold. The authors developed MWCNTs reinforced Nafion composite membrane for the H_2O_2 fuel cell using ball-milling and solution-casting method. The as-prepared Nafion/CNTs composite membrane had uniform dispersion of MWCNTs within the Nafion polymer matrix. The addition of 1 wt% of CNTs resulted in excellent mechanical strength of the composite membrane. The composite membrane exhibited decreased dimensional change compared to the commercial Nafion NRE-212 membrane. However, the performance of the CNTs reinforced Nafion and the commercial Nafion NRE-212 membrane in fuel cell tests was almost similar [35], which revealed that the addition of CNTs did not improve the proton conductivity of the Nafion membrane. Also, the ball-milling method followed by solution-casting used for the preparation of Nafion/CNTs composite membrane is expensive and not suitable for mass production [36,38].

Thomassin and coworkers [36] used melt mixing technique for MWCNT modification of Nafion membranes. Since no electrical conductivity was observed in the Nafion membrane for MWCNT content lower than 2 wt%, the short circuit risk was minimized. The improved dispersion of MWCNTs within the Nafion membranes resulted in drastically reduction in methanol permeability of about 60%, with only a slight reduction in the ionic conductivity of the Nafion membrane. The ionic conductivity of Nafion membranes was later enhanced by the functionalization of MWCNTs by carboxylic acid groups. The Nafion membrane with 2 wt% of modified-MWCNT content exhibited a 160% increase in Young's modulus as compared to pure Nafion membrane [36].

Wang *et al.* [38] used a more convenient and effective solution-cast method for the preparation of MWCNTs reinforced Nafion membrane. MWCNTs were oxidized by H_2O_2 and subsequently an improved solution-casting method by sodium hydroxide (NaOH) addition in the MWCNTs/Nafion/*N,N*-dimethylacetamide (DMAC) solution was used to prepare MWCNTs reinforced Nafion membranes. The addition of NaOH and the oxidation of MWCNTs greatly improved the dispersion quality of MWCNTs in Nafion/DMAC solution. The observed better dispersion of oxidized MWCNTs in the Nafion membrane could be attributed to the interactions of –OH with –COOH groups of the oxidized MWCNTs and the sodium sulfonate groups of the Nafion. The incorporation of MWCNTs improved the tensile strength of the reinforced membrane by 54%. The polarization curves of the fuel cells with NaOH-MWCNTs/Nafion reinforced membranes with different MWCNTs contents (from 1 to 4 wt%) measured at 80 °C and under fully humidified conditions indicated slightly better fuel cell performance of reinforced membranes than commercial NRE-212 membrane. The increase in the MWCNTs content resulted in reduced fuel cell performance owing to the minor reduction in the proton conductivity of the membrane by substituting polymer with MWCNTs. The authors concluded from the analysis of the comparison of the mechanical properties, proton conductivities and cell performance of the composite membranes that the content of MWCNTs in NaOH-MWCNTs/Nafion reinforced membrane should not exceed 3 wt% [38].

Several studies have focused on the modification of aromatic hydrocarbon based polymers using CNTs for membranes in fuel cells. Kang *et al.* [39] synthesized high molecular-weight poly(2,5-benzimidazole) (ABPBI)/CNT composite membranes via *in situ* polymerization of the AB-monomer in the presence of SWCNT or MWCNT in a mildly acidic polyphosphoric acid. The authors observed uniform dispersion of SWCNTs and MWCNTs throughout the PBI polymer matrix. The incorporation of CNTs into the ABPBI polymer matrix significantly improved the tensile properties of the ABPBI polymer. The composite membranes exhibited toughness of ~200 MPa which was close to that of the nature's toughest spider silk, ~215 MPa. The electrical conductivities of ABPBI films improved by ~19 and 52,700 times by the modification with SWCNT and MWCNT, respectively. The composite membranes of ABPBI with SWCNT and MWCNT without acid impregnation exhibited proton conducting nature with maximum conductivities of 0.018 and 0.017 Scm^{-1} respectively at 130 °C and 50% RH [39].

Guerrero Moreno *et al.* [40] evaluated the composite membranes of PBI and non-functionalized MWCNTs (PBI-CNT) as membrane electrolytes in PEMFC. At 180 °C, the proton conductivities of the PBI doped with H_3PO_4 (PBIPA) and the doped PBI-CNT (PBICNTPA) membranes were measured to be 6.3×10^{-2} and 7.4×10^{-2} Scm^{-1} respectively. The better phosphoric acid retention in the PBICNTPA composite membrane due to the incorporation of CNTs in PBI resulted in increased proton conductivity of the PBIPA membrane. At 180 °C, the single fuel cell having Pt catalyzed hydrogen fed anode and a similar oxygen cathode using PBICNTPA as membrane electrolyte without feed gas humidification showed higher open circuit voltage (OCV) (0.96 versus 0.8 V) and higher maximum power density (174.5 versus 153 $mWcm^{-2}$) than that using a PBIPA membrane, respectively. It was also observed that the OCV of the composite membrane did not change in a second cell test, whereas the OCV of the PBIPA membrane dropped. The polarization curves of fuel cells using PBIPA and PBICNTPA membranes at varying temperatures suggested enhanced fuel cell performance with rise in temperature for both PBIPA and PBICNTPA membranes. The comparison of the fuel cell performance using PBICNTPA membrane and PBI only membrane indicates that the use of CNTs reduces gas permeation through the composite membrane and minimizes power losses in fuel cells. On comparing the observed cell performance with that from previous works up to 300 mA cm^{-2}, it was observed that PBICNT membrane exhibited superior cell performance [41-43]. The thermal oxidative stability and the spatial thermal distribution of the PBIPA membrane also enhanced with the incorporation of CNTs owing to the enhanced thermal properties of the composite membrane. PBICNTPA membrane with 1 wt% CNT exhibited 32% greater tensile strength and 147% greater Young's modulus compared to the corresponding values for a pristine PBI membrane. The mechanical properties of PBIPA and PBICNTPA membranes compared to earlier reports revealed that the values of tensile strength and Young's Modulus for these membranes were comparable or greater than the reported values [41-46]. The enhancement of the mechanical stability and the proton conductivity of the PBI membrane by the incorporation of CNTs indicate the potential of PBI-CNT membrane as electrolyte in PEMFCs [40].

5.2.2 Polymer/Functionalized CNTs Composite Membranes

The dispersion of CNTs in polymer membranes is crucial for obtaining improved membrane properties. The functionalization of CNTs has proved to be an effective approach in the preparation of polymer/CNT composites for improving the dispersion ability of CNTs within the polymer matrix. The presence of CNTs may result in the disruption of the separation of hydrophilic and hydrophobic polymer phases, thus, leading to very little effect on proton conductivity, however, this could be resolved by introducing the sulfonic acid groups on CNTs.

Kannan *et al.* [47] observed improved conductivity of protons in Nafion-based membranes through the incorporation of SWCNTs pre-functionalized with sulfonic acid groups (s-SWCNTs) into the Nafion matrix. The sulfonic acid groups in the SWCNT connect the hydrophobic regions of the membrane and, thus, form a channel-like network for proton transport through the membrane, which thereby increases its proton conductivity. The s-SWCNT-functionalized Nafion membrane exhibited superior fuel cell performance and mechanical stability to Nafion 1135 in H_2/O_2 fuel cells [47]. The same group [48] modified Nafion membranes using s-MWCNTs so as to manipulate the hydrophilic domain size of Nafion membranes for improving the proton conductivity. The hydrophilic domain size showed remarkable increase with increase in s-MWCNT loading from 0.01% to 0.05% and decreased further at CNT loadings higher than 0.05%. The authors reported that the increased ionic cluster size in s-MWCNT-modified Nafion membrane increased the proton conductivity. The single cell measurements of the Nafion S-MWCNT composite 0.05% membranes with maximum ionic cluster size of ~70 Å showed maximum power density of 380 mW cm^{-2}, which was greater than those of recast Nafion (230 mW cm^{-2}) and Nafion 115 with ionic cluster size of ~50 Å (250 mW cm^{-2}) membranes. Figure 5.3 also shows the single-cell polarization plots obtained with membrane electrode assemble (MEAs) using commercial Nafion, Nafion recast, and composite membranes at 70 °C along with that of the 0.05% Nafion composite with unfunctionalized pristine MWCNT (NapM 0.05%) [48]. These measurements indicated that nanotube-tailored Nafion membranes enhanced the performance of the H_2/O_2 fuel cell.

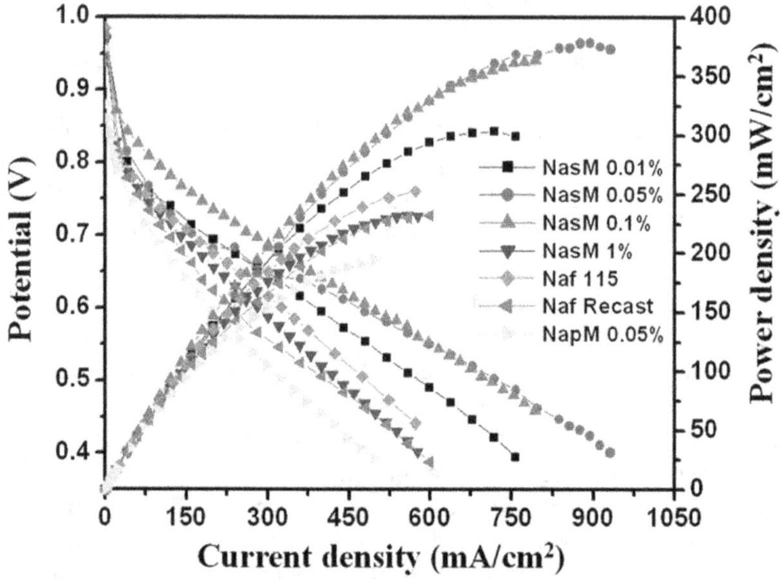

Figure 5.3 Fuel cell single-cell polarization plots of MEAs with Nafion and composite membranes at 70 °C. Reproduced from Reference 48 with permission from American Chemical Society.

Cele *et al.* [49] investigated the proton conductivity, thermal and mechanical stability of CNT based Nafion composite membranes using pure CNTs (pCNTs), oxidized CNTs (oCNTs) and amine functionalized CNTs (fCNTs) prepared by melt-mixing-compression-moulding process to analyze the influence of surface oxidation and functionalization of CNTs in Nafion membranes. The addition of 1 wt% of CNTs improved the thermal stability of all composite membranes. Composite membranes containing 1 wt% of oCNTs showed drastic improvements in thermo-mechanical stability than that of pure Nafion and other CNT-containing Nafion composite membranes, due to high compatibility of oCNT outer surfaces with Nafion matrix. The DC electrical conductivity of the CNT-containing Nafion composite membrane improved with the use of pCNT filler in Nafion, whereas the proton conductivity decreased with the addition of CNTs. Therefore, the authors suggested the need for reduction of filler amount to maintain low electrical conductivity and to maximize the proton conductivity of Nafion membrane. Out of the various

CNT-based Nafion composite membranes, oCNTs-containing Nafion composite membrane was claimed as the best polymer electrolyte membrane for fuel cells, as the surface oxidation of CNTs in Nafion membranes improved the properties of pure Nafion, compared to that of other Nafion composite membranes with pure CNTs and amine functionalized CNTs [49].

Functionalized CNTs/Nafion nanocomposite membranes synthesized by surfactant-assisted method exhibited highly enhanced dispersion of CNTs in the nanocomposite membranes [50]. The reinforcement of 1 wt% sulfonated CNTs (s-CNTs) and oxidized CNTs (ox-CNTs) into Nafion membrane by solution casting method lowered the methanol uptake and enhanced the thermal property of cast Nafion, while maintaining the ion exchange capacity. These results indicate that s-CNTs or ox-CNTs/Nafion nanocomposite membranes prepared by surfactant-assisted method could be used as PEMs for fuel cells. Wu *et al.* [46] observed improved ionic conductivity and fuel cell performance in KOH-doped PBI-CNT composite membranes by adding <1% of functionalized MWCNTs via solution casting method, revealing the potential of PBI-CNT composite membranes in high temperature alkaline direct methanol fuel cell (ADMFC) applications. Single cell evaluation using 6 M KOH as anode fuel and humidified oxygen as cathode oxidant generated the maximum power density of 104.7 mW cm^{-2} at 90 °C under 2 M methanol feed concentration.

A self-humidifying composite membrane with a layer of Pt/CNTs dispersed Nafion sandwiched between two outer layers of plain Nafion was developed by Liu *et al.* [51] using solution casting method. The Pt/CNTs were observed to be uniformly dispersed in the composite membrane. The N-Pt/CNTs-N membrane-based PEMFC using dry and humidified H_2 and O_2 reactants exhibited better performance compared to N-CNTs-N and Nafion NRE-212 membrane. With dry reactants, PEMFC containing N-Pt/CNTs-N membrane exhibited nearly 90% of the performance achieved with humidified reactants whereas the Nafion NRE-212 membrane exhibited only 72% of the performance displayed with humidified reactants. The superior performance of the composite membrane could be attributed to the catalytic recombination of the reactants permeating from the anode to the cathode which resulted in the production of water. In addition to this, the Pt/CNTs in the composite membrane could improve the mechanical strength of the Nafion membrane. The short circuiting issue of the electrically conductive Pt/CNTs was avoided by the two plain Nafion layers on the outer surface. The authors claimed that the reinforced composite membrane could overcome the limitations associated with Nafion/CNTs membrane by incorporating the self-humidifying layer in the membrane [51].

The polymer-functionalized CNTs are also reported to have good compatibility with the polymer matrix so as to avoid any possible micro-phase separation in the nanocomposites [52-54]. Several research studies have reported the use of polymer-functionalized CNTs as membrane electrolytes for FCs. For instance, Chen *et al.* [55] synthesized proton-conducting membrane by blending polysiloxane-functionalized MWCNTs with Nafion. The polysiloxane-functionalized MWCNTs were synthesized by the authors by covalently grafting hydrophilic layers composed of poly(oxyalkylene)diamines (EO) and tetraethyl orthosilicate-reinforced polysiloxane in a layer-by-layer manner onto the nanotube walls. The electron conduction between the CNTs was effectively prohibited by the polysiloxane layer on the CNT sidewalls which acted as an electrical resistance layer for MWCNTs. The addition of polysiloxane layer onto the CNT sidewalls allowed uninterrupted pathways for fast proton conduction (2.8×10^{-2} S cm^{-1} at 30 °C) through Nafion membrane *via* the ionic interactions between the amino groups of the CNTs and the sulfonated groups of Nafion. The fast proton conduction was maintained even at high temperatures (6.3×10^{-2} S cm^{-1} at 130 °C). These results suggested the potential use of functionalized CNT/Nafion nanocomposites as proton exchange electrolyte materials for high temperature fuel cell applications [55]. Figure 5.4 shows the TEM images of the pristine CNTs and CNT-EO. The polymer shell around the individual nanotubes was observed to be thin and uniform, thus, confirming uniform surface modification.

Liu *et al.* [56] reported ozone-mediated synthesis of PEMs using Nafion-functionalized MWCNTs (MWCNT-Nafion), in an effort to impart the high sulfonic acid content of Nafion to CNTs. The authors observed high compatibility of MWCNT-Nafion nanohybrid with the Nafion matrix through the Nafion chains attached to the MWNTs. At 60 °C temperature, the resulting composite membrane with 0.05 wt% MWCNT-Nafion loading (N/MN-0.05) exhibited an increase in mechanical strength and proton conductivity of about 1.5 times and 5 times as compared to pristine Nafion. The current density at 0.6 V and the maximum power density were observed to be 1556 mA cm^{-2} and 650 mW cm^{-2}, respectively, for N/MN-0.05 composite membrane, which are 1.5 times greater than those observed using pristine Nafion membrane. The high proton conductivity of Nafion/MWCNT-Nafion composite membrane consequently yielded high performance for PEMFC application. The superior thermal stability,

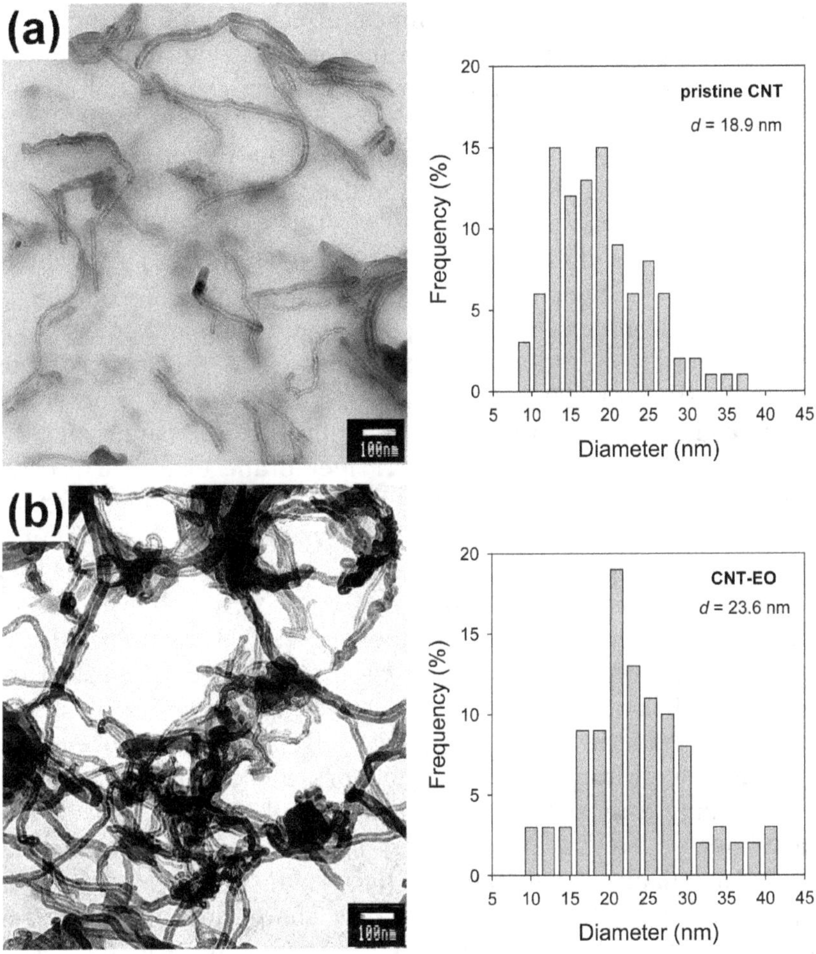

Figure 5.4 TEM images and average size distribution of (a) pristine CNTs and (b) modified CNTs CNT-EO. Reproduced from Reference 55 with permission from American Chemical Society.

improved mechanical strength and good single cell performance of the N/MN-0.05 composite membrane indicated that Nafion/MWCNT-Nafion composite membranes are promising candidates as PEMs in fuel cell applications [56]. In another study,

Hasani-Sadrabadi *et al.* [57] fabricated nanocomposite membranes by incorporating polybenzimidazole-decorated CNT (PBI-CNT) doped with phosphotungstic acid (PWA) into the Nafion matrix. The PWA-doped PBI-decorated CNT/Nafion composite membrane exhibited good proton conductivity at elevated temperatures and low humidity. Moreover, the nanocomposite membrane had improved power generation capacity of 386 mW cm^{-2}, at low humidity and high temperatures than the commercial Nafion commercial Nafion 117 membrane of 73 mW cm^{-2} [57].

Composite membranes of functionalized CNTs with other polymer electrolytes have also been reported. Kannan *et al.* [58] described the ability of phosphonated MWCNTs (P-MWCNTs) in tuning the properties of PBI-based membrane electrolytes for FC applications. Single-cell evaluation of composite membrane exhibited a 50% and 40% enhancement in proton conductivity and power density compared to a PA-doped PBI membrane. The fuel cell maximum power densities of 780 mW cm^{-2} and 600 mW cm^{-2} were observed for the composite and pristine PBI membrane under similar conditions. The PA-doped PBI/P-MWCNT membrane exhibited greatly improved ultimate strength compared to PA-doped PBI membrane because of the added MWCNTs in the polymer matrix. The high proton conductivity and good mechanical stability indicate the potential of phosphonic acid-functionalized MWCNT based hybrid membranes as polymer electrolytes for fuel cells [58]. In an earlier study, Kannan *et al.* [42] also employed composite membrane based on PA-doped PBI and P-MWCNTs for high temperature PEMFCs. The improvement in the proton conductivity was mainly because of the formation of proton conducting network in the composite membrane arising from the increase in the net phosphonic acid by the phosphonated groups on MWCNTs. Thus, the incorporation of P-MWCNTs enhanced the fuel cell performance of P-MWCNT/PA-doped PBI composite membranes. The composite membrane reduced the activation energy of the oxygen reduction reaction (ORR). The proton conductivity, stability and overall performance of composite membranes were observed to be greatly enhanced compared with PBI membranes. The optimum P-MWCNT doping level was reported to be 1 wt%. The acid leaching from the polymer matrix was found to be reduced by the use of composite membranes [42].

Composite membranes consisting of CNTs modified with sulfonic acid or PtRu nanoparticles and sulfonated poly(arylene sulfone) (sPAS) were synthesized by Joo *et al.* [59] for DMFC by solution casting. The functionalized CNTs were uniformly dispersed within the sPAS matrix. The obtained SO_3CNT-sPAS or PtRu/CNT-sPAS composite membranes exhibited improved tensile strength and toughness compared to the pure sPAS membrane. The addition of functionalized CNTs to sPAS matrix resulted in improved ionic conductivity and reduced methanol permeability compared with a sPAS membrane. This was due to the fact that sPAS/CNT composite membranes contained smaller and uniform ion clusters compared to the sPAS membrane. The single cell DMFC tests demonstrated that PtRu/CNT-sPAS membrane resulted in low ohmic resistance and high OCV compared with the neat sPAS membrane, which ultimately leads to high power density [59]. In another study, Tripathi *et al.* [60] developed nanocomposites based on sulfonated poly(ether ether ketone) (SPEEK) polymer and covalently functionalized MWCNTs obtained by grafting MWCNTs with sulfonic acid groups. The authors investigated the influence of f-MWCNT content on membrane properties for enabling hydrogen or alcohol fuel cell applications. The introduction of f-MWCNTs in the SPEEK membrane with 5 to 10 wt% content resulted improvement of proton conductivity, a jump in electrical conductivity, an increase in mechanical stability, a decrease in swelling, a decrease in water permeability down to 1/3rd of that of plain SPEEK membrane and also a decrease in methanol permeability by 1/20th of that of plain SPEEK membrane. The increase in the bound water with increasing content of f-MWCNTs in membrane favored the operation of hydrogen fuel cells above 100 °C and under low external humidity conditions [60].

Suryani *et al.* [41] reported ozone mediated synthesis of PBI- and Nafion-functionalized MWCNTs (MWCNT- PBI and MWCNT-Nafion) and investigated the effect of different functionalization of MWCNTs on membrane properties. Both MWCNT-PBI and MWCNT-Nafion were used as additives in the preparation of nanocomposite membranes for PEMFC. The high compatibility between PBI matrix and functionalized-MWCNTs was evident from the morphological analysis of the PBI/MWCNT nanocomposite membranes. The thermal and mechanical properties of the nanocomposite membranes were observed to be enhanced by the addition of the functionalized MWCNTs. The authors observed the formation of proton conducting pathways in composite membranes, and significant increase in the proton conductivities of the membranes by PBI functionalization of MWCNT. The PA-doped PBI/MWCNT nanocomposite membranes containing 0.2 wt% of MWCNT–Nafion (N-0.2) and 0.2 wt%

MWCNT–PBI (P-0.2) exhibited proton conductivities of 0.05 and 0.08 S cm^{-1} at 160 °C under anhydrous conditions, as compared to the proton conductivity of the PA-doped PBI membrane (0.03 S cm^{-1}). At 150 °C, maximum power densities of 700 and 600 mW cm^{-2} were observed for N-0.2 and P-0.2, respectively, which were greater than that measured with pristine PBI (530 mW cm^{-2}). The composite membranes exhibited improved proton conductivities and superior fuel cell performance compared to pristine PBI membrane. Nafion functionalization of MWCNTs produced ionically cross-linked structures in the nanocomposite membranes, which enhanced the compatibility between PEM and electrocatalyst, and ultimately led to efficient fuel cell performance due to the improved interfacial contact of membrane electrode assembly [41].

Composite membranes consisting of PBI and functionalized MWNTs have been investigated as PEMs [43]. The functionalization of MWNTs with sodium poly(4-styrene sulfonate) and imidazole provided remarkably improved mechanical reinforcement of the composite membranes due to much better compatibility with PBI. The proton conductivities measured at 160 °C under anhydrous condition for the MWNT-poly(NaSS)/PBI and MWNT-imidazole/PBI composite membranes were up to 5.1×10^{-2} and 4.3×10^{-2} S cm^{-1} respectively, which were higher than the pristine PBI membrane (2.8×10^{-2} S cm^{-1}). At 170 °C, the fuel cell performance of the MWNT-poly(NaSS)/PBI and MWNT-imidazole/PBI composite membranes was also observed to be greatly improved compared to the pristine PBI. In an study, Gong *et al.* [61] demonstrated the use of a super proton conductor, boron phosphate (BPO_4) to functionalize CNTs. The BPO_4 coated CNTs prepared by polydopamine-assisted sol-gel method were used as additive to modify SPEEK. The use of polydopamine enabled uniform adhesion of BPO_4 nanoparticles on to CNTs and the BPO_4 coating layer on CNTs reduced the short-circuiting risks of CNTs. The SPEEK/BPO_4@CNT composite membranes exhibited significantly improved tensile properties, dimensional and thermal stability compared to pure SPEEK membrane due to the homogeneous dispersion of BPO_4@CNTs in the SPEEK polymer matrix. The newly formed proton conducting channels in the SPEEK/ BPO_4@CNT composite membranes yielded improved proton conductivity compared with pure SPEEK membrane (45% and 150% at 20 °C and at 80 °C, respectively). The peak power density released by the H_2/O_2 fuel cell using SPEEK/ BPO_4@CNT composite membrane was much higher than achieved by using pure

SPEEK membrane. These results suggested that composites of SPEEK and proton conductor-functionalized CNTs possessed great potential as high performance PEM materials [61]

Jana *et al.* [62] investigated the effect of varying proportions of functionalized multi-walled carbon nanotubes (f-MWNTs) on the DMFC performance properties of polyethylene co-acrylic acid-based ionomer (Lotek 4200) based membranes. The authors observed that the proton conductivity and the methanol permeability of the membranes were strongly influenced by the introduction of f-MWNTs, as the incorporation of f-MWNTs in polymer matrix modified the polymer structure. At 1.0 M methanol feed concentration, it was observed that the DMFC performance of the composite membranes was comparable to that of a Nafion117 and the methanol crossover flux was 2.5-fold lower than that of Nafion117. In single-cell tests, the DMFC performance of composite membrane was superior to Nafion117 up to 10.0 M methanol concentration [62]. These results indicated that the functionalized CNTs can be used as nanofillers for the modification of polymer electrolyte membranes for achieving desired performance in fuel cells.

5.2.3 Polymer/Graphene Oxide Composite Membranes

GO has been used as nanofiller in polymer composite membranes due to its unique properties such as presence of oxygen rich functional groups, high surface area, high mechanical strength, poor electrical conductivity and exceptional structural resilience [63,64]. The addition of GO as additive to polymer membranes greatly enhances their proton conductivity and methanol-blocking efficiency, thus, GO/polymer composite membranes have the potential to be employed as one of the best PEMs for fuel cells. Kumar *et al.* [65] observed high ionic conductivities of 4.1×10^{-2} S cm^{-1} to 8.2×10^{-2} S cm^{-1} at temperatures of 25–90 °C, for a free standing GO paper membrane (with thickness of 100 μm), and underlined the potential of GO for use as PEM for DMFC application. The acidic functional groups, hydrogen bonds, and connecting oxygen atoms in the GO structure resulted in abundant proton conducting paths, which ultimately led to high proton conductivity. However, the methanol permeability of the GO paper membrane was observed to be significantly higher than that of Nafion 115 membrane, which resulted in poor DMFC performance despite its improved proton conductivity. By underlining the strong potential of GO as a filler in polymer matrix, the study suggested that the future work on GO based composite membranes with binder could provide great mechanical strength, enhanced proton conductivity and good flexibility with reduced methanol permeability for DMFCs.

Ionic conducting composite membranes of GO and poly (ethylene oxide) (PEO) were reported to be desired membrane electrolyte for low-temperature PEMFC application, without any modification [66]. The reason behind the ionic conductivity of the membrane is the release of protons from the COOH groups on GO sheets in the PEO polymer matrix. It was observed that the ionic conductivity of the PEO/GO membrane increased from 0.086 to 0.134 S cm^{-1} with increase in temperature from 25 to 60 °C, at 100% RH. Above 60 °C, the mechanical properties of the composite membrane got deteriorated. The PEO/GO membrane released a maximum power density of 53 mW cm^{-2} in a single cell operated at around 60 °C.

Composite membranes based on Nafion and GO as PEMs for fuel cell applications have also been discussed. Specifically, the Nafion/GO composite has been exploited as polymer electrolyte membrane for DMFC applications [67]. The interaction of the non-polar Nafion backbone with the hydrophobic GO structure and that of the polar ionic clusters of Nafion with the hydrophilic GO functional groups leads to the modification of the microstructures of the Nafion domains. The incorporation of GO into Nafion matrix favorably altered the physical and transport properties of Nafion for fuel cells. The incorporation of GO enhanced the mechanical properties of the Nafion due to strong interfacial interactions. Due to the same reason, the composite membrane also exhibited higher thermal stability than pure Nafion. The Nafion/GO composite membranes had a 14% increase in tensile strength and a 26% increase in Young's modulus as compared the Nafion membrane. The prepared Nafion/GO composite membranes with 0.5 wt% loading of GO sheets exhibited a 40% reduction in methanol crossover compared to the value for pristine Nafion membrane. The proton conductivity of the composite membrane reduced to ~98.1% of that of pure Nafion. The addition of 0.5 wt% GO in Nafion 112 membrane increased the selectivity (ratio of proton conductivity to methanol permeability) by ~150%, due to the reduced size of the ionic channels and the presence of hydrophilic functional groups of GO. The DMFC with Nafion/GO composite membrane exhibited remarkable improvement in performance with power densities much higher than earlier observed for Nafion-based DMFC. Also, the composite membrane with

5 M methanol had ~3 times higher power density than Nafion and ~14 times higher than that of the composite membrane at 1M methanol feed concentration, which indicated that the modification of ionic clusters of Nafion with GO nanofiller suppressed the methanol crossover even at high methanol concentrations, thus, resulting in the observed behavior [67].

Wang *et al.* [68] prepared GO/Nafion composite membranes to provide mechanical reinforcement to Nafion and optimized the GO loading based on fuel cell performance and mechanical properties of membrane. The addition of GO improved the tensile strength and dimensional stability of Nafion significantly. The mechanical reinforcement and dimensional stability provided by the individually dispersed GO sheets in Nafion resin enhanced the integrity of the membrane electrode assembly within the fuel cell and provided better fuel cell durability. GO/Nafion composite membrane with 3 wt% exhibited superior mechanical properties as compared to the recast Nafion, but the fuel cell performance was similar to that of recast Nafion. Apart from providing mechanical reinforcement, the use of GO as additive in Nafion preserved the low electrical conductivity of Nafion due to its electrically insulating nature [68]. GO/Nafion composites fabricated by LbL assembly were also applied as membrane electrolytes in DMFCs [69]. GO/Nafion composite membranes had lower uptake of water than recast Nafion 117 membrane. The ion exchange capacity (IEC) of composite membrane was only slightly higher than that of the recast Nafion 117 membrane. The enhanced IEC was due to the presence of carboxyl and hydroxyl functional groups at the GO sheet edges which released H^+ upon humidification. The oxidative stability of Nafion 117 membrane was improved remarkably by the GO coating on the Nafion surface. GO/Nafion composite membranes exhibited proton conductivities lower than that of Nafion 117 membrane owing to their decreased water uptake arising from the shrinkage in the proton diffusion channels. The methanol permeability of the composite membranes was much lower than that measured for pristine Nafion membrane under similar conditions. This observation was due to the reduced swelling ratio of the composite membranes which blocked the passage of methanol molecules through the proton conducting channels. The lowest methanol permeability (6.7×10^{-8} cm^2 s^{-1}), achieved for the composite membrane with 80-layered GO nanosheets, was about two orders of magnitude lower than that of pristine Nafion membrane (1.82×10^{-6} cm^2 s^{-1}) at 30 °C. The composite membranes exhibited decreasing methanol permeability with increase in the number of GO layers, which indicated the advantage of modifying Nafion with GO. The comparison of the performance of all composite membranes with that of the recast Nafion 117 membrane was also performed by defining relative selectivity (RS) as the ratio of the selectivity of the composite membranes to that of the Nafion 117. The RS of the GO/Nafion composite membrane with 80-layered GO nanosheets was 5-fold greater than that of the Nafion 117 membrane and also RS was found to be increasing with the increase in the number of GO layers. The observed dramatic suppression of methanol crossover by the modification with GO nanosheets, thus, guarantee the potential of composite membranes based on GO and polymer for future DMFCs. The single cell using GO/Nafion composite membrane with 50-layered GO achieved a maximum power density of 64.38 mW cm^{-2} at 60 °C and ambient pressure with RH of 100% under high methanol concentration of 5 M, whereas for the cell utilizing recast Nafion 117 membrane, a maximum power density of 41.60 Mw cm^{-2} was obtained, which indicated the superior fuel cell performance of GO/Nafion composite membranes compared to recast Nafion 117 membrane [69].

Several reports have revealed the application of GO as an excellent methanol-blocking film on Nafion surface for DMFCs. GO-laminated Nafion membrane was observed to be beneficial as PEM for DMFC operating under high methanol feed concentrations, compared to GO-dispersed poly(vinyl alcohol) (PVA) composite membrane [70]. Lin *et al.* [70] fabricated a dual layer membrane by laminating the Nafion 115 with a highly ordered and parallel oriented GO paper by vacuum filtration, through transfer printing followed by hot-pressing. The 1 μm thick, 2D GO paper adhered to the Nafion base membrane acted as efficient methanol barrier under high methanol feed concentrations. The methanol permeability of the GO-Nafion nanocomposite membrane was observed to be approximately 70% lower than that of Nafion 115, whereas the proton conductivity decreased by 22% by GO-lamination of Nafion 115. The ratio of proton conductivity to methanol crossover of GO-laminated Nafion 115 was observed to be 40% higher than Nafion 115. The GO-Nafion nanocomposite membrane was observed to be superior to neat Nafion membrane for DMFC during operation under 8 M methanol feed concentration [70].

Sulfonated polyimide (SPI)/GO composite membranes as proton exchange membranes were fabricated by He *et al.* [71] by incorporating GO with different sheet sizes into SPI. The SPI/GO composite membranes exhibited an increase in mechanical properties and proton conductivities, along with a decrease in methanol permeabilities by the addition of GO into SPI. It was observed that the size of GO sheets affected the properties of the composite membranes. SPI/GO composite membrane with the smallest size of GO achieved the best proton conductivity, lowest methanol permeability and significantly enhanced fuel cell performance compared to that of pure SPI and other SPI/GO composite membranes, at 0.5 wt% loading of GO in SPI. The observed outstanding properties of composites by the incorporation of smaller size GO could be ascribed to the formation of highly uniform microstructure and continuous proton conducting pathways due to strong hydrogen bonding between SPI and the smaller sized GO. The DMFC with SPI/GO composite membrane containing 0.5% GO exhibited a 1.4-fold higher power density compared with that of the SPI membrane at 25 °C [71].

PA-doped PBI/GO nanocomposite membranes were prepared by Uregen *et al.* [72] by solution blending method for high temperature fuel cell applications. The proton conductivity and acid retention properties of PBI/GO nanocomposite membrane were improved by the dispersion of GO in PBI polymer matrix. The PBI/GO composite membrane exhibited the highest proton conductivity for 2 wt% GO loading, which was 79% higher than that of a neat PBI membrane at 180 °C. In high temperature PEMFC tests, the PBI/GO composite membrane with 2 wt% GO content exhibited better performance than pure PBI membrane under 165 °C and non-humidifying conditions, with peak power densities of 0.31 and 0.38 W cm^{-2} obtained for composite and neat PBI membranes respectively. The enhanced proton conductivity of the nanocomposite membrane was mainly attributed to the facile proton hopping induced by the acidic functional groups (epoxy oxygen and carboxylic acid groups) of GO and also the interconnection of proton conducting channels in PBI membrane by GO. However, at temperatures above 165 °C, it was observed that the conductivity decreased due to the degradation of acidic functional groups in GO. These results demonstrate the strong potential of utilizing PBI/GO composites as electrolyte materials for high temperature PEMFCs [72].

5.2.4 Polymer/Functionalized Graphene Oxide Membranes

Polymer electrolyte/functionalized GO composite membranes have also been reported for use in fuel cell applications. Choi *et al.* [73] reported the incorporation of sulfonated GOs (SGOs) into Nafion matrix (Figure 5.5) and observed that the transport properties of SGO/Nafion composite membranes were superior to those of pristine Nafion and GO-incorporated nanocomposites. The physical and chemical modification of ion-conducting channels through chemical tuning of the graphene functionality controlled the confinement of bound water within the ionic channels and, thus, resulted in selectively facilitated transport behavior of SGO/Nafion composite membranes, as compared to pristine Nafion and GO-incorporated nanocomposites. The composite membrane exhibited enhanced proton conductivity (~66%) with low activation energy, and reduced methanol permeability (~35%) at high temperature for high performance DMFCs. In another study, composites of Nafion with sulfonic acid-functionalized GO were also presented as novel PEMs for high temperature PEM fuel cell applications due to the exhibited high proton conductivity and fuel cell performance [74]. At 120 °C with 25% humidity, the proton conductivity and fuel cell performance was observed to be 4 times greater than that of recast Nafion. Figure 5.6 demonstrates the polarization and power density curves for 10 wt% functionalized GO/Nafion and recast Nafion membranes.

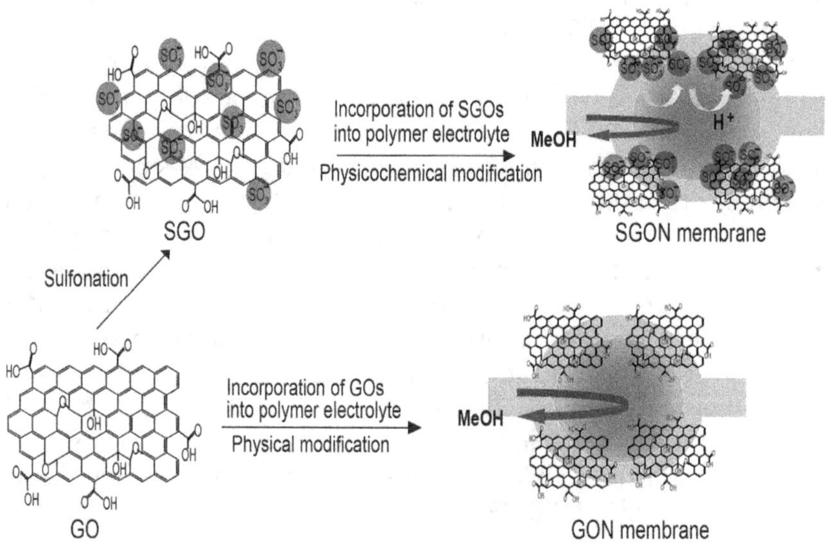

Figure 5.5 Schematic of generating tunable transport properties using functionalized graphene. Reproduced from Reference 73 with permission from American Chemical Society.

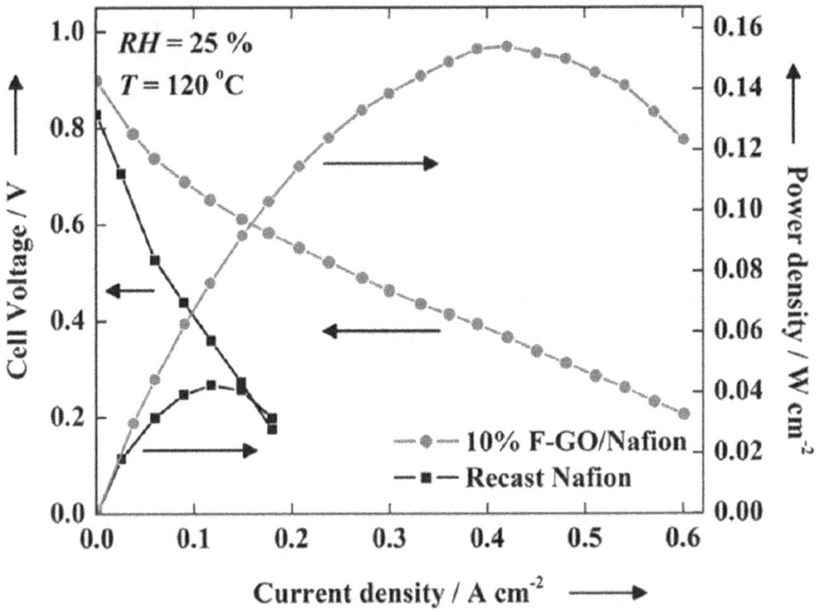

Figure 5.6 Polarization and power density curves for 10 wt% functionalized GO/Nafion and recast Nafion membranes. Reproduced from Reference 74 with permission from American Chemical Society.

In a recent study, Sahu *et al.* [75] employed sulfonic acid-functionalized graphene (S-graphene) as filler in Nafion for the preparation of composite membrane for PEMFCs under low RH. The incorporation of S-graphene in Nafion increased the sulfonate moieties per unit volume of each graphene domain and, thus, reconstructed the hydrophilic domains of Nafion, thereby providing fast transport of protons through the composite membrane at low RH. The high density of sulfonic acid groups on S-graphene dispersed in Nafion and the high surface area of S-graphene resulted in significantly enhanced IEC and proton conductivity of Nafion-S-graphene, which ultimately led to better PEMFC performance of composite membrane at low RH than that attained by employing pure Nafion. It was observed that the proton conductivity of Nafion-S-graphene (1%) composite membrane was five times higher that of the pristine Nafion membrane, at 20 % RH. The comparison of mechanical properties of Nafion-S-graphene composite membrane with 1% S-graphene content and pristine Nafion membrane exhibited higher mechanical stability of composite membrane. Nafion-S-graphene-based PEMFC delivered a peak power density of 300 mW cm^{-2} at 70

°C under 20% RH and ambient pressure, whereas a pristine recast Nafion-based PEMFC released a peak power density of only 220 mW cm^{-2} (Figure 5.7). The high surface roughness of the Nafion-S-graphene resulting from the high surface area and the crumpling nature of S-graphene increased the amount of water uptake and improved its compatibility with electrodes during the fabrication of membrane electrode assemblies. Therefore, Nafion–S-graphene hybrid membranes are promising for applications at low RH and elevated temperature PEMFCs [75].

Enotiadis *et al.* [76] also prepared GO/Nafion composite membranes with GO carrying hydrophilic functional groups such as $-NH_2$, -OH, $-SO_3H$ and observed that water retention and proton conduction of Nafion was favorably improved by the addition of organo-functionalized GO materials. The nanocomposite membranes were observed to be much stronger and could withstand higher temperatures compared to that of Nafion. In another study, Lee *et al.* [77] also explored GO as potential filler in low humidifying PEMFCs. Higher water uptake was observed in Nafion/GO composite membrane as compared to casting Nafion due to the hydrophilic nature of GO and the uptake increased with GO content. Nafion/Pt-G composite membrane obtained via *in-situ* deposition of Pt nanoparticles onto GO using microwave method exhibited lower water uptake than that of Nafion/GO membrane as GO reduced to hydrophobic graphene during Pt deposition. The water uptake of Nafion/Pt-G composite membrane also decreased with Pt-G content. However, the proton conductivities of the composite membranes followed the opposite trend. The proton conductivities of Nafion/GO composite membrane were lower than those of the Nafion/Pt-G, irrespective of the content of GO or Pt-G. The reason behind the higher proton conductivity of Nafion/Pt-G composite membrane was the presence of electronically conductive Pt nanoparticles. Nafion/GO composite membranes with 0.5 wt% and 3.0 wt% filler also exhibited reduced proton conductivities compared with those of pristine Nafion membrane due to blocking of the ionic clusters in Nafion by GO. It is noteworthy that incorporating 4.5 wt% of GO in Nafion led to higher proton conductivity than Nafion membrane owing to the fact that the abundant oxygen rich functional groups facilitated proton transfer channels in Nafion/GO membrane and, thus, overcame the blockage of Nafion's ionic clusters by GO. Significant increase in tensile strength was observed for both Nafion based composite membranes, thus, allowing their use at high temperature. DMFCs with

Nafion/GO composite membrane with 0.5 wt%, 3.0 wt% and 4.5 wt% GO showed superior performance under various RHs (100%, 60% and 40%) compared to that with pristine casting Nafion membrane. The cell performance of Nafion/Pt-G composite membrane was found to be inferior to that of Nafion/GO composite membrane under similar conditions, due to the poor water retention of graphene and the loss of electrons caused by the electrical network formation by Pt nanoparticles on GO. The measured OCVs of Nafion/Pt-G and Nafion/GO composite membranes experienced no significant variation with RH down to 40% [77].

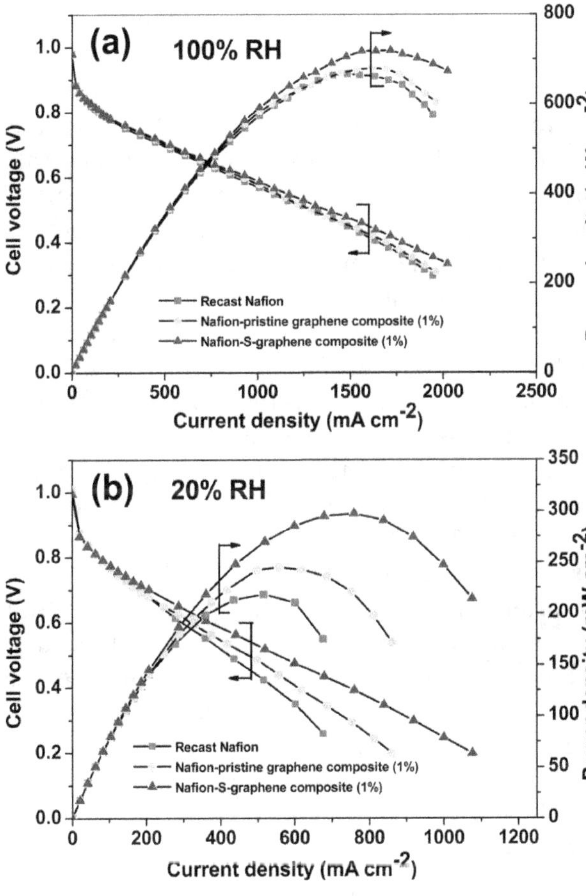

Figure 5.7 Performance of H_2/O_2 PEFC with recast Nafion, Nafion-graphene, and Nafion-S-graphene composite membranes at (a) 100% RH and (b) 20% RH at 70 °C under atmospheric pressure. Reproduced from Reference 75 with permission from American Chemical Society.

Xu *et al.* [78] prepared PBI/GO and PBI/SGO composite membranes by adding graphite oxide and sulfonated graphite oxide into PBI membranes with low phosphoric loadings (1.8-2.0 PRU) for high temperature PEMFCs. The PBI/GO and PBI/SGO composite membranes with low content of H_3PO_4 exhibited higher proton conductivity than that of the PBI/PA membrane with high phosphoric loadings. The hydrogen bonding in GO resulted in high proton conductivity of the composite membranes. The authors reported that both PBI/GO and PBI/SGO composite membranes generated superior fuel cell performance and higher peak power densities than that of the PBI/PA membranes. Therefore, PBI/SGO composite membranes with low PA loadings could be used as potential PEMs for high temperature operation [78]. In another study, Liu *et al.* [79] reported nanohybrid membranes based on SGO nanosheets and chitosan (CS). The uniform dispersion of SGO nanosheets in the chitosan membrane resulted in improved thermal and mechanical stabilities of nanohybrid membrane due to the inhibition of mobility of CS chains through the strong electrostatic interactions between the acid group of SGO and $-NH_2$ group of CS matrix. The addition of SGO generated acid-base pairs at the CS–SGO interface, which served as proton-hopping locations. The high surface area, nanosheet structure and better dispersion of SGO in CS matrix allowed the formation of continuous channels for efficient proton migration to the nanohybrid membrane, thus, resulting in significantly enhanced proton conductivity under both hydrated (100%) and anhydrous (0% RH) conditions. It was observed that the incorporation of 2.0% SGO could provide a 122.5% increase of hydrated conductivity and 90.7% increase of anhydrous conductivity when compared with CS control membrane. The lower water uptake of SGO-filled membranes compared to GO-filled membranes reduced the area swelling, which resulted in reinforced structural stabilities of nanohybrid membranes and thus, making them suitable for practical fuel cell application. The maximum current density and the maximum power density of a H_2/O_2 fuel cell based on SGO-filled CS membrane increased by 58% and 64% with 2 wt% of SGO at 120 °C and under anhydrous conditions [79].

Gahlot *et al.* [80] prepared SGO/SPES nanocomposite membranes by incorporating various concentrations of SGO into sulfonated poly(ether sulfone) (SPES). The SGO sheets were observed to uniformly distribute throughout the SPES matrix. The IEC, proton conductivity, water retention, selectivity, methanol crossover resistance and mechanical strength of SPES were improved by the incorporation of SGO in SPES (Figure 5.8). The strong interfacial interactions between SGO and SPES matrix prevented the formation of hydrophilic channels in the nanocomposite membrane, which led to reduced methanol permeability and higher proton conductivity. The SGO/SPES nanocomposite with 5% SGO possessed larger methanol crossover resistance and maximum proton conductivity. The observed good proton conductivity of the nanocomposite membrane at temperatures varying from 30 to 90 °C with low activation energy suggested the applicability of SGO/SPES membrane as PEM for DMFCs at high temperature [80].

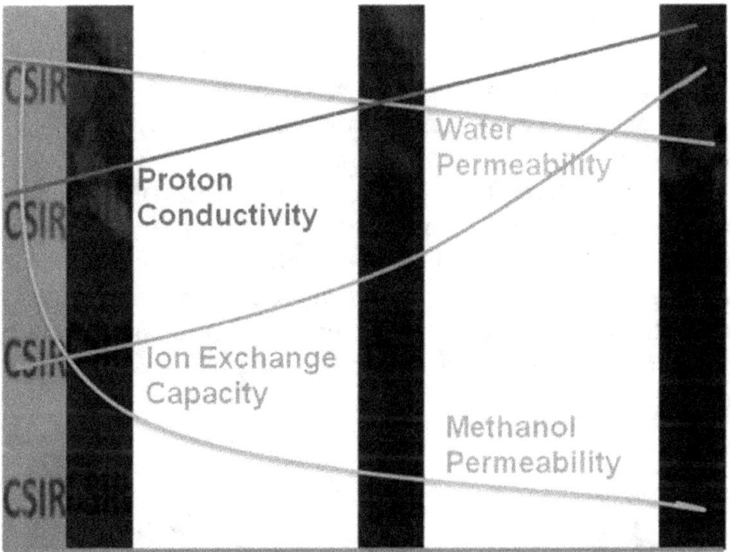

Figure 5.8 Effect of addition of SGO on various properties of SPES membrane. Reproduced from Reference 80 with permission from American Chemical Society.

Kumar *et al.* [81] also observed that the composite membrane of SPEEK with SGO exhibited improved proton conductivity as compared to the neat SPEEK membrane due to the interconnection of proton transport networks within the SPEEK membrane by the incorporated SGO and creation of more proton conducting pathways enabled by presence of water between GO layers. The morphological analysis confirmed the uniform distribution of SGO sheets in the SPEEK structure. At 30% RH, the composite membrane exhibited a proton conductivity of 0.055 S cm^{-1} which was higher than that of a neat SPEEK membrane (0.015 S cm^{-1}). The excellent fuel cell performance of SGO/SPEEK composite membrane compared to neat SPEEK membrane indicates the suitability of using SGO/SPEEK as alternative low-cost PEM for fuel cells [81].

In another study, Pandey *et al.* [82] reported that the SPI/sulfonated propylsilane graphene oxide (SPSGO) composite membrane containing 8 wt% of SPSGO (SPI/SPSGO-8) exhibited significant improvement in proton conductivity (nearly equal to that of Nafion 117) as well as thermal and mechanical stability, along with 10-fold enhanced bound water content. The incorporation of the multi-functionalized SPSGO in the SPI matrix also promoted internal self-humidification and reduced the methanol crossover as compared to the pristine SPI membrane. The membrane properties desirable for DMFC application were observed to be greatly influenced by the loading of acid functionalized GO in the SPI matrix (Figure 5.9). A maximum power density of 75.06 mW cm^{-2} was exhibited by SPI/SPSGO-8 in single-cell DMFC tests, as compared to the commercial Nafion 117 membrane (62.40 mW cm^{-2}) under 2M ethanol fuel at 70 °C. The strong H-bonding between the multifunctional groups, the presence of hydrophobic graphene sheets and polymer chains contributed to the proton conducting property by providing a suitable design of proton conducting channels in the membrane matrix.

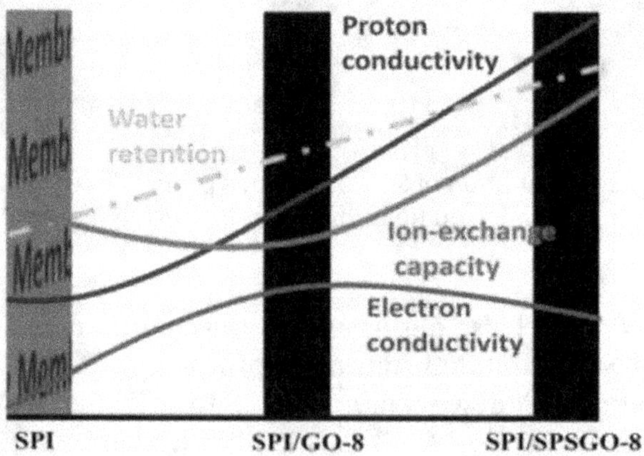

Figure 5.9 Effect of GO and SPSGO loading on SPI membrane properties. Reproduced from Reference 82 with permission from American Chemical Society.

Graphite oxide/polybenzimidazole containing tertiary butyl group (GO/BuIPBI) and isocyanate modified graphite oxide/BuIPBI (iGO/ BuIPBI) composites were explored as PEMs for high temperature fuel cell applications [83]. It was observed that the incorporation of 1% isocyanate functionalized GO could enhance the chemical stability of the BuIPBI matrix due to hydrogen bonding between the iGO filler and BuIPBI. The GO/BuIPBI and iGO/BuIPBI composite membranes doped with PA had higher proton conductivities than neat BuIPBI membrane and exhibited increasing trend with increase in temperature. The enhanced proton conductivities of the composite membranes could be ascribed to the proton transport through the membrane induced by the well-connected proton conduction pathways provided by the GO. However, as the temperature was increased as high as 140 °C, it was observed that the conductivities of the composite membranes decreased due to the destruction of carboxyl and epoxy functional groups in GO. The authors also observed that the proton conductivity was influenced by the PA doping level in the membrane. The iGO/BuIPBI membrane with 10 wt% of iGO in BuIPBI and with 76.6% PA doping had the highest proton conductivity of 0.027 S cm^{-1} at 140 °C without humidity, compared to that of BuIPBI and GO/BuIPBI [83].

Recently, Shukla *et al.* [84] prepared sulfonated graphene nanoribbons (sGNR) by unzipping and sulfonation of MWCNTs by a hydrothermal synthetic route which had better dispersion and functional compatibility with SPEEK polymer matrix. Nanocomposite PEMs of SPEEK with sGNR exhibited improvement in water uptake, ion exchange capacity and proton conductivity. The SPEEK/sGNR (0.1 wt%) composite membrane displayed improved mechanical stability than the pristine SPEEK membrane. The SPEEK/sGNR composite membrane had current density of 840 mA cm^{-2} at 0.6 V and maximum power density of 660 mW cm^{-2} compared to the current density of 480 mA cm^{-2} at 0.6 V and maximum power density of 331 mW cm^{-2} for pristine SPEEK membrane in fuel cell tests. The SPEEK/sGNR nanocomposite membranes were also observed to be highly durable. The observed results indicated better fuel cell performance and durability of SPEEK/sGNR composite membranes. The authors suggested that sGNR could be used as effective additives in different polymer matrices so as to achieve durable and cheaper nanocomposites as PEMs for fuel cells [84]. In another study, SPI membrane, incorporating 0.5 wt% GO/poly(sodium-4-styrenesulfonate) modified graphene (PSS-G) was observed to exhibit an improvement in proton conductivity of 490.3% and a drastically decreased methanol permeability of 508.9%, with ~25 times increase in selectivity (proton conductivity/methanol permeability) at 80 °C. The SPI/PSS-G composite membrane also had enhanced tensile strength of 134.4% at 30 °C [85].

Ye *et al.* [86] prepared PEM based on protic ionic liquids (PILs) composite membranes and ionic liquid polymer modified graphene sheets (denoted as PIL(NTFSI)-G) using solution blending method. The sulfonated polyimide (SPI)/PIL(NTFSI)-G/PIL composite membranes with 0.5 wt% graphene loading exhibited enhanced ionic conductivity of 7.5×10^{-3} S cm^{-1}, nearly four times higher than that of a pure SPI/PIL membrane at 160 °C. GO incorporation also resulted in PIL cost-saving of 20%. The Young's modulus and tensile strength of the composite membrane exhibited an increase of 127% and 345% respectively with 0.9 wt% addition of graphene, in comparison with the plain SPI/PIL membrane. The 3D network of homogeneously distributed graphene sheets throughout the polymer matrix in the SPI/PIL(NTFSI)-G/PIL composite membrane facilitated high-level mechanical reinforcement through enhanced nanofiller-polymer adhesion. The homogeneously dispersed graphene nanosheets in SPI matrix created 3D ion transport channels throughout the membrane, which resulted in high ionic conductivity [86].

In a similar study, He *et al.* [87] demonstrated the preparation of highly conductive nanocomposite membrane by incorporating polydopamine-modified graphene oxide (DGO) sheets into SPEEK matrix [87]. The strong electrostatic interaction between the $-NH_2$ and –NH– groups of DGO sheets and the $-SO_3H$ groups of SPEEK chains generated acid-base pairs with closely linked proton donor and acceptor and allowed facile proton transport between the donor and acceptor in the nanocomposite membrane. The proton conductivity of SPEEK/DGO nanocomposite membrane was observed to be much higher than SPEEK membrane under both hydrated and anhydrous conditions. The enhanced proton conductivity of the composite membrane led to much higher H_2/O_2 cell performance, with an increase in maximum current density of 47% and increase in maximum power density of 38 % under 120 °C and anhydrous conditions by the addition of 5 wt% of DGO to SPEEK. SPEEK/DGO composite membrane exhibited enhanced thermal stability due to the tuning of the nanophase-separated structure and inhibition of the chain motion and thermal decomposition of the composite membrane due to strong electrostatic interactions of DGO with the SPEEK polymer matrix. The strong mutual interactions between the DGO and SPEEK also resulted in an increase in mechanical stability of SPEEK/DGO composites. The tensile strength and the Young's modulus of the composite reached up to 57.5 MPa and 896.1 MPa by 2.5% DGO loading. These findings indicated that SPEEK/DGO nanocomposite membranes could be employed as promising PEMs for practical fuel cell applications under high temperature and anhydrous condition [87].

Mishra *et al.* [88] reported that the presence of both SPEEK and GO in Nafion membrane enhanced the membrane's proton conductivity. Nafion containing 1wt% SPEEK and 0.75wt% of highly oxidized GO (HGO) exhibited maximum proton conductivity of 322.2 mS cm^{-1} at 100% RH and 90 ^{o}C, as compared to that of pure Nafion (198 mS cm^{-1}). Nafion-HGO (NHG) composite membranes had the highest storage modulus among virgin Nafion and other composite membranes such as Nafion-SPEEK (NS), Nafion-GO (NG), Nafion-SPEEK-GO (NSG) and Nafion-SPEEK-HGO (NSHG) in the temperature range of 50-150 ^{o}C, which could be attributed to the increase in the secondary interactions of HGO with Nafion arising from higher degree of oxidation and smaller size of GO. The chemical stability of Nafion membrane was observed to be enhanced by the addition of GO, SPEEK or SPEEK-GO. The storage modulus decreased with the addition of highly sulfonated PEEK to NG and NHG composite membranes owing to the high stiffness of GO and HGO as compared to SPEEK. Benefiting from the enhanced proton conductivities and water uptake, NSHG composite membrane with 0.75 wt% of GO and HGO exhibited higher power density and current density than virgin Nafion and other composite membranes, which provided superior fuel cell performance [88]. In another study, Yuan *et al.* [89] reported the LbL assembly of polyelectrolyte poly(diallyldimethylammonium chloride) (PDDA) and GO multilayer film on Nafion membranes and demonstrated the capability of Nafion-PDDA-GO as an efficient methanol-blocking membrane for DMFC. The composite membrane suppressed the methanol crossover with 67% decrease in methanol diffusion coefficient as compared to the pristine Nafion membrane, and also enhanced membrane strength. The methanol crossover limiting current density across the Nafion membrane with PDDA-GO bilayer structure was about $1/3^{rd}$ of pristine Nafion membrane, which demonstrated that the dense and uniform bilayer structure on Nafion reduced the methanol crossover by about 63%. The passive DMFC using Nafion-PDDA-GO composite membrane exhibited higher energy efficiency and enhanced power density as compared to the pristine Nafion membrane [89].

Nanocomposite anion-exchange membranes (AEMs) were prepared by dispersing GO functionalized with PDDA into polyvinyl alcohol/PDDA semi-interpenetrating polymer networks (PVA/PDDA SIPNs) via solvent casting followed by thermal cross-linking [90]. The OH- conductivity and the mechanical stability of the membranes prepared by physical cross-linking was higher than that of the chemically cross-linked ones. The optimal conditions for fabricating nanocomposite AEMs were determined by studying the effects of PVA/PDDA ratio, FGO content, cross-linking method and temperature on the material properties of nanocomposite membranes. The OH- conductivity reached the highest value (12.1 mS cm^{-1} at 30 ^{o}C and 21 mS cm^{-1} at 80 ^{o}C) for the PVA/PDDA membrane having a PVA/PDDA weight ratio of 70/30 by the inclusion of 20 wt% FGO. The high OH- conductivity of the nanocomposite membrane at high FGO content (20 wt%) was due to the transfer of OH- through the membrane facilitated by the high-water uptake of the membrane having hydrophilic FGO with high surface area. The hydrogen fuel cell performance of the nanocomposite membrane was observed to be at the maximum with 10 wt% FGO. The use of FGO as filler in the PVA/PDDA membrane also improved the thermo-mechanical stability of the nanocomposite membrane at 100% RH. The FGO based SIPNs synthesized in an environmentally friendly approach utilizing deionized water as solvent are promising membranes for AEM fuel cells.

Zhao *et al.* [91] reported the use of sulfonated polymer brush functionalized GO (SP-GO) as filler in SPEEK polymer matrix. The effect of the SP-GO integrated into SPEEK membrane on membrane properties and single PEMFC performance was studied. SP-GO fillers were observed to be more uniformly distributed within SPEEK than GO due to the hydrogen bonding between the sulfonic acid groups of SP-GO and polar groups of SPEEK, which resulted in improved interfacial compatibility between GO and SPEEK. The uniformly dispersed SP-GO connected the ionic clusters in SPEEK and formed wide and well-connected proton-transfer pathways along the SPEEK/SP-GO interface through interfacial interactions. Hence, SP-GO-filled composite membranes achieved enhanced proton conductivities and reduced activation energy for proton conduction as compared to the SPEEK control membrane and the GO-filled membrane. The proton conductivity of the composite membrane exhibited further increase with increase in the polymer brush length and the content of SP-GO filler. The composite membrane with 10 wt% SSGO had a 95.5% increase in proton conductivity at 65 °C under hydrated condition and a 178% increase in conductivity at 150 °C under anhydrous conditions, compared to SPEEK control membrane. The addition of SP-GO fillers in SPEEK polymer also lowered the H_2 permeability of the composite membranes. The improved proton conductivity and the low H_2 permeability of the composite membranes led to higher PEMFC performance than Nafion under anhydrous conditions and high temperature [91].

In another study, Chu *et al.* [92] explored the potential of PBI/zwitterion-coated GO (ZC-GO) membranes for DMFC applications, with ZC-GO containing ammonium and sulfonic acid groups as filler in PBI. The authors investigated the influence of ZC-GO content on the properties of composite membranes for DMFCs by varying the amount of ZC-GO incorporated into PBI. The membrane properties such as water uptake, proton conductivity, methanol permeability, and swelling ratio of the PBI/ZC-GO increased with increase in the amount of ZC-GO. PBI/ZC-GO composite membrane with 25 wt% ZC-GO exhibited a higher water uptake and a lower swelling ratio compared with Nafion 117 membrane, due to the restricted hydrophilic domains owing to the hydrogen bonding between ZC-GO and PBI. The methanol permeability of PBI/ZC-GO with 25 wt% ZC-GO was observed to be higher than that of pure PBI membrane due to the formation of proton transport channels through the membrane by the adsorbed water, but was much lower than that of Nafion 117 membrane or other hybrid membranes under similar conditions. The increase in ZC-GO content facilitated transfer of more protons due to the formation of hydrogen bonds between PBI and ZC-GO. The increase in proton mobility in the composite membrane with increasing ZC-GO content resulted from the extension of the number of available proton exchange sites per cluster. The conductivities of PBI/ZC-GO membranes with different ZC-GO percentage exhibited increase in conductivities with increase in temperature. The reason behind this phenomenon was the faster migration of ions and increased diffusivity at higher temperatures. The PBI/ZC-GO composite membrane with 25 wt% ZC-GO exhibited the highest proton conductivity of 4.12×10^{-2} S cm^{-1} at 90 °C at 100% RH. Recently, Ye *et al.* [93] used simple blending to incorporate graphene nanosheets into poly(vinyl alcohol) (PVA) and investigated the transport properties of the prepared graphene/PVA nanocomposites. The continuous and well-connected ionic channels introduced in the graphene/PVA composite membrane by the uniformly dispersed graphene nanosheets in the PVA matrix improved the transport properties of the composite membrane. As a result, the graphene/PVA composite membranes exhibited an increase in ionic conductivity of ~126% and a decrease in methanol permeability of ~55% with a 0.7 wt% graphene loading. The significant improvement of the transport properties observed for the graphene/PVA composite membranes resulted in better cell performance than the PVA membrane, with ~148% improved maximum power density at 60 °C. The addition of graphene into PVA also resulted in enhanced mechanical properties (~73% improvement in tensile strength) at graphene loading of 1.4 wt%, owing to interfacial interactions between graphene and PVA matrix through H-bonding.

Kim *et al.* [94] prepared phosphotungstic acid (PW) coupled GO-Nafion (Nafion/PW-mGO) membrane and investigated PEFCs utilizing (Nafion/PW-mGO) composite membrane at elevated temperature and low RH operating conditions. The composite membrane exhibited improved proton conductivity compared to pristine and recast Nafion membranes. Nafion/PW-mGO composite membrane provided maximum power density, greater than those of the Nafion-212 and recast Nafion membranes at 80 °C and under low RH. Maximum power density delivered by a Nafion/PW-mGO membrane operated at 80 °C and under low RH was found to be about 4-fold higher than the Nafion-212 membrane. The fuel cell performance of the Nafion/PW-mGO composite membrane was remarkably better than those of the Nafion-212 and recast Nafion membranes under low RH. The higher PEFC performance at elevated temperature was ascribed to the hygroscopic nature and strong acid strength of PW which allowed water retention in the composite membrane through hydrogen bonding with protons and facile hopping of protons, thus, resulting in increased proton transfer. Yang *et al.* [95] also prepared novel composite membranes of PBI with triazole modified GO (MGO) for high temperature PEMFCs. The composite membrane exhibited simultaneous improvement in both tensile strength and proton conductivity with the incorporation of well-dispersed triazole functionalized GO. PA-doped PBI/MGO membrane had high proton conductivity of 0.135 S cm^{-1} at 180 °C. The presence of triazole groups in MGO provided facile proton hopping pathways through hydrogen bonds with PA, which led to superior proton conduction through the hybrid membrane at high temperature. The fuel cell performance of PBI/MGO membrane was remarkably better than that of the PBI membrane with the same acid doping level. For example, at 180 °C, the maximum power densities of 537 mW cm^{-2} and 506 mW cm^{-2} were obtained with fuel cells based on composite and pristine membranes, respectively [95].

Bai *et al.* [96] described highly conductive PEMs based on chitosan and phosphorylated graphene oxide (PGO). The authors synthesized GO with phosphorylated polymeric layer by distillation-precipitation polymerization and subsequently PGO nanosheets were embedded into the CS matrix. The thermal and mechanical stabilities of PGO-filled membranes were observed to be higher than that of CS and GO-filled membranes due to the strong electrostatic interactions between the PA groups of PGO and the amine groups of CS in PGO-filled membranes. The attractive interactions between the PA groups on PGO surface and the CS chains resulted in better dispersion of PGO nanosheets in the CS membrane and, thus, facilitated efficient proton migration pathways along PGO surface. The nanohybrid membrane exhibited improved proton conductivities under both hydrated and anhydrous conditions. At 160 °C and under anhydrous conditions, the conductivity achieved for the nanohybrid membrane with 2.5% PGO was 22.2 and 95.5 times higher than those of CS control and Nafion membranes respectively. This superior proton conductivity of the nanohybrid membrane led to remarkably higher H_2/O_2 cell performance compared to CS control and GO-filled membranes at 0% RH.

Pandey and Shahi [97] prepared SPI/sulphonated imidized GO (SIGO) composite membranes for PEFCs. SPI/SIGO composite membranes exhibited enhanced mechanical, thermal and oxidative stabilities. The bound water content in PEM matrix increased with SIGO content and, thus, composite membranes had improved water-retention properties (with bound water 5.12% higher than that of Nafion 117 membrane) and high proton conduction ability. The proton conductivity of pristine PBI membrane increased after the incorporation of SIGO in SPI, due to the inter-connected proton conducting channels formed by the acidic groups and was observed to be comparatively higher than Nafion 117 membrane at 30 °C. The proton-transfer activation energy was improved by the presence of SIGO. The integration of covalently bonded SIGO in polymer matrix resulted in narrow hydrophilic-hydrophobic phase separation in

SPI/SIGO composite and SPI/SIGO composite membrane exhibited lower methanol permeability than both pristine SPI and Nafion 117 membrane at 30 °C. The composite membrane exhibited high maximum power density in comparison with pristine PBI and Nafion 117 membrane in single-cell DMFC performance at 70 °C. These results demonstrated that the developed SPI/SIGO composite membrane could be utilized as suitable materials for high performance DMFC application.

Dai *et al.* [98] introduced octadecylamine-functionalized GO (GOA) and partially quaternized GOA (GOAN) into cross-linked quaternized poly(styrene-*b*-isobutylene-*b*-styrene) (QSIBS) to prepare cross-linked QSIBS/GOA and QSIBS/GOAN composite membranes. The authors evaluated these hybrid composite membranes as anion exchange polymer electrolytes in DMFCs. The introduction of GOA into QSIBS improved the ionic conductivity of QSIBS, the ionic conductivity of which was much lower than that of Nafion 115. The QSIBS/GOA composite membranes exhibited an increasing ionic conductivity with temperature due to the increase in the mobility of both ionic and polymeric chains. The QSIBS/GOA membrane exhibited 138% increase in ionic conductivity compared to that of the pure QSIBS with the increase in GOA content up to 0.50 wt%. However, further increase in GOA amount reduced the ionic conductivity to 37% with 1 wt% GOA. The decreased conductivity beyond 0.5 wt% GOA loading could be ascribed to the "barrier effect" arising from the aggregation of the GOA in the hybrid membrane. The GOA integrated QSIBS membrane exhibited a 32.1% decrease in methanol crossover compared to that for a pure QSIBS at 0.5 wt% GOA loading. The QSIBS/GOAN composite membrane containing 0.5 wt% GOAN loading exhibited higher ionic conductivity than that of QSIBS/GOA, but was comparable to that of Nafion 115. Methanol permeabilities of hybrid composite membrane were observed to be much lower compared to Nafion 115. QSIBS/GOAN with 0.5 wt% filler loading possessed the highest selectivity (ratio of ionic conductivity to methanol permeability), which was about 12 times greater than that of the Nafion 115 membrane. The excellent performance was attributed to the barrier effect of GOAN, which favored the methanol crossover, and also due to the interconnected ion transport channels between GOAN and QSIBS by the hydrophilic groups in the GOAN, which favored the ionic transport. These results demonstrated that developed QSIBS/GOAN system could be utilized as promising AEMs for DMFCs [98].

5.3 Conclusion and Future Prospects

Fuel cells are preferred over other energy producing devices for stationary, transportation and marine applications due to their advantages such as high energy density, low polluting emissions, silent operation, simple maintenance and longer operating durations than batteries. Perfluorosulfonic acid based polymers are the most promising candidates as proton exchange membrane electrolytes because of their excellent proton conducting property, good thermal, mechanical, physical and chemical stabilities at moderate temperatures and in a wide range of relative humidities. The high cost, high methanol permeability, low proton conductivity and loss of mechanical stability at high temperature and under low humidity practically limited the applications of PFSA polymers as proton conduction membranes for fuel cells. The modification of Nafion membrane with various materials has proved to be a promising method in attaining desired properties such as high proton conductivities, low methanol permeability and good mechanical properties. Specifically, the incorporation of nanocarbon materials into polymer matrices can remarkably improve their applicability as PEMs.

CNTs and GO are advanced nanomaterials that could be used as additives in the fabrication of electrolyte membranes for potential fuel cell applications. CNT/polymer composites were employed to improve the mechanical properties and reduce the methanol crossover with negligible effect on the proton transport behavior. The integration of GO into polymer matrices has shown to enhance the mechanical properties, proton transport, methanol blocking properties of PEMs and hence can be used as remarkable membranes for fuel cells. The homogenous dispersion of the filler in host polymeric matrices and strong interfacial interactions of the filler with the polymer matrix are important factors determining the performance of polymer/nanocarbon composite membranes. The modification of nanocarbon materials with different functional groups leads to a better compatibility between the nanocarbon materials and polymer matrix. It was revealed that the use of functionalized nanocarbon materials as filler in polymeric membranes could greatly improve the properties of state-of-the-art polymer membranes. Various kinds of functionalized CNTs and GO were explored and employed in the fabrication of hybrid membranes. The addition of sulfonic or phosphonic acid groups to CNTs and GO resulted in an increase in the proton conductivity of membrane electrolytes. CNTs and GO can be more effectively solubilized through functionalization with matrix polymer and polymer/nanocarbon composite membranes based on matrix polymer-functionalized nanocarbon materials have shown supreme properties such as improved proton conductivity, alleviated methanol crossover effect and good mechanical strength even at high temperature and low humidity conditions, which could ultimately lead to a better fuel cell performance compared to the currently used PEM electrolytes. Still further research is needed to improve the membrane properties so as to obtain desirable electrolyte membranes.

Future research efforts in this area will focus on the preparation of high performance hybrid membranes by the incorporation of modified nanocarbon materials into polymer electrolytes and the performance evaluation of the fuel cells based on the newly developed PEMs. The mass production of fuel cells and potential utilization of newly developed hybrid composite membranes based on modified nanocarbon materials can revolutionize the field of electricity in the near future.

References

1. Li, X., Fields, L., and Way, G. (2006) Principles of fuel cells. *Platinum Metals Review,* **50**, 200-201.
2. Peighambardoust, S. J., Rowshanzamir, S., and Amjadi, M. (2010) Review of the proton exchange membranes for fuel cell applications. *International Journal of Hydrogen Energy,* **35**, 9349-9384.
3. Barbir, F., and Gomez, T. (1997) Efficiency and economics of proton exchange membrane (PEM) fuel cells. *International Journal of Hydrogen Energy,* **22**, 1027-1037.
4. Kamarudin, S. K., Daud, W. R. W., Ho, S. L., and Hasran, U. A. (2007) Overview on the challenges and developments of micro-direct methanol fuel cells (DMFC). *Journal of Power Sources,* **163**, 743-754.
5. Kamarudin, S. K., Achmad, F., and Daud, W. R. W. (2009) Overview on the application of direct methanol fuel cell (DMFC) for portable electronic devices. *International Journal of Hydrogen Energy,* **34**, 6902-6916.
6. Ismail, A., Kamarudin, S. K., Daud, W. R. W., Masdar, S., and Yosfiah, M. R. (2011) Mass and heat transport in direct methanol fuel cells. *Journal of Power Sources,* **196**, 9847-9855.
7. Rikukawa, M., and Sanui, K. (2000) Proton-conducting polymer electrolyte membranes based on hydrocarbon polymers. *Progress*

in Polymer Science, **25**, 1463-1502.

8. Ye, Y.-S., Rick, J., and Hwang, B.-J. (2012) Water soluble polymers as proton exchange membranes for fuel cells. *Polymers,* **4**(2), 913-963.
9. Kerres, J. A. (2001) Development of ionomer membranes for fuel cells. *Journal of Membrane Science,* **185**, 3-27.
10. Mauritz, K. A., and Moore, R. B. (2004) State of understanding of Nafion. *Chemical Reviews,* **104**, 4535-4586.
11. Appleby, A. J., and Foulkes, F. R. (1989) *Fuel Cell Handbook,* Van Nostrand Reinhold, USA.
12. Surampudi, S., Narayanan, S. R., Vamos, E., Frank, H., Halpert, G., LaConti, A., Kosek, J., Prakash, G. K. S., and Olah, G. A. (1994) Advances in direct oxidation methanol fuel cells. *Journal of Power Sources,* **47**, 377-385.
13. Ren, X., Wilson, M. S., and Gottesfeld, S. (1996) High performance direct methanol polymer electrolyte fuel cells. *Journal of The Electrochemical Society,* **143**, L12-L15.
14. Antonucci, P. L., Arico, A. S., Creti, P., Ramunni, E., and Antonucci, V. (1999) Investigation of a direct methanol fuel cell based on a composite Nafion®-silica electrolyte for high temperature operation. *Solid State Ionics,* **125**, 431-437.
15. Arico, A. S., Srinivasan, S., and Antonucci, V. (2001) DMFCs: From fundamental aspects to technology development. *Fuel Cells,* **1**, 133-161.
16. Song, S., Zhou, W., Liang, Z., Cai, R., Sun, G., Xin, Q., Stergiopoulos, V., and Tsiakaras, P. (2005) The effect of methanol and ethanol crossover on the performance of PtRu/C-based anode DAFCs. *Applied Catalysis B: Environmental,* **55**, 65-72.
17. Kang, K., Park, S., Gwak, G., Jo, A., Kim, M., Lim, Y.-D., Kim, W.-G., Hong, T., Kim, D., and Ju, H. (2014) Effect of variation of hydrophobicity of anode diffusion media along the through-plane direction in direct methanol fuel cells. *International Journal of Hydrogen Energy,* **39**, 1564-1570.
18. Li, Q., He, R., Jensen, J. O., and Bjerrum, N. J. (2004) PBI-based polymer membranes for high temperature fuel cells - Preparation, characterization and fuel cell demonstration. *Fuel Cells,* **4**, 147-159.
19. Qingfeng, L., Hjuler, H. A., and Bjerrum, N. J. (2001) Phosphoric acid doped polybenzimidazole membranes: Physiochemical characterization and fuel cell applications. *Journal of Applied Electrochemistry,* **31**, 773-779.
20. Iijima, S. (1991) Helical microtubules of graphitic carbon. *Nature,* **354**, 56-58.
21. Baughman, R. H., Zakhidov, A. A., and de Heer, W. A. (2002) Carbon nanotubes - the route toward applications. *Science,* **297**, 787-792.
22. Tsukagoshi, K., Yoneya, N., Uryu, S., Aoyagi, Y., Kanda, A., Ootuka, Y.,

and Alphenaar, B. W. (2002) Carbon nanotube devices for nanoelectronics. *Physica B: Condensed Matter,* **323**, 107-114.
23. Saino Hanna, V., Remya, N., Baiju, G. N., Hanajiri, T., Maekawa, T., Yoshida, Y., and Kumar, D. S. (2010) Sensors based on carbon nanotubes and their applications: A review. *Current Nanoscience,* **6**, 331-346.
24. De Volder, M. F. L., Tawfick, S. H., Baughman, R. H., and Hart, A. J. (2013) Carbon nanotubes: Present and future commercial applications. *Science,* **339**, 535-539.
25. Lee, Y. H., An, K. H., Lim, S. C., Kim, W. S., Jeong, H. J., Doh, C.-H., and Moon, S.-I. (2002) Applications of carbon nanotubes to energy storage devices. *New Diamond & Frontier Carbon Technology,* **12**, 209-228.
26. Ajayan, P. M., and Zhou, O. Z. (2001) Applications of carbon nanotubes. In: *Carbon Nanotubes: Synthesis, Structure, Properties, and Applications*, Dresselhaus, M. S., Dresselhaus, G., and Avouris, P. (eds.), Springer, Germany, pp. 391-425.
27. Wang, C., Waje, M., Wang, X., Tang, J. M., Haddon, R. C., and Yushan (2004) Proton exchange membrane fuel cells with carbon nanotube based electrodes. *Nano Letters,* **4**, 345-348.
28. Danilov, M. O., Kolbasov, G. Y., and Melezhyk, A. V. (2008) Electrodes for fuel cells based on carbon nanotubes and catalysts. In: *Carbon Nanomaterials in Clean Energy Hydrogen Systems*, Baranowski, B., Zaginaichenko, S. Y., Schur, D. V., Skorokhod, V. V., and Veziroglu, A. (eds.), Springer, Netherlands, pp. 279-284.
29. Khantimerov, S. M., Kukovitsky, E. F., Sainov, N. A., and Suleimanov, N. M. (2013) Fuel cell electrodes based on carbon nanotube/metallic nanoparticles hybrids formed on porous stainless steel pellets. *International Journal of Chemical Engineering,* 2013, ID 157098.
30. Luo, C., Xie, H., Wang, Q., Luo, G., and Liu, C. (2015) A review of the application and performance of carbon nanotubes in fuel cells. *Journal of Nanomaterials,* 2015, ID 560392.
31. Wang, W., Ciselli, P., Kuznetsov, E., Peijs, T., and Barber, A. H. (2008) Effective reinforcement in carbon nanotube–polymer composites. *Philosophical Transactions of the Royal Society A: Mathematical, Physical and Engineering Sciences,* **366**, 1613-1626.
32. Coleman, J. N., Khan, U., Blau, W. J., and Gun'ko, Y. K. (2006) Small but strong: A review of the mechanical properties of carbon nanotube-polymer composites. *Carbon,* **44**, 1624-1652.
33. Coleman, J. N., Khan, U., and Gun'ko, Y. K. (2006) Mechanical reinforcement of polymers using carbon nanotubes. *Advanced Materials,* **18**, 689-706.
34. Arash, B., Wang, Q., and Varadan, V. K. (2014) Mechanical properties of carbon nanotube/polymer composites. *Scientific Reports,* 4,

ID 6479.

35. Liu, Y.-H., Yi, B., Shao, Z.-G., Xing, D., and Zhang, H. (2006) Carbon nanotubes reinforced Nafion composite membrane for fuel cell applications. *Electrochemical and Solid-State Letters,* **9**, A356-A359.
36. Thomassin, J.-M., Kollar, J., Caldarella, G., Germain, A., Jerome, R., and Detrembleur, C. (2007) Beneficial effect of carbon nanotubes on the performances of Nafion membranes in fuel cell applications. *Journal of Membrane Science,* **303**, 252-257.
37. Jana, R. N., and Bhunia, H. (2008) Thermal stability and proton conductivity of silane based nanostructured composite membranes. *Solid State Ionics,* **178**, 1872-1878.
38. Wang, L., Xing, D. M., Zhang, H. M., Yu, H. M., Liu, Y. H., and Yi, B. L. (2008) MWCNTs reinforced Nafion® membrane prepared by a novel solution-cast method for PEMFC. *Journal of Power Sources,* **176**, 270-275.
39. Kang, J.-Y., Eo, S.-M., Jeon, I.-Y., Choi, Y. S., Tan, L.-S., and Baek, J.-B. (2010) Multifunctional poly(2,5-benzimidazole)/carbon nanotube composite films. *Journal of Polymer Science Part A: Polymer Chemistry,* **48**, 1067-1078.
40. Guerrero Moreno, N., Gervasio, D., Godinez Garcia, A., and Perez Robles, J. F. (2015) Polybenzimidazole-multiwall carbon nanotubes composite membranes for polymer electrolyte membrane fuel cells. *Journal of Power Sources,* **300**, 229-237.
41. Suryani, Chang, C.-M., Liu, Y.-L., and Lee, Y. M. (2011) Polybenzimidazole membranes modified with polyelectrolyte-functionalized multiwalled carbon nanotubes for proton exchange membrane fuel cells. *Journal of Materials Chemistry,* **21**, 7480-7486.
42. Kannan, R., Kagalwala, H. N., Chaudhari, H. D., Kharul, U. K., Kurungot, S., and Pillai, V. K. (2011) Improved performance of phosphonated carbon nanotube-polybenzimidazole composite membranes in proton exchange membrane fuel cells. *Journal of Materials Chemistry,* **21**, 7223-7231.
43. Jheng, L.-c., Huang, C.-y., and Hsu, S. L.-c. (2013) Sulfonated MWNT and imidazole functionalized MWNT/polybenzimidazole composite membranes for high-temperature proton exchange membrane fuel cells. *International Journal of Hydrogen Energy,* **38**, 1524-1534.
44. Zhang, L., Ni, Q.-Q., Shiga, A., Natsuki, T., and Fu, Y. (2011) Preparation of polybenzimidazole/functionalized carbon nanotube nanocomposite films for use as protective coatings. *Polymer Engineering & Science,* **51**, 1525-1532.
45. Shao, H., Shi, Z., Fang, J., and Yin, J. (2009) One pot synthesis of multiwalled carbon nanotubes reinforced polybenzimidazole hybrids: Preparation, characterization and properties. *Polymer,* **50**, 5987-5995.

46. Wu, J.-F., Lo, C.-F., Li, L.-Y., Li, H.-Y., Chang, C.-M., Liao, K.-S., Hu, C.-C., Liu, Y.-L., and Lue, S. J. (2014) Thermally stable polybenzimidazole/carbon nano-tube composites for alkaline direct methanol fuel cell applications. *Journal of Power Sources,* **246**, 39-48.
47. Kannan, R., Kakade, B. A., and Pillai, V. K. (2008) Polymer electrolyte fuel cells using Nafion-based composite membranes with functionalized carbon nanotubes. *Angewandte Chemie International Edition,* **47**, 2653-2656.
48. Kannan, R., Parthasarathy, M., Maraveedu, S. U., Kurungot, S., and Pillai, V. K. (2009) Domain size manipulation of perflouorinated polymer electrolytes by sulfonic acid-functionalized MWCNTs to enhance fuel cell performance. *Langmuir,* **25**, 8299-8305.
49. Cele, N. P., Sinha Ray, S., Pillai, S. K., Ndwandwe, M., Nonjola, S., Sikhwivhilu, L., and Mathe, " M. K. (2010) Carbon nanotubes based Nafion composite membranes for fuel cell applications. *Fuel Cells,* **10**, 64-71.
50. Kim, Y. H., Sayeed, M. A., Lee, H. K., Park, Y., Gopalan, A. I., Lee, K.-P., and Choi, S.-J. (2012) Functionalized Carbon Nanotubes Reinforced Polymer Electrolyte Membranes Prepared by a Surfactant-assisted Method for Fuel Cell Applications. *2012 12th IEEE International Conference on Nanotechnology (IEEE-NANO),* UK, pp. 1-2.
51. Liu, Y.-H., Yi, B., Shao, Z.-G., Wang, L., Xing, D., and Zhang, H. (2007) Pt/CNTs-Nafion reinforced and self-humidifying composite membrane for PEMFC applications. *Journal of Power Sources,* **163**, 807-813.
52. Lin, Y., Zhou, B., Shiral Fernando, K. A., Liu, P., Allard, L. F., and Sun, Y.-P. (2003) Polymeric carbon nanocomposites from carbon nanotubes functionalized with matrix polymer. *Macromolecules,* **36**, 7199-7204.
53. Lin, Y., Meziani, M. J., and Sun, Y.-P. (2007) Functionalized carbon nanotubes for polymeric nanocomposites. *Journal of Materials Chemistry,* **17**, 1143-1148.
54. Chang, C.-M., and Liu, Y.-L. (2010) Functionalization of multi-walled carbon nanotubes with non-reactive polymers through an ozone-mediated process for the preparation of a wide range of high performance polymer/carbon nanotube composites. *Carbon,* **48**, 1289-1297.
55. Chen, W.-F., Wu, J.-S., and Kuo, P.-L. (2008) Poly(oxyalkylene)diamine-functionalized carbon nanotube/perfluorosulfonated polymer composites: Synthesis, water state, and conductivity. *Chemistry of Materials,* **20**, 5756-5767.
56. Liu, Y.-L., Su, Y.-H., Chang, C.-M., Suryani, Wang, D.-M., and Lai, J.-Y. (2010) Preparation and applications of Nafion-functionalized multiwalled carbon nanotubes for proton exchange membrane fuel cells. *Journal of Materials Chemistry,* **20**, 4409-4416.

57. Hasani-Sadrabadi, M. M., Dashtimoghadam, E., Majedi, F. S., Moaddel, H., Bertsch, A., and Renaud, P. (2013) Superacid-doped polybenzimidazole-decorated carbon nanotubes: a novel high-performance proton exchange nanocomposite membrane. *Nanoscale,* **5**, 11710-11717.
58. Kannan, R., Aher, P. P., Palaniselvam, T., Kurungot, S., Kharul, U. K., and Pillai, V. K. (2010) Artificially designed membranes using phosphonated multiwall carbon nanotube-polybenzimidazole composites for polymer electrolyte fuel cells. *The Journal of Physical Chemistry Letters,* **1**, 2109-2113.
59. Joo, S. H., Pak, C., Kim, E. A., Lee, Y. H., Chang, H., Seung, D., Choi, Y. S., Park, J.-B., and Kim, T. K. (2008) Functionalized carbon nanotube-poly(arylene sulfone) composite membranes for direct methanol fuel cells with enhanced performance. *Journal of Power Sources,* **180**, 63-70.
60. Tripathi, B. P., Schieda, M., Shahi, V. K., and Nunes, S. P. (2011) Nanostructured membranes and electrodes with sulfonic acid functionalized carbon nanotubes. *Journal of Power Sources,* **196**, 911-919.
61. Gong, C., Zheng, X., Liu, H., Wang, G., Cheng, F., Zheng, G., Wen, S., Law, W.-C., Tsui, C.-P., and Tang, C.-Y. (2016) A new strategy for designing high-performance sulfonated poly(ether ether ketone) polymer electrolyte membranes using inorganic proton conductor-functionalized carbon nanotubes. *Journal of Power Sources,* **325**, 453-464.
62. Jana, R. N., Maity, B., Mallick, S., Majumdar, A., and Singh, P. (2015) Nanostructured ionomeric membranes for direct methanol fuel cell. *Indian Chemical Engineer,* **57**, 103-114.
63. Dreyer, D. R., Park, S., Bielawski, C. W., and Ruoff, R. S. (2010) The chemistry of graphene oxide. *Chemical Society Reviews,* **39**, 228-240.
64. Zhu, J., Liu, F., Mahmood, N., and Hou, Y. (2015) Graphene polymer nanocomposites for fuel cells. In: *Graphene-Based Polymer Nanocomposites in Electronics,* Sadasivuni, K. K., Ponnamma, D., Kim, J., and Thomas, S. (eds.), Springer International Publishing, Switzerland, pp. 91-130.
65. Kumar, R., Mamlouk, M., and Scott, K. (2011) A graphite oxide paper polymer electrolyte for direct methanol fuel cells. *International Journal of Electrochemistry,* **2011**, ID 434186.
66. Cao, Y.-C., Xu, C., Wu, X., Wang, X., Xing, L., and Scott, K. (2011) A poly (ethylene oxide)/graphene oxide electrolyte membrane for low temperature polymer fuel cells. *Journal of Power Sources,* **196**, 8377-8382.
67. Choi, B. G., Huh, Y. S., Park, Y. C., Jung, D. H., Hong, W. H., and Park, H. (2012) Enhanced transport properties in polymer electrolyte

composite membranes with graphene oxide sheets. *Carbon,* **50**, 5395-5402.

68. Wang, L., Kang, J., Nam, J.-D., Suhr, J., Prasad, A. K., and Advani, S. G. (2015) Composite membrane based on graphene oxide sheets and Nafion for polymer electrolyte membrane fuel cells. *ECS Electrochemistry Letters,* **4**(1), F1-F4.
69. Sha Wang, L., Nan Lai, A., Xiao Lin, C., Gen Zhang, Q., Mei Zhu, A., and Lin Liu, Q. (2015) Orderly sandwich-shaped graphene oxide/Nafion composite membranes for direct methanol fuel cells. *Journal of Membrane Science,* **492**, 58-66.
70. Lin, C. W., and Lu, Y. S. (2013) Highly ordered graphene oxide paper laminated with a Nafion membrane for direct methanol fuel cells. *Journal of Power Sources,* **237**, 187-194.
71. He, Y., Tong, C., Geng, L., Liu, L., and Lu, C. (2014) Enhanced performance of the sulfonated polyimide proton exchange membranes by graphene oxide: Size effect of graphene oxide. *Journal of Membrane Science,* **458**, 36-46.
72. Uregen, N., Pehlivanoglu, K., Ozdemir, Y., and Devrim, Y. (2017) Development of polybenzimidazole/graphene oxide composite membranes for high temperature PEM fuel cells. *International Journal of Hydrogen Energy,* **42**(4), 2636-2647.
73. Choi, B. G., Hong, J., Park, Y. C., Jung, D. H., Hong, W. H., Hammond, P. T., and Park, H. (2011) Innovative polymer nanocomposite electrolytes: Nanoscale manipulation of ion channels by functionalized graphenes. *ACS Nano,* **5**, 5167-5174.
74. Zarrin, H., Higgins, D., Jun, Y., Chen, Z., and Fowler, M. (2011) Functionalized graphene oxide nanocomposite membrane for low humidity and high temperature proton exchange membrane fuel cells. *The Journal of Physical Chemistry C,* **115**, 20774-20781.
75. Sahu, A. K., Ketpang, K., Shanmugam, S., Kwon, O., Lee, S., and Kim, H. (2016) Sulfonated graphene–Nafion composite membranes for polymer electrolyte fuel cells operating under reduced relative humidity. *The Journal of Physical Chemistry C,* **120**, 15855-15866.
76. Enotiadis, A., Angjeli, K., Baldino, N., Nicotera, I., and Gournis, D. (2012) Graphene-based Nafion nanocomposite membranes: Enhanced proton transport and water retention by novel organo-functionalized graphene oxide nanosheets. *Small,* **8**, 3338-3349.
77. Lee, D. C., Yang, H. N., Park, S. H., and Kim, W. J. (2014) Nafion/graphene oxide composite membranes for low humidifying polymer electrolyte membrane fuel cell. *Journal of Membrane Science,* **452**, 20-28.
78. Xu, C., Cao, Y., Kumar, R., Wu, X., Wang, X., and Scott, K. (2011) A polybenzimidazole/sulfonated graphite oxide composite membrane for high temperature polymer electrolyte membrane fuel cells. *Journal of Materials Chemistry,* **21**, 11359-11364.

79. Liu, Y., Wang, J., Zhang, H., Ma, C., Liu, J., Cao, S., and Zhang, X. (2014) Enhancement of proton conductivity of chitosan membrane enabled by sulfonated graphene oxide under both hydrated and anhydrous conditions. *Journal of Power Sources,* **269**, 898-911.
80. Gahlot, S., Sharma, P. P., Kulshrestha, V., and Jha, P. K. (2014) SGO/SPES-based highly conducting polymer electrolyte membranes for fuel cell application. *ACS Applied Materials & Interfaces,* **6**, 5595-5601.
81. Kumar, R., Mamlouk, M., and Scott, K. (2014) Sulfonated polyether ether ketone - sulfonated graphene oxide composite membranes for polymer electrolyte fuel cells. *RSC Advances,* **4**, 617-623.
82. Pandey, R. P., Thakur, A. K., and Shahi, V. K. (2014) Sulfonated polyimide/acid-functionalized graphene oxide composite polymer electrolyte membranes with improved proton conductivity and water-retention properties. *ACS Applied Materials & Interfaces,* **6**, 16993-17002.
83. Xue, C., Zou, J., Sun, Z., Wang, F., Han, K., and Zhu, H. (2014) Graphite oxide/functionalized graphene oxide and polybenzimidazole composite membranes for high temperature proton exchange membrane fuel cells. *International Journal of Hydrogen Energy,* **39**, 7931-7939.
84. Shukla, A., Bhat, S. D., and Pillai, V. K. (2016) Simultaneous unzipping and sulfonation of multi-walled carbon nanotubes to sulfonated graphene nanoribbons for nanocomposite membranes in polymer electrolyte fuel cells. *Journal of Membrane Science,* **520**, 657-670.
85. Tseng, C.-Y., Ye, Y.-S., Cheng, M.-Y., Kao, K.-Y., Shen, W.-C., Rick, J., Chen, J.-C., and Hwang, B.-J. (2011) Sulfonated polyimide proton exchange membranes with graphene oxide show improved proton conductivity, methanol crossover impedance, and mechanical properties. *Advanced Energy Materials,* **1**, 1220-1224.
86. Ye, Y.-S., Tseng, C.-Y., Shen, W.-C., Wang, J.-S., Chen, K.-J., Cheng, M.-Y., Rick, J., Huang, Y.-J., Chang, F.-C., and Hwang, B.-J. (2011) A new graphene-modified protic ionic liquid-based composite membrane for solid polymer electrolytes. *Journal of Materials Chemistry,* **21**, 10448-10453.
87. He, Y., Wang, J., Zhang, H., Zhang, T., Zhang, B., Cao, S., and Liu, J. (2014) Polydopamine-modified graphene oxide nanocomposite membrane for proton exchange membrane fuel cell under anhydrous conditions. *Journal of Materials Chemistry A,* **2**, 9548-9558.
88. Mishra, A. K., Kim, N. H., Jung, D., and Lee, J. H. (2014) Enhanced mechanical properties and proton conductivity of Nafion–SPEEK–GO composite membranes for fuel cell applications. *Journal of Membrane Science,* **458**, 128-135.
89. Yuan, T., Pu, L., Huang, Q., Zhang, H., Li, X., and Yang, H. (2014) An

effective methanol-blocking membrane modified with graphene oxide nanosheets for passive direct methanol fuel cells. *Electrochimica Acta,* **117**, 393-397.

90. Movil, O., Frank, L., and Staser, J. A. (2015) Graphene oxide-polymer nanocomposite anion-exchange membranes. *Journal of The Electrochemical Society,* **162**, F419-F426.
91. Zhao, L., Li, Y., Zhang, H., Wu, W., Liu, J., and Wang, J. (2015) Constructing proton-conductive highways within an ionomer membrane by embedding sulfonated polymer brush modified graphene oxide. *Journal of Power Sources,* **286**, 445-457.
92. Chu, F., Lin, B., Feng, T., Wang, C., Zhang, S., Yuan, N., Liu, Z., and Ding, J. (2015) Zwitterion-coated graphene-oxide-doped composite membranes for proton exchange membrane applications. *Journal of Membrane Science,* **496**, 31-38.
93. Ye, Y.-S., Cheng, M.-Y., Xie, X.-L., Rick, J., Huang, Y.-J., Chang, F.-C., and Hwang, B.-J. (2013) Alkali doped polyvinyl alcohol/graphene electrolyte for direct methanol alkaline fuel cells. *Journal of Power Sources,* **239**, 424-432.
94. Kim, Y., Ketpang, K., Jaritphun, S., Park, J. S., and Shanmugam, S. (2015) A polyoxometalate coupled graphene oxide-Nafion composite membrane for fuel cells operating at low relative humidity. *Journal of Materials Chemistry A,* **3**, 8148-8155.
95. Yang, J., Liu, C., Gao, L., Wang, J., Xu, Y., and He, R. (2015) Novel composite membranes of triazole modified graphene oxide and polybenzimidazole for high temperature polymer electrolyte membrane fuel cell applications. *RSC Advances,* **5**, 101049-101054.
96. Bai, H., Li, Y., Zhang, H., Chen, H., Wu, W., Wang, J., and Liu, J. (2015) Anhydrous proton exchange membranes comprising of chitosan and phosphorylated graphene oxide for elevated temperature fuel cells. *Journal of Membrane Science,* **495**, 48-60.
97. Pandey, R. P., and Shahi, V. K. (2015) Sulphonated imidized graphene oxide (SIGO) based polymer electrolyte membrane for improved water retention, stability and proton conductivity. *Journal of Power Sources,* **299**, 104-113.
98. Dai, P., Mo, Z.-H., Xu, R.-W., Zhang, S., Lin, X., Lin, W.-F., and Wu, Y.-X. (2016) Development of a cross-linked quaternized poly(styrene-b-isobutylene-b-styrene)/graphene oxide composite anion exchange membrane for direct alkaline methanol fuel cell application. *RSC Advances,* **6**, 52122-52130.

6

Energy Storage Devices

6.1 Introduction

The ever-growing energy consumption and consequential diminishing of fossil fuels have triggered the need to develop effective, economic and environmentally friendly energy conversion and storage technologies, especially using renewable energy sources. In this respect, various energy storage devices such as supercapacitors, lithium-ion batteries (LIBs), lithium-sulfur batteries (LSBs), sodium-ion batteries (SIBs), etc., have gained preferable recognition for diverse applications [1-4]. Considerable advancements have also been attained in the recent years, especially related to energy density, power density and longevity, to develop high performance energy storage devices with good flexibility, light weight and modulating morphological features for distinct applications in a large number of diverse fields [5,6].

Supercapacitors or electrochemical capacitors/ultracapacitors are a special class of energy storage devices displaying outstanding power density, prolonged cycle life, ability to handle elevated power rates, rapid charging-discharging, charge propagation dynamics, pulse power supply, cryogenic workability, notable proficiency to cover the power/energy breach, working safety, low maintenance need, etc. [4,7-10]. These remarkable features of supercapacitors drive their applications in hybrid electrical vehicles, memory backup appliances, sustained power supplies, sizeable industrial apparatus, renewable energy power plants, portable implements, emergency systems, flash photography, breaking systems, military implements, wearable electronics like watches, smart wristbands, displays, cloths, bags, health monitoring implements, etc. [6,11-17]. Based on the energy storage mechanism or cell design, supercapacitors are sorted into three groups, namely, electric double layer capacitors (EDLCs), pseudocapacitors and hybrid capacitors [4,9,17,18]. In the case of EDLCs, the energy storage mechanism is related to the charge build-up at the boundaries of electrodes and

Anish M. Varghese and Vikas Mittal
Khalifa University of Science and Technology, Abu Dhabi, UAE

electrolytes. The substantial storage capacity of these supercapacitors can be achieved through the use of electrode materials with considerable surface area together with appreciable nanoscopic porosity and electrical conductivity, e.g. nano-porous carbonaceous materials such as graphene, carbon nanotubes and activated carbon [6,15,18-23]. The energy storage mechanism of pseudocapacitors is based on the faradic redox reactions, which are reversible as well as continuous in nature. In contrast to EDLCs, pseudocapacitors exhibit greater specific capacitance and energy density. For this purpose, electrodes of various nanostructured frameworks based on conducting polymers, functionalized carbonaceous materials as well as oxides and hydroxides of various metals have received wide acceptance [6,7,15,18,19,21,24]. Hybrid capacitors are desirable class of supercapacitors developed through the integration of EDLC electrode and pseudocapacitor/lithium-ion cell electrode to overcome the difficulties associated with individual EDLC and pseudocapacitor systems and, thereby, to obtain good mechanical permanence and improved cycle life [4, 5,25,26].

Lithium-ion batteries (LIBs) are one of the rechargeable energy storage devices with significant energy density owing to their considerable charge accumulation on the electrode surfaces resulting from the movements of active Li ions [25,27,28]. LIBs have high potential for wide range of energy storage applications such as hybrid/full electric vehicles, electric power implements, portable electronic devices like mobile phones, camcorders, computers, etc., due to their remarkable energy density, cyclability, nominal capacity loss, efficiency, appreciable operating voltage and operational safety [1,2,29-36]. However, the power density of LIBs is lower as compared to supercapacitors [25], thus, significant attempts have been made to enrich it without affecting energy density and prolonged cyclability.

Sodium-ion batteries (SIBs) are advanced class of high performance rechargeable energy storage devices with features such as easy availability, non-poisonous and satisfactory redox potential of sodium together with compatibility with Li. These intelligent energy storage devices display considerable potential for use in hybrid/full electric vehicles, power tools, etc. [2,37-39]. Lately, energy storage devices based on lithium-sulfur batteries and lithium-air batteries have been introduced, which exhibit benefits such as capacity, working voltage, cost and safety [40-42]. Figure 6.1 demonstrates the major categories of energy storage devices and their commonly used electrode structures. To date, many research studies have been carried out to further enrich the performance of these smart energy storage devices. The present review focuses on some of the recent advancements in the development of these functional energy storage devices.

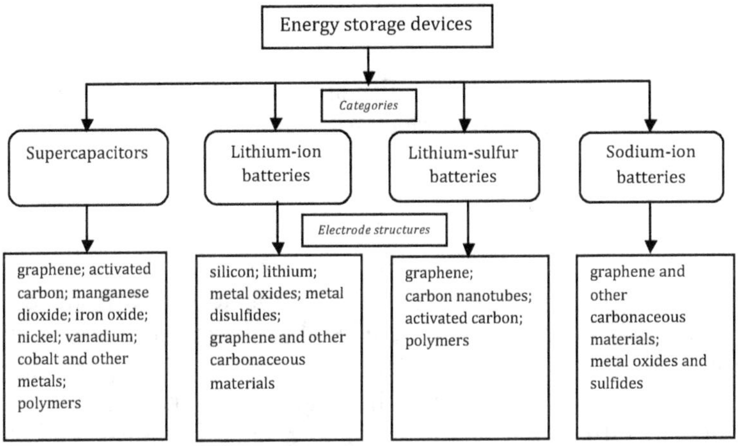

Figure 6.1 Major categories of energy storage devices and their electrode materials.

6.2 Supercapacitors

6.2.1 Graphene Based Supercapacitors

Due to the outstanding physical, chemical, thermal and mechanical performance, along with substantial surface area and porosity, graphene has received significant research interest for diverse supercapacitor applications [5]. Wang *et al.* [43] examined the electrochemical properties of graphene necessary for supercapacitor electrodes. Graphene exhibited effective specific capacitance of 205 F/g with capacitance retention of 90% over 1200 charging-discharging cycles. Further, energy density and power density of 29 Wh/kg and 10 kW/kg were observed respectively for graphene based electrode structure. In another study, Liu *et al.* [44] reported appreciable energy densities of about 86 and 136 Wh/kg at ambient temperature and 80 °C respectively for graphene sheets. Lv *et al.* [45] also reported the potential of graphene derived at low-temperature for efficient energy storage applications owing to higher specific capacitance compared to high-temperature exfoliation. In another study, Yan *et al.* [46] attained significant specific capacitance of 456 F/g for low temperature functionalized graphene nano-sheets with the help of a template, $Mg(OH)_2$.

Lee *et al.* [47] studied the electrochemical properties of supercapacitor electrode made of mesoporous graphene nanotubes. Specific capacitance of 206 F/g along with high cyclability of 96% over 10000 cycles were noted. In another study, Xu *et al.* [48] observed high specific capacitance for self-convened structure based on graphene hydrogel. Also, good electrical conductivity along with excellent mechanical and thermal stability were observed. To develop supercapacitor electrode structure exhibiting both transparency and stretchability, Chen *et al.* [49] detailed the use of wrinkled graphene. In another study, Yoo *et al.* [50] also reported in-plane design of graphene and enhancement in supercapacitor performance.

Chemically modified graphene exhibited a specific capacitance of about 135 F/g in aqueous electrolyte and 99 F/g in organic solvents [51]. Also, the chemically modified graphene displayed stabilized performance towards voltage changes owing to appreciable electrical conductivity. Mishra and Ramaprabhu [52] observed higher specific capacitance of 375 F/g for polyaniline functionalized graphene compared to RuO_2, TiO_2 and Fe_3O_4 functionalized graphene. With an aim to generate enhanced electrochemical properties of graphene, Kim *et al.* [53] carried out poly(ionic liquid) modification of graphene. As a result, modified graphene based electrode with an energy density of 6.5 Wh/kg and a power density of 2.4 kW/kg was obtained. In addition, stable electrochemical properties were observed up to 3.5 V. In another study, Lee *et al.* [54] reported restacking hindered graphene through the inclusion of melamine monomers for energy storage performance as supercapacitor electrodes. The resultant electrode material displayed specific capacitance of about 210 F/g, which remained unaltered over 20000 charging-discharging cycles. In another study, the electrochemical properties of chemically modified graphene were enriched through the development of 3D macroporous structure and further improvement in performance was also achieved by introducing MnO_2 [55]. The resultant electrode structure comprising of 3D graphene/MnO_2 hybrid system displayed superior electrochemical properties such as high specific capacitance of 389 F/g, energy density of 44 Wh/kg and power density of 25 kW/kg, together with appreciable stability and cyclability.

Chemical doping is one of the useful routes to enhance the electrochemical properties of graphene based supercapacitors. In a recent study, Han *et al.* [56] reported the generation of boron-doped graphene to widen the potential of graphene based supercapacitors. Boron-doped graphene based electrode architecture possessed improved specific capacitance of 200 F/g with an excellent cyclability over 4500 cycles on account of its enhanced surface area. In another study, Paek *et al.* [57] acquired improvement in electrochemical capacitance of graphene through N-doping due to the modification in its molecular architecture. Ma *et al.* [58] also reported excellent electrochemical capacitance of 618 F/g for Fe_2O_3 decorated N-doped graphene based electrode structure. Sui *et al.* [59] reported specific capacitance of 223 F/g with a prolonged cycling stability for electrode structure fabricated using aerogels of N-doped graphene possessing considerable porosity.

Pham *et al.* [60] reported the fabrication of 3D nanostructured supercapacitor electrode using CNT impregnated graphene. The hybrid system offered autonomous capability, flexibility and excellent electrical conductivity along with ultrahigh energy density. In another study, Yu and Dai [61] developed a composite system composed of polyethyleneimine functionalized graphene and CNTs with a specific capacitance of 120 F/g for supercapacitor applications. Lei *et al.* [62] reported specific capacitance of 193 F/g for supercapacitor electrode made of intercalated composite structure based on graphene sheets and MnO_2 decorated CNT. The schematic of the composite synthesis route is shown in Figure 6.2. Also, the resultant electrode structure exhibited high rate capability and good

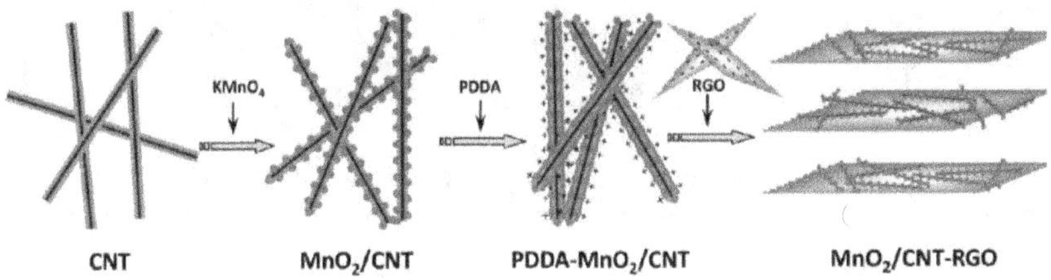

Figure 6.2 Schematic showing the synthesis route for graphene/MnO_2 decorated graphene composite. Reproduced from Reference 62 with permission from American Chemical Society.

cyclability. In another study, You *et al.* [63] reported outstanding specific capacitance, cyclability and significant rate capability for electrode structure made of 3D highly ordered porous structure based on ternary composite containing mesoporous carbon, CNTs and graphene.

Gao *et al.* [64] reported the use of graphene nano-sheet hybrid system of Ni^{2+}/Al^{3+} double hydroxide layered structure (GNS/LDH) for supercapacitor electrode with outstanding specific capacitance of 782 F/g (Figure 6.3). About 38% and 23% increase in specific capacitance was observed after 50 and 200 cycles respectively. In another study, Yang *et al.* [65] obtained specific capacitance of up to 1562 F/g for electrode based on hybrid system of reduced GO, Ni^{2+}/Al^{3+} double hydroxide layered structure and MWCNTs. Further, the system displayed good cycle life permanence. In another recent study, Huang *et at.* [66] reported considerable enhancement in specific capacitance for Co/Al layered double hydroxides through the inclusion of GO nano-sheets.

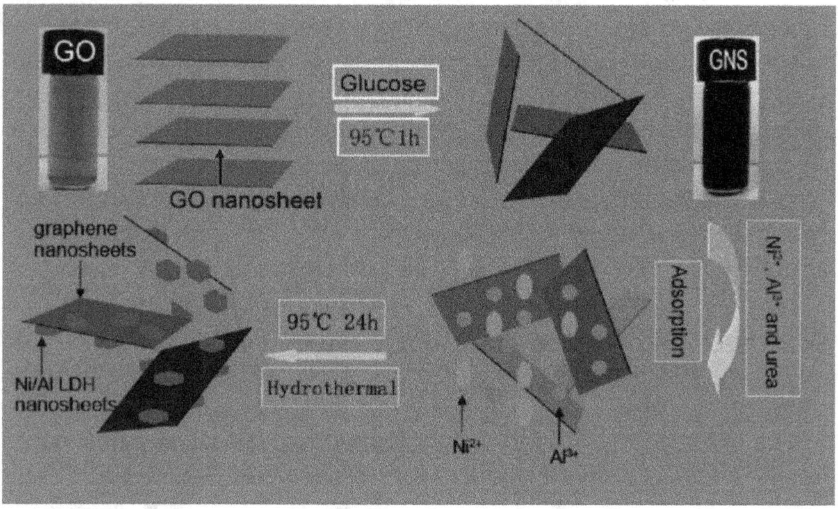

Figure 6.3 Schematic of the formation process of the hybrid GNS/LDH composite. Reproduced from Reference 64 with permission from American Chemical Society.

With an aim to achieve improved supercapacitor performance, Peng *et al.* [67] reported an electrode based on 2D planar nanostructures composed of graphene nano-sheets and MnO_2 nano-sheets. He *et al.* [68] examined the possibility of graphene/MnO_2 hybrid 3D framework for energy storage applications as supercapacitor electrode. The resultant electrode material possessed a specific capacitance of 130 F/g together with low density. In another study, Wu *et al.* [69] observed specific energy of 30.4 Wh/kg and about 79% capacitance recuperation over 1000 charging-discharging cycles for supercapacitor electrode based on graphene composite with MnO_2 nanowires. Lee *et al.* [70] also developed graphene/Mn_3O_4 composite based supercapacitor electrode with appreciable cyclability. The resultant electrode material displayed substantial specific capacitance along with rapid charging-discharging capability, thus, confirming its advanced performance.

The performance of graphene/MnO_2 framework was improved by applying a wrapping of CNTs or conducting polymers [71]. The electrode architecture with 3D conductive wrapping resulted in enhanced capacitance of about 380 F/g with more than 95% capacitance recuperation after 3000 charging-discharging cycles. For fabricating flexible supercapacitor electrode structure, Jin *et al.* [72] used graphene decorated CNT/MnO_2 hybrid displaying an optimal specific capacitance of 487 F/g and specific energy of 25 Wh/kg [72]. Chen *et al.* [73] reported ultrahigh specific capacitance of 1985 F/g and pronounced energy density of 54 Wh/kg for supercapacitor electrode based on ternary composite spheres containing graphene, MnO_2 and $Ni(OH)_2$. The observed performance was attributed to the presence of high surface area and hierarchical arrangement of porous and spherical structures resulting from the synergistic effect of the components.

In the study by Dong *et al.* [74], composite of 3D graphene and cobalt oxide was demonstrated to have high potential for supercapacitor applications due to its higher specific capacitance of about 1100 F/g under 10 A/g and significant cyclability. Graphene/$Ni(OH)_2$ hybrid based supercapacitor electrode was observed by Wang *et al.* [75] to exhibit a specific capacitance as high as 1335 F/g. Ratha and Rout [76] reported the use of a composite of graphene and tungsten disulphide with a layered architecture to generate high performance supercapacitor electrode with a good specific capacitance of 350 F/g, outstanding cyclability and adequate energy density. To enhance the specific capacitance of metal-organic frameworks (MOFs), Choi *et al.* [77] also reported the development of nano-crystalline structure by doping MOFs nanocrystals with graphene. The resultant supercapacitor generated greater stability up to 10000 cycles.

Table 6.1 demonstrates the comparison of specific capacitance of various graphene based supercapacitor electrode structures.

Table 6.1 Comparison of specific capacitance of various graphene based supercapacitor electrodes

Authors	Electrode structure	Specific capacitance (F/g)
Yan *et al.* [46]	Low temperature functionalized graphene nanosheets	456
Lee *et al.* [47]	Mesoporous graphene nanotubes	206
Mishra *et al.* [52]	Polyaniline functionalized graphene	375
Lee *et al.* [54]	Restacking hindered graphene	210
Choi *et al.* [55]	3D graphene/MnO_2 hybrid	389
Han *et al.* [56]	Boron-doped graphene	200
Ma *et al.* [58]	Fe_2O_3 decorated N-doped graphene	618
Sui *et al.* [59]	N-doped graphene aerogels	223
Yu *et al.* [61]	Polyethyleneimine functionalized graphene/CNT composite	120
Lei *et al.* [62]	Graphene sheets/MnO_2 decorated CNT composite	193
Gao *et al.* [64]	Graphene sheet/Ni^{2+}/Al^{3+} LDH composite	782
Yang *et al.* [65]	Reduced GO/Ni^{2+}/Al^{3+} LDH/MWCNT composite	1562
He *et al.* [68]	Graphene/MnO_2 hybrid	130
Yu *et al.* [71]	CNT or conducting polymer covered graphene/MnO_2 framework	380
Jin *et al.* [72]	Graphene decorated CNT/MnO_2 hybrid	487
Chen *et al.* [73]	Graphene/MnO_2/$Ni(OH)_2$ ternary composite spheres	1985
Dong *et al.* [74]	3D graphene/cobalt oxide composite	1100
Wang *et al.* [75]	Graphene/$Ni(OH)_2$ composite	1335
Ratha *et al.* [76]	Graphene/tungsten disulphide composite	350

6.2.2 Activated Carbon Based Supercapacitors

Activated carbon has displayed high potential in energy storage applications due to its unique electrochemical characteristics. Kim *et al.* [78] used graphene sourced carbon with 3D pore architecture and substantial surface area to fabricate supercapacitor electrode. The resultant 3D pore architecture generated energy density of 74

Wh/kg, along with power density of 338 kW/kg and specific capacitance as high as 174 F/g. Zhang *et al.* [79] reported an outstanding power density of 500 kW/kg for graphene derived chemically activated carbon owing to substantial surface area and electrical conductivity. Wang *et al.* [80] made use of hemp fiber sourced carbon nano-sheets having network structure for effective supercapacitor electrode applications due to their high surface area, mesoporosity and electrical conductivity. The authors observed specific energy of 40 Wh/kg along with a specific capacitance of 140 F/g for the developed electrode.

The practicability of activated carbon nanofibers for supercapacitor electrode applications was also confirmed by Barranco *et al.* [81] as specific capacitance of 255 F/g in acidic electrolyte along with prolonged cycle stability were observed. Dai *et al.* [82] reported a specific capacitance of 253 F/g along with high rate capacity and capacitance retention over 5000 cycles for carbon nano-spheres having hollow mesoporous structure and extremely small thickness. Sevilla and Fuertes [83] reported the use of carbon nano-sheets with porous network for designing supercapacitor electrode structure. Specific energy of about 30 Wh/kg with a specific power of 13 kW/kg were observed. Hulicova-Jurkacova *et al.* [84] observed considerable permanence in electrochemical properties for electrode structure based on microporous activated carbon with high amount of phosphorous.

In another study, Wickramaratne *et al.* [85] reported specific capacitance of about 388 F/g together with a long cycle life for porous N-doped carbon spheres. In another study, Chen *et al.* [86] observed excellent supercapacitor properties for porous N-doped carbon nanofibers. Hou *et al.* [87] also reported the use of silk sourced N-doped carbon nano-sheets with a 3D highly ordered porous framework for diverse supercapacitor applications owing to high specific capacitance of 242 F/g, along with high energy density of 102 Wh/kg and an outstanding capacitance recuperation of 91% over 1000 charging-recharging cycles. In another study, Tan *et al.* [88] observed a specific capacitance of 248 F/g along with excellent cyclability over 5000 charging-recharging cycles for supercapacitor electrode made of polyaniline sourced N-doped graphitic carbon nano-cages.

Lu *et al.* [89] reported the use of CNT/activated carbon hybrid as supercapacitor electrode possessing specific energy and specific power of 50 Wh/kg and 22 kW/kg respectively, along with an appreciable specific capacitance of 188 F/g. For designing supercapacitor electrode with substantial power density, Hahm *et al.* [90] utilized 3D framework based electrode composed of nanocomposite of CNTs and carbon nano-cups by combining anodization and chemical vapor deposition techniques (Figure 6.4). The observed increase in the capacitance of the electrode was attributed to high surface area of the hybrid structure.

Shi *et al.* [91] synthesized supercapacitor electrode based on nitrogen impregnated hybrid structure of activated carbon decorated MWCNTs. Specific capacitance of 103 F/g along with substantial cyclability were observed. Du *et al.* [92] fabricated activated carbon/Fe_3O_4 hybrid based supercapacitor electrode with a specific capacitance of 38 F/g in KOH electrolyte solution at 0.5 mA/cm^2 current density. Only 18% loss in inceptive capacitance after 500 cycles was observed.

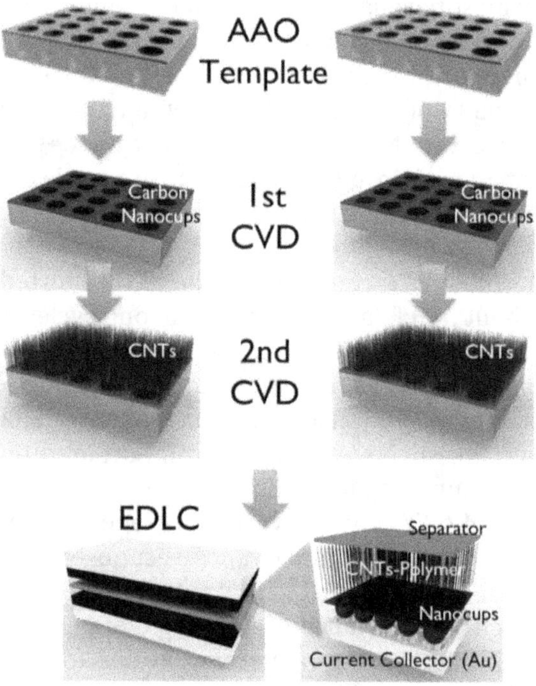

Figure 6.4 Schematic illustrating of the fabrication process (AAO: anodized aluminum oxide; CVD: chemical vapor deposition and EDLC: electrochemical double layer capacitors). Reproduced from Reference 90 with permission from American Chemical Society.

6.2.3 Metal Based Supercapacitors

MnO_2 Based Supercapacitors

In order to develop vast range of high performance supercapacitor materials, MnO_2 has received immense research interest due to its substantial energy storage capacity together with low cost. In the last few years, many studies have also been carried out to enrich the electrical or ionic conductivities of MnO_2 based electrodes, thereby, extending the application range. Qu *et al.* [93] reported the use of MnO_2 nano-rods for the synthesis of supercapacitor cathode with a capacitance of 201 F/g. Energy density and power density of 17 Wh/kg and 2 kW/kg were observed respectively for MnO_2/activated carbon hybrid. Wang *et al.* [94] reported the generation of 3D mesoporous MnO_2 nano-architectures possessing high surface area and specific capacitance of about 200 F/g for supercapacitor electrode application. In another study, Yu *et al.* [95] reported specific capacitance of 120 F/g for MnO_2 with highly oriented 3D architecture.

Yan *et al.* [96] reported a method to coat amorphous MnO_2 on crystalline tin oxide nano-wires for generating supercapacitor electrodes. High specific capacitance of 800 F/g was noted along with 99% capacitance retention over 2000 charging-recharging cycles. In another study, MnO_2/graphene composite exhibited excellent specific capacitance of 465 F/g along with high specific energy of 26 Wh/kg [97]. Li *et al.* [98] also reported specific capacitance of 516 F/g together with a prolonged life cycle for electrode structure based on nanocomposite composed of mesoporous MnO_2 and carbon aerogel. Reddy *et al.* [99] reported the generation of substantial power density of 33 kW/kg for electrode structure based on coaxial Au-MnO_2/CNT nanocomposite arrays. Significant specific energy and rate capability were observed by Yang *et al.* [100] for flexible electrode structure based on MnO_2 nanowires and Fe_2O_3 nanotubes.

For generating supercapacitor electrode with effective rate capability, Jiang *et al.* [101] reported the use of 3D nano-framework based on MnO_2/CNT/reduced GO ternary hybrid. Improved electrochemical properties for the resulting electrode structure were observed (Figure 6.5). Hu *et al.* [102] reported the suitability of MnO_2/CNT/cotton textile ternary composite framework for wearable supercapacitor application as negative electrode. Specific capacitance of 410 F/g was noted for the resultant electrode material. In another study, Hou *et al.* [103] fabricated supercapacitor electrode based on ternary composite containing MnO_2, CNT and poly(3,4-ethylenedioxythiophene)-poly(styrenesulfonate). Appreciable specific capacitance of 427 F/g with capacitance recuperation of 99% over 1000 charging-recharging cycles was noted for the resultant electrode material. This was attributed to the synergistic electrical conducting effect of CNT and conducting polymer in the composition. Yang *et al.* [104] also detailed the fabrication of high performance electrode by the use core-shell nano-cables constituted of hydrogenated ZnO as core and ZnO enriched MnO_2 as shell over carbon cloth. The flexible free-standing electrode displayed a high specific capacitance of 1261 F/g.

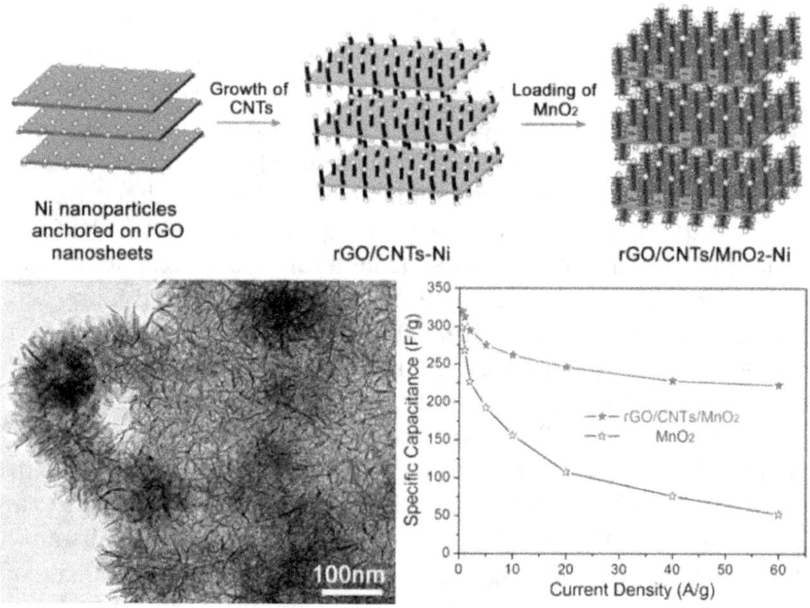

Figure 6.5 Preparation of the synthesis and performance of the hybrid elected. Reproduced from Reference 101 with permission from American Chemical Society.

Nickel Based Supercapacitors

Nickel (Ni) based materials exhibit attractive properties like high storage capacity and outstanding thermal permanence. For instance, Kim *et al.* [105] observed excellent electrochemical performance useful for supercapacitor application for electrode based on 3D NiO nano-framework exhibiting large pore volume. Cai *et al.* [106] reported $NiMoO_4$ nano-spheres possessing hierarchical architecture along with high surface area and electrical conductivity for high performance supercapacitor electrodes. The electrode structure demonstrated specific capacitance and specific energy of up to 974 F/g and 20 Wh/kg respectively along with specific power of 2100 W/kg. Also, capacitance retention over 2000 cycles was observed for the electrode material. Li *et al.* [107] developed Ni-Co nano-sheets with nanostructured pores and extremely small thickness for use in supercapacitors. The resultant supercapacitor displayed energy density of 41 Wh/kg, together with superior specific capacitance of 1304 F/g and excellent cycle life. In another study, Chen *et al.* [108] reported the applicability of 3D nanostructured nickel cobalt sulfide ternary nano-sheet arrays for supercapacitor electrodes. Optimal specific capacitance of 1418 F/g along with specific energy of 60 Wh/kg and prolonged cycle life were observed.

Ji *et al.* [109] reported high power density of 44 kW/kg for electrode structure based on hybrid system containing nano-porous $Ni(OH)_2$ grafted graphitic foam. Zhang *et al.* [110] also reported outstanding electrochemical performance for $Ni(OH)_2$/reduced GO composite film based supercapacitor electrode structure. The resultant composite architecture exhibited high specific capacitance of 1404 F/g as well as superlative cycle life and rate performance. In another study, Dubal *et al.* [111] reported the generation of flexible electrode material from nanostructured composite composed of $Ni(OH)_2$ decorated MWCNTs. Ultrahigh specific capacitance of 1487 F/g together with good cyclability, rate capacity and energy as well as power density were reported.

Many other Ni based materials have been introduced as high performance supercapacitor electrode materials, including $NiCo_2O_4$ porous nanostructure [112], $NiMnO_4$ nano-sheets with mesoporous architecture [113], $NiCo_2O_4$ with size adjustable nano-coral structure [114], $NiCo_2S_4$ mesoporous nano-sheets [115], $NiCo_2S_4$ mesoporous nanoparticles [116], $NiCo_2O_4$ microspheres [117], $NiCo_2O_4$ sub-microspheres [17], morphology controlled nickel cobalt oxide [118], $NiCoO_2$ nanotubes containing nano-sheets [119], $NiCoMnO_2$ mesoporous nanotubes containing nano-sheets [120], $Co_xNi_{1-x}(OH)_2$ nanoparticles [121], $Co_{1-x}Ni_x(OH)_2$ nano-flakes [122], $Co_xNi_{1-x}(OH)_2$ nano-sheets [123], $NiCo_2O_4$ spinel [124], NiCoS spinel [125], $NiCo_2O_4$ nano-sheet arrays decorated Ni foam [126], $NiCo_2S_4$ nanotube arrays decorated Ni foam [127], $NiCo_2S_4$ nanotube arrays decorated Ni foam [128], $NiCo_2O_4$ nano-sheets decorated 3D porous Ni film [129], $NiCo_2S_4$ arrays decorated Ni foam [130], Ni-Co-Mn oxide nano-sheets decorated Ni foam [131], $NiCo_2O_4$ nano-sheets decorated carbon fiber paper [132], $NiCo_2O_4$ nano-needles decorated Ni/graphene hybrid foam with mesoporous 3D structure [133], NiP/MnO_2 hybrid with ordered mesoporous structure [134], Ni-Co_2O_4/3D graphene foam nanocomposite [135], $NiCo_2O_4$/reduced GO nanocomposite [29], $NiCo_2O_4$ nanowire/N-doped graphene hybrid [136], NiO/carbon mesoporous hybrid [137], $Ni(OH)_2$ nano-flakes decorated ZnO nanowires [138], Fe-doped $NiCo_2O_4$ micro-spheres decorated nano-meshes [139], silver nanowire/$Ni(OH)_2$ core/shell hybrid [140], $NiCo_2O_4/NiMoO_4$ hybrid core/shell nano-sheet arrays decorated Ni foam [141], $NiMoO_4/Co_3O_4$ nano-hybrid with 3D structure decorated Ni foam [142], $NiCo_2S_4/MnO_2$ core/shell mesoporous ordered hybrid arrays decorated Ni foam [143], NiCoS/graphene hybrid ordered mesoporous arrays decorated Ni foam [144], $NiCo_2O_4$ nano-belt/graphene nano-sheet hybrid [145], $NiCo_2S_4$/carbon core/shell hybrid arrays decorated Ni foam [146], $NiMoO_4/MnO_2$ hybrid ordered nano-sheet arrays decorated Ti mesh [147], Ni-$Co(OH)_2$ nano-sphere/$NiCo_2O_4$ nanowire hybrid decorated Ni foam [148] and $NiCo_2O_4$/graphene hybrid on Ni foam [149].

Vanadium Based Supercapacitors

Supercapacitor electrodes based on vanadium pentoxide (V_2O_5) nano-porous framework with simplistic ion affinity exhibited specific capacitance of 315 F/g together with superior cyclability [150]. To achieve enhanced electrochemical performance, Zhu *et al.* [151] made use of 3D V_2O_5 frameworks with extremely small wall thickness as well as high surface area and porosity (Figure 6.6). About 451 F/g specific capacitance together with high capacitance stability of greater than 90% over 4000 charging-discharging cycles were observed. The resultant electrode system exhibited energy density of about 107 Wh/kg, along with a power density of 9.4 kW/kg. In another study, Butt *et al.* [152] reported specific capacitance of 360 F/g and cycling stability for supercapacitor electrode based on ZnV_2O_4 nano-spheres with ordered morphology.

Feng *et al.* [153] reported VS_2 nano-sheets with extremely thin layered architecture for supercapacitor electrode applications. The resultant electrode structure possessed a specific capacitance of

4760 $\mu F/cm^2$ and outstanding cycling stability over 1000 cycles. In another study, Lu *et al.* [154] reported the fabrication of supercapacitor with high energy density through the use of VN porous nano-wires based anode structure. Also, many other vanadium based materials have been developed as high performance supercapacitor electrodes, including VN nanofibers [155], hierarchical VN nanostructures [156], V_2O_5/GO hybrid nanofibers [157], V_2O_5 xerogel/graphene sheet hybrid [158], V_2O_5 nanofiber/exfoliated graphene nanocomposite [159] and VN/N-doped graphene hybrid [160].

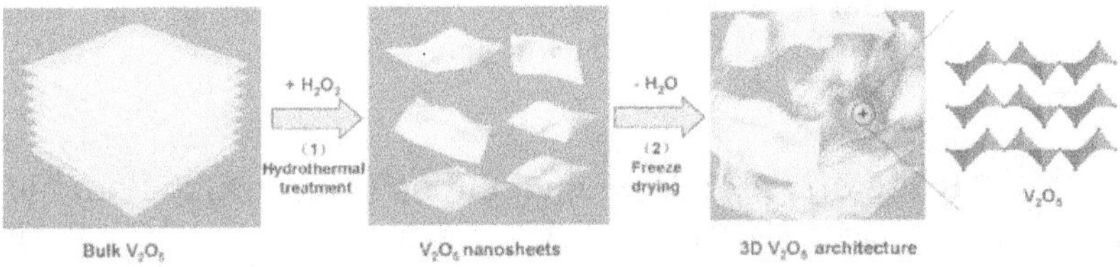

Figure 6.6 Schematic of the fabrication steps of the 3D V_2O_5 architectures. Reproduced from Reference 151 with permission from American Chemical Society.

Cobalt Based Supercapacitors

Owing to their unique features, such as high surface area, superior redox characteristics and modulating size and shape, cobalt (Co) based materials find suitability in a variety of applications as high performance supercapacitor electrodes. Meher *et al.* [161] reported specific capacitance as high as 548 F/g along with outstanding recuperation in capacitance of about 99% over 2000 cycles for Co_3O_4 layered nanostructure due to the facile OH^- ions capture as shown in Figure 6.7.

In another study, Deori *et al.* [162] reported superior electrochemical properties for Co_3O_4 nanoparticles with hexagonal platelet morphology compared to spherical morphology. To enrich the rate capability of supercapacitor electrode developed using Co_3O_4 nanosheets, Rakhi *et al.* [163] also carried out RuO_2 decoration. The resultant electrode structure containing RuO_2 decorated Co_3O_4 nanosheets exhibited a high specific capacitance of 905 F/g along with considerable capacitance retention of greater than 96% for 5000 cycles.

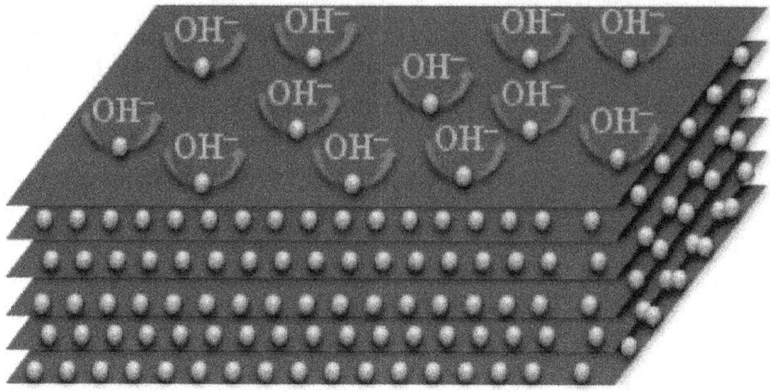

Figure 6.7 Schematic showing the facile OH^- ions capture on multilayered nanostructure of Co_3O_4. Reproduced from Reference 161 with permission from American Chemical Society.

Pendashteh *et al.* [164] developed $CuCo_2O_4$ nanowires with hierarchical mesoporous structure as supercapacitor electrode material. Superior electrochemical properties such as excellent specific capacitance (1210 F/g), specific energy (42.8 Wh/kg), specific power (15 kW/kg) as well as cycling stability were observed. In another study, Huang *et al.* [165] prepared cobalt sulfide with 2D layered structure for supercapacitor electrode application and observed an ultrahigh specific capacitance of 1314 F/g together with appreciable cycling stability. Also, other Co based materials such as CoS nanotubes [166], CoS ordered 3D nanostructure [167], $CoMoO_4$ nano-rods [168], $CoMoO_4$ monoclinic nanostructures [21], spinel $CoMn_2O_4$ nanofibers [169], CoS nanotube/graphene hybrid [170] and $CoMoO_4$/graphene hybrid [171] were reported to have immense potential as supercapacitor electrodes.

Wang *et al.* [172] prepared nano-cable using Co_3O_4 coated MWCNTs for generating effective supercapacitor cathode. The specific capacitance of 590 F/g of the cathode structure remained unaltered after 2000 charging-discharging cycles. With an objective to fabricate supercapacitor electrode with high storage capability, Yang *et al.* [173] made use of highly ordered network framework based on Co_3O_4 nano-net strengthened by carbon fiber paper. High specific capacitance of about 1124 F/g together with effective cyclability were reported.

Salunkhe *et al.* [174] developed nanocomposite based on metal-organic framework sourced Co_3O_4 and carbon exhibiting 3D nanoporous network structure for supercapacitor application. The electrode material exhibited energy density of 36 Wh/kg together with good rate performance and cyclic stability over 2000 charging-discharging cycles. In another study, Xia *et al.* [175] reported the generation of a supercapacitor electrode displaying efficient electrochemical performance by the use of integrated composite of 3D foam with nanowire framework containing Co_3O_4 as core and poly (3,4-ethylenedioxythiophene)/MnO_2 composite as shell.

Iron Based Supercapacitors

The potential of nano-flakes of $FeCo_2O_4$ for the supercapacitor electrode fabrication was revealed by Mohamed *et al.* [176], as specific capacitance of about 433 F/g along with powerful cyclability after 2500 cycles were recorded. Long *et al.* [177] reported specific capacitance of 720 F/g for electrode structure based on iron nanosheets decorated graphene sheets. Guan *et al.* [178] reported Fe_2O_3 decoration over graphite foam/CNTs composite with an objective to acquire supercapacitance properties. The electrode structure exhibited specific energy of about 75 Wh/kg with an outstanding specific power of 1400 W/kg. Also, capacitance retention of more than 95% over 50000 cycles was accomplished.

Supercapacitors Based on Other Metals

Ma *et al.* [179] studied the suitability of hierarchical MoS_2 nanospheres for supercapacitor electrode and observed effective cycle life and specific capacitance of 145 F/g. To enrich the electrochemical properties of MoS_2 based supercapacitor electrode structure, many MoS_2 composites have also been developed, including MoS_2/carbon hybrid [180], N-doped carbon decorated MoS_2 [181], MoS_2/MWCNT composite [182] and MoS_2/carbon aerogel hybrid [183].

Lu *et al.* [184] reported considerable specific capacitance of 3.25 mF/cm^2 and a prolonged cycle life for electrode structure based on hydrogenated TiO_2 nanotube arrays. In another study, Lu *et al.* [185] also reported outstanding cycling stability for nanowire arrays of TiN. Shang *et al.* [186] designed coaxial nanostructured arrays based on $Ni_xCo_{2x}(OH)_{6x}$ decorated TiN nanotubes for high performance supercapacitor electrode application. The generation of an ultrahigh specific capacitance of 2543 F/g together with capacitance retention of about 94% after 5000 charging-discharging cycles were observed for the developed electrode.

Zhang *et al.* [187] synthesized RuO_2 nanotubes for supercapacitor applications and reported specific capacitance of 860 F/g together with high cyclability. In another study, specific capacitance of 953 F/g was reported by Bi *et al.* [188] for nanocomposite containing RuO_2 and CNTs.

6.2.4 Polymer Based Supercapacitors

Polyaniline Based Supercapacitors

Nanofibers of hollow polyaniline with a specific capacitance of 601 F/g exhibited feasibility for use as supercapacitor electrodes owing to the improved ion distribution resulting from the synergistic effect of sequential porous architecture, small wall thickness and hollow framework [189]. Fan *et al.* [190] reported the development of supercapacitor electrode based on polyaniline hollow sphere composite with graphene. The resultant electrode system exhibited superior electrochemical properties such as specific capacitance of 614 F/g and capacitance recuperation of 90% over 500 cycles.

Wang *et al.* [191] described the use of polyaniline coated graphene paper as supercapacitor electrode with sufficient flexibility and self-supporting nature. Specific capacitance of 233 F/g together with mechanical strength of 12.6 MPa were observed. In another study, Wang *et al.* [192] obtained specific capacitance of 590 F/g for supercapacitor electrode based on hierarchical framework of polyaniline nanowire/graphene nanocomposite, which remained unaltered over 200 charging-discharging cycles. Wu *et al.* [193] developed polyaniline nanofiber composite with chemically converted graphene exhibiting substantial flexibility and electrical conductivity for smart supercapacitor electrode application. Specific capacitance of 210 F/g along with good cycling stability and rate capacity were reported.

Jianhua *et al.* [194] reported efficient electrochemical properties such as optimal specific capacitance of 580 F/g and capacitance retention of 96% for polyaniline nanofiber/graphene composite interconnected through amide linkages. Zhang and Zhao [195] observed specific capacitance of 361 F/g for polyaniline/graphene nanocomposite compared to poly(3,4-ethylenedioxythiophene) and polypyrrole based graphene composites. In another study, De Oliveira *et al.* [196] generated p-doped polyaniline composite with carboxylated graphene with a specific capacitance of 278 F/g for supercapacitor electrode applications. Li *et al.* [197] reported specific capacitance of 257 F/g for nanocomposite composed of polyaniline nanoparticles and graphene sheets. Kumar *et al.* [198] also reported specific capacitance of 250 F/g together with effective cyclability for polyaniline composite with reduced GO, also exhibiting substantial electrical conductivity of 8.66 S/cm. Figure 6.8 shows the schematic of the grafting mechanism for the polyaniline/reduced GO composite synthesis.

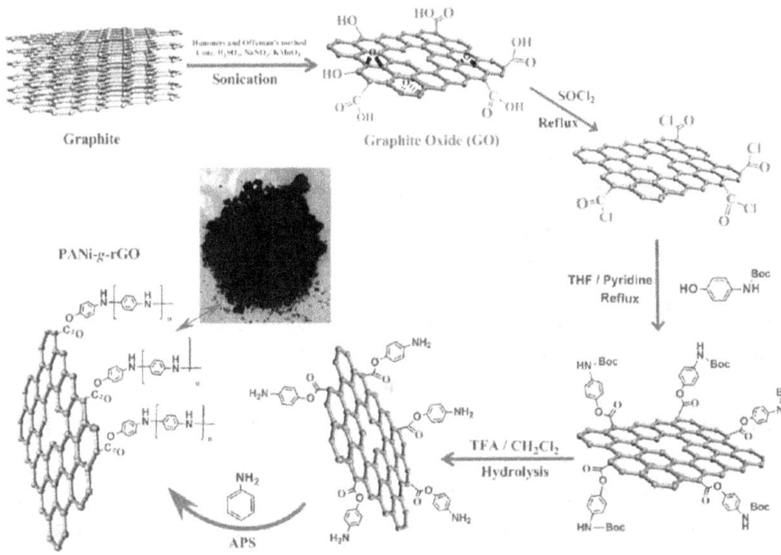

Figure 6.8 Schematic of the grafting mechanism for the polyaniline/ reduced GO composite synthesis. Reproduced from Reference 198 with permission from American Chemical Society.

Lai *et al.* [199] analyzed the effect of various surface modifications of graphene on the specific capacitance of polyaniline. Supercapacitor electrode made of polyaniline composite with amine functionalized graphene demonstrated greater increment in capacitance over GO, pristine graphene and N-doped graphene based composites (Figure 6.9). Moreover, the capacitance of amine functionalized graphene based polyaniline composite remained unaltered after 680 electrochemical cycles. Xu *et al.* [200] reported effective specific capacitance, together with appreciable stability of the electrode framework based on ordered nanocomposite containing polyaniline nanowires and GO sheets. In another study, Xu *et al.* [201] also reported specific capacitance of about 450 F/g for polyaniline decorated carbon/GO nanocomposite film exhibiting 3D ordered framework.

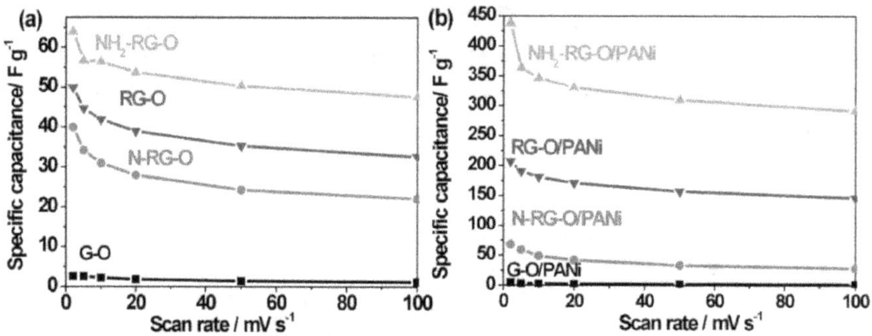

Figure 6.9 Specific capacitance of (a) various graphene and (b) polyaniline/graphene composites. Reproduced from Reference 199 with permission from American Chemical Society.

Meng *et al.* [202] developed polyaniline/CNT composite as effective electrode material for wearable supercapacitor applications. Specific capacitance of about 350 F/g was observed under flexible conditions. Furthermore, the composite system presented high extent of cyclability and low capacitance loss. In another study, Cheng *et al.* [203] synthesized polyaniline laminated etched-carbon fabric based supercapacitor electrode possessing specific capacitance of 673 F/g. Wang *et al.* [204] also observed specific capacitance of 273 F/g along with appreciable electrical conductivity of 5.1 S/cm for polyaniline composite with bacterial sourced cellulose nanofibers based electrode structure. In addition, polyaniline composites with SnO_2 nanoparticles [205], N-doped CNTs [206], black TiO_2 [207], MoS_2 [208], graphene nano-scroll [209], carbon nanofibers [210], graphene/carbon fibers [211], CNT/cellulose nanofibers [212], MnO_2/graphene [213], N-doped graphene/nickel ferrite [214], graphene/$MnFe_2O_4$ [215] and graphene/SnO_2 [216] have been reported to have high potential as supercapacitor electrodes.

Polypyrrole Based Supercapacitors

Zhang *et al.* [217] observed specific capacitance as high as 500 F/g for layered composite structure containing polypyrrole and GO. Davies *et al.* [218] reported the generation of flexible supercapacitor electrode using polypyrrole coated graphene film. The flexible electrode structure exhibited an outstanding energy density of 33 Wh/kg, along with appreciable power density of 1184 W/kg. Also, a specific capacitance of 237 F/g was exhibited. Zhang *et al.* [219] also attained specific capacitance of 480 F/g for electrode structure based on polypyrrole nanofiber composite with graphene. In another study, Biswas and Drzal [220] utilized multilayered nanocomposite architecture composed of polypyrrole nanowires and graphene nano-sheets for generating superior supercapacitor electrode structure, which exhibited specific capacitance of 165 F/g as well as long cycle life.

3D arrays of polypyrrole/RuO_2 hybrid displaying highly oriented structure were used by Zhang *et al.* [221] to fabricate high performance supercapacitor electrode. High specific capacitance of around 302 F/g along with adequate cyclability was reported. In another work, Li *et al.* [222] observed substantial specific capacitance and cycling stability for supercapacitor electrode architecture containing CNT as core, polypyrrole as transitional layer and MnO_2 as superficial layer.

In addition, the supercapacitor electrode materials based on a variety of other polypyrrole based composites with MoS_2 [223], GO nano-ribbons [224], vertically oriented CNTs [225] and graphene/Ag nanoparticles [226] have been demonstrated as favorable materials owing to efficient electrochemical performance.

Poly(3,4-ethylenedioxythiophene) Based Supercapacitors

D'Arcy *et al.* [227] generated poly(3,4-ethylenedioxythiophene) (PEDOT) nanofibers and observed excellent supercapacitor performance owing to their appreciable electrical conductivity. Aradilla *et al.* [228] also reported enhanced electrochemical capacitance for supercapacitor electrode structure constituted of PEDOT and poly(*N*-methylpyrrole) multilayered system. In another study, Liu and Lee [229] used 1D nanomaterials based on PEDOT and MnO_2 for developing high performance energy storage device possessing effective electrochemical performance and mechanical properties. Tang *et al.* [230] also reported the practicability of flexible and low cost interconnected PEDOT/MnO_2/graphite composite to serve as intelligent supercapacitor electrode. Specific capacitance of 196 F/g along with notable energy density were observed for the developed electrode.

6.3 Lithium-Ion Batteries (LIBs)

6.3.1 Silicon Based LIB Anodes

Silicon has received considerable interest as LIB anode material owing to significant specific capacity compared to other materials. Also, silicon based materials offer other advantages including comparatively smaller operational potential, environmental friendliness and easy availability. Capacity reduction of the anode materials resulting from volume change upon lithiation-delithiation actions has also been studied in depth. In this respect, the application of stable surface coatings on Si anodes has presented immense potential. For instance, Lu *et al.* [231] reported the use of nonfilling carbon coating on Si micro-particle porous framework. Ultrahigh reversible specific capacity of 1500 mAh/g together with cyclability after 1000 cycles were noted for the resultant composite framework. This was attributed to the reduced contact area between carbon coated electrode and electrolyte through the reduced electrolyte penetration. In another study, Jia *et al.* [232] also observed reversible specific capacitance of about 1500 mAh/g after 100 charging-discharging cycles for surface carbon coated hierarchical 3D mesoporous Si based LIB anode structure.

Peled *et al.* [233] reported Si nanowires with 3D architecture as anode. High storage capacity of 457 mAh/cm^2 along with outstanding reversible capacity, cyclability and almost 100% current efficiency were observed. In another study, Teki *et al.* [234] also observed reversible storage capacity and prolonged cyclability for LIB anode based on both Si nanowire and nano-rod arrays. Sourice *et al.* [235] reported specific capacitance and reversible capacitance in the range of 2400 mAh/g and 500 mAh/g respectively after 500 cycles for carbon coated Si nanoparticles. Gohier *et al.* [236] reported rate capability of 1900 mAh/g and storage capacity of 3000 mAh/g for ordered nano-architecture composed of Si and aligned CNTs. Jung *et al.* [237] revealed graphene covered Si based metallic glass alloy as LIB anode with good specific capacity and rate capability. Liu *et al.* [238] also used naturally rolled Si/reduced GO layered nanostructure as high performance LIB anode with prolonged cycle life for up to 2000 charging-discharging cycles (Figure 6.10). He *et al.* [239] reported enhancement in cyclability of patterned Si through the application of alumina coating. Also, this approach was helpful to improve the fracture resistance of Si based LIB anodes. In another study, Han *et al.* [240] obtained efficient performance of LIB anode based on Si nanoparticles through the application of metal-organic framework (ZIF-8) coating. Yu *et al.* [241] developed thin sheet LIB anode with favorable stress relaxation by depositing Si ribbon arrays on polydimethylsiloxane layer for stretchable applications.

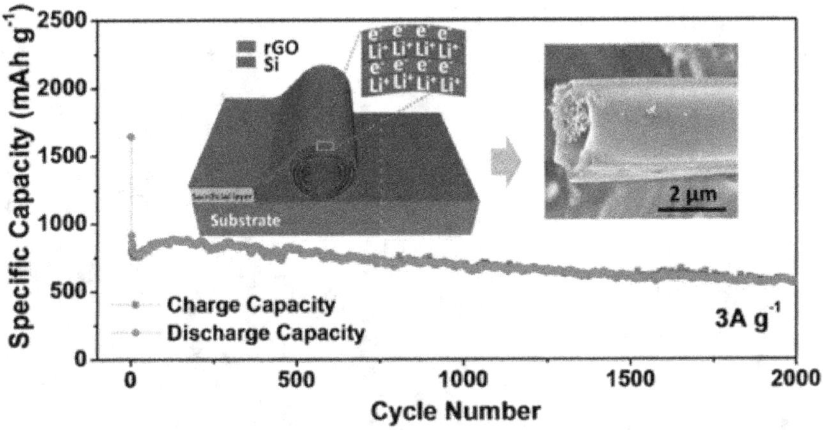

Figure 6.10 Performance of the LIB anode. Reproduced from Reference 238 with permission from American Chemical Society.

With an aim to enhance electrochemical performance, Deng *et al.* [242] developed nano-membrane based electrode structure comprised of C/Si/C trilayer with naturally rolled up tubular configuration. The tubular composite structure demonstrated high specific capacitance of 2000 mAh/g as well as reversible capacitance of 1000 mAh/g for 300 charging-discharging cycles. In another study, Liu *et al.* [243] generated C/Si/C trilayer nanotube arrays with substantially high electrochemical capacitance of 2200 mAh/g, consistent Coulombic efficiency and improved cycling stability (Figure 6.11). In another study, Jia *et al.* [244] developed LIB anode structure based on $NiSi_2$/Si/carbon ternary composite. Specific capacitance of 1272 mAh/g after 200 cycles was reported for the anode structure, along with superior rate capability.

Cao *et al.* [245] developed LIB anode based on Cu/Si core/shell nano-cable arrays and their Al_2O_3 decorated analog for effective electrochemical performance. Both anode structures presented outstanding electrochemical capacity and cyclability, with the one containing Al_2O_3 surface coating exhibiting higher rate capability. In another study, Hu *et al.* [246] demonstrated the generation of nanocomposite composed of Si as core and SiO_x/C hybrid as shell for generating LIB anode. The resultant composite exhibited reversible specific capacity of 1100 mAh/g along with good rate capability and long cycle life.

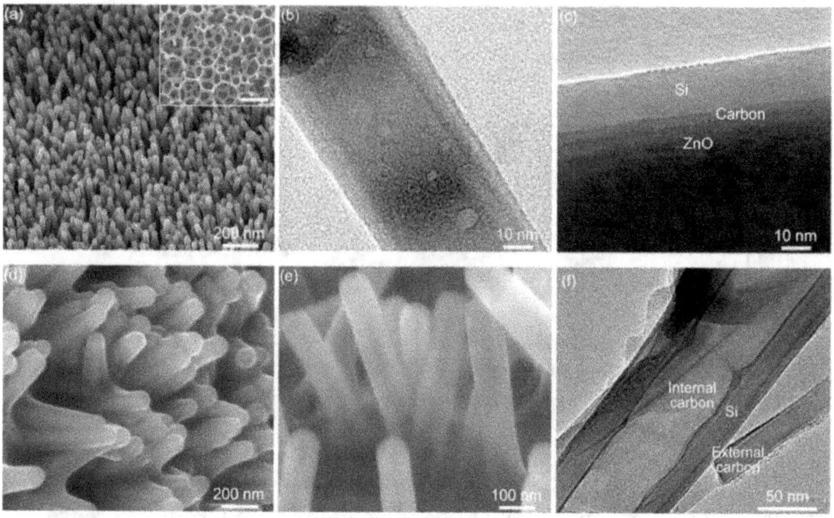

Figure 6.11 (a) Scanning electron image of the ZnO nano-rods grown on the Ni foam, (b) transmission image of a single carbon-coated ZnO nano-rod after carbonization of carbon precursor-coated ZnO nano-rods, (c) image of a ZnO@C@Si nano-rod, (d) image of the ZnO@C@Si@C nano-rod array, (e) image of the C@Si@C nanotube array after etching the ZnO core and (f) TEM image of C@Si@C nanotube. Reproduced from Reference 243 with permission from American Chemical Society.

In order to improve the electrolyte interface stability and cyclability of Si based LIB anode structures, Park *et al.* [247] used pyrene based methacrylate coating. This approach was useful to intensify storage and rate capacity of Si through the side chain conducting nature of the coating. The electrochemical performance of the anode structure was further enriched through the addition of ethylene oxide. In another study, Zhao *et al.* [248] used 1-flourodecane reduction and corresponding coating formation on the surface of Li_xSi nanoparticles to achieve improved electrochemical performance. Also, Ryou *et al.* [249] used mussel inspired binder to enhance the electrochemical properties of Si nanoparticles and their composite with graphite. In another study, Schroder *et al.* [250] discussed the use of fluoroethylene carbonate along with Si as efficient LIB anode. Liu *et al.* [251] reported the potential of PFFOMB (poly(9,9-dioctylfluorene-co-fluorenone-co-methylbenzoic acid)) as effective binder for Si nanoparticles to enable high performance LIB anode. Table 6.2 demonstrates the comparison of storage capacity of various silicon based LIB anode structures.

Table 6.2 Comparison of storage capacity of various silicon based LIB anode structures

Authors	LIB anode structure	Storage capacity
Lu *et al.* [231]	Nonfilling carbon coated porous Si microparticles	1500 mAh/g
Jia *et al.* [232]	Hierarchical 3D mesoporous Si	1500 mAh/g
Peled *et al.* [233]	3D Si nanowires	457 mAh/cm^2
Sourice *et al.* [235]	Carbon coated Si nanoparticles	2400 mAh/g
Gohier *et al.* [236]	Hierarchical Si/aligned CNT composite	3000 mAh/g
He *et al.* [239]	Al coated patterned Si	1125 mAh/g
Han *et al.* [240]	Si/MOF nanocomposite	1050 mAh/g
Deng *et al.* [242]	C/Si/C trilayer nanocomposite	2000 mAh/g
Liu *et al.* [243]	C/Si/C trilayer nanotube arrays	2200 mAh/g
Jia *et al.* [244]	$NiSi_2$/Si/carbon ternary composite	1272 mAh/g
Cao *et al.* [245]	Cu/Si core/shell nanocable arrays	1890 mAh/g
	Cu/Si/Al_2O_3 nanocable arrays	1820 mAh/g
Hu *et al.* [246]	Si/SiO_x-C core/shell hybrid	1100 mAh/g

6.3.2 Lithium Based LIB Electrodes

As an efficient electrode structure of LIBs, lithium (Li) presents considerable interest due to unique features such as high specific capacity, low redox potential and light weight. For instance, Liang *et al.* [252] reported the application of oxidized polyacrylonitrile nanofiber with 3D architecture on Li anode structure, which demonstrated sturdy electrochemical cycling along with chemical stability. An improved Coulombic efficiency of 97% after 120 charging-discharging cycles was reported. Shen *et al.* [253] observed rate performance of 121 mAh/g under 30 C for LIB anode based on hydrogenated nanowire arrays of $Li_4Ti_5O_{12}$. Also, good storage capacity and long cycle life were noted. In another study, Pohjalainen *et al.* [254] reported the potential of $Li_4Ti_5O_{12}$ nanoparticles with small primary size as LIB anode. Wang *et al.* [255] developed LIB anode using highly ordered nanocrystals containing $Li_4Ti_5O_{12}$ spheres, which exhibited rate performance of 149 mAh/g at 10 C with superior cyclic stability. This was due to the higher electrical conductivity of 3D interconnected architecture and corresponding rapid flow of ions and electrons through it.

Zhang *et al.* [256] used carbon coated Li_3VO_4 as LIB anode to prevent Li_3VO_4 dendrites growth and, thereby, to maintain efficient ion transport. As a result, improved reversible storage capacity of up to 450 mAh/g with superior cyclability were accomplished. In another study, Wang *et al.* [257] designed hierarchical mesoporous nanostructure composed of $LiFePO_4$ nanoparticles and carbon as LIB cathode. High Coulombic efficiency along with good specific capacitance and long cycle life were reported.

6.3.3 Metal Disulfide Based LIB Anodes

Molybdenum Disulfide Based LIB Anodes

Molybdenum disulfide (MoS_2), a category of transition metal dichalcogenide, is an attractive class of LIB electrode materials due to interconnected structure-property correlations responsible for effective energy storage. However, the considerable restacking and subsequent lowering in electrochemical performance of MoS_2 limit its applications. In order to overcome this drawback, various studies have been reported. For instance, Zhang *et al.* [258] developed hierarchical 3D MoS_2 nano-spheres (Figure 6.12) with an outstanding reversible specific capacitance of 1009 mAh/g over 500 cycles for enabling high performance LIB anode structure. Jiang *et al.* [259] detailed the fabrication of effective LIB anode using intercalated 2D nano-architecture containing single-layer MoS_2 and mesoporous carbon. The anode structure had reversible specific capacity of 1113 mAh/g and rate performance of 943 mAh/g together with good cycling stability.

Yu *et al.* [260] developed MoS_2 nano-sheet/N-doped carbon shell composite with hollow architecture for LIB anode application. The resultant anode structure displayed 1000 mAh/g storage capacity along with cyclability over 200 electrochemical cycles and excellent rate capability. Zhou *et al.* [261] synthesized 3D architecture containing few-layer MoS_2 nano-sheets grafted carbon nano-sheet with a reversible specific capacitance of 676 mAh/g and good cyclability for LIB anode application. Xiong *et al.* [262] also discussed the use of LIB anode material composed of 3D graphene/MoS_2 nano-flowers hybrid on carbon fiber textile. The resultant anode material exhibited storage capacity of 1255 mAh/g and high cyclability owing to its electrical conductivity.

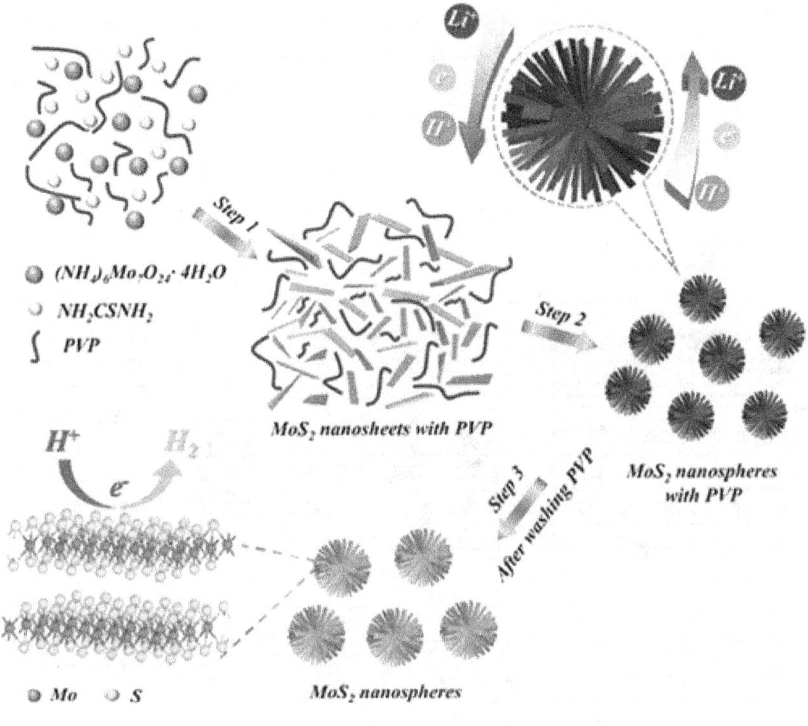

Figure 6.12 Schematic illustration for the generation of 3D radially oriented MoS_2 nano-spheres. Reproduced from Reference 258 with permission from American Chemical Society.

Other Metal Disulfides Based LIB Anodes

Owing to its high specific capacity, electrical conductivity and low cost, CoS_2 is also considered as a beneficial LIB anode material. Recently, Wang *et al.* [263] designed LIB anode structure with outstanding cyclic performance using composite composed of CoS_2 nanoparticles and nano-MOF sourced N-enriched carbon. Specific capacitance as high as 560 mAh/g with minimal capacitance fading upon faster lithiation/delithiation cycles were observed. Balogun *et al.* [264] reported the synthesis of LIB anode structure based on composite of porous VN nanowires and SnS_2 nano-sheets (Figure 6.13). Appreciable reversible specific capacitance of up to 819 mAh/g together with significant capacitance retention and mechanical longevity were observed. In another study, Fang *et al.* [265] used highly oriented composite containing VS_2 and graphene nanosheets as LIB anode with reversible storage capacity as high as 528 mAh/g, together with outstanding cycle life stability.

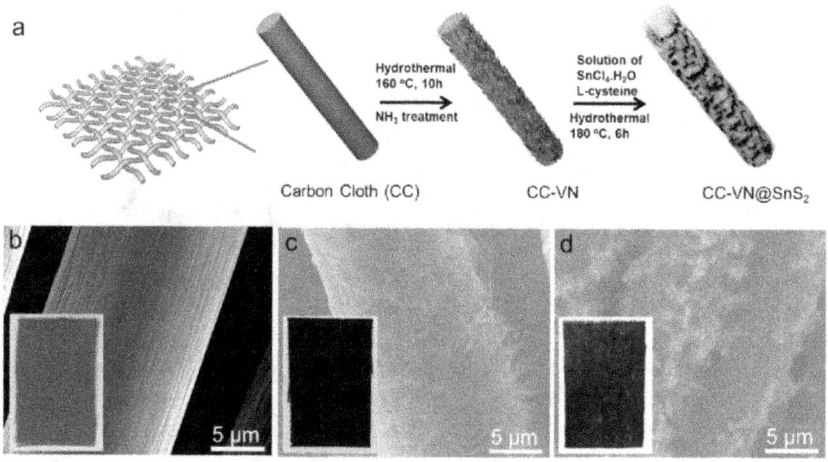

Figure 6.13 (a) Schematic displaying the synthesis routes for vanadium nitride (VN)/SnS_2 nanocomposite. Scanning electron images of (b) CC; c) CC-VN and c) CC-VN/SnS_2 showing the morphological changes during each growth stages of nanocomposite. Reproduced from Reference 264 with permission from American Chemical Society.

6.3.4 Metal Oxide Based LIB Anodes

Cobalt Oxide Based LIB Anodes

Cao *et al.* [266] reported reversible storage capacity of 1516 mAh/g with prolonged cycle life for LIB anode made up of CoO nanowire clusters. The electrochemical performance of these clusters was further enriched through the use of SiO_2 coating. Chen *et al.* [267] also developed high performance anode using Co_3O_4 nano-flakes with network mesoporous structure displaying a substantial storage capacity of 883 mAh/g with effective cycling permanence.

Gu *et al.* [268] reported the use of Co_3O_4 nanoparticle decorated CNT arrays with mesoporous structure for LIB anode application. High cycling stability along with good rate capability and storage capacity were observed for the anode structure. Huang *et al.* [269] used MWCNT/Co_3O_4 hybrid with highly oriented architecture for LIB anode. The resultant hybrid structure exhibited specific storage capacity of 813 mAh/g for 100 electrochemical cycles, as well as effective rate performance and cyclability. Li *et al.* [270] also achieved high specific capacity of 1184 mAh/g along with superior rate capacity and cyclability for Co_3O_4 nanowire array/Ge core/shell composite.

Iron Oxide Based LIB Anodes

Iron oxide is considered as promising LIB anode material due to its specific capacitance, corrosion resistance, availability and low cost. Cao *et al.* [271] used porous α-Fe_2O_3 nano-sheet with 3D highly oriented framework for LIB anode fabrication and a reversible specific capacity of 878 mAh/g for 1000 electrochemical cycles was noted. In another study, Cho *et al.* [272] designed composite nanofibers with bubble nano-rod framework containing hollow Fe_2O_3 nanospheres and carbon for generating long term high performance LIB anode structure. Zhang *et al.* [273] designed an anode structure comprising of Fe_3O_4/Fe_3C core/shell hybrid nanoparticles as yolks and carbon nano-spindles as shells. The resultant composite structure displayed cycling stability and high reversible storage capacitance of 1128 mAh/g. Zou *et al.* [274] used MOF derived hollow ZnO/$ZnFe_2O_4$/C hybrid with hierarchical porous architecture possessing outstanding electrochemical properties for generating LIB anode structure.

Tin Dioxide Based LIB Anodes

Park *et al.* [275] fabricated 1D SnO_2 nanowires with an aim to develop LIB anode structure and observed superior electrochemical performance. Zhou *et al.* [276] synthesized SnO_2/carbon core/shell composite fiber with 1D hollow architecture as LIB anode with specific capacity as high as 1002 mAh/g, along with cyclability and superior rate performance. In another study, Liang *et al.* [277] reported the use of SnO_2 coated carbon hollow particles as LIB anode and the achievement of superior electrochemical properties.

Titanium Dioxide Based LIB Anodes

Liu *et al.* [278] observed superior electrochemical performance for LIB anode based on ultrathin TiO_2-B nano-sheets exhibiting oriented porous structure owing to the faster ion flow and corresponding electrode reactions. In another study, Liu *et al.* [279] revealed the feasibility of mesoporous TiO_2-B microspheres for high performance LIB anode application with outstanding rate capability. Li *et al.* [280] designed a composite based on mesoporous anatase TiO_2 nano-spheres and graphene nano-sheets for enabling superior LIB anode structure with effective electrochemical features. In another study, Liu *et al.* [281] also reported high as well as stable electrochemical properties for mesoporous hollow spheres based on TiO_2/graphitic carbon hybrid.

Other Metal Oxides Based LIB Electrodes

Sun *et al.* [282] used naturally rolled-up NiO nano-membranes with 3D curved structure possessing high rate performance together with reversible storage capacity and long cycle life for use as efficient LIB anode. Ding *et al.* [283] reported beneficial LIB performance such as reversible storage capacity of up to 326 mAh/g and adequate cycling stability for oriented 3D V_6O_{13} nano-textiles based cathode structure (Figure 6.14). In another study, Zhang *et al.* [284] reported the generation of LIB anode structure by the use of ball-in-ball hollow $ZnMn_2O_4$ microspheres with improved storage capacity, effective rate performance and prolonged cycle life. Petnikota *et al.* [285] reported high reversible storage capacity of 1052 mAh/g and long cyclability for LIB anode based on nanostructured MgO/few-layer graphene composite. In another study, Wang *et al.* [286] observed reversible storage capacity of 1068 mAh/g along with good rate capability and capacitance retention for porous N-doped carbon laminated MnO_x based LIB anode structure.

6.3.5 Graphene Based LIB Anodes

With an objective to design high performance LIB anode, Hu *et al.* [287] developed graphene nano-capsules with branched architecture. High reversible storage capacity of 1373 mAh/g, efficient rate performance as well as cyclability were accomplished. In another study, Huang *et al.* [288] reported the development of β-$Co(OH)_2$

nano-sheet covered graphene with outstanding rate capacity and superior cycling stability over 5000 cycles for high performance LIB anode structure.

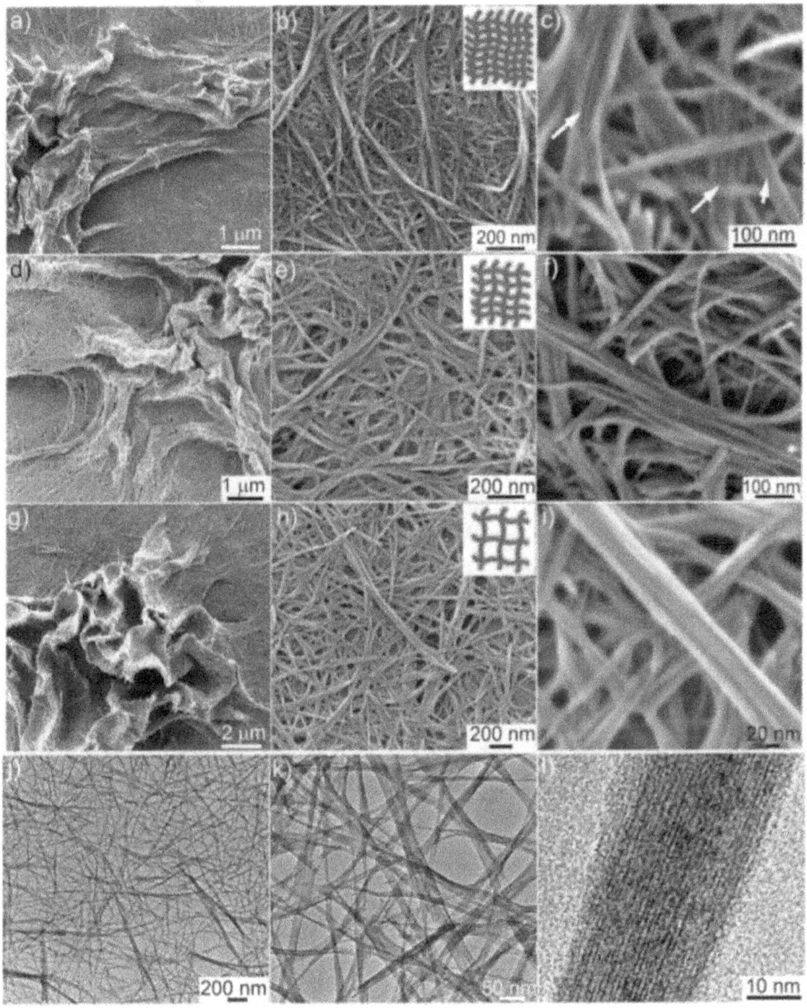

Figure 6.14 Scanning electron images of V_6O_{13} nano-textiles synthesized using $VOSO_4$ precursor with different concentrations. Reproduced from Reference 283 with permission from American Chemical Society.

Zhong *et al.* [289] fabricated LIB anode structure based on hierarchical composite composed of $MnCO_3$ and large area graphene. The resultant anode structure exhibited superior power performance, good storage capacity and significant cyclability due to high surface area of graphene. Liu *et al.* [290] reported a composite of graphene and mesoporous SnO_2 as promising LIB anode material owing to the accomplishment of reversible capacity of 1354 mAh/g under 100 mA/g over 50 charging-discharging cycles, along with pronounced rate capability of 664 mAh/g under 2 A/g. Recently, Jacob *et al.* [291] also reported binary Cu/ZnO doped graphene nanocomposites as effective anode materials for Li-ion batteries. Transmission electron microscopy analysis exhibited the distribution of Cu/ZnO nanoparticles over the graphene sheets with size in the range 10-50 nm. From the electrochemical analysis, it was inferred that the hybrid Cu-ZnO-G-2 (generated using 0.05 M Cu precursor and 0.1M ZnO precursor) exhibited a stable and specific capacity of 630 mAh/g at a current rate of 100 mA/g with almost 95% capacity retention for up to 100 cycles (Figure 6.15).

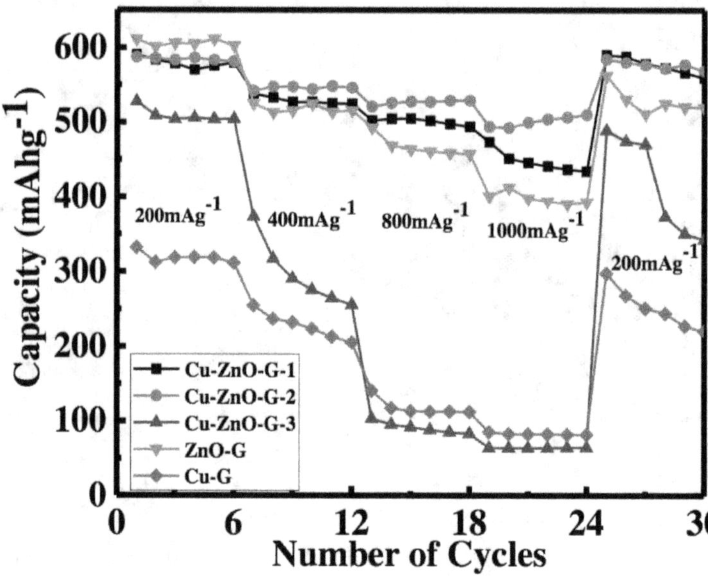

Figure 6.15 Rate capability behavior of Cu/ZnO/graphene, Cu-graphene and ZnO/graphene nanocomposites. Reproduced from Reference 291 with permission from Elsevier.

Evanoff *et al.* [292] reported high storage capacity of 2000 mAh/g along with excellent cycling permanence and Coulombic efficiency of about 99% for LIB anode made of porous composite nanostructure composed of nano-silicon decorated graphene. Zeng *et al.* [293] detailed the fabrication of binder-free LIB anode using graphene hybrid aerogel with spinel-type metal oxide nanoparticles. Appreciable rate capability and improved cycling stability for the resultant anode structure were noted. Guo *et al.* [294] reported the synthesis of highly oriented 3D graphene/PEDOT/MnO_2 ternary composite with substantial electrochemical properties appropriate for application as LIB anode.

In another study, Gao *et al.* [295] reported high specific capacity of up to 1710 mAh/g with rate capability and capacitance retention for ternary composite structure comprising of activated cotton textile (ACT) with porous tubular fibers embedded with NiS_2 nanobowls and wrapped with conductive graphene sheets. In another study, Dong *et al.* [296] also generated dually fixed SnO_2 nanoparticles on graphene nano-sheets for use as LIB anode structure. Efficient electrochemical performance of the developed anode was re-recorded.

6.3.6 Other Carbonaceous Materials Based LIB Anodes

Hou *et al.* [87] reported the development of LIB anode with reversible specific capacity of 1865 mAh/g from silk sourced N-doped carbon possessing oriented porous structure and substantial surface area. In another study, Qie *et al.* [297] fabricated smart LIB anode using N-doped carbon nanofiber network. High reversible specific capacity of 943 mAh/g as well as notable rate performance were acquired. Liu *et al.* [298] utilized Sn nanoparticles supported spherical carbon web as effective LIB anode material. The composite exhibited specific capacitance of about 494 mAh/g with superior rate capability and cycling stability.

Huang *et al.* [299] observed prolonged cycling life for LIB anode based on highly oriented composite of Sn nanoparticles and CNTs. In a recent study, Guo *et al.* [300] observed outstanding rate performance and cyclability for LIB anode based on mesoporous carbon/MWCNT composite. In another study, Wu *et al.* [301] reported the use of 3D hollow nanostructure composed of CoS coated porous carbon/CNT hybrid as effective LIB anode with reversible storage capacity of up to 1668 mAh/g and high rate capability of up to 1038 mAh/g.

Thus, an extensive range of studies has been carried out to develop a large number of advanced materials useful as LIB electrodes.

6.4 Sodium-Ion Batteries (SIBs)

6.4.1 Graphene Based SIB Electrodes

Xu *et al.* [302] discussed the use of 3D N-doped graphene foam for SIB anode application owing to its substantial surface area, 3D mesoporosity, hierarchical porosity, lattice space and enriched N content. The resultant anode structure displayed reversible storage capacity of 853 mAh/g along with appreciable rate capability of 594 mAh/g and cycling stability. Zhang *et al.* [303] used ultrathin graphene nanocomposite of Sb as high performance SIB anode and noted superior reversible storage capacitance of 612 mAh/g. In another study, Sun *et al.* [304] reported VS_4 grafted graphene hybrid as LIB anode (Figure 6.16) with excellent rate capacity as well as storage capacity of 362 mAh/g. Also, Hu *et al.* [305] discussed the viability of multilayer graphene composite with antimony for use as SIB anode. Liu *et al.* [291] reported satisfactory electrochemical properties for SIB anode comprising of TiO_2 nanoparticles decorated graphene sheets.

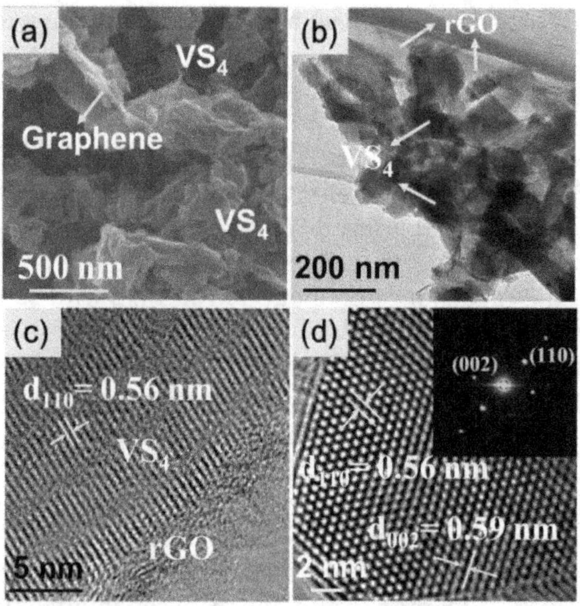

Figure 6.16 (a) Scanning and (b) transmission electron images of VS_4/rGO, high resolution images of (c) VS_4/rGO and (d) VS_4. Reproduced from Reference 304 with permission from American Chemical Society.

Xu *et al.* [306] reported reversible storage capacitance of 640 mAh/g and cycling permanence for SIB anode comprised of hollow MnS microsphere decorated graphene sheets. In another study, Ma *et al.* [307] observed attractive SIB anode features for SnS_2/graphene nanocomposite. Choi *et al.* [308] developed SIB anode based on 3D microspheres of MoS_2/graphene composite and observed excellent reversible storage capacity and Coulombic efficiency of about 100%. This was attributed to the rapid ion/electron transport through the resultant anode structure in correspondence to the minimized MoS_2 restacking. In another study, Xie *et al.* [309] observed suitability of MoS_2/graphene composite with 2D hetero-interfaces for effective SIB anode application. Wu *et al.* [310] observed substantial rate capacity and cyclability of LIB anode through the usage of nanostructured composite comprising of porous $NaTi_2(PO_4)_3$ nanoparticles and 3D graphene web. Rui *et al.* [311] also reported the applicability of 3D porous hybrid composed of carbon coated $Na_3V_2(PO_4)_3$ and graphene for superior cathode structure for LIB application.

6.4.2 Other Carbonaceous Materials Based SIB Electrodes

Hou *et al.* [312] reported exceptional specific capacity, cycle life and rate performance for 3D porous carbon architecture derived from carbon quantum dots. Luo *et al.* [313] observed enhancement in cyclic stability and Coulombic efficiency through the use of hard carbon based anode system. At the same time, SIB anode structure made of soft carbon presented substantial rate capacity [314]. Liu *et al.* [315] used mesoporous N-enriched carbon with network framework as effective SIB anode structure. Also, porous N-doped carbon nano-sheets exhibited substantial potential for effective application as SIB anode [316]. Liu *et al.* [317] reported the development of efficient SIB anode structure through the introduction of Sn nano-dots on porous matrix of N-doped carbon nanofibers. In another study, Jian *et al.* [318] observed improved rate performance and cyclic stability for potassium impregnated carbon based SIB anode. Xu *et al.* [319] also revealed the potential of mesoporous composite comprised of carbon and graphene as effective anode material.

Liu *et al.* [298] used Sn nanoparticles supported spherical carbon web as high quality SIB anode. The resultant anode structure displayed storage capacity of about 494 mAh/g, efficient rate capability and prolonged cyclability. In another study, Wang *et al.* [320] revealed the potential of SnS_2/C hybrid nano-spheres to generate superior SIB anodes. The system exhibited reversible storage capacity of 660 mAh/g over 100 charging-discharging cycles along with high cyclability. Kalubarme *et al.* [321] employed carbon encapsulated SnO_2 as SIB anode to enable efficient electrochemical properties. The composite system exhibited reversible specific capacity of 514 mAh/g, along with effective cycling consistency as well as rate capacity. Wang *et al.* [322] observed optimal reversible specific capacitance of about 520 mAh/g with prolonged cycle life for SIB anode based on MoS_2/C hybrid nano-spheres. Also, Coulombic efficiency of about 100% was acquired for the system. Jian *et al.* [323] reported the practicability of $Na_3V_2(PO_4)_3$/carbon nano-hybrid for desirable SIB cathode applications. In another study, Jiang *et al.* [324] achieved prolonged cycle performance and outstanding rate capacity for mesoporous carbon composite with carbon coated $Na_3V_2(PO_4)_3$ nanoparticles.

Li *et al.* [325] also reported the potential of phosphorene for applications in novel electronic devices. Song *et al.* [326] described the generation of SIB anode structure using composite system containing red phosphorus and carboxylated CNTs together with a binder based on crosslinked polymer. High reversible storage capacity of 1586 mAh/g for 100 electrochemical sequences together with high capacitance retention and Coulombic efficiency were observed (Figure 6.17).

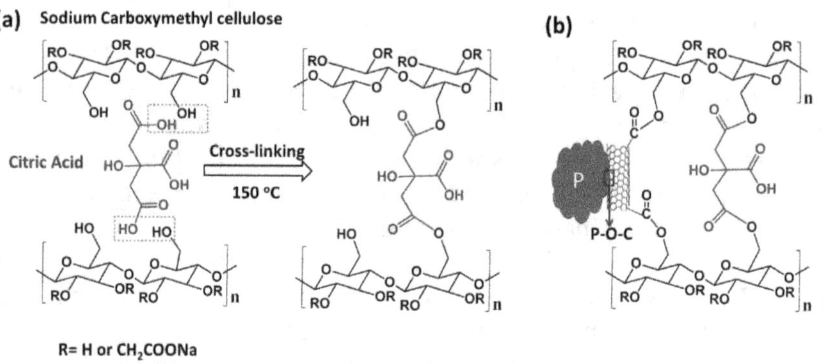

Figure 6.17 (a) Crosslinked polymer binder generated by thermal-induced condensation of carboxymethyl cellulose and citric acid (c-NaCMC-CA) and (b) representation of the interaction between P-CNT hybrid and c-NaCMC-CA binder. Reproduced from Reference 326 with permission from American Chemical Society.

6.4.3 Metal Based SIB Electrodes

Su *et al.* [327] reported efficient electrochemical properties for SIB anode generated from hollow nano-spheres of anatase TiO_2 compared to rutile and amorphous forms. With an objective to achieve desirable electrochemical performance for SIB anode structure, Xu *et al.* [328] also fabricated hierarchical 3D Ni/TiO_2 nano-arrays. Reversible storage capacity in the range of 200 mAh/g with effective capacitance retention were noted for the SIB anode. The practicability of MoS_2 nano-flowers displaying substantial interlayer free volume for generating superior SIB anode structure was discussed by Hu *et al.* [329].

6.5 Lithium-Sulfur Batteries (LSBs)

6.5.1 Graphene Based LSB Electrodes

Zhou *et al.* [330] reported the development of high performance LSB cathode structure through the impregnation of S on graphene foam. In another study, Zhou *et al.* [331] also detailed the generation of graphene/S layered framework for LSB cathode with excellent electrochemical properties. Wang *et al.* [332] acquired high reversible storage capacity for cathode system based on thin macroporous nanocomposite containing nano-S and graphene. In another study, Wang *et al.* [333] also fabricated thin LiS_2/graphene based effective free-standing LSB cathode displaying superior rate performance and prolonged cyclability. Song *et al.* [334] reported high reversible storage capacity of about 1000 mAh/g for LSB through the use of N-doped wrinkled graphene sheets as S cathode host. With an aim to achieve high performance LSB cathode structure, Xie *et al.* [335] also synthesized percolative B-doped graphene aerogel with 3D network structure. The developed cathode system presented reversible storage capacitance in the range of 994 mAh/g along with significant rate capacity and cyclability.

Li *et al.* [336] synthesized vertically oriented S/graphene hybrid nano-walls for LSB cathode application (Figure 6.18). The cathode structure exhibited reversible storage capacitance of about 1261 mAh/g together with effective rate capability and cycle life. Zhu *et al.* [337] revealed the possibility of CNTs/graphene hollow nanosphere scaffolds for use as LSB cathode host. In another study, Chen *et al.* [338] reported the use of CNT/graphene/S hybrid system for pliable LSB cathode application and observed reversible storage capacitance as high as 912 mAh/g together with high rate capability. Wu *et al.* [339] acquired outstanding LSB performance for cathode structure based on graphene covered by Li_2S nanoparticles, thus, protected from undesirable electrolyte interactions. High storage capacity of about 1040 mAh/g was accomplished for the developed material. Recently, Wang *et al.* [340] observed efficient LSB performances for cathode structure comprising of Li_2S/C/3D graphene ternary nanocomposite.

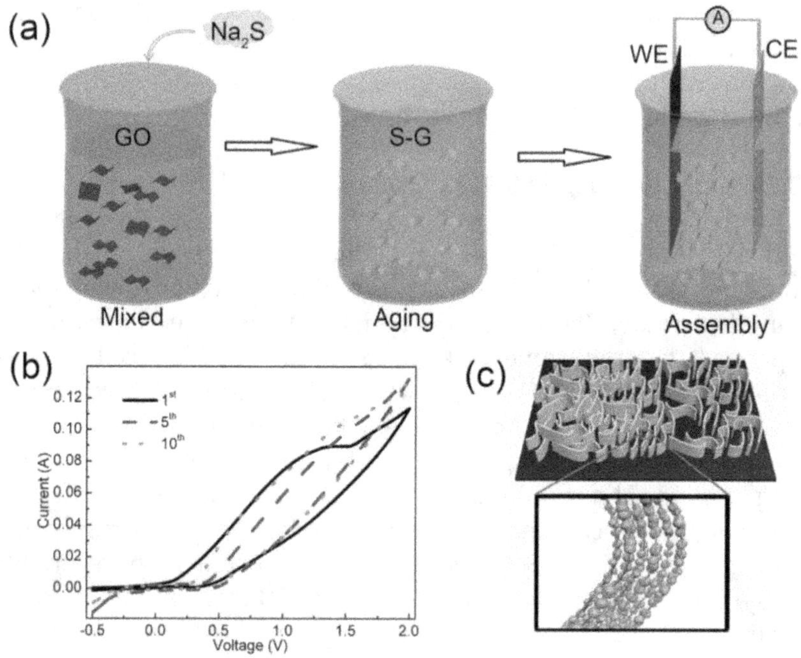

Figure 6.18 Schematic of the fabrication of vertically aligned S-G nano-walls on substrates and representative cyclic voltammograms employed during the electrochemical assembly processes. Reproduced from Reference 336 with permission from American Chemical Society.

6.5.2 Carbon Nanotubes Based LSB Electrodes

Sun *et al.* [341] observed impressive electrochemical properties in the case of cathode structure based on S/percolative CNT hybrid. In another study, Guo *et al.* [342] attained specific storage capacity of about 1106 mAh/g for P infused CNTs based cathode host. Song *et al.* [343] reported high reversible storage capacity of about 1200 mAh/g over 200 electrochemical cycles for lithium-sulfur battery cathode based on composite of N-doped carbon spheres and CNTs. Fan *et al.* [344] used CNT/$NiFe_2O_4$/S ternary nanocomposite structure as cathode for high performance LSB. The cathode system exhibited reversible storage capacity of 1350 mAh/g under 0.1 C, along with impressive rate capacity, cyclability and Coulombic efficiency.

6.5.3 Other Carbonaceous Materials Based LSB Electrodes

Strubel *et al.* [345] reported high reversible storage capacitance of 1200 mAh/g and effective rate capacity for LSB cathode based on composite containing S and highly oriented percolative carbon. In another study, Du *et al.* [346] achieved reversible storage capacity of 1115 mAh/g and rate capability of 551 mAh/g for LSB cathode fabricated using 3D nanocarbon/S hybrid. Sun *et al.* [347] also described the use of hollow carbon spheres with core-shell morphology along with S as an effective cathode structure. In another study, Zhou *et al.* [348] studied the feasibility of hollow structured N-doped carbon nano-spheres as hosts for generating LSB cathode. Schneider *et al.* [349] also reported highly oriented N-rich carbon as successful S cathode host. Li *et al.* [350] generated LSB cathode based on S composite with MOF derived N-doped carbon and observed impressive electrochemical performance including 937 mAh/g reversible storage capacity, 632 mAh/g rate capability and 100% Coulombic efficiency. Pang *et al.* [351] acquired considerable enhancement in cycling stability of LSB through the application of N and S enriched percolative carbon as cathode.

Lu *et al.* [231] introduced carbon laminated percolative Si microparticles as cathode structure with superior LSB performance such as storage capability of about 1500 mAh/g and prolonged cyclability. Hwa *et al.* [352] utilized carbon laminated nano-spheres of Li_2S/GO hybrid as cathode for efficient LSB performance. Zeng *et al.* [353] noticed remarkable electrochemical features for LSB architecture containing percolative $S_{1-x}Se_x$/carbon nanofiber hybrid based pliable cathode. Li *et al.* [354] reported the generation of MnO_2 nano-sheet decorated hollow carbon nanofibers as S cathode host for LSB. Reversible storage capacity of 1161 mAh/g together with appreciable cyclability were obtained. In another study, Niu *et al.*

[355] reported improved electrochemical performance in the case of $Ni(OH)_2$ coated S/carbon black composite based LSB cathode.

6.5.4 Polymers Based LSB Electrodes

Wei *et al.* [356] designed S/polyacrylonitrile composite based LSB cathode structure and observed high reversible storage capacity and 100% Coulombic efficiency. In another study, Konarov *et al.* [357] reported the use of composite composed of S and dehydrogenated polyacrylonitrile as cathode material possessing high reversible storage capacity of about 1050 mAh/g with outstanding consistency. Li *et al.* [40] reported effective cyclic performance, storage capacitance, rate capacity and Coulombic efficacy for LSB cathode developed using S/polyacrylonitrile/graphene ternary nanocomposite. Hu *et al.* [358] evaluated the usefulness of polyaniline coated nanospheres of S/carbon hybrid for LSB cathode structure. In another study, Xie *et al.* [359] obtained efficient energy performance as high as 400 Wh/kg and reversible storage capacity of about 1142 mAh/g for S/polypyrrole core/shell hybrid spheres based LSB cathode.

6.6 Conclusions

A large variety of energy storage devices have been developed in the recent years with an aim to overcome challenges associated with rising energy needs. The present study reviewed the recent advancements gained for these energy storage devices, especially supercapacitors, lithium-ion batteries, lithium-sulfur batteries and sodium-ion batteries in terms of storage capacity, energy density, power density, rate capability, capacitance retention, cycling stability, Coulombic efficiency, etc. Energy storage devices based on materials such as graphene, carbon, metals, metal oxides and polymers were observed to display remarkable electrochemical properties owing to the beneficial features of these materials, which can be synergistically combined with each other to enhance performance.

References

1. Kalra, V., Basu, S., and Smith, W. A. (2016) Special issue on recent advances in energy storage and conversion devices. *Chemical Engineering Science,* **154**, 1-2.
2. Ma, X., Luo, W., Yan, M., He, L., and Mai, L. (2016) In situ characteri-

zation of electrochemical processes in one dimensional nanomaterials for energy storages devices. *Nano Energy,* **24**, 165-188.
3. Yang, Z., Zhang, J., Kintner-Meyer, M. C., Lu, X., Choi, D., Lemmon, J. P., and Liu, J. (2011) Electrochemical energy storage for green grid. *Chemical Reviews,* **111**(5), 3577-3613.
4. Gonzalez, A., Goikolea, E., Barrena, J. A., and Mysyk, R. (2016) Review on supercapacitors: technologies and materials. *Renewable and Sustainable Energy Reviews,* **58**, 1189-1206.
5. Lv, W., Li, Z., Deng, Y., Yang, Q.-H., and Kang, F. (2016) Graphene-based materials for electrochemical energy storage devices: opportunities and challenges. *Energy Storage Materials,* **2**, 107-138.
6. Jeong, H. T., Kim, B. C., Gorkin, R., Higgins, M, J., and Wallace, G. G. (2014) Capacitive behavior of latex/single-wall carbon nanotube stretchable electrodes. *Electrochimica Acta,* **137**, 372-380.
7. Wang, G., Zhang, L., and Zhang, J. (2012) A review of electrode materials for electrochemical supercapacitors. *Chemical Society Reviews,* **41**, 797-828.
8. Sk, M. M., Yue, C. Y., Ghosh, K., and Jena, R. K. (2016) Review on advances in porous nanostructured nickel oxides and their composite electrodes for high-performance supercapacitors. *Journal of Power Sources,* **308**, 121-140.
9. Simon, P., and Gogotsi, Y. (2008) Materials for electrochemical capacitors. *Nature Materials,* **7**, 845-854.
10. Liu, C., Li, F., Ma, L. P., and Cheng, H. M. (2010) Advanced materials for energy storage. *Advanced Materials,* **22**(8), E28-E62.
11. Kandalkar, S., Dhawale, D., Kim, C.-K., and Lokhande, C. (2010) Chemical synthesis of cobalt oxide thin film electrode for supercapacitor application. *Synthetic Metals,* **160**, 1299-1302.
12. Zhi, M., Xiang, C., Li, J., Li, M., and Wu, N. (2013) Nanostructured carbon–metal oxide composite electrodes for supercapacitors: a review. *Nanoscale,* **5**, 72-88.
13. Lv, T., Yao, Y., Li, N., and Chen, T. (2016) Wearable fiber-shaped energy conversion and storage devices based on aligned carbon nanotubes. *Nano Today,* **11**, 644-660.
14. Wu, H., Zhang, Y., Cheng, L., Zheng, L., Li, Y., Yuan, W., and Yuan, X. (2016) Graphene based architectures for electrochemical capacitors. *Energy Storage Materials,* **5**, 8-32.
15. Meng, X., Feng, M., Zhang, H., Ma, Z., and Zhang, C. (2017) Solvothermal synthesis of cobalt/nickel layered double hydroxides for energy storage devices. *Journal of Alloys and Compounds,* **695**, 3522-3529.
16. Nithya, V., and Arul, N. S. (2016) Review on α-Fe_2O_3 based negative electrode for high performance supercapacitors. *Journal of Power Sources,* **327**, 297-318.
17. Zhu, Y., Wang, J., Wu, Z., Jing, M., Hou, H., Jia, X., and Ji, X. (2015) An

electrochemical exploration of hollow $NiCo_2O_4$ submicrospheres and its capacitive performances. *Journal of Power Sources,* **287**, 307-315.

18. Shi, F., Xie, D., Zhong, Y., Wang, D., Xia, X., Gu, C., Wang, X., and Tu, J. (2016) Self-supporting hierarchical rGO@Ni nanosheet@Co_3O_4 nanowire array and its application in high-rate batteries. *Journal of Power Sources,* **327**, 281-288.
19. Pandolfo, A., and Hollenkamp, A. (2006) Carbon properties and their role in supercapacitors. *Journal of Power Sources,* 157, 11-27.
20. Ke, Q., and Wang, J. (2016) Graphene-based materials for supercapacitor electrodes - A review. *Journal of Materiomics,* **2**, 37-54.
21. Veerasubramani, G. K., Krishnamoorthy, K., Radhakrishnan, S., Kim, N.-J., and Kim, S. J. (2014) Synthesis, characterization, and electrochemical properties of $CoMoO_4$ nanostructures. *International Journal of Hydrogen Energy,* **39**, 5186-5193.
22. Zhang, L. L., and Zhao, X. (2009) Carbon-based materials as supercapacitor electrodes. *Chemical Society Reviews,* **38**, 2520-2531.
23. Bose, S., Kuila, T., Mishra, A. K., Rajasekar, R., Kim, N. H., and Lee, J. H. (2012) Carbon-based nanostructured materials and their composites as supercapacitor electrodes. *Journal of Materials Chemistry,* **22**, 767-784.
24. Zhao, X., Sanchez, B. M., Dobson, P. J., and Grant, P. S. (2011) The role of nanomaterials in redox-based supercapacitors for next generation energy storage devices. *Nanoscale,* **3**, 839-855.
25. Cericola, D., Novak, P., Wokaun, A., and Kotz, R. (2011) Mixed bi-material electrodes based on $LiMn_2O_4$ and activated carbon for hybrid electrochemical energy storage devices. *Electrochimica Acta,* **56**, 8403-8411.
26. Cericola, D., Ruch, P., Kotz, R., Novak, P., and Wokaun, A. (2010) Characterization of bi-material electrodes for electrochemical hybrid energy storage devices. *Electrochemistry Communications,* **12**, 812-815.
27. Winter, M., and Brodd, R. J. (2004) What are batteries, fuel cells, and supercapacitors? *Chemical Reviews,* **104**, 4245-4270.
28. Winter, M., Besenhard, J. O., Spahr, M. E., and Novak, P. (1998) Insertion electrode materials for rechargeable lithium batteries. *Advanced Materials,* **10**, 725-763.
29. Shen, J., Li, X., Li, N., and Ye, M. (2014) Facile synthesis of $NiCo_2O_4$-reduced graphene oxide nanocomposites with improved electrochemical properties. *Electrochimica Acta,* **141**, 126-133.
30. Palacin, M. R. (2009) Recent advances in rechargeable battery materials: a chemist's perspective. *Chemical Society Reviews,* **38**, 2565-2575.
31. Nitta, N., Wu, F., Lee, J. T., and Yushin, G. (2015) Li-ion battery materials: present and future. *Materials Today,* **18**, 252-264.

32. Wu, D., Shi, C., Huang, S., Qiu, X., Wang, H., Zhan, Z., Zhang, P., Zhao, J., Sun, D., and Lin, L. (2015) Electrospun nanofibers for sandwiched polyimide/poly (vinylidene fluoride)/polyimide separators with the thermal shutdown function. *Electrochimica Acta,* **176**, 727-734.
33. Bresser, D., Kim, G.-T., Binetti, E., Striccoli, M., Comparelli, R., Seidel, S., Ozkaya, D., Copley, M., Bishop, P., Paillard, E., and Passerini, S. (2015) Transforming anatase TiO_2 nanorods into ultrafine nanoparticles for advanced electrochemical performance. *Journal of Power Sources,* **294**, 406-413.
34. Li, Y., Qian, K., He, Y.-B., Kaneti, Y. V., Liu, D., Luo, D., Li, H., Li, B., and Kang, F. (2017) Study on the reversible capacity loss of layered oxide cathode during low-temperature operation. *Journal of Power Sources,* **342**, 24-30.
35. Zhang, Z., Sui, G., Bi, H., and Yang, X. (2015) Radiation-crosslinked nanofiber membranes with well-designed core–shell structure for high performance of gel polymer electrolytes. *Journal of Membrane Science,* **492**, 77-87.
36. Goriparti, S., Miele, E., De Angelis, F., Di Fabrizio, E., Zaccaria, R. P., and Capiglia, C. (2014) Review on recent progress of nanostructured anode materials for Li-ion batteries. *Journal of Power Sources,* **257**, 421-443.
37. Palomares, V., Serras, P., Villaluenga, I., Hueso, K. B., Carretero-González, J., and Rojo, T. (2012) Na-ion batteries, recent advances and present challenges to become low cost energy storage systems. *Energy & Environmental Science,* **5**, 5884-5901.
38. Hong, S. Y., Kim, Y., Park, Y., Choi, A., Choi, N.-S., and Lee, K. T. (2013) Charge carriers in rechargeable batteries: Na ions vs. Li ions. *Energy & Environmental Science,* **6**, 2067-2081.
39. Hasa, I., Dou, X., Buchholz, D., Shao-Horn, Y., Hassoun, J., Passerini, S., and Scrosati, B. (2016) A sodium-ion battery exploiting layered oxide cathode, graphite anode and glyme-based electrolyte. *Journal of Power Sources,* **310**, 26-31.
40. Li, J., Li, K., Li, M., Gosselink, D., Zhang, Y., and Chen, P. (2014) A sulfur–polyacrylonitrile/graphene composite cathode for lithium batteries with excellent cyclability. *Journal of Power Sources,* **252**, 107-112.
41. Ji, X., and Nazar, L. F. (2010) Advances in Li–S batteries. *Journal of Materials Chemistry,* **20**, 9821-9826.
42. Bruce, P. G., Freunberger, S. A., Hardwick, L. J., and Tarascon, J.-M. (2012) Li-O_2 and Li-S batteries with high energy storage. *Nature Materials,* **11**, 19-29.
43. Wang, Y., Shi, Z., Huang, Y., Ma, Y., Wang, C., Chen, M., and Chen, Y. (2009) Supercapacitor devices based on graphene materials. *The Journal of Physical Chemistry C,* **113**, 13103-13107.

44. Liu, C., Yu, Z., Neff, D., Zhamu, A., and Jang, B. Z. (2010) Graphene-based supercapacitor with an ultrahigh energy density. *Nano Letters,* **10**, 4863-4868.
45. Lv, W., Tang, D.-M., He, Y.-B., You, C.-H., Shi, Z.-Q., Chen, X.-C., Chen, C.-M., Hou, P.-X., Liu, C., and Yang, Q.-H. (2009) Low-temperature exfoliated graphenes: vacuum-promoted exfoliation and electrochemical energy storage. *ACS Nano,* **3**, 3730-3736.
46. Yan, J., Wang, Q., Wei, T., Jiang, L., Zhang, M., Jing, X., and Fan, Z. (2014) Template-assisted low temperature synthesis of functionalized graphene for ultrahigh volumetric performance supercapacitors. *ACS Nano,* **8**, 4720-4729.
47. Lee, J.-S., Kim, S.-I., Yoon, J.-C., and Jang, J.-H. (2013) Chemical vapor deposition of mesoporous graphene nanoballs for supercapacitor. *ACS Nano,* **7**, 6047-6055.
48. Xu, Y., Sheng, K., Li, C., and Shi, G. (2010) Self-assembled graphene hydrogel via a one-step hydrothermal process. *ACS Nano,* **4**, 4324-4330.
49. Chen, T., Xue, Y., Roy, A. K., and Dai, L. (2013) Transparent and stretchable high-performance supercapacitors based on wrinkled graphene electrodes. *ACS Nano,* **8**, 1039-1046.
50. Yoo, J. J., Balakrishnan, K., Huang, J., Meunier, V., Sumpter, B. G., Srivastava, A., Conway, M., Mohana Reddy, A. L., Yu, J., Vajtai, R., and Ajayan, P. M. (2011) Ultrathin planar graphene supercapacitors. *Nano Letters,* **11**, 1423-1427.
51. Stoller, M. D., Park, S., Zhu, Y., An, J., and Ruoff, R. S. (2008) Graphene-based ultracapacitors. *Nano Letters,* **8**, 3498-3502.
52. Mishra, A. K., and Ramaprabhu, S. (2011) Functionalized graphene-based nanocomposites for supercapacitor application. *The Journal of Physical Chemistry C,* **115**, 14006-14013.
53. Kim, T. Y., Lee, H. W., Stoller, M., Dreyer, D. R., Bielawski, C. W., Ruoff, R. S., and Suh, K. S. (2010) High-performance supercapacitors based on poly (ionic liquid)-modified graphene electrodes. *ACS Nano,* **5**, 436-442.
54. Lee, J. H., Park, N., Kim, B. G., Jung, D. S., Im, K., Hur, J., and Choi, J. W. (2013) Restacking-inhibited 3D reduced graphene oxide for high performance supercapacitor electrodes. *ACS Nano,* **7**, 9366-9374.
55. Choi, B. G., Yang, M., Hong, W. H., Choi, J. W., and Huh, Y. S. (2012) 3D macroporous graphene frameworks for supercapacitors with high energy and power densities. *ACS Nano,* **6**, 4020-4028.
56. Han, J., Zhang, L. L., Lee, S., Oh, J., Lee, K.-S., Potts, J. R., Ji, J., Zhao, X., Ruoff, R. S., and Park, S. (2012) Generation of B-doped graphene nanoplatelets using a solution process and their supercapacitor applications. *ACS Nano,* **7**, 19-26.
57. Paek, E., Pak, A. J., Kweon, K. E., and Hwang, G. S. (2013) On the ori-

gin of the enhanced supercapacitor performance of nitrogen-doped graphene. *The Journal of Physical Chemistry C,* **117**, 5610-5616.

58. Ma, Z., Huang, X., Dou, S., Wu, J., and Wang, S. (2014) One-pot synthesis of Fe_2O_3 nanoparticles on nitrogen-doped graphene as advanced supercapacitor electrode materials. *The Journal of Physical Chemistry C,* **118**, 17231-17239.
59. Sui, Z.-Y., Meng, Y.-N., Xiao, P.-W., Zhao, Z.-Q., Wei, Z.-X., and Han B.-H. (2015) Nitrogen-doped graphene aerogels as efficient supercapacitor electrodes and gas adsorbents. *ACS Applied Materials & Interfaces,* **7**, 1431-1438.
60. Pham, D. T., Lee, T. H., Luong, D. H., Yao, F., Ghosh, A., Le, V. T., Kim, T. H., Li, B., Chang, J., and Lee, Y. H. (2015) Carbon nanotube-bridged graphene 3D building blocks for ultrafast compact supercapacitors. *ACS Nano,* **9**, 2018-2027.
61. Yu, D., and Dai, L. (2009) Self-assembled graphene/carbon nanotube hybrid films for supercapacitors. *The Journal of Physical Chemistry Letters,* **1**, 467-470.
62. Lei, Z., Shi, F., and Lu. L. (2012) Incorporation of MnO_2-coated carbon nanotubes between graphene sheets as supercapacitor electrode. *ACS Applied Materials & Interfaces,* **4**, 1058-1064.
63. You, B., Jiang, J., and Fan, S. (2014) Three-dimensional hierarchically porous all-carbon foams for supercapacitor. *ACS Applied Materials & Interfaces,* **6**, 15302-15308.
64. Gao, Z., Wang, J., Li, Z., Yang, W., Wang, B., Hou, M., He, Y., Liu, Q., Mann, T., Yang, P., Zhang, M., and Liu, L. (2011) Graphene nanosheet/Ni^{2+}/Al^{3+} layered double-hydroxide composite as a novel electrode for a supercapacitor. *Chemistry of Materials,* **23**, 3509-3516.
65. Yang, W., Gao, Z., Wang, J., Ma, J., Zhang, M., and Liu, L. (2013) Solvothermal one-step synthesis of Ni-Al layered double hydroxide/carbon nanotube/reduced graphene oxide sheet ternary nanocomposite with ultrahigh capacitance for supercapacitors. *ACS Applied Materials & Interfaces,* **5**, 5443-5454.
66. Huang, S., Zhu, G.-N., Zhang, C., Tjiu, W. W., Xia, Y.-Y., and Liu, T. (2012) Immobilization of Co-Al layered double hydroxides on graphene oxide nanosheets: growth mechanism and supercapacitor studies. *ACS Applied Materials & Interfaces,* **4**, 2242-2249.
67. Peng, L., Peng, X., Liu, B., Wu, C., Xie, Y., and Yu, G. (2013) Ultrathin two-dimensional MnO_2/graphene hybrid nanostructures for high-performance, flexible planar supercapacitors. *Nano Letters,* **13**, 2151-2157.
68. He, Y., Chen, W., Li, X., Zhang, Z., Fu, J., Zhao, C., and Xie, E. (2012) Freestanding three-dimensional graphene/MnO_2 composite networks as ultralight and flexible supercapacitor electrodes. *ACS Na-*

no, **7**, 174-182.

69. Wu, Z.-S., Ren, W., Wang, D.-W., Li, F., Liu, B., and Cheng, H.-M. (2010) High-energy MnO_2 nanowire/graphene and graphene asymmetric electrochemical capacitors. *ACS Nano,* **4**, 5835-5842.
70. Lee, J. W., Hall, A. S., Kim, J.-D., and Mallouk, T. E. (2012) A facile and template-free hydrothermal synthesis of Mn_3O_4 nanorods on graphene sheets for supercapacitor electrodes with long cycle stability. *Chemistry of Materials,* **24**, 1158-1164.
71. Yu, G., Hu, L., Liu, N., Wang, H., Vosgueritchian, M., Yang, Y., Cui, Y., and Bao, Z. (2011) Enhancing the supercapacitor performance of graphene/MnO_2 nanostructured electrodes by conductive wrapping. *Nano Letters,* **11**, 4438-4442.
72. Jin, Y., Chen, H., Chen, M., Liu, N., and Li, Q. (2013) Graphene-patched CNT/MnO_2 nanocomposite papers for the electrode of high-performance flexible asymmetric supercapacitors. *ACS Applied Materials & Interfaces,* **5**, 3408-3416.
73. Chen, H., Zhou, S., and Wu, L. (2014) Porous nickel hydroxide-manganese dioxide-reduced graphene oxide ternary hybrid spheres as excellent supercapacitor electrode materials. *ACS Applied Materials & Interfaces,* **6**, 8621-8630.
74. Dong, X.-C., Xu, H., Wang, X.-W., Huang, Y.-X., Chan-Park, M. B., Zhang, H., Wang, L.-H., Huang, W., and Chen, P. (2012) 3D graphene–cobalt oxide electrode for high-performance supercapacitor and enzymeless glucose detection. *ACS Nano,* **6**, 3206-3213.
75. Wang, H., Casalongue, H. S., Liang, Y., and Dai, H. (2010) $Ni(OH)_2$ nanoplates grown on graphene as advanced electrochemical pseudocapacitor materials. *Journal of the American Chemical Society,* **132**, 7472-7477.
76. Ratha, S., and Rout, C. S. (2013) Supercapacitor electrodes based on layered tungsten disulfide-reduced graphene oxide hybrids synthesized by a facile hydrothermal method. *ACS Applied Materials & Interfaces,* **5**, 11427-11433.
77. Choi, K. M., Jeong, H. M., Park, J. H., Zhang, Y.-B., Kang, J. K., and Yaghi, O. M. (2014) Supercapacitors of nanocrystalline metal-organic frameworks. *ACS Nano,* **8**, 7451-7457.
78. Kim, T., Jung, G., Yoo, S., Suh, K. S., and Ruoff, R. S. (2013) Activated graphene-based carbons as supercapacitor electrodes with macro- and mesopores. *ACS Nano,* **7**, 6899-6905.
79. Zhang, L. L., Zhao, X., Stoller, M. D., Zhu, Y., Ji, H., Murali, S., Wu, Y., Perales, S., Clevenger, B., and Ruoff, R. S. (2012) Highly conductive and porous activated reduced graphene oxide films for high-power supercapacitors. *Nano Letters,* **12**, 1806-1812.
80. Wang, H., Xu, Z., Kohandehghan, A., Li, Z., Cui, K., Tan, X., Stephenson, T. J., King'ondu, C. K., Holt, C. M., Olsen, B. C., Tak, J. K., Harfield, D., Anyia, A. O., and Mitlin, D. (2013) Interconnected carbon

nanosheets derived from hemp for ultrafast supercapacitors with high energy. *ACS Nano,* **7**, 5131-5141.

81. Barranco, V., Lillo-Rodenas, M., Linares-Solano, A., Oya, A., Pico, F., Ibanez, J., Agullo-Rueda, F., Amarilla, J. M., and Rojo, J. M. (2010) Amorphous carbon nanofibers and their activated carbon nanofibers as supercapacitor electrodes. *The Journal of Physical Chemistry C,* **114**(22), 10302-10307.
82. Dai, Y., Jiang, H., Hu, Y., Fu, Y., and Li, C. (2014) Controlled synthesis of ultrathin hollow mesoporous carbon nanospheres for supercapacitor applications. *Industrial & Engineering Chemistry Research,* **53**, 3125-3130.
83. Sevilla, M., and Fuertes, A. B. (2014) Direct synthesis of highly porous interconnected carbon nanosheets and their application as high-performance supercapacitors. *ACS Nano,* **8**, 5069-5078.
84. Hulicova-Jurcakova, D., Puziy, A. M., Poddubnaya, O. I., Suárez-García, F., Tascon, J. M., and Lu, G. Q. (2009) Highly stable performance of supercapacitors from phosphorus-enriched carbons. *Journal of the American Chemical Society,* **131**, 5026-5027.
85. Wickramaratne, N. P., Xu, J., Wang, M., Zhu, L., Dai, L., and Jaroniec, M. (2014) Nitrogen enriched porous carbon spheres: Attractive materials for supercapacitor electrodes and CO_2 adsorption. *Chemistry of Materials,* **26**, 2820-2828.
86. Chen, L.-F., Zhang, X.-D., Liang, H.-W., Kong, M., Guan, Q.-F., Chen, P., Wu, Z.-Y., and Yu, S.-H. (2012) Synthesis of nitrogen-doped porous carbon nanofibers as an efficient electrode material for supercapacitors. *ACS Nano,* **6**, 7092-7102.
87. Hou, J., Cao, C., Idrees, F., and Ma, X. (2015) Hierarchical porous nitrogen-doped carbon nanosheets derived from silk for ultrahigh-capacity battery anodes and supercapacitors. *ACS Nano,* **9**, 2556-2564.
88. Tan, Y., Xu, C., Chen, G., Liu, Z., Ma, M., Xie, Q., Zheng, N., and Yao, S. (2013) Synthesis of ultrathin nitrogen-doped graphitic carbon nanocages as advanced electrode materials for supercapacitor. *ACS Applied Materials & Interfaces,* **5**, 2241-2248.
89. Lu, W., Hartman, R., Qu, L., and Dai, L. (2011) Nanocomposite electrodes for high-performance supercapacitors. *The Journal of Physical Chemistry Letters,* **2**, 655-660.
90. Hahm, M. G., Leela Mohana Reddy, A., Cole, D. P., Rivera, M., Vento, J. A., Nam, J., Jung, H. Y., Kim, Y. L., Narayanan, N. T., Hashim, D., Galande, C., Jung, Y. J., Bundy, M., Karna, S., Ajayan, P. M., and Vajtai, R. P. (2012) Carbon nanotube–nanocup hybrid structures for high power supercapacitor applications. *Nano Letters,* **12**(11), 5616-5621.
91. Shi, K., Ren, M., and Zhitomirsky, I. (2014) Activated carbon-coated carbon nanotubes for energy storage in supercapacitors and ca-

pacitive water purification. *ACS Sustainable Chemistry & Engineering,* **2**, 1289-1298.

92. Du, X., Wang, C., Chen, M., Jiao, Y., and Wang, J. (2009) Electrochemical performances of nanoparticle Fe3O4/activated carbon supercapacitor using KOH electrolyte solution. *The Journal of Physical Chemistry C,* **113**, 2643-2646.
93. Qu, Q., Zhang, P., Wang, B., Chen, Y., Tian, S., Wu, Y., and Holze, R. (2009) Electrochemical performance of MnO_2 nanorods in neutral aqueous electrolytes as a cathode for asymmetric supercapacitors. *The Journal of Physical Chemistry C,* **113**, 14020-14027.
94. Wang, Y.-T., Lu, A.-H., Zhang, H.-L., and Li, W.-C. (2011) Synthesis of nanostructured mesoporous manganese oxides with three-dimensional frameworks and their application in supercapacitors. *The Journal of Physical Chemistry C,* **115**, 5413-5421.
95. Yu, P., Zhang, X., Wang, D., Wang, L., and Ma, Y. (2008) Shape-controlled synthesis of 3D hierarchical MnO_2 nanostructures for electrochemical supercapacitors. *Crystal Growth and Design,* **9**, 528-533.
96. Yan, J., Khoo, E., Sumboja, A., and Lee, P. S. (2010) Facile coating of manganese oxide on tin oxide nanowires with high-performance capacitive behavior. *ACS Nano,* **4**, 4247-4255.
97. Xin, G., Wang, Y., Zhang, J., Jia, S., Zang, J., and Wang, Y. (2015) A self-supporting graphene/MnO_2 composite for high-performance supercapacitors. *International Journal of Hydrogen Energy,* **40**, 10176-10184.
98. Li, G.-R., Feng, Z.-P., Ou, Y.-N., Wu, D., Fu, R., and Tong, Y.-X. (2010) Mesoporous MnO_2/carbon aerogel composites as promising electrode materials for high-performance supercapacitors. *Langmuir,* **26**, 2209-2213.
99. Reddy, A. L. M., Shaijumon, M. M., Gowda, S. R., and Ajayan, P. M. (2009) Multisegmented Au-MnO_2/carbon nanotube hybrid coaxial arrays for high-power supercapacitor applications. *The Journal of Physical Chemistry C,* **114**, 658-663.
100. Yang, P., Ding, Y., Lin, Z., Chen, Z., Li, Y., Qiang, P., Ebrahimi, M., Mai, W., Wong, C. P., and Wang, Z. L. (2014) Low-cost high-performance solid-state asymmetric supercapacitors based on MnO_2 nanowires and Fe_2O_3 nanotubes. *Nano Letters,* **14**, 731-736.
101. Jiang, H., Dai, Y., Hu, Y., Chen, W., and Li. C. (2013) Nanostructured ternary nanocomposite of rGO/CNTs/MnO_2 for high-rate supercapacitors. *ACS Sustainable Chemistry & Engineering,* **2**, 70-74.
102. Hu, L., Chen, W., Xie, X., Liu, N., Yang, Y., Wu, H., Yao, Y., Pasta, M., Alshareef, H. N., and Cui, Y. (2011) Symmetrical MnO_2-carbon nanotube–textile nanostructures for wearable pseudocapacitors with high mass loading. *ACS Nano,* **5**, 8904-8913.
103. Hou, Y., Cheng, Y., Hobson, T., and Liu, J. (2010) Design and synthe-

sis of hierarchical MnO_2 nanospheres/carbon nanotubes/conducting polymer ternary composite for high performance electrochemical electrodes. *Nano Letters,* **10**, 2727-2733.

104. Yang, P., Xiao, X., Li, Y., Ding, Y., Qiang, P., Tan, X., Mai, W., Lin, Z., Wu, W., Li, T., Jin, H., Liu, P., Zhou, J., Wong, C. P., and Wang, Z. L. (2013) Hydrogenated ZnO core-shell nanocables for flexible supercapacitors and self-powered systems. *ACS Nano,* **7**, 2617-2626.
105. Kim, S.-I., Lee, J.-S., Ahn, H.-J., Song, H.-K., and Jang, J.-H. (2013) Facile route to an efficient NiO supercapacitor with a three-dimensional nanonetwork morphology. *ACS Applied Materials & Interfaces,* **5**, 1596-1603.
106. Cai, D., Wang, D., Liu, B., Wang, Y., Liu, Y., Wang, L., Li, H., Huang, H., Li, Q., and Wang, T. (2013) Comparison of the electrochemical performance of $NiMoO_4$ nanorods and hierarchical nanospheres for supercapacitor applications. *ACS Applied Materials & Interfaces,* **5**, 12905-12910.
107. Li, X., Li, Q., Wu, Y., Rui, M., and Zeng, H. (2015) Two-dimensional, porous nickel-cobalt sulfide for high-performance asymmetric supercapacitors. *ACS Applied Materials & Interfaces,* **7**, 19316-19323.
108. Chen, W., Xia, C., and Alshareef, H. N. (2014) One-step electrodeposited nickel cobalt sulfide nanosheet arrays for high-performance asymmetric supercapacitors. *ACS Nano,* 8, 9531-9541.
109. Ji, J., Zhang, L. L., Ji, H., Li, Y., Zhao, X., Bai, X., Fan, X., Zhang, F., and Ruoff, R. S. (2013) Nanoporous $Ni(OH)_2$ thin film on 3D ultrathin-graphite foam for asymmetric supercapacitor. *ACS Nano,* **7**, 6237-6243.
110. Zhang, H., Zhang, X., Zhang, D., Sun, X., Lin, H., Wang, C., and Ma, Y. (2012) One-step electrophoretic deposition of reduced graphene oxide and $Ni(OH)_2$ composite films for controlled syntheses supercapacitor electrodes. *The Journal of Physical Chemistry B,* **117**, 1616-1627.
111. Dubal, D. P., Gund, G. S., Lokhande, C. D., and Holze, R. (2013) Decoration of sponge like $Ni(OH)_2$ nanoparticles onto MWCNTs using an easily manipulated chemical protocol for supercapacitors. *ACS Applied Materials & Interfaces,* **5**, 2446-2454.
112. Chen, H., Jiang, J., Zhang, L., Qi, T., Xia, D., and Wan. H. (2014) Failely synthesized porous $NiCo_2O_4$ flowerlike nanostructure for high-rate supercapacitors. *Journal of Power Sources,* **248**, 28-36.
113. Cai. D., Liu, B., Wang, D., Liu, Y., Wang, L., Li, H., Wang, Y., Wang, C., Li, Q., and Wang, T. (2014) Enhanced performance of supercapacitors with ultrathin mesoporous $NiMoO_4$ nanosheets. *Electrochimica Acta,* **125**, 294-301.
114. Tao, Y., Ruiyi, L., Zaijun, L., and Yinjun, F. (2014) A facile and scalable strategy for synthesis of size-tunable $NiCo_2O_4$ with nanocoral-

like architecture for high-performance supercapacitors. *Electrochimica Acta,* **134**, 384-392.

115. Wu, Z., Pu, X., Ji, X., Zhu, Y., Jing, M., Chen, Q., and Jiao, F. (2015) High energy density asymmetric supercapacitors from mesoporous $NiCo_2S_4$ nanosheets. *Electrochimica Acta,* **174**, 238-245.
116. Zhu, Y., Wu, Z., Jing, M., Yang, X., Song, W., and Ji, X. (2015) Mesoporous $NiCo_2S_4$ nanoparticles as high-performance electrode materials for supercapacitors. *Journal of Power Sources,* **273**, 584-590.
117. Kuang, M., Zhang, W., Guo, X. L., Yu, L., and Zhang, Y. X. (2014) Template-free and large-scale synthesis of hierarchical dandelion-like $NiCo_2O_4$ microspheres for high-performance supercapacitors. *Ceramics International,* **40**, 10005-10011.
118. Zhao, Y., Zhang, X., He, J., Zhang, L., Xia, M., and Gao, F. (2015) Morphology controlled synthesis of nickel cobalt oxide for supercapacitor application with enhanced cycling stability. *Electrochimica Acta,* **174**, 51-56.
119. Xu, X., Zhou, H., Ding, S., Li, J., Li, B., and Yu. D. (2014) The facile synthesis of hierarchical $NiCoO_2$ nanotubes comprised ultrathin nanosheets for supercapacitors. *Journal of Power Sources,* **267**, 641-647.
120. Wang, H., Fu, Y., Wang, X., Gao, J., Zhang, Y., and Zhao, Q. (2015) Preparation and performance of tubular nanoflaky (Ni, Co, Mn) oxides with hierarchical mesoporous structure. *Journal of Alloys and Compounds,* **639**, 352-358.
121. Xie, L., Hu, Z., Lv, C., Sun, G., Wang, J., Li, Y., He, H., Wang, J., and Li, K. (2012) Co_xNi_{1-x} double hydroxide nanoparticles with ultrahigh specific capacitances as supercapacitor electrode materials. *Electrochimica Acta,* **78**, 205-211.
122. Kulkarni, S., Jagadale, A., Kumbhar, V., Bulakhe, R., Joshi, S., and Lokhande, C. (2013) Potentiodynamic deposition of composition influenced $Co_{1-x}Ni_x$ LDHs thin film electrode for redox supercapacitors. *International Journal of Hydrogen Energy,* **38**, 4046-4053.
123. Chen, G., Liaw, S. S., Li, B., Xu, Y., Dunwell, M., Deng, S., Fan, H., and Luo, H. (2014) Microwave-assisted synthesis of hybrid $Co_xNi_{1-x}(OH)_2$ nanosheets: Tuning the composition for high performance supercapacitor. *Journal of Power Sources,* **251**, 338-343.
124. Zhu, Y., Ji, X., Wu, Z., Song, W., Hou, H., Wu, Z., He, X., Chen, Q., and Banks, C. E. (2014) Spinel $NiCo_2O_4$ for use as a high-performance supercapacitor electrode material: understanding of its electrochemical properties. *Journal of Power Sources,* **267**, 888-900.
125. Tang, Y., Chen, T., Yu, S., Qiao, Y., Mu, S., Zhang, S., Zhao, Y., Hou, L., Huang, W., and Gao, F. (2015) A highly electronic conductive cobalt nickel sulphide dendrite/quasi-spherical nanocomposite for a supercapacitor electrode with ultrahigh areal specific capacitance. *Journal of Power Sources,* **295**, 314-322.

126. Cai, D., Xiao, S., Wang, D., Liu, B., Wang, L., Liu, Y., Li, H., Wang, Y., Li, Q., and Wang, T. (2014) Morphology controlled synthesis of $NiCo_2O_4$ nanosheet array nanostructures on nickel foam and their application for pseudocapacitors. *Electrochimica Acta,* **142**, 118-124.
127. Cai, D., Wang, D., Wang, C., Liu, B., Wang, L., Liu, Y., Li, Q., and Wang, T. (2015) Construction of desirable $NiCo_2S_4$ nanotube arrays on nickel foam substrate for pseudocapacitors with enhanced performance. *Electrochimica Acta,* **151**, 35-41.
128. Chen, H., Jiang, J., Zhang, L., Xia, D., Zhao, Y., Guo, D., Qi, T., and Wan, H. (2014) In situ growth of $NiCo_2S_4$ nanotube arrays on Ni foam for supercapacitors: maximizing utilization efficiency at high mass loading to achieve ultrahigh areal pseudocapacitance. *Journal of Power Sources,* **254**, 249-257.
129. Wang, T., Guo, Y., Zhao, B., Yu, S., Yang, H.-P., Lu, D., Fu, X.-Z., Sun, R., and Wong, C.-P. (2015) $NiCo_2O_4$ nanosheets in-situ grown on three dimensional porous Ni film current collectors as integrated electrodes for high-performance supercapacitors. *Journal of Power Sources,* **286**, 371-379.
130. Yang, Z., Zhu, X., Wang, K., Ma, G., Cheng, H., and Xu, F. (2015) Preparation of $NiCo_2S_4$ flaky arrays on Ni foam as binder-free supercapacitor electrode. *Applied Surface Science,* **347**, 690-695.
131. Chen, S., Yang, G., and Zheng, H. (2016) Aligned Ni-Co-Mn oxide nanosheets grown on conductive substrates as binder-free electrodes for high capacity electrochemical energy storage devices. *Electrochimica Acta,* **220**, 296-303.
132. Deng, F., Yu, L., Cheng, G., Lin, T., Sun, M., Ye, F., and Li, Y. (2014) Synthesis of ultrathin mesoporous $NiCo_2O_4$ nanosheets on carbon fiber paper as integrated high-performance electrodes for supercapacitors. *Journal of Power Sources,* **251**, 202-207.
133. Yu, M., Chen, J., Liu, J., Li, S., Ma, Y., Zhang, J., and An, J. (2015) Mesoporous $NiCo_2O_4$ nanoneedles grown on 3D graphene-nickel foam for supercapacitor and methanol electro-oxidation. *Electrochimica Acta,* **151**, 99-108.
134. Li, X., Ding, R., Shi, W., Xu, Q., Wang, L., Jiang, H., Yang, Z., and Liu, E. (2017) Hierarchical mesoporous Ni-P@MnO_2 composite for high performance supercapacitors. *Materials Letters,* **187**, 144-147.
135. Zhang, C., Kuila, T., Kim, N. H., Lee, S. H., and Lee, J. H. (2015) Facile preparation of flower-like $NiCo_2O_4$/three dimensional graphene foam hybrid for high performance supercapacitor electrodes. *Carbon,* **89**, 328-339.
136. Yu, M., Chen, J., Ma, Y., Zhang, J., Liu, J., Li, S., and An, J. (2014) Hydrothermal synthesis of $NiCo_2O_4$ nanowires/nitrogen-doped graphene for high-performance supercapacitor. *Applied Surface Science,* **314**, 1000-1006.

137. Tao, K., Li, P., Kang, L., Li, X., Zhou, Q., Dong, L., and Liang, W. (2015) Facile and low-cost combustion-synthesized amorphous mesoporous NiO/carbon as high mass-loading pseudocapacitor materials. *Journal of Power Sources,* **293**, 23-32.
138. Lo, I.-H., Wang, J.-Y., Huang, K.-Y., Huang, J.-H., and Kang, W. P. (2016) Synthesis of $Ni(OH)_2$ nanoflakes on ZnO nanowires by pulse electrodeposition for high-performance supercapacitors. *Journal of Power Sources,* **308**, 29-36.
139. Liu, L., Zhang, H., Fang, L., Mu, Y., and Wang, Y. (2016) Facile preparation of novel dandelion-like Fe-doped $NiCo_2O_4$ microspheres@ nanomeshes for excellent capacitive property in asymmetric supercapacitors. *Journal of Power Sources,* **327**, 135-144.
140. Yuksel, R., Coskun, S., Kalay, Y. E., and Unalan, H. E. (2016) Flexible, silver nanowire network nickel hydroxide core-shell electrodes for supercapacitors. *Journal of Power Sources,* **328**, 167-173.
141. Xue, W.-d., Wang, W.-j., Fu, Y.-f., He, D.-x, Zeng, F.-y., and Zhao, R. (2017) Rational synthesis of honeycomb-like $NiCo_2O_4$@$NiMoO_4$ core/shell nanofilm arrays on Ni foam for high-performance supercapacitors. *Materials Letters,* **186**, 34-37.
142. Ma, X.-J., Kong, L.-B., Zhang, W.-B., Liu, M.-C., Luo, Y.-C., and Kang, L. (2014) Design and synthesis of 3D Co_3O_4@$MMoO_4$ (M=Ni, Co) nanocomposites as high-performance supercapacitor electrodes. *Electrochimica Acta,* **130**, 660-669.
143. Shim, J.-J. (2015) In situ growth of hierarchical mesoporous $NiCo_2S_4$@ MnO_2 arrays on nickel foam for high-performance supercapacitors. *Electrochimica Acta,* **166**, 302-309.
144. Lamiel, C., and Shim, J.-J. (2015) Hierarchical mesoporous graphene@ Ni-Co-S arrays on nickel foam for high-performance supercapacitors. *Electrochimica Acta,* **161**, 351-357.
145. Xu, J., Li, L., Gao, P., Yu, L., Chen, Y., Yang, P., Gai, S., and Yang, P. (2015) Facile preparation of $NiCo_2O_4$ nanobelt/graphene composite for electrochemical capacitor application. *Electrochimica Acta,* **166**, 206-214.
146. Wang, H., Wang, C., Qing, C., Sun, D., Wang, B., Qu, G., Sun, M., and Tang, Y. (2015) Construction of carbon-nickel cobalt sulphide hetero-structured arrays on nickel foam for high performance asymmetric supercapacitors. *Electrochimica Acta,* **174**, 1104-1112.
147. Chen, H., Yu, L., Zhang, J. M., and Liu, C. P. (2016) Construction of hierarchical $NiMoO_4$@ MnO_2 nanosheet arrays on titanium mesh for supercapacitor electrodes. *Ceramics International,* **42**, 18058-18063.
148. Gong, X., Cheng, J., Liu, F., Zhang, L., and Zhang, X. (2014) Nickel–Cobalt hydroxide microspheres electrodepositioned on nickel cobaltite nanowires grown on Ni foam for high-performance pseudocapacitors. *Journal of Power Sources,* **267**, 610-616.

149. Shim, J.-J. (2015) Three-dimensional nickel foam/graphene/ $NiCo_2O_4$ as high-performance electrodes for supercapacitors. *Journal of Power Sources,* **273**, 110-117.
150. Saravanakumar, B., Purushothaman, K. K., and Muralidharan, G. (2012) Interconnected V_2O_5 nanoporous network for high-performance supercapacitors. *ACS Applied Materials & Interfaces,* **4**, 4484-4490.
151. Zhu, J., Cao, L., Wu, Y., Gong, Y., Liu, Z., Hoster, H. E., Zhang, Y., Zhang, S., Yang, S., Yan, Q., Ajayan, P. M., and Vajtai, R. (2013) Building 3D structures of vanadium pentoxide nanosheets and application as electrodes in supercapacitors. *Nano Letters,* **13**, 5408-5413.
152. Butt, F. K., Tahir, M., Cao, C., Idrees, F., Ahmed, R., Khan, W. S., Ali, Z., Mahmood, N., Tanveer, M., Mahmood, A., and Aslam, I. (2014) Synthesis of novel ZnV_2O_4 hierarchical nanospheres and their applications as electrochemical supercapacitor and hydrogen storage material. *ACS Applied Materials & Interfaces,* **6**, 13635-13641.
153. Feng, J., Sun, X., Wu, C., Peng, L., Lin, C., Hu, S., Yang, J., and Xie, Y. (2011) Metallic few-layered VS_2 ultrathin nanosheets: high two-dimensional conductivity for in-plane supercapacitors. *Journal of the American Chemical Society,* **133**, 17832-17838.
154. Lu, X., Yu, M., Zhai, T., Wang, G., Xie, S., Liu, T., Liang, C., Tong, Y., and Li, Y. (2013) High energy density asymmetric quasi-solid-state supercapacitor based on porous vanadium nitride nanowire anode. *Nano Letters,* **13**, 2628-2633.
155. Xu, Y., Wang, J., Shen, L., Dou, H., and Zhang, X. (2015) One-dimensional vanadium nitride nanofibers fabricated by electrospinning for supercapacitors. *Electrochimica Acta,* **173**, 680-686.
156. Wang, B., Chen, Z., Lu, G., Wang, T., and Ge, Y. (2016) Exploring electrolyte preference of vanadium nitride supercapacitor electrodes. *Materials Research Bulletin,* **76**, 37-40.
157. Thangappan, R., Kalaiselvam, S., Elayaperumal, A., and Jayavel, R. (2014) Synthesis of graphene oxide/vanadium pentoxide composite nanofibers by electrospinning for supercapacitor applications. *Solid State Ionics,* **268**, 321-325.
158. Xu, J., Sun, H., Li, Z., Lu, S., Zhang, X., Jiang, S., Zhu, Q., and Zakharova, G. S. (2014) Synthesis and electrochemical properties of graphene/V_2O_5 xerogels nanocomposites as supercapacitor electrodes. *Solid State Ionics,* **262**, 234-237.
159. Choudhury, A., Bonso, J. S., Wunch, M., Yang, K. S., Ferraris, J. P., and Yang, D. J. (2015) In-situ synthesis of vanadium pentoxide nanofibre/exfoliated graphene nanohybrid and its supercapacitor applications. *Journal of Power Sources,* **287**, 283-290.
160. Balamurugan, J., Karthikeyan, G., Thanh, T. D., Kim, N. H., and Lee, J. H. (2016) Facile synthesis of vanadium nitride/nitrogen-doped graphene composite as stable high performance anode materials

for supercapacitors. *Journal of Power Sources,* **308**, 149-157.
161. Meher, S. K., and Rao, G. R. (2011) Ultralayered Co_3O_4 for high-performance supercapacitor applications. *The Journal of Physical Chemistry C,* **115**, 15646-15654.
162. Deori, K., Ujjain, S. K., Sharma, R. K., and Deka, S. (2013) Morphology controlled synthesis of nanoporous Co3O4 nanostructures and their charge storage characteristics in supercapacitors. *ACS Applied Materials & Interfaces,* **5**, 10665-10672.
163. Rakhi, R., Chen, W., Hedhili, M. N., Cha, D., and Alshareef, H. N. (2014) Enhanced rate performance of mesoporous Co_3O_4 nanosheet supercapacitor electrodes by hydrous RuO_2 nanoparticle decoration. *ACS Applied Materials & Interfaces,* **6**, 4196-4206.
164. Pendashteh, A., Moosavifard, S. E., Rahmanifar, M. S., Wang, Y., El-Kady, M. F., Kaner, R. B., and Mousavi, M. F. (2015) Highly ordered mesoporous $CuCo_2O_4$ nanowires, a promising solution for high-performance supercapacitors. *Chemistry of Materials,* **27**, 3919-3926.
165. Huang, K.-J., Zhang, J.-Z., Shi, G.-W., and Liu, Y.-M. (2014) One-step hydrothermal synthesis of two-dimensional cobalt sulfide for high-performance supercapacitors. *Materials Letters,* **131**, 45-48.
166. Wan, H., Ji, X., Jiang, J., Yu, J., Miao, L., Zhang, L., Bie, S., Chen, H., and Ruan, Y. (2013) Hydrothermal synthesis of cobalt sulfide nanotubes: The size control and its application in supercapacitors. *Journal of Power Sources,* **243**, 396-402.
167. Luo, F., Li, J., Yuan, H., and Xiao, D. (2014) Rapid synthesis of three-dimensional flower-like cobalt sulfide hierarchitectures by microwave assisted heating method for high-performance supercapacitors. *Electrochimica Acta,* **123**, 183-189.
168. Liu, M.-C., Kong, L.-B., Lu, C., Li, X.-M., Luo, Y.-C., and Kang, L. (2013) Facile fabrication of $CoMoO_4$ nanorods as electrode material for electrochemical capacitors. *Materials Letters,* **94**, 197-200.
169. Bhagwan, J., Sivasankaran, V., Yadav, K., and Sharma, Y. (2016) Porous, one-dimensional and high aspect ratio nanofibric network of cobalt manganese oxide as a high performance material for aqueous and solid-state supercapacitor (2 V). *Journal of Power Sources,* **327**, 29-37.
170. Wan, H.-Z., Jiang, J.-J., Yu, J.-W., Ruan, Y.-J., Peng, L., Zhang, L., Chen, H.-C., and Bie, S.-W. (2014) Cobalt sulfide nanotube arrays grown on FTO and graphene membranes for high-performance supercapacitor application. *Applied Surface Science,* **311**, 793-798.
171. Xia, X., Lei, W., Hao, Q., Wang, W., and Wang, X. (2013) One-step synthesis of $CoMoO_4$/graphene composites with enhanced electrochemical properties for supercapacitors. *Electrochimica Acta,* **99**, 253-261.
172. Wang, X., Li, M., Chang, Z., Yang, Y., Wu, Y., and Liu, X. (2015) Co_3O_4

@ MWCNT nanocable as cathode with superior electrochemical performance for supercapacitors. *ACS Applied Materials & Interfaces,* **7**, 2280-2285.

173. Yang, L., Cheng, S., Ding, Y., Zhu, X., Wang, Z. L., and Liu, M. (2011) Hierarchical network architectures of carbon fiber paper supported cobalt oxide nanonet for high-capacity pseudocapacitors. *Nano Letters,* **12**, 321-325.
174. Salunkhe, R. R., Tang, J., Kamachi, Y., Nakato, T., Kim, J. H., and Yamauchi, Y. (2015) Asymmetric supercapacitors using 3D nanoporous carbon and cobalt oxide electrodes synthesized from a single metal–organic framework. *ACS Nano,* **9**, 6288-6296.
175. Xia, X., Chao, D., Fan, Z., Guan, C., Cao, X., Zhang, H., and Fan, H. J. (2014) A new type of porous graphite foams and their integrated composites with oxide/polymer core/shell nanowires for supercapacitors: structural design, fabrication, and full supercapacitor demonstrations. *Nano Letters,* **14**, 1651-1658.
176. Mohamed, S. G., Chen, C.-J., Chen, C. K., Hu, S.-F., and Liu, R.-S. (2014) High-performance lithium-ion battery and symmetric supercapacitors based on $FeCo_2O_4$ nanoflakes electrodes. *ACS Applied Materials & Interfaces,* **6**, 22701-22708.
177. Long, C., Wei, T., Yan, J., Jiang, L., and Fan, Z. (2013) Supercapacitors based on graphene-supported iron nanosheets as negative electrode materials. *ACS Nano,* **7**, 11325-11332.
178. Guan, C., Liu, J., Wang, Y., Mao, L., Fan, Z., Shen, Z., Zhang, H., and Wang, J. (2015) Iron oxide-decorated carbon for supercapacitor anodes with ultrahigh energy density and outstanding cycling stability. *ACS Nano,* **9**, 5198-5207.
179. Ma, L., Xu, L.-M., Zhou, X.-P., and Xu, X.-Y. (2014) Biopolymer-assisted hydrothermal synthesis of flower-like MoS_2 microspheres and their supercapacitive properties. *Materials Letters,* **132**, 291-294.
180. Fan, L.-Q., Liu, G.-J., Zhang, C.-Y., Wu, J.-H., and Wei, Y.-L. (2015) Facile one-step hydrothermal preparation of molybdenum disulfide/carbon composite for use in supercapacitor. *International Journal of Hydrogen Energy,* **40**, 10150-10157.
181. Yang, M., Hwang, S.-K., Jeong, J.-M., Huh, Y. S., and Choi, B. G. (2015) Nitrogen-doped carbon-coated molybdenum disulfide nanosheets for high-performance supercapacitor. *Synthetic Metals,* **209**, 528-533.
182. Huang, K. J., Wang, L., Zhang, J.-Z., Wang, L.-L., and Mo, Y.-P. (2014) One-step preparation of layered molybdenum disulfide/multi-walled carbon nanotube composites for enhanced performance supercapacitor. *Energy,* **67**, 234-240.
183. Huang, K.-J., Wang, L., Zhang, J.-Z., and Xing, K. (2015) Synthesis of molybdenum disulfide/carbon aerogel composites for supercapac-

itors electrode material application. *Journal of Electroanalytical Chemistry,* **752**, 33-40.

184. Lu, X., Wang, G., Zhai, T., Yu, M., Gan, J., Tong, Y., and Li, Y. (2012) Hydrogenated TiO_2 nanotube arrays for supercapacitors. *Nano Letters,* **12**, 1690-1696.
185. Lu, X., Wang, G., Zhai, T., Yu, M., Xie, S., Ling, Y., Liang, C., Tong, Y., and Li, Y. (2012) Stabilized TiN nanowire arrays for high-performance and flexible supercapacitors. *Nano Letters,* **12**, 5376-5381.
186. Shang, C., Dong, S., Wang, S., Xiao, D., Han, P., Wang, X., Gu, L., and Cui. G. (2013) Coaxial $Ni_xCo_{2x}(OH)_{6x}$/TiN Nanotube arrays as supercapacitor electrodes. *ACS Nano,* **7**, 5430-5436.
187. Zhang, J., Ma, J., Zhang, L. L., Guo, P., Jiang, J., and Zhao, X. (2010) Template synthesis of tubular ruthenium oxides for supercapacitor applications. *The Journal of Physical Chemistry C,* **114**, 13608-13613.
188. Bi, R.-R., Wu, X.-L., Cao, F.-F., Jiang, L.-Y., Guo, Y.-G., and Wan, L.-J. (2010) Highly dispersed RuO_2 nanoparticles on carbon nanotubes: facile synthesis and enhanced supercapacitance performance. *The Journal of Physical Chemistry C,* **114**, 2448-2451.
189. Miao, Y.-E., Fan, W., Chen, D., and Liu, T. (2013) High-performance supercapacitors based on hollow polyaniline nanofibers by electrospinning. *ACS Applied Materials & Interfaces,* **5**, 4423-4428.
190. Fan, W., Zhang, C., Tjiu, W. W., Pramoda, K. P., He, C., and Liu, T. (2013) Graphene-wrapped polyaniline hollow spheres as novel hybrid electrode materials for supercapacitor applications. *ACS Applied Materials & Interfaces,* **5**, 3382-3391.
191. Wang, D.-W., Li, F., Zhao, J., Ren, W., Chen, Z.-G., Tan, J., Wu, Z.-S., Gentle, I., Lu, G. Q., and Cheng, H.-M. (2009) Fabrication of graphene/polyaniline composite paper via in situ anodic electropolymerization for high-performance flexible electrode. *ACS Nano,* **3**, 1745-1752.
192. Wang, L., Ye, Y., Lu, X., Wen, Z., Li, Z., Hou, H., and Song, Y. (2013) Hierarchical nanocomposites of polyaniline nanowire arrays on reduced graphene oxide sheets for supercapacitors. *Scientific Reports,* **3**, 3568.
193. Wu, Q., Xu, Y., Yao, Z., Liu, A., and Shi, G. (2010) Supercapacitors based on flexible graphene/polyaniline nanofiber composite films. *ACS Nano,* **4**, 1963-1970.
194. Jianhua, L., Junwei, A., Yecheng, Z., Yuxiao, M., Mengliu, L., Mei, Y., and Songmei, L. (2012) Preparation of an amide group-connected graphene–polyaniline nanofiber hybrid and its application in supercapacitors. *ACS Applied Materials & Interfaces,* **4**, 2870-2876.
195. Zhang, J., and Zhao, X. (2012) Conducting polymers directly coated on reduced graphene oxide sheets as high-performance superca-

pacitor electrodes. *The Journal of Physical Chemistry C,* **116**, 5420-5426.

196. De Oliveira, H. P., Sydlik, S. A., and Swager, T. M. (2013) Supercapacitors from free-standing polypyrrole/graphene nanocomposites. *The Journal of Physical Chemistry C,* **117**, 10270-10276.
197. Li, Z.-F., Zhang, H., Liu, Q., Sun, L., Stanciu, L., and Xie, J. (2013) Fabrication of high-surface-area graphene/polyaniline nanocomposites and their application in supercapacitors. *ACS Applied Materials & Interfaces,* **5**, 2685-2691.
198. Kumar, N. A., Choi, H.-J., Shin, Y. R., Chang, D. W., Dai, L., and Baek, J.-B. (2012) Polyaniline-grafted reduced graphene oxide for efficient electrochemical supercapacitors. *ACS Nano,* **6**, 1715-1723.
199. Lai, L., Yang, H., Wang, L., Teh, B. K., Zhong, J., Chou, H., Chen, L., Chen, W., Shen, Z., Ruoff, R. S., and Lin, J. (2012) Preparation of supercapacitor electrodes through selection of graphene surface functionalities. *ACS Nano,* **6**, 5941-5951.
200. Xu, J., Wang, K., Zu, S.-Z., Han, B.-H., and Wei, Z. (2010) Hierarchical nanocomposites of polyaniline nanowire arrays on graphene oxide sheets with synergistic effect for energy storage. *ACS Nano,* **4**, 5019-5026.
201. Xu, D., Xu, Q., Wang, K., Chen, J., and Chen, Z. (2013) Fabrication of free-standing hierarchical carbon nanofiber/graphene oxide/polyaniline films for supercapacitors. *ACS Applied Materials & Interfaces,* **6**, 200-209.
202. Meng, C., Liu, C., Chen, L., Hu, C., and Fan, S. (2010) Highly flexible and all-solid-state paperlike polymer supercapacitors. *Nano Letters,* **10**, 4025-4031.
203. Cheng, Q., Tang, J., Ma, J., Zhang, H., Shinya, N., and Qin, L.-C. (2011) Polyaniline-coated electro-etched carbon fiber cloth electrodes for supercapacitors. *The Journal of Physical Chemistry C,* **115**, 23584-23590.
204. Wang, H., Zhu, E., Yang, J., Zhou, P., Sun, D., and Tang, W. (2012) Bacterial cellulose nanofiber-supported polyaniline nanocomposites with flake-shaped morphology as supercapacitor electrodes. *The Journal of Physical Chemistry C,* **116**, 13013-13019.
205. Luo, Z., Zhu, Y., Liu, E., Hu, T., Li, Z., Liu, T., and Song, L. (2014) Synthesis of polyaniline/SnO_2 nanocomposite and its improved electrochemical performance. *Materials Research Bulletin,* **60**, 105-110.
206. Dubal, D. P., Chodankar, N. R., Caban-Huertas, Z., Wolfart, F., Vidotti, M., Holze, R., Lokhande, C. D., and Gomez-Romero, P. (2016) Synthetic approach from polypyrrole nanotubes to nitrogen doped pyrolyzed carbon nanotubes for asymmetric supercapacitors. *Journal of Power Sources,* **308**, 158-165.
207. Chen, J., Xia, Z., Li, H., Li, Q., and Zhang, Y. (2015) Preparation of hi-

ghly capacitive polyaniline/black TiO_2 nanotubes as supercapacitor electrode by hydrogenation and electrochemical deposition. *Electrochimica Acta,* **166**, 174-182.

208. Huang, K.-J., Wang, L., Liu, Y.-J., Wang, H.-B., Liu, Y.-M., and Wang, L.-L. (2013) Synthesis of polyaniline/2-dimensional graphene analog MoS_2 composites for high-performance supercapacitor. *Electrochimica Acta,* **109**, 587-594.
209. Zheng, B.-n., and Gao, C. (2016) Preparation of graphene nanoscroll/polyaniline composites and their use in high performance supercapacitors. *New Carbon Materials,* **31**, 315-320.
210. Tran, C., Singhal, R., Lawrence, D., and Kalra, V. (2015) Polyaniline-coated freestanding porous carbon nanofibers as efficient hybrid electrodes for supercapacitors. *Journal of Power Sources,* **293**, 373-379.
211. Jiang, X., Cao, Y., Li, P., Wei, J., Wang, K., Wu, D., and Zhu, H. (2015) Polyaniline/graphene/carbon fiber ternary composites as supercapacitor electrodes. *Materials Letters,* **140**, 43-47.
212. Yang, C., and Li, D. (2015) Flexible and foldable supercapacitor electrodes from the porous 3D network of cellulose nanofibers, carbon nanotubes and polyaniline. *Materials Letters,* **155**, 78-81.
213. Yu, L., Gan, M., Ma, L., Huang, H., Hu, H., Li, Y., Tu, Y., Ge, C., Yang, F., and Yan J. (2014) Facile synthesis of MnO_2/polyaniline nanorod arrays based on graphene and its electrochemical performance. *Synthetic Metals,* **198**, 167-174.
214. Wang, W., Hao, Q., Lei, W., Xia, X., and Wang, X. (2014) Ternary nitrogen-doped graphene/nickel ferrite/polyaniline nanocomposites for high-performance supercapacitors. *Journal of Power Sources,* **269**, 250-259.
215. Sankar, K. V., and Selvan, R. K. (2015) The ternary $MnFe_2O_4$/ graphene/polyaniline hybrid composite as negative electrode for supercapacitors. *Journal of Power Sources,* **275**, 399-407.
216. Jin, Y., and Jia, M. (2015) Design and synthesis of nanostructured graphene-SnO_2-polyaniline ternary composite and their excellent supercapacitor performance. *Colloids and Surfaces A: Physicochemical and Engineering Aspects,* **464**, 17-25.
217. Zhang, L. L., Zhao, S., Tian, X. N., and Zhao, X. (2010) Layered graphene oxide nanostructures with sandwiched conducting polymers as supercapacitor electrodes. *Langmuir,* **26**, 17624-17628.
218. Davies, A., Audette, P., Farrow, B., Hassan, F., Chen, Z., Choi, J.-Y., and Yu, A. (2011) Graphene-based flexible supercapacitors: pulse-electropolymerization of polypyrrole on free-standing graphene films. *The Journal of Physical Chemistry C,* **115**, 17612-17620.
219. Zhang, K., Zhang, L. L., Zhao, X., and Wu, J. (2010) Graphene/polyaniline nanofiber composites as supercapacitor electrodes. *Chemistry of Materials,* **22**, 1392-1401.

220. Biswas, S., and Drzal, L. T. (2010) Multilayered nanoarchitecture of graphene nanosheets and polypyrrole nanowires for high performance supercapacitor electrodes. *Chemistry of Materials,* **22**, 5667-5671.
221. Zang, J., Bao, S.-J., Li, C. M., Bian, H., Cui, X., Bao, Q., Sun, C. Q., Guo, J., and Lian, K. (2008) Well-aligned cone-shaped nanostructure of polypyrrole/RuO_2 and its electrochemical supercapacitor. *The Journal of Physical Chemistry C,* **112**, 14843-14847.
222. Li, P., Yang, Y., Shi, E., Shen, Q., Shang, Y., Wu, S., Wei, J., Wang, K., Zhu, H., Yuan, Q., Cao, A., and Wu, D. (2014) Core-double-shell, carbon nanotube@ polypyrrole@ MnO_2 sponge as freestanding, compressible supercapacitor electrode. *ACS Applied Materials & Interfaces,* **6**, 5228-5234.
223. Ma, G., Peng, H., Mu, J., Huang, H., Zhou, X., and Lei, Z. (2013) In situ intercalative polymerization of pyrrole in graphene analogue of MoS_2 as advanced electrode material in supercapacitor. *Journal of Power Sources,* **229**, 72-78.
224. Hsu, F.-H., and Wu, T.-M. (2014) Enhanced capacitance of one-dimensional polypyrrole/graphene oxide nanoribbon nanocomposite as electrode material for high performance supercapacitors. *Synthetic metals,* **198**, 188-195.
225. Warren, R., Sammoura, F., Teh, K. S., Kozinda, A., Zang, X., and Lin, L. (2015) Electrochemically synthesized and vertically aligned carbon nanotube–polypyrrole nanolayers for high energy storage devices. *Sensors and Actuators A: Physical,* **231**, 65-73.
226. Kalambate, P. K., Dar, R. A., Karna, S. P., and Srivastava, A. K. (2015) High performance supercapacitor based on graphene-silver nanoparticles-polypyrrole nanocomposite coated on glassy carbon electrode. *Journal of Power Sources,* **276**, 262-270.
227. D'Arcy, J. M., El-Kady, M. F., Khine, P. P., Zhang, L., Lee, S. H., Davis, N. R., Liu, D. S., Yeung, M. T., Kim, S. Y., Turner, C. L., Lech, A. T., Hammond, P. T., and Kaner, R. B. (2014) Vapor-phase polymerization of nanofibrillar poly (3, 4-ethylenedioxythiophene) for supercapacitors. *ACS Nano,* **8**, 1500-1510.
228. Aradilla, D., Estrany, F., and Alemán, C. (2011) Symmetric supercapacitors based on multilayers of conducting polymers. *The Journal of Physical Chemistry C,* **115**, 8430-8438.
229. Liu, R., and Lee, S. B. (2008) MnO_2/poly (3,4-ethylenedioxythiophene) coaxial nanowires by one-step coelectrodeposition for electrochemical energy storage. *Journal of the American Chemical Society,* **130**, 2942-2943.
230. Tang, P., Han, L., and Zhang L. (2014) Facile synthesis of graphite/PEDOT/MnO_2 composites on commercial supercapacitor separator membranes as flexible and high-performance supercapacitor electrodes. *ACS Applied Materials & Interfaces,* **6**, 10506-10515.

231. Lu, Z., Liu, N., Lee, H.-W., Zhao, J., Li, W., Li, Y., and Cui, Y. (2015) Nonfilling carbon coating of porous silicon micrometer-sized particles for high-performance lithium battery anodes. *ACS Nano,* **9**, 2540-2547.
232. Jia, H., Gao, P., Yang, J., Wang, J., Nuli, Y., and Yang, Z. (2011) Novel Three-Dimensional Mesoporous Silicon for High Power Lithium-Ion Battery Anode Material. *Advanced Energy Materials,* **1**, 1036-1039.
233. Peled, E., Patolsky, F., Golodnitsky, D., Freedman, K., Davidi, G., and Schneier, D. (2015) Tissue-like silicon nanowires-based three-dimensional anodes for high-capacity lithium ion batteries. *Nano Letters,* **15**, 3907-3916.
234. Teki, R., Datta, M. K., Krishnan, R., Parker, T. C., Lu, T. M., Kumta, P. N., and Koratkar, N. (2009) Nanostructured silicon anodes for lithium ion rechargeable batteries. *Small,* **5**, 2236-2242.
235. Sourice, J., Quinsac, A., Leconte, Y., Sublemontier, O., Porcher, W., Haon, C., Bordes, A., De Vito, E., Boulineau, A., and Jouanneau Si Larbi, S. v., Herlin-Boime, N., and Reynaud, C. (2015) One-step synthesis of Si@C nanoparticles by laser pyrolysis: High-capacity anode material for lithium-ion batteries. *ACS Applied Materials & Interfaces,* **7**(12), 6637-6644.
236. Gohier, A., Laik, B., Kim, K. H., Maurice, J. L., Pereira-Ramos, J. P., Cojocaru, C. S., and Van, P. T. (2012) High-rate capability silicon decorated vertically aligned carbon nanotubes for Li-ion batteries. *Advanced Materials,* **24**, 2592-2597.
237. Jung, J.-W., Ryu, W.-H., Shin, J., Park, K., and Kim, I.-D. (2015) Glassy metal alloy nanofiber anodes employing graphene wrapping layer: toward ultralong-cycle-life lithium-ion batteries. *ACS Nano,* **9**, 6717-6727.
238. Liu, X., Zhang, J., Si, W., Xi, L., Eichler, B., Yan, C., and Schmidt, O. G. (2015) Sandwich nanoarchitecture of Si/reduced graphene oxide bilayer nanomembranes for Li-ion batteries with long cycle life. *ACS Nano,* **9**, 1198-1205.
239. He, Y., Yu, X., Wang, Y., Li, H., and Huang, X. (2011) Alumina-Coated Patterned Amorphous Silicon as the Anode for a Lithium-Ion Battery with High Coulombic Efficiency. *Advanced Materials,* **23**, 4938-4941.
240. Han, Y., Qi, P., Feng, X., Li, S., Fu, X., Li, H., Chen, Y., Zhou, J., Li, X., and Wang, B. (2015) In situ growth of MOFs on the surface of si nanoparticles for highly efficient lithium storage: Si@MOF nanocomposites as anode materials for lithium-ion batteries. *ACS Applied Materials & Interfaces,* **7**, 2178-2182.
241. Yu, C., Li, X., Ma, T., Rong, J., Zhang, R., Shaffer, J., An, Y., Liu, Q., Wei, B., and Jiang, H. (2012) Silicon Thin Films as Anodes for High-Performance Lithium-Ion Batteries with Effective Stress Rel-

axation. *Advanced Energy Materials,* **2**, 68-73.

242. Deng, J., Ji, H., Yan, C., Zhang, J., Si, W., Baunack, S., Oswald, S., Mei, Y., and Schmidt, O. G. (2013) Naturally rolled-up C/Si/C trilayer nanomembranes as stable anodes for lithium-ion batteries with remarkable cycling performance, *Angewandte Chemie International Edition,* **52**, 2326-2330.
243. Liu, J., Li, N., Goodman, M. D., Zhang, H. G., Epstein, E. S., Huang, B., Pan, Z., Kim, J., Choi, J. H., Huang, X., Liu, J., Hsia, K. J., Dillon, S. J., and Braun, P. V. (2015) Mechanically and chemically robust sandwich-structured C@Si@C nanotube array Li-ion battery anodes. *ACS Nano,* **9**(2), 1985-1994.
244. Jia, H., Stock, C., Kloepsch, R., He, X., Badillo, J. P., Fromm, O., Vortmann, B., Winter, M., and Placke, T. (2015) Facile synthesis and lithium storage properties of a porous $NiSi_2$/Si/carbon composite anode material for lithium-ion batteries. *ACS Applied Materials & Interfaces,* **7**, 1508-1515.
245. Cao, F. F., Deng, J. W., Xin, S., Ji, H. X., Schmidt, O. G., Wan, L. J., and Guo, Y. G. (2011) Cu-Si Nanocable arrays as high-rate anode materials for lithium-ion batteries. *Advanced Materials,* **23**, 4415-4420.
246. Hu, Y. S., Demir-Cakan, R., Titirici, M. M., Muller, J. O., Schlogl, R., Antonietti, M., and Maier, J. (2008) Superior storage performance of a Si@SiO_x/C nanocomposite as anode material for lithium-ion batteries. *Angewandte Chemie International Edition,* **47**, 1645-1649.
247. Park, S.-J., Zhao, H., Ai, G., Wang, C., Song, X., Yuca, N., Battaglia, V. S., Yang, W., and Liu, G. (2015) Side-chain conducting and phase-separated polymeric binders for high-performance silicon anodes in lithium-ion batteries. *Journal of the American Chemical Society,* **137**, 2565-2571.
248. Zhao, J., Lu, Z., Wang, H., Liu, W., Lee, H.-W., Yan, K., Zhuo, D., Lin, D., Liu, N., and Cui, Y. (2015) Artificial solid electrolyte interphase-protected Li x Si nanoparticles: An efficient and stable prelithiation reagent for lithium-ion batteries. *Journal of the American Chemical Society,* **137**, 8372-8375.
249. Ryou, M. H., Kim, J., Lee, I., Kim, S., Jeong, Y. K., Hong, S., Ryu, J. H., Kim, T. S., Park, J. K., and Lee, H. (2013) Mussel-inspired adhesive binders for high-performance silicon nanoparticle anodes in lithium-ion batteries. *Advanced Materials,* **25**, 1571-1576.
250. Schroder, K., Alvarado, J., Yersak, T. A., Li, J., Dudney, N., Webb, L. J., Meng, Y. S., and Stevenson, K. J. (2015) The effect of fluoroethylene carbonate as an additive on the solid electrolyte interphase on silicon lithium-ion electrodes. *Chemistry of Materials,* **27**, 5531-5542.
251. Liu, G., Xun, S., Vukmirovic, N., Song, X., Olalde-Velasco, P., Zheng, H., Battaglia, V. S., Wang, L., and Yang, W. (2011) Polymers with tailored electronic structure for high capacity lithium battery electro-

des. *Advanced Materials,* **23**, 4679-4683.

252. Liang, Z., Zheng, G., Liu, C., Liu, N., Li, W., Yan, K., Yao, H., Hsu, P.-C., Chu, S., and Cui, Y. (2015) Polymer nanofiber-guided uniform lithium deposition for battery electrodes. *Nano Letters,* **15**, 2910-2916.
253. Shen, L., Uchaker, E., Zhang, X., and Cao, G. (2012) Hydrogenated $Li_4Ti_5O_{12}$ nanowire arrays for high rate lithium ion batteries. *Advanced Materials,* **24**, 6502-6506.
254. Pohjalainen, E., Rauhala, T., Valkeapaa, M., Kallioinen, J., and Kallio, T. (2015) Effect of $Li_4Ti_5O_{12}$ particle size on the performance of lithium ion battery electrodes at high C-rates and low temperatures. *The Journal of Physical Chemistry C,* **119**, 2277-2283.
255. Wang, C., Wang, S., He, Y.-B., Tang, L., Han, C., Yang, C., Wagemaker, M., Li, B., Yang, Q.-H., Kim, J.-K., and Kang, F. (2015) Combining fast Li-ion battery cycling with large volumetric energy density: grain boundary induced high electronic and ionic conductivity in $Li_4Ti_5O_{12}$ spheres of densely packed nanocrystallites. *Chemistry of Materials,* **27**, 5647-5656.
256. Zhang, C., Song, H., Liu, C., Liu, Y., Zhang, C., Nan, X., and Cao, G. (2015) Fast and reversible Li ion insertion in carbon-encapsulated Li_3VO_4 as anode for lithium-ion battery. *Advanced Functional Materials,* **25**, 3497-3504.
257. Wang, G., Liu, H., Liu, J., Qiao, S., Lu, G. M., Munroe, P., and Ahn, H. (2010) Mesoporous $LiFePO_4$/C nanocomposite cathode materials for high power lithium ion batteries with superior performance. *Advanced Materials,* **22**, 4944-4948.
258. Zhang, S., Chowdari, B., Wen, Z., Jin, J., and Yang, J. (2015) Constructing highly oriented configuration by few-layer MoS_2: toward high-performance lithium-ion batteries and hydrogen evolution reactions. *ACS Nano,* **9**, 12464-12472.
259. Jiang, H., Ren, D., Wang, H., Hu, Y., Guo, S., Yuan, H., Hu, P., Zhang, L., and Li, C. (2015) 2D monolayer MoS_2-carbon interoverlapped superstructure: Engineering ideal atomic interface for lithium ion storage. *Advanced Materials,* **27**, 3687-3695.
260. Yu, X. Y., Hu, H., Wang, Y., Chen, H., and Lou, X. W. D., (2015) Ultrathin MoS_2 nanosheets supported on N-doped carbon nanoboxes with enhanced lithium storage and electrocatalytic properties. *Angewandte Chemie International Edition,* **54**, 7395-7398.
261. Zhou, J., Qin, J., Zhang, X., Shi, C., Liu, E., Li, J., Zhao, N., and He, C. (2015) 2D space-confined synthesis of few-layer MoS_2 anchored on carbon nanosheet for lithium-ion battery anode. *ACS Nano,* **9**, 3837-3848.
262. Xiong, F., Cai, Z., Qu, L., Zhang, P., Yuan, Z., Asare, O. K., Xu, W., Lin, C., and Mai, L. (2015) Three-dimensional crumpled reduced graphene oxide/MoS_2 nanoflowers: a stable anode for lithium-ion batteries. *ACS Applied Materials & Interfaces,* **7**, 12625-12630.

263. Wang, Q., Zou, R., Xia, W., Ma, J., Qiu, B., Mahmood, A., Zhao, R., Yang, Y., Xia, D., and Xu, Q. (2015) Facile synthesis of ultrasmall CoS_2 nanoparticles within thin N-doped porous carbon shell for high performance lithium-ion batteries. *Small,* **11**, 2511-2517.
264. Balogun, M.-S., Qiu, W., Jian, J., Huang, Y., Luo, Y., Yang, H., Liang, C., Lu, X., and Tong, Y. (2015) Vanadium nitride nanowire supported SnS2 nanosheets with high reversible capacity as anode material for lithium ion batteries. *ACS Applied Materials & Interfaces,* **7**, 23205-23215.
265. Fang, W., Zhao, H., Xie, Y., Fang, J., Xu, J., and Chen, Z. (2015) Facile hydrothermal synthesis of VS_2/graphene nanocomposites with superior high-rate capability as lithium-ion battery cathodes. *ACS Applied Materials & Interfaces,* **7**, 13044-13052.
266. Cao, K., Jiao, L., Liu, Y., Liu, H., Wang, Y., and Yuan, H. (2015) Ultra-high capacity lithium-ion batteries with hierarchical CoO nanowire clusters as binder free electrodes. *Advanced Functional Materials,* **25**, 1082-1089.
267. Chen, S., Zhao, Y., Sun, B., Ao, Z., Xie, X., Wei, Y., and Wang, G. (2015) Microwave-assisted synthesis of mesoporous Co_3O_4 nanoflakes for applications in lithium ion batteries and oxygen evolution reactions. *ACS Applied Materials & Interfaces,* **7**, 3306-3313.
268. Gu, D., Li, W., Wang, F., Bongard, H., Spliethoff, B., Schmidt, W., Weidenthaler, C., Xia, Y., Zhao, D., and Schuth, F. (2015) Controllable synthesis of mesoporous peapod-like Co_3O_4@carbon nanotube arrays for high-performance lithium-ion batteries. *Angewandte Chemie International Edition,* **54**, 7060-7064.
269. Huang, G., Zhang, F., Du, X., Qin, Y., Yin, D., and Wang, L. (2015) Metal organic frameworks route to in situ insertion of multiwalled carbon nanotubes in Co_3O_4 polyhedra as anode materials for lithium-ion batteries. *ACS Nano,* **9**, 1592-1599.
270. Li, X., Yang, Z., Fu, Y., Qiao, L., Li, D., Yue, H., and He, D. (2015) Germanium anode with excellent lithium storage performance in a germanium/lithium–cobalt oxide lithium-ion battery. *ACS Nano,* **9**, 1858-1867.
271. Cao, K., Jiao, L., Liu, H., Liu, Y., Wang, Y., Guo, Z., and Yuan, H. (2015) 3D hierarchical porous α-Fe_2O_3 nanosheets for high-performance lithium-ion batteries. *Advanced Energy Materials,* **5**(4), 1401421.
272. Cho, J. S., Hong, Y. J., and Kang, Y. C. (2015) Design and synthesis of bubble-nanorod-structured Fe_2O_3-carbon nanofibers as advanced anode material for Li-ion batteries. *ACS Nano,* **9**, 4026-4035.
273. Zhang, J., Wang, K., Xu, Q., Zhou, Y., Cheng, F., and Guo, S. (2015) Beyond yolk–shell nanoparticles: Fe_3O_4@ Fe_3C core@ shell nanoparticles as yolks and carbon nanospindles as shells for efficient lithium ion storage. *ACS Nano,* **9**, 3369-3376.
274. Zou, F., Hu, X., Li, Z., Qie, L., Hu, C., Zeng, R., Jiang, Y., and Huang, Y.

(2014) MOF-derived porous $ZnO/ZnFe_2O_4/C$ octahedra with hollow interiors for high-rate lithium-ion batteries. *Advanced Materials,* **26**, 6622-6628.

275. Park, M. S., Wang, G. X., Kang, Y. M., Wexler, D., Dou S. X., and Liu, H. K. (2007) Preparation and electrochemical properties of SnO_2 nanowires for application in lithium-ion batteries. *Angewandte Chemie,* **119**, 764-767.
276. Zhou, D., Song, W.-L., and Fan, L.-Z. (2015) Hollow core-shell SnO_2/C fibers as highly stable anodes for lithium-ion batteries. *ACS Applied Materials & Interfaces,* **7**, 21472-21478.
277. Liang, J., Yu, X. Y., Zhou, H., Wu, H. B., Ding, S., and Lou, X. W. D. (2014) Bowl-like SnO_2@carbon hollow particles as an advanced anode material for lithium-ion batteries. *Angewandte Chemie International Edition,* **53**, 12803-12807.
278. Liu, S., Jia, H., Han, L., Wang, J., Gao, P., Xu, D., Yang, J., and Che, S. (2012) Nanosheet-constructed porous TiO_2-B for advanced lithium ion batteries. *Advanced Materials,* **24**, 3201-3204.
279. Liu, H., Bi, Z., Sun, X. G., Unocic, R. R., Paranthaman, M. P., Dai, S., and Brown, G. M. (2011) Mesoporous TiO_2-B microspheres with superior rate performance for lithium ion batteries. *Advanced Materials,* **23**, 3450-3454.
280. Li, N., Liu, G., Zhen, C., Li, F., Zhang, L., and Cheng, H. M. (2011) Battery performance and photocatalytic activity of mesoporous anatase TiO_2 nanospheres/graphene composites by template-free self-assembly. *Advanced Functional Materials,* **21**, 1717-1722.
281. Liu, H., Li, W., Shen, D., Zhao, D., and Wang, G. (2015) Graphitic carbon conformal coating of mesoporous TiO_2 hollow spheres for high-performance lithium ion battery anodes. *Journal of the American Chemical Society,* **137**, 13161-13166.
282. Sun, X., Yan, C., Chen, Y., Si, W., Deng, J., Oswald, S., Liu, L., and Schmidt, O. G. (2014) Three-dimensionally "curved" NiO nanomembranes as ultrahigh rate capability anodes for Li-ion batteries with long cycle lifetimes. *Advanced Energy Materials,* **4**(4), 1300912.
283. Ding, Y.-L., Wen, Y., Wu, C., van Aken, P. A., Maier, J., and Yu, Y. (2015) 3D V_6O_{13} nanotextiles assembled from interconnected nanogrooves as cathode materials for high-energy lithium ion batteries. *Nano Letters,* **15**, 1388-1394.
284. Zhang, G., Yu, L., Wu, H. B., Hoster, H. E., and Lou, X. W. D. (2012) Formation of $ZnMn_2O_4$ ball-in-ball hollow microspheres as a high-performance anode for lithium-Ion batteries. *Advanced Materials,* **24**, 4609-4613.
285. Petnikota, S., Rotte, N. K., Reddy, M., Srikanth, V. V., and Chowdari, B. (2015) MgO-decorated few-layered graphene as an anode for Li-ion batteries. *ACS Applied Materials & Interfaces,* **7**, 2301-2309.

286. Wang, J., Zhang, C., and Kang, F. (2015) Nitrogen-enriched porous carbon coating for manganese oxide nanostructures toward high-performance lithium-ion batteries. *ACS Applied Materials & Interfaces,* **7**, 9185-9194.
287. Hu, C., Lv, L., Xue, J., Ye, M., Wang, L., and Qu, L. (2015) Branched graphene nanocapsules for anode material of lithium-ion batteries. *Chemistry of Materials,* **27**, 5253-5260.
288. Huang, X. l., Wang, R. z., Xu, D., Wang, Z. l., Wang, H. g., Xu, J. j., Wu, Z., Liu, Q. c., Zhang, Y., and Zhang, X. b. (2013) Homogeneous CoO on graphene for binder-free and ultralong-life lithium ion batteries. *Advanced Functional Materials,* **23**, 4345-4353.
289. Zhong, Y., Yang, M., Zhou, X., Luo, Y., Wei, J., and Zhou, Z. (2015) Orderly packed anodes for high-power lithium-ion batteries with super-long cycle life: Rational design of $MnCO_3$/large-area graphene composites. *Advanced Materials,* **27**, 806-812.
290. Liu, X., Zhong, X., Yang, Z., Pan, F., Gu, L., and Yu, Y. (2015) Gram-scale synthesis of graphene-mesoporous SnO 2 composite as anode for lithium-ion batteries. *Electrochimica Acta,* **152**, 178-186.
291. Jacob. L., Prasanna, K., Vengatesan, M. R., Santhoshkumar, P., Lee, C. W., and Mittal, V. (2018) Binary Cu/ZnO decorated graphene nanocomposites as an efficient anode for lithium ion batteries. *Journal of Industrial and Engineering Chemistry,* **59**, 108-114
292. Evanoff, K., Magasinski, A., Yang, J., and Yushin, G. (2011) Nanosilicon-coated graphene granules as anodes for Li-ion batteries. *Advanced Energy Materials,* **1**, 495-498.
293. Zeng, G., Shi, N., Hess, M., Chen, X., Cheng, W., Fan, T., and Niederberger, M. (2015) A general method of fabricating flexible spinel-type oxide/reduced graphene oxide nanocomposite aerogels as advanced anodes for lithium-ion batteries. *ACS Nano,* **9**, 4227-4235.
294. Guo, C. X., Wang, M., Chen, T., Lou, X. W., and Li, C. M. (2011) A hierarchically nanostructured composite of MnO_2/conjugated polymer/graphene for high-performance lithium ion batteries. *Advanced Energy Materials,* **1**, 736-741.
295. Gao, Z., Song, N., Zhang, Y., and Li, X. (2015) Cotton-textile-enabled, flexible lithium-ion batteries with enhanced capacity and extended lifespan. *Nano Letters,* **15**, 8194-8203.
296. Dong, Y., Zhao, Z., Wang, Z., Liu, Y., Wang, X., and Qiu, J. (2015) Dually fixed SnO_2 nanoparticles on graphene nanosheets by polyaniline coating for superior lithium storage. *ACS Applied Materials & Interfaces,* **7**, 2444-2451.
297. Qie, L., Chen, W. M., Wang, Z. H., Shao, Q. G., Li, X., Yuan, L. X., Hu, X. L., Zhang, W. X., and Huang, Y. H. (2012) Nitrogen-doped porous carbon nanofiber webs as anodes for lithium ion batteries with a superhigh capacity and rate capability. *Advanced Materials,* **24**,

2047-2050.
298. Liu, Y., Zhang, N., Jiao, L., Tao, Z., and Chen, J. (2015) Ultrasmall Sn nanoparticles embedded in carbon as high-performance anode for sodium-ion batteries. *Advanced Functional Materials,* **25**, 214-220.
299. Huang, X., Cui, S., Chang, J., Hallac, P. B., Fell, C. R., Luo, Y., Metz, B., Jiang, J., Hurley, P. T., and Chen, J. (2015) A hierarchical tin/carbon composite as an anode for lithium-ion batteries with a long cycle life. *Angewandte Chemie International Edition,* **54**, 1490-1493.
300. Guo, B., Wang, X., Fulvio, P. F., Chi, M., Mahurin, S. M., Sun, X. G., and Dai, S. (2011) Soft-templated mesoporous carbon-carbon nanotube composites for high performance lithium-ion batteries. *Advanced Materials,* **23**, 4661-4666.
301. Wu, R., Wang, D. P., Rui, X., Liu, B., Zhou, K., Law, A. W., Yan, Q., Wei, J., and Chen, Z. (2015) In-situ formation of hollow hybrids composed of cobalt sulfides embedded within porous carbon polyhedra/carbon nanotubes for high-performance lithium-ion batteries. *Advanced Materials,* **27**, 3038-3044.
302. Xu, J., Wang, M., Wickramaratne, N. P., Jaroniec, M., Dou, S., and Dai, L. (2015) High-performance sodium ion batteries based on a 3D anode from nitrogen-doped graphene foams. *Advanced Materials,* **27**, 2042-2048.
303. Zhang, W., Liu, Y., Chen, C., Li, Z., Huang, Y., and Hu, X. (2015) Flexible and binder-free electrodes of Sb/rGO and $Na_3V_2(PO_4)_3$/rGO nanocomposites for sodium-ion batteries. *Small,* **11**, 3822-3829.
304. Sun, R., Wei, Q., Li, Q., Luo, W., An, Q., Sheng, J., Wang, D., Chen, W., and Mai, L. (2015) Vanadium sulfide on reduced graphene oxide layer as a promising anode for sodium ion battery. *ACS Applied Materials & Interfaces,* **7**, 20902-20908.
305. Hu, L., Zhu, X., Du, Y., Li, Y., Zhou, X., and Bao, J. (2015) A chemically coupled antimony/multilayer graphene hybrid as a high-performance anode for sodium-ion batteries. *Chemistry of Materials,* **27**, 8138-8145.
306. Xu, X., Ji, S., Gu, M., and Liu, J. (2015) In situ synthesis of MnS hollow microspheres on reduced graphene oxide sheets as high-capacity and long-life anodes for Li-and Na-ion batteries. *ACS Applied Materials & Interfaces,* **7**, 20957-20964.
307. Ma, C., Xu, J., Alvarado, J., Qu, B., Somerville, J., Lee, J. Y., and Meng, Y. S. (2015) Investigating the energy storage mechanism of SnS_2-rGO composite anode for advanced Na-ion batteries. *Chemistry of Materials,* **27**, 5633-5640.
308. Choi, S. H., Ko, Y. N., Lee, J. K., and Kang, Y. C. (2015) 3D MoS_2–graphene microspheres consisting of multiple nanospheres with superior sodium ion storage properties. *Advanced Functional Materials,* **25**, 1780-1788.
309. Xie, X., Ao, Z., Su, D., Zhang, J., and Wang, G. (2015) MoS_2/graphene

composite anodes with enhanced performance for sodium-ion batteries: The role of the two-dimensional heterointerface. *Advanced Functional Materials,* **25**, 1393-1403.

310. Wu, C., Kopold, P., Ding, Y.-L., van Aken, P. A., Maier, J., and Yu, Y. (2015) Synthesizing porous $NaTi_2(PO_4)_3$ nanoparticles embedded in 3D graphene networks for high-rate and long cycle-life sodium electrodes. *ACS Nano,* **9**, 6610-6618.
311. Rui, X., Sun, W., Wu, C., Yu, Y., and Yan, Q. (2015) An advanced sodium-ion battery composed of carbon coated $Na_3V_2(PO_4)_3$ in a porous graphene network. *Advanced Materials,* **27**, 6670-6676.
312. Hou, H., Banks, C. E., Jing, M., Zhang, Y., and Ji, X. (2015) Carbon quantum dots and their derivative 3D porous carbon frameworks for sodium-ion batteries with ultralong cycle life. *Advanced Materials,* **27**, 7861-7866.
313. Luo, W., Bommier, C., Jian, Z., Li, X., Carter, R., Vail, S., Lu, Y., Lee, J.-J., and Ji, X. (2015) Low-surface-area hard carbon anode for Na-ion batteries via graphene oxide as a dehydration agent. *ACS Applied Materials & Interfaces,* **7**, 2626-2631.
314. Luo, W., Jian, Z., Xing, Z., Wang, W., Bommier, C., Lerner, M. M., and Ji, X. (2015) Electrochemically expandable soft carbon as anodes for Na-ion batteries. *ACS Central Science,* **1**, 516-522.
315. Liu, H., Jia, M., Sun, N., Cao, B., Chen, R., Zhu, Q., Wu, F., Qiao, N., and Xu, B. (2015) Nitrogen-rich mesoporous carbon as anode material for high-performance sodium-ion batteries. *ACS Applied Materials & Interfaces,* **7**, 27124-27130.
316. Wang, H. g., Wu, Z., Meng, F. l., Ma, D. l., Huang, X. l., Wang, L. m., and Zhang, X. b. (2013) Nitrogen-doped porous carbon nanosheets as low-cost, high-performance anode material for sodium-ion batteries. *ChemSusChem,* **6**, 56-60.
317. Liu, Y., Zhang, N., Jiao, L., and Chen, J. (2015) Tin nanodots encapsulated in porous nitrogen-doped carbon nanofibers as a free-standing anode for advanced sodium-ion batteries. *Advanced Materials,* **27**, 6702-6707.
318. Jian, Z., Luo, W., and Ji, X. (2015) Carbon electrodes for K-ion batteries. *Journal of the American Chemical Society,* **137**, 11566-11569.
319. Xu, Y., Zhu, Y., Liu, Y., and Wang, C. (2013) Electrochemical performance of porous carbon/tin composite anodes for sodium-ion and lithium-ion batteries. *Advanced Energy Materials,* **3**, 128-133.
320. Wang, J., Luo, C., Mao, J., Zhu, Y., Fan, X., Gao, T., Mignerey, A. C., and Wang, C. (2015) Solid-state fabrication of SnS_2/C nanospheres for high-performance sodium ion battery anode. *ACS Applied Materials & Interfaces,* **7**, 11476-11481.
321. Kalubarme, R. S., Lee, J.-Y., and Park, C.-J. (2015) Carbon encapsulated tin oxide nanocomposites: an efficient anode for high per-

formance sodium-ion batteries. *ACS Applied Materials & Interfaces,* **7**, 17226-17237.

322. Wang, J., Luo, C., Gao, T., Langrock, A., Mignerey, A. C., and Wang, C. (2015) An advanced MoS_2/carbon anode for high-performance sodium-ion batteries. *Small.* **11**, 473-481.
323. Jian, Z., Han, W., Lu, X., Yang, H., Hu, Y. S., Zhou, J., Zhou, Z., Li, J., Chen, W., Chen, D., and Chen, L. (2013) Superior electrochemical performance and storage mechanism of $Na_3V_2(PO_4)_3$ cathode for room-temperature sodium-ion batteries. *Advanced Energy Materials,* **3**, 156-160.
324. Jiang, Y., Yang, Z., Li, W., Zeng, L., Pan, F., Wang, M., Wei, X., Hu, G., Gu, L., and Yu, Y. (2015) Nanoconfined carbon-coated $Na_3V_2(PO_4)_3$ particles in mesoporous carbon enabling ultralong cycle life for sodium-ion batteries. *Advanced Energy Materials,* **5**.
325. Li, W., Yang, Y., Zhang, G., and Zhang, Y.-W. (2015) Ultrafast and directional diffusion of lithium in phosphorene for high-performance lithium-ion battery. *Nano Letters,* **15**, 1691-1697.
326. Song, J., Yu, Z., Gordin, M. L., Li, X., Peng, H., and Wang, D. (2015) Advanced sodium ion battery anode constructed via chemical bonding between phosphorus, carbon nanotube, and cross-linked polymer binder. *ACS Nano,* **9**, 11933-11941.
327. Su, D., Dou, S., and Wang, G. (2015) Anatase TiO_2: Better anode material than amorphous and rutile phases of TiO_2 for Na-ion batteries. *Chemistry of Materials,* **27**, 6022-6029.
328. Xu, Y., Zhou, M., Wen, L., Wang, C., Zhao, H., Mi, Y., Liang, L., Fu, Q., Wu, M., and Lei, Y. (2015) Highly ordered three-dimensional Ni-TiO_2 nanoarrays as sodium ion battery anodes. *Chemistry of Materials,* **27**, 4274-4280.
329. Hu, Z., Wang, L., Zhang, K., Wang, J., Cheng, F., Tao, Z., and Chen, J. (2014) MoS_2 nanoflowers with expanded interlayers as high-performance anodes for sodium-ion batteries. *Angewandte Chemie,* **126**, 13008-13012.
330. Zhou, G., Li, L., Ma, C., Wang, S., Shi, Y., Koratkar, N., Ren, W., Li, F., and Cheng, H.-M. (2015) A graphene foam electrode with high sulfur loading for flexible and high energy Li-S batteries. *Nano Energy,* **11**, 356-365.
331. Zhou, G., Pei, S., Li, L., Wang, D. W., Wang, S., Huang, K., Yin, L. C., Li, F., and Cheng, H. M. (2014) A graphene–pure-sulfur sandwich structure for ultrafast, long-life lithium–sulfur batteries. *Advanced Materials,* **26**, 625-631.
332. Wang, C., Wang, X., Wang Y., Chen, J., Zhou, H., and Huang, Y. (2015) Macroporous free-standing nano-sulfur/reduced graphene oxide paper as stable cathode for lithium -sulfur battery. *Nano Energy,* **11**, 678-686.
333. Wang, C., Wang, X., Yang, Y., Kushima, A., Chen, J., Huang, Y., and Li,

J. (2015) Slurryless Li_2S/reduced graphene oxide cathode paper for high-performance lithium sulfur battery. *Nano Letters,* **15**, 1796-1802.

334. Song, J., Yu, Z., Gordin, M. L., and Wang, D. (2016) Advanced sulfur cathode enabled by highly crumpled nitrogen-doped graphene sheets for high-energy-density lithium–sulfur batteries. *Nano Letters,* **16**, 864-870.
335. Xie, Y., Meng, Z., Cai, T., and. Han, W.-Q. (2015) Effect of boron-doping on the graphene aerogel used as cathode for the lithium–sulfur battery. *ACS Applied Materials & Interfaces,* **7**, 25202-25210.
336. Li, B., Li, S., Liu, J., Wang, B., and Yang, S. (2015) Vertically aligned sulfur–graphene nanowalls on substrates for ultrafast lithium–sulfur batteries. *Nano Letters,* **15**, 3073-3079.
337. Zhu, L., Peng, H.-J., Liang, J., Huang, J.-Q., Chen, C.-M., Guo, X., Zhu, W., Li, P., and Zhang, Q. (2015) Interconnected carbon nanotube/graphene nanosphere scaffolds as free-standing paper electrode for high-rate and ultra-stable lithium–sulfur batteries. *Nano Energy,* **11**, 746-755.
338. Chen, Y., Lu, S., Wu, X., and Liu, J. (2015) Flexible carbon nanotube-graphene/sulfur composite film: free-standing cathode for high-performance lithium/sulfur batteries. *The Journal of Physical Chemistry C,* **119**, 10288-10294.
339. Wu F., Lee, J. T., Zhao, E., Zhang, B., and Yushin, G. (2015) Graphene-Li_2S-carbon nanocomposite for lithium-sulfur batteries. *ACS Nano,* **10**, 1333-1340.
340. Wang, D., Xie, D., Yang, T., Zhong, Y., Wang, X., Xia, X., Gu, C., and Tu, J. (2016) Li_2S@C composite incorporated into 3D reduced graphene oxide as a cathode material for lithium-sulfur batteries. *Journal of Power Sources,* **313**, 233-239.
341. Sun, L., Wang, D., Luo, Y., Wang, K., Kong, W., Wu, Y., Zhang, L., Jiang, K., Li, Q., and Zhang, Y. (2015) Sulfur embedded in a mesoporous carbon nanotube network as a binder-free electrode for high-performance lithium–sulfur batteries. *ACS Nano,* **10**, 1300-1308.
342. Guo, M.-q., Huang, J.-q., Kong, X.-y., Peng, H.-j., Shui, H., Qian, F.-y., Zhu, L., Zhu, W.-c., and Zhang, Q. (2016) Hydrothermal synthesis of porous phosphorus-doped carbon nanotubes and their use in the oxygen reduction reaction and lithium-sulfur batteries. *New Carbon Materials,* **31**, 352-362.
343. Song, J., Gordin, M. L., Xu, T., Chen, S., Yu, Z., Sohn, H., Lu, J., Ren, Y., Duan, Y., and Wang, D. (2015) Strong lithium polysulfide chemisorption on electroactive sites of nitrogen-doped carbon composites for high-performance lithium-sulfur battery cathodes. *Angewandte Chemie International Edition,* **54**, 4325-4329.
344. Fan, Q., Liu, W., Weng, Z., Sun, Y., and Wang, H. (2015) Ternary hybrid material for high-performance lithium-sulfur battery. *Journal*

of the American Chemical Society, **137**, 12946-12953.

345. Strubel, P., Thieme, S., Biemelt, T., Helmer, A., Oschatz, M., Bruckner, J., Althues, H., and Kaskel, S. (2015) ZnO hard templating for synthesis of hierarchical porous carbons with tailored porosity and high performance in lithium-sulfur battery. *Advanced Functional Materials,* **25**, 287-297.
346. Du, W.-C., Yin, Y.-X., Zeng, X.-X., Shi, J.-L., Zhang, S.-F., Wan, L.-J., and Guo, Y.-G. (2015) Wet chemistry synthesis of multidimensional nanocarbon-sulfur hybrid materials with ultrahigh sulfur loading for lithium-sulfur batteries. *ACS Applied Materials & Interfaces,* **8**, 3584-3590.
347. Sun, Q., He, B., Zhang, X.-Q., and Lu, A.-H. (2015) Engineering of hollow core-shell interlinked carbon spheres for highly stable lithium-sulfur batteries. *ACS Nano,* **9**, 8504-8513.
348. Zhou, W., Wang, C., Zhang, Q., Abruña, H. D., He. Y., Wang, J., Mao, S. X., and Xiao, X. (2015) Tailoring pore size of nitrogen-doped hollow carbon nanospheres for confining sulfur in lithium-sulfur batteries. *Advanced Energy Materials,* **5**(16), 1401752.
349. Schneider, A., Weidmann, C., Suchomski, C., Sommer, H., Janek, J. r., and Brezesinski, T. (2015) Ionic liquid-derived nitrogen-enriched carbon/sulfur composite cathodes with hierarchical microstructure - A step toward durable high-energy and high-performance lithium-sulfur batteries. *Chemistry of Materials,* **27**, 1674-1683.
350. Li, Z., and Yin, L. (2015) Nitrogen-doped MOF-derived micropores carbon as immobilizer for small sulfur molecules as a cathode for lithium sulfur batteries with excellent electrochemical performance. *ACS Applied Materials & Interfaces,* **7**, 4029-4038.
351. Pang, Q., Tang, J., Huang, H., Liang, X., Hart, C., Tam, K. C., and. Nazar, L. F. (2015) A nitrogen and sulfur dual-doped carbon derived from polyrhodanine@cellulose for advanced lithium-sulfur batteries. *Advanced Materials,* **27**, 6021-6028.
352. Hwa, Y., Zhao, J., and Cairns, E. J. (2015) Lithium sulfide (Li_2S)/graphene oxide nanospheres with conformal carbon coating as a high-rate, long-life cathode for Li/S cells. *Nano Letters,* **15**, 3479-3486.
353. Zeng, L., Yao, Y., Shi, J., Jiang, Y., Li, W., Gu, L., and Yu, Y. (2016) A flexible $S_{1-x}Se_x$@ porous carbon nanofibers ($x \leq 0.1$) thin film with high performance for Li-S batteries and room-temperature Na-S batteries. *Energy Storage Materials,* **5**, 50-57.
354. Li, Z., Zhang, J., and Lou, X. W. D. (2015) Hollow carbon nanofibers filled with MnO_2 nanosheets as efficient sulfur hosts for lithium-sulfur batteries. *Angewandte Chemie International Edition,* **54**, 12886-12890.
355. Niu, X.-Q., Wang, X.-L., Xie, D., Wang, D.-H., Zhang, Y.-D., Li, Y., Yu, T., and Tu, J.-P. (2015) Nickel hydroxide-modified sulfur/carbon

composite as a high-performance cathode material for lithium sulfur battery. *ACS Applied Materials & Interfaces,* **7**, 16715-16722.

356. Wei, S., Ma, L., Hendrickson, K. E., Tu, Z., and Archer, L. A. (2015) Metal-sulfur battery cathodes based on PAN-sulfur composites. *Journal of the American Chemical Society,* **137**, 12143-12152.
357. Konarov, A., Gosselink, D., Doan, T. N. L., Zhang, Y., Zhao, Y., and Chen, P. (2014) Simple, scalable, and economical preparation of sulfur–PAN composite cathodes for Li/S batteries. *Journal of Power Sources,* **259**, 183-187.
358. Hu, H., Cheng, H., Liu, Z., Li, G., Zhu, Q., and Yu, Y. (2015) In situ polymerized PAN-assisted S/C nanosphere with enhanced high-power performance as cathode for lithium/sulfur batteries. *Nano Letters,* **15**, 5116-5123.
359. Xie, Y., Zhao, H., Cheng, H., Hu, C., Fang, W., Fang, J., Xu, J., and Chen, Z. (2016) Facile large-scale synthesis of core–shell structured sulfur@ polypyrrole composite and its application in lithium–sulfur batteries with high energy density. *Applied Energy,* **175**, 522-528.

7

Nano-catalysts: Advances in Synthesis and Applications in the Field of Energy

7.1 Introduction

Globally, catalyst synthesis solely accounts around US$10 billion in sales in different chemical processes [1]. Catalysis is a process during which the catalyst lowers the activation energy and increases the rate of a reaction without affecting the state of equilibrium of the reaction and regenerates at the last step of reaction. The field of catalysis has been developed in many directions like electro (e.g., ethanol powered fuel cells), photo (e.g. self-cleaning glass), enzymatic (e.g. Baeyer–Villiger oxidation), homogeneous (e.g. hydroformylation) and heterogeneous (e.g. automotive catalytic converter) [2]. Catalysts play vital role in many industrial processes such as olefin polymerization, hydrogenation, cracking, reforming, gas synthesis, fermentation, etc. Since the discovery of catalysts, the improvement in the catalytic activity and performance is continuously under focus in order to reduce the major cost associated with the large-scale chemical processes [3]. This includes many aspects of the catalytic process such as elimination of catalyst poisoning, re-generation, enhancing the useful lifetime, etc.

Nano-scale materials refer to those which have at least one external dimension or internal structure between 1-100 nanometers. These materials may have different physico-chemical properties than coarser materials of same type and can be used to produce state-of-the-art technologies and end-products [4]. Example of such technologies and end-products containing novel nanomaterials are clinical neuroscience [5,6], bio-labelling [7], rechargeable batteries [8], antimicrobial surfaces [9], tissue engineering and regenerative medicine [10], etc. With the advent of nanotechnology in technological and industrial areas, nano-science has also made considerable and revolutionary advances in the field of catalysis (late 1990s) for different applications. Nano-catalysts refer to nano-materials that deal with the

Ali U. Chaudhry and Vikas Mittal*
Khalifa University of Science and Technology, Abu Dhabi, UAE
**Current address: Texas A&M University, Qatar*

unconventional material properties for innovative applications. According to recent reports, nano-catalyst is a "*hierarchical system of basic structural units (active sites) with defined assembling strategies for multiple length scales*" [11,12]. Nano-catalysts are used in several environmental, chemical and process industries such as energy conversion and conservation [13], pharmaceutical, production of chemicals and petrochemicals, control emission of hazardous gases like CO, NO [14], etc. Considering the benefits of catalysts, there is always a need to develop catalysts with improved efficiency, recycling, selectivity and stability. Nano-science has made it possible to produce nano-catalysts with high activity and selectivity which is highly unlikely to be achieved with micro dimensional materials [2]. At nano scale, different chemical and physical properties of materials may appear due to different extent of delocalized electrons. Also, the presence of large intrinsic surface reactivity (catalytic activity) owing to smaller size makes nano-catalysts promising candidates for catalysis. For instance, Figure 7.1 demonstrates the TEM image of multi-armed nano-star colloidal platinum nano-catalyst particles [15]. For the synthesis of nano-star catalyst particles, the authors employed tetrahedral platinum nanocrystals as seeds, which avoided the need of organic solvents, templates, substrates, etc.

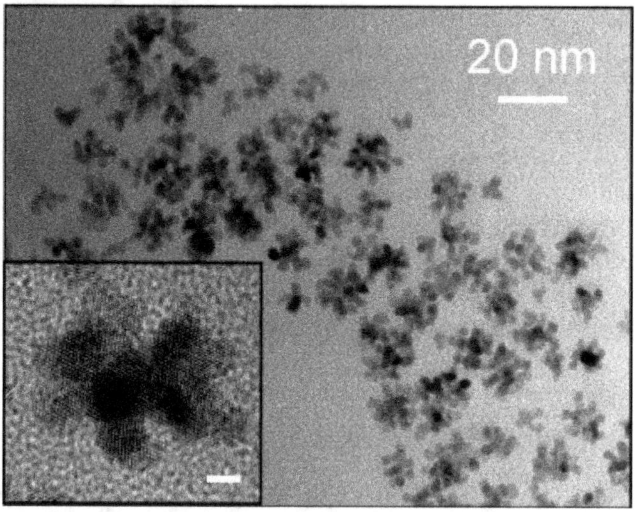

Figure 7.1 TEM image of the multi-armed nano-star colloidal platinum nano-catalyst particles. Inset represents the HRTEM image of the nano-star displaying multiple arms (scale bar 2 nm). Reproduced from Reference 15 with permission from American Chemical Society.

7.2 Applications of Nano-catalysts in the Field of Energy

In the twenty-first century, the nano-catalysts have gained significant research attention for exploring large scale applications, especially in the field of energy due to ecological and environmental friendly operation and low cost synthesis. Also, the depleting fossil fuel resources indicate the need to reassess our energy needs and optimize the efficiency of existing processes [16]. According to a recent report, around 85% of the energy used globally comes from fossil fuels which in turn discharges 30 billion tons of carbon dioxide [17]. The grand challenge of clean environment and efficiently control processes may be met earnestly with the help of precisely controlled composition and nano-catalyst structure [18]. In addition, the implications of nanostructure materials should not only encircle the efficient conversion (clean) of fossil fuel to useful products, but should also lead to a vital role in developing new catalytic routes for renewable energy. Development of new catalytic routes for non-conventional energy is specifically important to open new avenues so as to reduce the reliance on the conventional fossil based fuels, however, the increment in the efficiency of the existing processes would equally benefit the process industries.

This review reports recent advances in nanostructured materials as catalysts for energy applications. The review covers not only the research and development of nano-catalysts in already developed processes for fossil fuel conversion, but also focuses on the alternative means of energy using nano-catalysts (Table 7.1). More specifically, the review focuses on some of the energy related applications mentioned in Table 7.1.

7.2.1 Oil Refining

Oil refinery uses various kind of processes including catalyzed reactions for the separation of individual compounds present in the crude oil. During these operations, many environmentally hazardous gases like CO_2, hydrogen sulfide, ammonia, etc., are liberated. Refineries have to treat/process all kind of harmful products to make them harmless for the environment. Moreover, in order to meet the current and future demands of market in terms of fuel quality and quantity, efficient conversion of heaviest compounds of crude oil is required. In these scenarios, the contributions by the functional nano-catalysts with increased activity and selectivity are expected to be beneficial, especially in following vital refining processes commonly employed in the oil industry [19]:

Hydrotreating

Hydrotreating (HT) is a major reductive process to remove oxygen, sulfur, nitrogen, halogen atoms and metals from the oil streams. Mainly, the catalysts containing hydrogen reduced palladium, platinum (low H_2S environments), sulfides of nickel, cobalt, tungsten or molybdenum metals supported on alumina, amorphous aluminosilicates or zeolites (X.Y or morenite) are used in the treatment process

Table 7.1 Application of nano-catalysis in the field of Energy [1,17,19-33]

	Catalytic Processes
Energy from Fossil & Renewable Resources	Hydrotreating
	Hydrodemetalation
	Catalytic reforming
	Fluid catalytic cracking
	Hydrocracking
	Isomerization
	Alkylation of isobutene with light alkenes
	Oligomerization
	Etherification
	Water gas shift and COS removal
	Steam reforming
	Methanol synthesis
	Methanol to hydrocarbons
	Fischer-Tropsch (FT) synthesis process
	Gas to liquid
	Oxidative coupling of methane
	Direct coal liquefaction
	Coal and carbon gasification
	Fuel cells
	Biomass gasification and syngas conversion
	Fast pyrolysis and bio-oil upgrading
	Liquid-phase/aqueous-phase catalytic processing
	Vegetable oil conversion to bio-fuel
	Photocatalytic water splitting
	Hydrogenation
	Reforming
	Dehydration
	Energy storage

[25,34]. Hydrodesulfurization (HDS) is one of the vital processes of hydrotreating which is used to reduce the sulfur compounds from the crude oil. The performance of the catalysts during the treatment process mainly depends on the support on which catalyst resides, sulfidation conditions and promoters. The main function of the support is to provide good dispersion of the active part of the catalysts whereas promoters improve the performance. In a recent report, trimetallic nano-catalysts composed of Ni–Mo–W supported on multiwalled carbon nanotubes (MWCNT) were synthesized using ultrasound assisted co-precipitation method. The high energy environment (ultrasound) was used to avoid the metallic agglomeration. The incorporation of MWCNT was also useful due to weak interaction of metallic part and nanotubes [35].

Nano-catalysts prepared with conventional methods result in samples with polydisperse particles and inhomogeneous compositions. Further, metallic sulfide results in low active site densities which are unable to generate ultralow level of sulfur in petroleum products. Previous reports showed that silica supported metallic phosphide showed excellent activity, selectivity and resistance to sulfur poisoning. Danforth *et al.* [36] prepared robust system of monodisperse nanoparticles composed of $Ni_{2-x}Co_xP$ and $Ni_{2-x}Fe_xP$ using solution-phase arrested precipitation. The nanoparticles were encapsulated in mesoporous silica ($mSiO_2$). Nano-catalyst composed of $Ni_{1.92}Co_{0.08}P@mSiO_2$ exhibited highest activity HDS on mass basis, whereas highest turnover frequency (TOF) was observed for $Ni_{2-x}Fe_xP@mSiO_2$ owing to increased electron donation from Fe to Ni than from Co to Ni [36].

Hydrodemetalation

Hydrodemetalation is among the basic steps of petroleum refining in which heteroatoms such as O, N, S and specifically metals are removed. With the reducing amount of light crude, refineries are now focusing on the heavier crude containing concentrated metals. The metals have damaging consequences on the catalysts and refinery equipment [28,37]. The fundamental process to remove the heteroatoms and metals is HM in the presence of catalyst and high hydrogen pressure. The heteroatoms usually convert to oxides (H_2O, SO_2) and hydrogen pnictides (NH_3), whereas the metals such as nickel and vanadium (as porphyrins and complexes) become sulfides. These sulfides usually deactivate the expensive catalyst by plugging the active sites on the catalyst surface and cause underutilization. The main catalysts of hydrodemetalation process are hydrogen (gas) and oxides of Co, Mo, etc., supported on γ-Al_2O_3 [19]. In order to overcome the problem of deactivation of catalyst due to diffusion limited (transport) reactions, Rao *et al.* [38] adopted a mathematical approach for the design of nano-catalyst with optimized broad pore network. This hierarchically structured catalyst (CoMo/AlO), using modeling, exhibited activity for a longer period of time as compared to optimized mesosphere industrial catalysts [38].

The use of nano-sized MoS_2 or MoS_2/NiS for the conversion of extra heavy crude oil and petroleum residue was also reported [39]. During this process V, Ni, and Mo metals were quantitatively passed into high boiling fractions.

Thus, the development of nanostructured nano-catalysts leads to process enhancement as well as alleviation of the existing challenges.

Catalytic Reforming

Catalytic reforming (CR) deals with the crude oil distillate in order to make high-octane reformates. The feed of CR is usually composed of aromatics (6-12 carbon atoms), naphthenes and paraffins. Generally, impregnated conventional catalysts based on platinum (noble metals) with promoters such as Re, Sn, and Ir and chloride alumina support are mainly used in this process [19,24,40]. The advances in the nanoparticles chemistry have made it possible to design nano-catalysts for CR process with improved resistance to deactivation, high catalytic activity and selectivity at high temperature [41]. In a recent research study, bimetallic nanoparticles based on Pt (catalyst) and Rh (promoter) supported on mesoporous silica were used for isomerization reaction. The study showed that turnover frequency (TOF) and selectivity were mainly affected by the change in composition and size of the nano-catalyst [42]. The high temperature usage of nano-catalyst is limited owing to unwanted sintering and Ostwald ripening due to non-equilibrium state of nano-catalyst and support. This problem can be solved by introducing isolate nanoparticles concept using core-shell structure approach [11]. Recently, An *et al.* [43] studied the effect of thermal stability of designed nano-catalyst and the role of the support on CR. The authors synthesized four different types of nano-catalysts from platinum nanoparticles (Figure 7.2) supported on mesoporous SiO_2 or TiO_2. The selectivity of the catalysts was highly improved due to improved charge transfer at the interface in case of Pt-TiO_2 for n-Hexane reforming reactions [43]. These findings underlined the role of nano-catalytic processes in enhancing the efficiency and yield of conventional processes.

Fluid Catalytic Cracking

Fluid catalytic cracking (FCC) is a major refinery process and used to convert heavy vacuum distillates (like vacuum gas oil), atmospheric and refinery residues into lighter ones like motor fuels and gasoline. The main catalysts of FCC are mainly composed of functional matrix (zeolite Y), binder (NaY) and kaolin clay. Owing to the new environmental regulations on the release of CO, NO_x and SO_x, there is need to develop new effective FCC catalysts [19,23]. The main challenge associated with the catalysts is limited transport phenomena (poor mass transfer) within the catalysts pore structures, as the zeolite

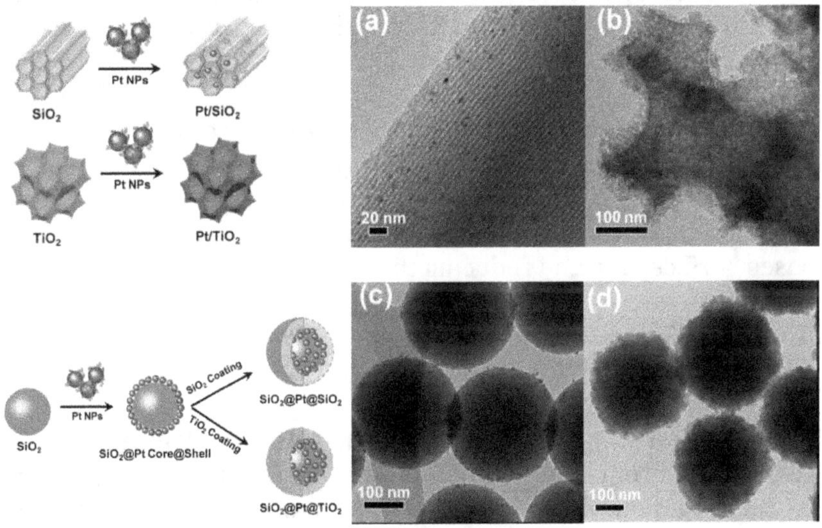

Figure 7.2 Preparation of supported Pt nanoparticle catalysts and sandwich-type Pt core@shell catalysts and TEM images of supported Pt nanoparticle catalysts (a,b) and sandwich catalysts (c,d): (a) Pt/SiO_2, (b) Pt/TiO_2, (c) SiO_2@Pt@SiO_2, and (d) SiO_2@Pt@TiO_2. Reproduced from Reference 43 with permission from American Chemical Society.

having porous structure less than 1 nm lead to limited diffusion reactions. Extensive research efforts have been dedicated to improve the pore accessibility through nano-sized zeolites [44-46]. Moreover, transport of reactants within catalyst pores can occur through single or combination of following three mechanisms or models, i.e., Knudsen diffusion [47], intercrystalline diffusion [48], and molecular diffusion [49]. Knudsen and molecular diffusion mechanism describe the trasnporatation of molecules within the micropores on the basis of molecular sizes and are often called rate-limiting step of a reaction [50,51]. On the other hand, intercrystalline diffusion is more active at smaller pore structure. Garcia-Martinez *et al.* [51] introduced a novel, well-controlled and direct surfactant-assisted synthesis method for controlled mesoporosity in Y zeolites (low Si/Al ratios) for FCC applications. The synthesized FCC catalysts (mesostructured zeolites) exhibited improved hydrothermal stability and selectivity in terms of useful fuels. In another research study, well-organized zeolitic nanocrystal aggregates were synthesized [52]. The aggregates system was composed of interconnected hierarchically micro–meso–macro porous. Catalytic testing validated their supremacy as the cracking catalysts compared to the conventional zeolite catalysts. Similar kind of structures were also reported in the other literature studies, such as delaminated zeolite [53] and stable single-unit-cell nanosheets of zeolite as nano-catalysts.

In another study, Dejhosseini *et al.* [54] also studied catalytic cracking reaction of heavy oil in the presence of cerium oxide nanocatalyst (Figure 7.3) in supercritical water. The authors concluded that the rate of conversion could be enhanced by increasing the exposed surface area and reducing the particle size of the catalyst.

Isomerization

Isomerization reaction in refinery industry deals with the skeleton isomerization of alkanes, alkenes, isomerization of double bond of alkenes, alkyl aromatics, saturated hydro-carbon (bi/poly-cyclic) and inter-conversion of alkyl cyclo-pentanes and cyclo-hexanes. In the various isomerization processes, two types of catalysts are mainly used, i.e., mono-functional acidic and bi-functional catalysts [19]. Mostly, mono-functional are consisted of Friedel–Crafts type of acid $AlCl_3$ with HCl whereas $HSbF_6$ is used as liquid acid catalyst. Bi-functional catalysts are advantageous, promising and mainly composed of combination of acid with a metal such as zeolites (acid forms) with platinum or palladium [26]. For the isomerization of hexane, nanocatalysts system comprising of Pd and zeolites HY, HZSM-5 and mixtures of zeolites with γ-Al_2O_3 were synthesized [55]. The increased activity was observed in case of Pd/HY due to improved dispersity and reduction in Pd cluster size. Recently Aghdam *et al.* [56] used membrane reactor packed with Pt/ZSM-5 nano-catalyst for the isomerization of n-C5. The membrane reactor containing nano-catalyts exhibited improved i-C5 yield due to selectivity of reactants in reaction zone. In another study, An *et al.* [57] synthesized colloidal Pt nanoparticles of different controlled sizes and supported onto ordered macroporous oxides such as SiO_2, Al_2O_3, TiO_2, Nb_2O_5, Ta_2O_5, and ZrO_2

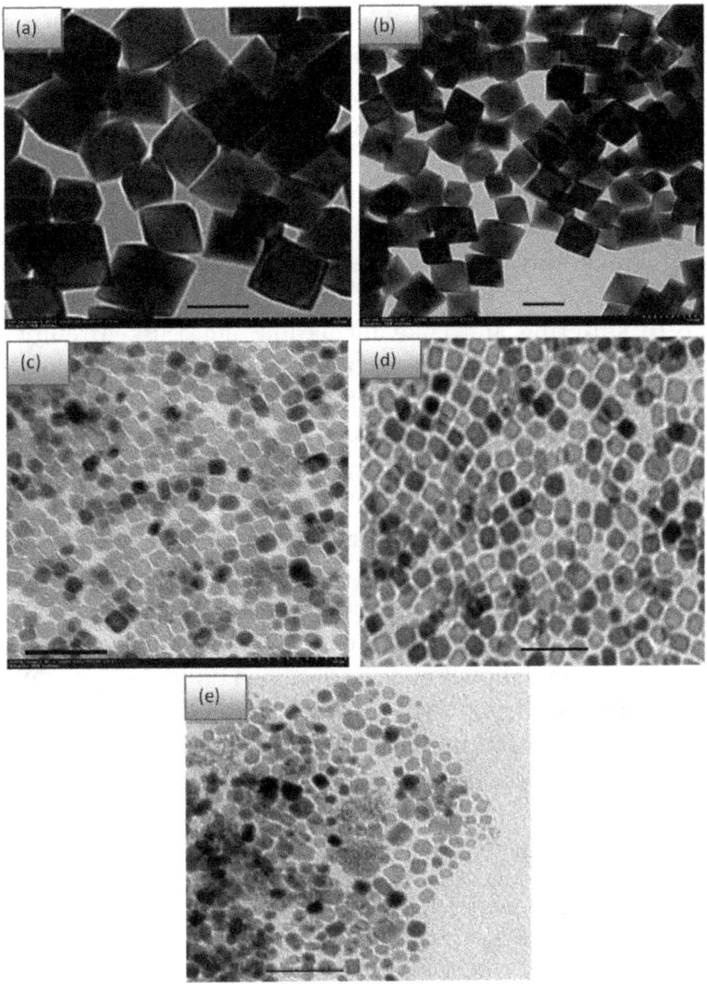

Figure 7.3 Transmission electron micrographs of the CeO_2 nanoparticles (a,b) without hexanoic acid, (c) with hexanoic acid, (d) after calcination at 300 °C, and (e) after calcination at 650 °C. Reproduced from Reference 54 with permission from American Chemical Society.

for n-hexane isomerization. The results revealed that Nb_2O_5 and Ta_2O_5 exhibited the highest selectivity when the nanoparticles size increased from 1.7 to 5.5 nm due to combined effect, i.e., charge transfer at Pt-oxide interfaces and smaller size of catalysts [57].

Steam Reforming

Steam reforming (SR), another key industrial process, is used to synthesize hydrogen and products like ammonia, methanol, CO-rich gas, etc., from steam, CO_2 and hydrocarbons. Main catalysts used of SR are usually based on Cu, Ni, Co, Fe Ru, Rh, etc., and perform better than non-metal catalysis [29]. There are certain disadvantages associated with the usage of catalysts for SR process as reported in literature such as reduction in the surface area due to sintering of particles, sulfur poisoning, carbon formation, etc. [19]. The requirements for good steam reforming catalysts were also reported like high selectivity to CO_2, minimum production of CO, low active energy, low cost, resistance to sintering and coke formation, etc. [58]. Attempt were made to improve the diffusion, i.e., the rate-limiting step, as well as to induce high activity and long-lasting stability of catalysts. Koo *et al.* [59] developed a monolith catalyst using locally developed coating technique, i.e., deposition-precipitation for hydrogen production. Nickel as nano-layer was deposited on Fe-Cr alloy flat rolled and corrugated metal strips using nickel nitrate as the precursor. The prepared catalysts exhibited improved activity and long-term stability due to high dispersion and large active surface area. Similarly for improved activity, stability, and reduced CO formation for steam reforming of methanol, novel nano-structured copper supported ZrO_2 was synthesized [60]. A polymer templating technique was used to control the morphology and porosity of the supported material. During catalysis, copper particles stayed apart due to ZrO_2 support, thus, resulting in improved sintering process and higher available surface area for carrying out the reaction.

Methanol Synthesis

According to a recent report, the global annual consumption of methanol was about of 65 million tons/year during 2013 [61]. Methanol is synthesized from synthesis gas, hydrogen mixture, CO and CO_2 using different technologies such as reforming of natural gas, adiabatic prefired, tubular, auto-thermal or oxygen-fired processes. Main catalysts for the process are composed of ZnO/Cr_2O_3, $Cu/ZnO/Al_2O_3$, Cu/ZrO_2, as well as Pd-based and sulfide-based. The problems associated with the catalyst have been reported in the literature such as sintering, poisoning of sulfur, chlorine, carbonyls of iron and by-product formation [19,30]. In order to improve the poor sintering behavior of Cu nano-catalyst and metal-support interaction, Berg *et al.* [62] synthesized Cu catalysts on SiO_2 via precipitation and incipient wetness impregnation method. The catalysts system created through precipitation exhibited entrapment of Cu particles in the plate-like silica structure which, in turn, limited the growth of Cu particles during catalysis [62]. Another attempt was made to avoid Ostwald ripening of nanoparticles during copper methanol catalysis by producing amine-functionalized support material (Figure 7.4) [63]. The improved sintering behavior was owing to the functionalization which, in turn, increased the distance between the Cu nanoparticles due to limiting transport of copper atoms over the support surface.

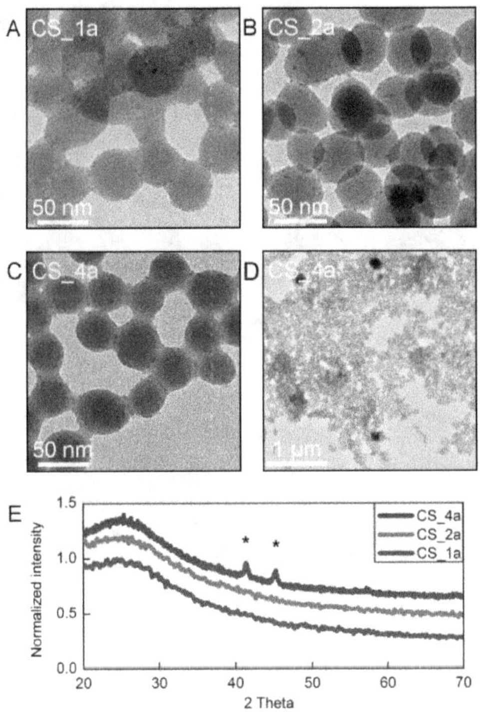

Figure 7.4 TEM images of various nanoparticles and X-ray diffractograms of the samples (peaks (*) correspond to CuO). Reproduced from Reference 63 with permission from American Chemical Society.

The function of a promoter element in main catalyst particle is to enhance or suppress the reactivity or selectivity for certain reaction. Recently, Kuld *et al.* [61] discussed the complex promotional interactions of ZnO with Cu during catalysis of methanol from synthesis gas. The authors showed with the help of experiments and density functional theory calculations that promotion of catalysis is due to Zn atoms migration in the Cu surface during catalysis [61].

Similarly, Berg *et al.* [64] also reported about the structure sensitivity of Cu and CuZn for methanol synthesis. It was revealed that activity of the catalyst was highly depending on structure. It was considerably decreased for smaller particles (8 nm) due to sole atomic configuration like step-edge sites, which is usually missing for smaller particles [64]. In another study, Rungtaweevoranit *et al.* [65] reported a catalyst system (100% selectivity and improved activity) consisting of Cu nanoparticles encapsulated in a metal-organic framework (MOF) for methanol conversion from CO_2 [65]. Figure 7.5

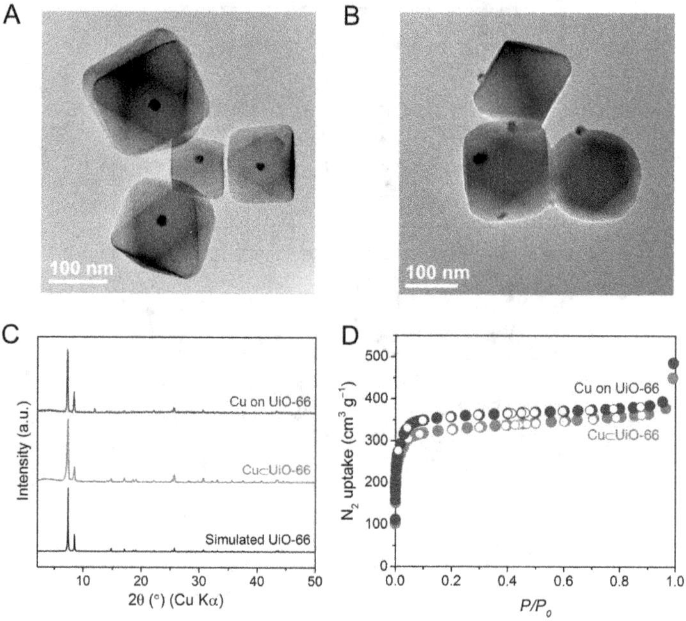

Figure 7.5 TEM images of (A) Cu⊂UiO-66 (single Cu NC inside UiO-66), (B) Cu on UiO-66, (C) XRD patterns in comparison with simulated pattern from single crystal X-ray diffraction data, and (D) N_2 adsorption–desorption isotherms at 77 K (adsorption and desorption points represented by closed circles and open circles, respectively). Reproduced from Reference 65 with permission from American Chemical Society.

shows the morphological features of the the Cu encapsulated in MOF, in comparison with the Cu on MOF. Ternary nano-sized $Cu/ZnO/Al_2O_3$ was also reported in another study, which was prepared from alternatives novel sol and colloidal primary solutions and exhibited improved catalytic activity [66].

Fischer–Tropsch Synthesis Processes

Fischer-Tropsch synthesis process (FT) is a highly exothermic reaction and used to produce wide-range of hydrocarbons such as high quality diesel fuel, gasoline and linear chemicals from catalytic conversion of mixtures of CO and H_2 (syngas, catalytic CO hydrogenation) through surface polymerization. The syngas is usually produced from coal or methane (non -crude oil feedstock) by treating with oxygen and steam at high temperatures. Main catalysts used in FT process are iron, nickel, cobalt, ruthenium or osmium [11,67]. Many attempts have been made to enhance the activity of the FT catalysts. Using promoter along with the main catalyst system is one of the method to increase the activity, selectivity, degree of reduction or the dispersion of the supported catalyst. Recently, Pt was used as promoter with Co/Nb_2O_5 catalyst system supported on γ-Al_2O_3- and α-Al_2O_3 [68]. Experimental data showed that Pt promoted the activity and active sites of catalyst system by a factor of 2.4 and 1.7 respectively. Similarly, different kinds of noble metals promoters such as Ag, Pt, Ru, and Re were added through (co-)impregnation to Co/TiO_2 catalysts system. Temperature programmed reduction (TPR) studies showed that Pt and Ru decreased the reduction temperatures of cobalt oxide more than Re and Ag [69].

An attempt was also made to improve the low porosity limits of promoter like niobia by synthesizing niobia modified silica by impregnation and subsequent use as support for Co and Pt-Co catalysts [70]. The selectivity of the niobia modified catalysts system was found to be pressure dependent due to higher intrinsic CO coverage. It was also found that combination of Nb nano-crystals and Pt increased the cobalt-weight normalized activity by a factor of 3-4 due to higher number of active sites and increased TOF. In another study, Wang *et al.* [71] reported platinum-modulated cobalt nano-catalysts for low-temperature Fischer–Tropsch process. The authors observed high activity of the catalyst for CO dissociation, which was attributed to the formation of Co overlayer structures on Pt NPs or Pt–Co alloy NPs. As observed in Figure 7.6, the higher activity of the supported

Co layer resulted from the formation of a more favorable transition state, which thus facilitates the C–O dissociation on supported Co layers.

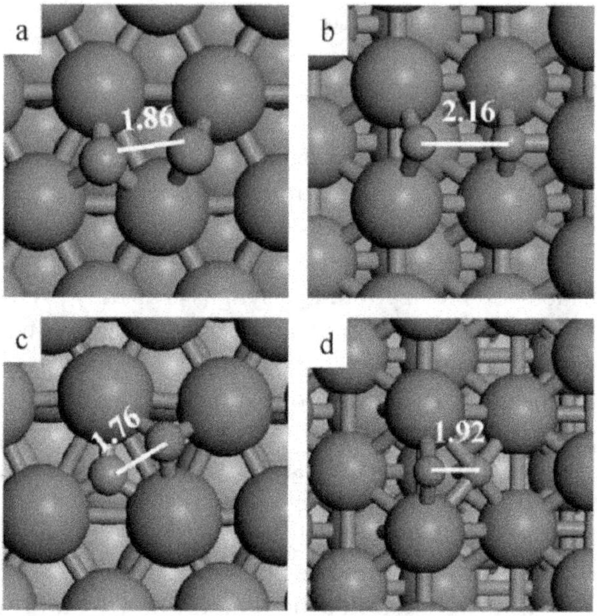

Figure 7.6 Optimized transition states for CO dissociation on: (a) Co(111), (b) Co(311), (c) 1 ML Co/Pt(111), and (d) 2 ML Co/Pt(311). Blue, orange, gray, and red spheres represent Co, Pt, C, and O atoms, respectively. Reproduced from Reference 71 with permission from American Chemical Society.

Usually, parameters associated with support in catalyst system like pore size, symmetry, geometry, connectivity and morphology, textures, specific surface area, stability, metal dispersion and reducibility play a vital role during catalysis. Use of ordered mesoporous substrates for FT catalysts have plenty of advantageous due to symmetry, long range order and narrow pore size distribution. Oschatz *et al.* [72] compared mesoporous ordered particles of silica, carbon and silicon carbide as supports for Fe-based Ft catalysts. The catalyst system was synthesized by encapsulating iron nanoparticles and sodium and sulphur (promoters) within the selected supports which, in turn, increased the stability of the system. The results showed that carbon had highly selective behavior for lower olefins and low methane production [72].

Etherification

Etherification is a useful exothermic and acid catalytic process in refining industry which is used to convert branched alkene into important products such as tertiary ethers (gasoline as methyl tert-butyl ether (MTBE), ethyl tert-butyl ether (ETBE) and tert-amyl methyl ether (TAME)), linear ethers (diesel fuel as dimethyl ether (DME), di-*n*-pentyl ether (DNPE)), etc. It is a simply the addition reaction of alcohol which is mainly catalyzed by acidic ion-exchange resins (sulfonated organic polymers), zeolites (molecular sieves), clays, solid hetro-polyacids, sulfided zirconia, etc. The main feedstock for etherification process is reactive streams of iso-butene or pentenes, or alcohol (methanol or ethanol). During catalysis, deactivation of catalysts usually happens due to poisoning of acidic sites by Na^{+2} and NH_4^+ ions, nitrogen containing compounds, dienes, metallic cations, etc. [31]. Other limitations of the catalysts have also been reported in the literature such as high solubility of the mineral acids and high temperature stability of hetero-polyacid. Yee *et al.* [73] discussed the catalysis of ETBE synthesis from tert-butyl alcohol (TBA) using functionalized multi-walled carbon nanotube (F-MCNT) having Lewis acid sites. The proposed mechanism of etherification was as follows: 1) creation of tert-butyl carbonium ion (TBC), 2) interaction of TBC ion with ethanol forming intermediate, 3) rearrangement of TBC to towards stable molecule of ETBE and 4) desorbance of ETBE molecule from the surface. The associated advantage of this process was to avoid the formation of side product like diethyl ether. The catalyst system improved degradation, reusability and regeneration behavior, selectivity and conversion (64% & 68%) at 3 wt% of catalytic loading [73]. Catalytic performance of ion-exchange resin during catalysis has critical importance for producing analogous ether. In a recent study sixteen sulfonic ion-exchange resins were evaluated for catalysis of iso-butene with four different alcohols, i.e., 1-butanol, 1-propanol, ethanol or methanol [74]. It was observed that reaction rates were higher as the chain length of the alcohol increased. It was also concluded that the highly acidic resins with inflexible morphology showed higher activity for the reaction.

Molecular sieves catalysts are known as "electron transfer stations" and can be used as alternatives to acidic catalysts due to many benefits, e.g., thermal stability, modification with different ions, amendable pore configuration and acidity and easy regeneration, even though these display a reduced activity at low temperatures.

These catalysts have also other properties like empty outermost (5d) orbital, different valency state and ability to form thermally stabilized structure. Modifications and treatments of zeolites with metallic ions, acidic leaching or steam treatment are also reported in literature for improved catalytic performance of these molecular sieves. Recently, Yan *et al.* [75] modified the Hβ zeolite with a combination of low steam treatment and lanthanum ion exchange. The comparative study with single modification process showed that combination modification processes exhibited best results in terms of etherfication activity and conversion. The combined modification process did not destroy the crystal phase and framework structure, while generated new Brønsted acid sites and enlarged pore structure, which was useful for the etherification [75]. Excellent activity of the sulfated zirconia is also reported for etherification by Jaworski *et al.* [76]. The authors reported the modification of zirconia with sulfuric acid which resulted in tetragonal phase of zirconia. The authors used the modified ZrO_2 for etherification of glycerol with benzyl alcohol at different temperatures (120-140° C), where the conversion grew with increasing sulfuric acid content and temperature.

7.2.2 Alternative/Renewable Energy

According to The International Energy Agency (IEA) World Energy Outlook 2009, the CO_2 target of 450 ppm up to year 2030 cannot be achieved without considerable involvement of renewable energy, and carbon capture and sequestration (CCS) [77]. Brining renewable energy into the main stream, thus, would be the significant challenge for the current century. Alternative or renewable energy refers to the energy which is evolved from sources other than earth's natural resources or does not damage the environment. In this category, mainstream technologies could be based on extraction of power from wind, hydro, etc., as well as the generation of energy from solar, geothermal, biomass, etc. In this respect, nano-catalysts due to their large surface to volume ratio and efficient catalytic capabilities have a pivotal role by improving the efficiency of clean energy production, energy conversion, energy storage and transport. Specifically, the role of nanotechnology in electro-catalysis of alternative resources can be beneficial in terms of improved efficiency and selectivity of current processes and permitting new technologies which are not viable till now [78].

Fuel Cells

Fuel cells are considered as ultra-low/zero noise and pollutant emission devices for the production of electrical energy directly converted from chemical energy. The main problem associated with the commercialization of fuel cells is the costly material used for manufacturing and its insufficient short life span of operation. Different kinds of low and high operating temperature fuel cells with different electrolytes have been commercialized/researched, like RMFC (reformed methanol), PEMFC (proton/polymer exchange/electrolyte membrane), UMFC (up-flow microbial), DMFC (direct methanol), MHFC (metal hydride), DCFC (direct carbon), MCFC (molten carbonate), PCFC (protonic ceramic), EGFC (electro-galvanic), SOFC (solid oxide), EBFC (enzymatic biofuel), PAFC (phosphoric acid), MFC (microbial), DBFC (direct boron-hydride), DFAFC (direct formic acid), AFC (alkaline), RFC (regenerative), TSOFC (tubular solid oxide), etc. The main working mechanism of the fuel cell depends on the reaction of positively charges H^+ ions with O_2 or some other oxidizing agent [33]. AFC is low operating temperature (80° C) FC and uses aqueous alkaline (30% KOH) solution in a porous matrix. The electrical efficiency and power density of cell is about 60% and 0.2-0.4 W/cm^2 respectively using H_2 and O_2 as fuel. The main catalyst reported in the literature is Pt as cathodic catalyst (oxygen reduction reaction (ORR)) and its foremost disadvantage is the high cost and scarcity. Bhandary *et al.* [79] proposed a green and cheaper method of synthesis of an electrocatalyst based on nano-flakes of AgCd alloy for ORR in AFC. The electro-activity of the catalyst was determined using cyclic voltammetry. The catalyst exhibited outstanding activity for ORR, producing <15% of peroxide in the process which facilitated it to tolerate its electrocatalytic activity for more than 3 hours in alkaline conditions. Similarly, Nagai *et al.* [80] proposed a complex system of non-precious metal electro-catalyst system based on perovskite type oxide-carbon mixture ($La_{0.6}Sr_{0.4}MnO_3$/C, $La_{0.6}Sr_{0.4}CoO_3$/C, and $La_{0.6}Sr_{0.4}Mn_{0.7}Co_{0.3}O_3$/C) and an iron phthalocyanine complex (Fe-Pc) for ORR in AFC. The activity of the catalysts were determined using rotating ring-disk electrode (RRDE) which showed improved ORR activities in all of the cases. Specifically, Fe-Pc/$La_{0.6}Sr_{0.4}Mn_{0.7}Co_{0.3}O_3$/C catalyst was observed to exhibit the maximum catalytic activity, which in comparison was similar to that of the Pt/C catalyst.

The other type of low operating temperature (60-90 °C) FC is PEMFC with high electrical efficiency between 50-60% and power density 0.5-1 W/cm^2 and appropriate for many applications. The main reactants are H_2, H_2 rich reformates with CO less than 30 ppm and oxygen/air. The main electrolyte is proton conducting polymer membrane whereas mostly Pt, Pt alloy or Pt/C are used as catalysts. The reactants are separated by the polymer membrane. Hydrogen and O_2 react at anode and cathodic side respectively and flow independently in the bipolar plate. The catalyst layer is loaded with supported metallic particles where the hydrogen and oxygen react to form water and byproducts. The electrons from H_2 dissociation flow through external circuit [81]. Along with many advantages like highest power density and fast start and stop sequence, catalyst poisoning (inactivation of catalytic sites) by CO is a major problem in commercialization of these promising PEMFCs. In order to overcome the drawback associated with the CO poisoning so as to increase the tolerance of the anodic catalyst, many efforts have been made. One of the solutions suggested the use of binary or ternary alloyed catalysts with Pt to improve the adsorption and oxidation of CO. Further, the cost of the Pt based catalysts in PEMFC can also be decreased by replacement with other cheaper metals or lower loadings. Kheradmandinia *et al.* [82] used low cost non-noble metals nano-catalysts composed of C/SnO_2, C/Ni and C/CoO. These nano-catalysts exhibited improved catalytic activities by lowering the potentials of CO electro-oxidation [33,82]. Another attempt was also made to reduce the cost of the Pt catalysts by developing nano-alloy Pt catalysts on high surface area carbon and Vulcun with different loadings of Pt, Ni and Co [83].

Another type of low operating temperature (20-90° C) and proton conducting polymer electrolyte membrane containing FC is DMFC where liquid methanol is utilized as direct fuel instead of H_2. The cell efficiency and power density is around 20-30% and 0.05-0.1 W/cm^2 respectively. The working of DMFC depends on the methanol oxidation at anode where generated protons transfer towards the cathode through a membrane followed by oxygen reduction at cathode. The electricity is produced through the transfer of electrons through an external circuit. Similar to the PEMFC, DMFCs also use carbon supported Pt metal based catalysts system [33]. In this regard, Pt has the same disadvantage of cost-effectiveness and much attention has been given to decrease the cost. The electro-catalysts should have good methanol tolerance, activity and selectivity for ORR. Along with this, other problems related to DEMFC have also been reported in the literature like lethargic kinetics of reactions at both electrodes, reduced durability, electrode flooding and methanol crossover. The overall performance of the DMFCs can be improved by modifying the structure of membrane electrode assembly (MEA), i.e., backing, diffusion and catalysts layers. Also, in DMFC, transport of methanol through membrane causes efficiency loss and drop in cell voltage owing to cathodic oxidation of methanol. To overcome this, studies have focused on replacement of Pt with other metals like Pd (low cost and activity for ORR). Efforts have been made to expand the electro-activity of Pd by using transition metals such as Co, Fe, and Ni as alloying agent. These studies have exhibited feasibility of Pd as electro-catalysts and have resolved the methanol crossover problem for DMFCs. For instance, Golmohammadi *et al.* [84] used Pd_3Co as nano-catalyst on binary carbon supports (MWCNTs and Vulcan) for fabrication of MEA for a passive DMFC and showed that the elector-catalytic properties for the ORR were best when MWCNTs and Vulcan at a mass ratio of 25:75 were used. Introduction of MWCNTs into the support provided high surface area, good electronic conductivity, and fast ORR kinetics as shown by improved electrode kinetic parameters [84].

In another study, Wang *et al.* [85] reported an electro-catalyst with low Pt mole fraction, consisting of intermetallic nanoparticles of Cu_3Pt for oxygen reduction reaction activity. A simple impregnation-reduction method was used for the catalyst synthesis. The authors used electrochemical and chemical dealloying methods for the synthesis of the catalyst particles. It was observed that the electrochemical dealloying method resulted in the formation of a core-shell morphology with 1 nm thin Pt skin and ordered Cu_3Pt core. On the other hand, the chemical dealloying method resulted in a spongy structure (Figure 7.7). The authors observed that both dealloying methods led to enhanced specific and mass activities toward the ORR. The mass activity was also observed to still enhance even after 5000 potential cycles, which indicated the functional nature of the developed electro-catalysts with improved ORR.

Vegetable Oil Conversion to Bio-fuel

The importance of energy from alternative resources is progressively vital owing to shrinking crude oil reserves and environmental hazards of liberated gases. Recently, vegetable oil (VO) conversion to biodiesel (BD) has fascinated the researchers and attracted significant attention. BD is the colloquial name for "fatty acid methyl ester" and usually produced by trans-esterification of VO and methanol in the presence of catalysts. It can also be produced by other methods like fermentation, ecofining, thermal and catalytic cracking [86]. Since it is synthesized from the renewable resources, a lesser amount of

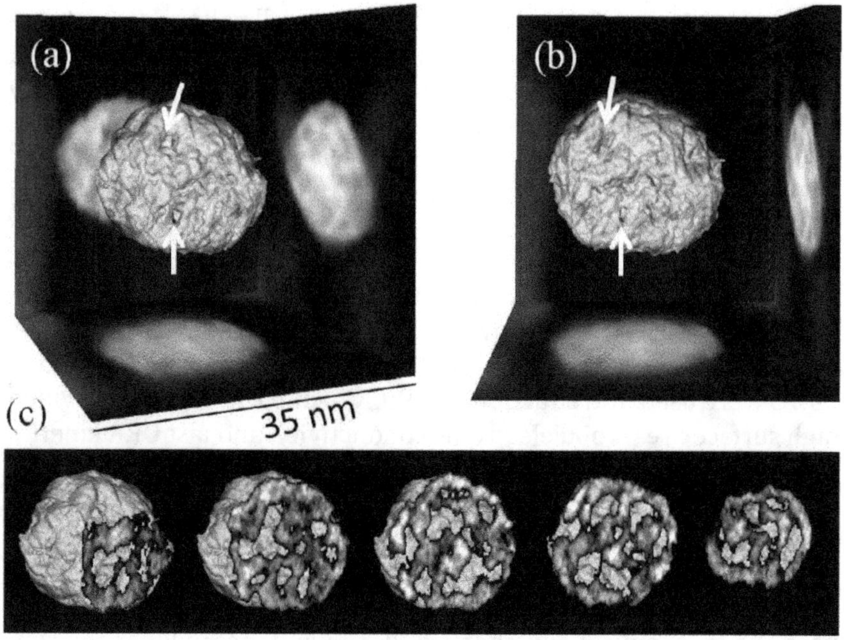

Figure 7.7 (a,b) Tomographic reconstruction of the spongy nanoparticle. The arrows indicate channels connecting to the exterior surfaces. (c) Consecutive slices through the surfaces to demonstrate the porous networks. Reproduced from Reference 85 with permission from American Chemical Society.

pollutants and hazardous gases (CO_2, sulfur and nitrogen) is released due to clean burning, biodegradability, nontoxicity, absence of sulfur and aromatic compounds, etc. VO represents a large family of resources with oil from many sources including canola, avocado, palm, mustard, soybean, olive sunflower, peanut rapeseed, recycled oil, etc. In the trans-esterification reaction with alcohol for the synthesis of BD, diglycerides are formed from triglycerides during the first stage, followed by the formation of monoglycerides. Finally, glycerol is synthesized producing one methyl ester molecule at every stage. The reactions are catalyzed by conventional homogeneous or heterogeneous based alkali (faster) or acidic (slower) based materials, enzymatic catalysts or supercritical alcoholysis conditions. The use of the conventional homogenous catalysts like soluble Na and KOH make this process longer in duration and due to improved solubility, the catalysts are difficult to be removed. Heterogeneous catalysts such as ZrO_2, titania, sodium hydroxide, sodium on alumina, potassium chloride, potassium carbonate, potassium nitrate, potassium iodide, potassium bromide, potassium fluoride have numerous benefits like regeneration and reuse of the catalyst. Different nanomaterials have also been employed for VO conversion to BD like calcium oxides, nano-γ-Al_2O_3, MgO, TiO_2, zeolites [87,88]. Verziu *et al.* [89] compared the performance of three dissimilar MgO nano-catalysts, different in terms of sizes and morphologies, for the conversion of sunflower and rapeseed oils. In order to prepare nanoparticles with different surfaces like (111) (110) and (100), different preparation methods were employed such hydrolysis, precipitation, and calcination. Under microwave conditions, the studies indicated that the activation temperature depended on the exposed face. It was lowest for (110) and (100) faces due to density of the basic sites [89]. Bet-Moushoul *et al.* [90] used five different types of gold nanoparticles supported calcium oxide-based nano-catalysts systems for biodiesel synthesis using trans-esterification of sunflower oil with methanol. Nano-gold deposited on CaO were prepared from AuNps solution using the impregnation method. The conversion process produced higher quality biodiesel and showed that the efficiency of the reaction was around 90–97% with 3.9-4.3 mg/kg amount of glycerol for all the samples. The reusability of the catalyst was also found to be improved as it could be re-employed up to ten times without impairment of activity [90].

Bio-gasoline (BG) is another kind of biofuel where exploration is being conducted in both academic and private sectors. Catalytic cracking process can be used to produce bio-based fuels like gasoline, kerosene and diesel from cooking or non-cooking oils at higher yields and comparatively lower temperature range. Common zeolites based catalysts are used for catalytic cracking due to their several advantages such as ordered structures and acidic sites. Ahmad *et al.* [86] prepared zeolite (ZSM-5) based nano-catalyst system loaded with different concentration of hetero-metallic (Fe–Zn–Cu) nano-oxides and used for the catalytic cracking of palm oil for the production of bio-gasoline. The synthesis of nano-catalysts initially started with the preparation of metallic organic frameworks with hetro-metallic complexes. Higher efficiency (56%) was obtained due to higher loading of loaded metal oxides catalysts [86].

Photo-catalytic Water Splitting

Photo-catalytic water splitting (PWS) is a promising technology to produce energy in the form of H_2 as fuel by artificial photosynthesis of water using artificial or natural light. PWS converts solar energy into chemical energy by photo-catalysis/photo-electro-catalysis in a photochemical/photo-electrochemical cell. The reactants (sunlight and H_2O) of PWS process are found as abundant resources on the globe. With increasing environmental concerns caused by fossil fuel burning, hydrogen as a fuel can be an alternative due to environmental friendliness and high energy yield. Among other water splitting methods like electrolysis, steam reforming of hydrocarbons, thermochemical and photobiological methods, PWS has numerous advantageous such as low cost, comparatively high efficiency, ability to separate hydrogen and oxygen streams, and appropriate technology for small scale production. The hydrogen production mechanism from H_2O splitting in 2:1 ratio is as follows: during initial stage, electron-hole pairs are generated after absorption of photons followed by migration of these charge carriers on the catalyst surface and finally electrons produce H_2 through water reduction, whereas holes produce oxygen by an oxidizing reaction [91]. Main catalysts reported in the literature for PWS are TiO_2, surface and structural modified TiO_2, metal sulfides and oxides, nitrides, oxy-sulfides and oxy-nitrides. The main principals associated with the use of photo-catalyst for PWS are their energy band gap which should be at least 1.23 eV and anti-photo-corrosion. The major challenge associated with the use of semiconductor as photo-catalyst is ultraviolet (UV) light sensitivity. Attempts have been made to improve the sensitivity and efficiency of the photo-catalysts for visible light response by developing visible-light-driven reactors. The use of conductive surface containing photo-catalyst particles can be useful for PWS process. Addition of reduced graphene oxide with photo-catalysts can create new paths for electron migration towards the electrode. Iwase *et al.* [92] fabricated composites of RGO and $CuGaS_2$ in order to achieve enhanced cathodic photo-current under sunlight irradiation. Composite electrode depicted improvements in terms of cathodic photo-current under visible light irradiation than the pristine photo-electrode. On coupling with a CoO_x loaded $BiVO_4$ based photo-anode, the composite electrode exhibited that hydrogen evolution can be achieved without applying external bias under visible irradiation [92]. Further, corrosion of the photo-catalysts is another major problem for photo-cells.

Hence, it is vital to develop photo-catalyst with improved corrosion resistance behavior. This can be achieved by either surface modification or doping with other materials more resistant to corrosion. For instance, nano-crystalline zinc oxide is a useful catalyst for PWS process owing to its relatively higher electron and holes mobility, and easy fabrication of nano-structures with large surface-to-volume ratios. The photo-degradation of ZnO in presence of aggressive solutions is a serious hurdle in using ZnO for PWS. Nano-coatings on ZnO electrodes of suitable material like TiO_2 could be beneficial for water splitting photo-anode. Liu *et al.* [93] stabilized the ZnO nanowire photo-anodes with ultrathin shell of TiO_2 through atomic layer deposition (ALD) (Figure 7.8). The chemically grown thin shells of TiO_2 were observed to exhibit stable operation for water splitting in aggressive solution with 25% higher water splitting activity [93].

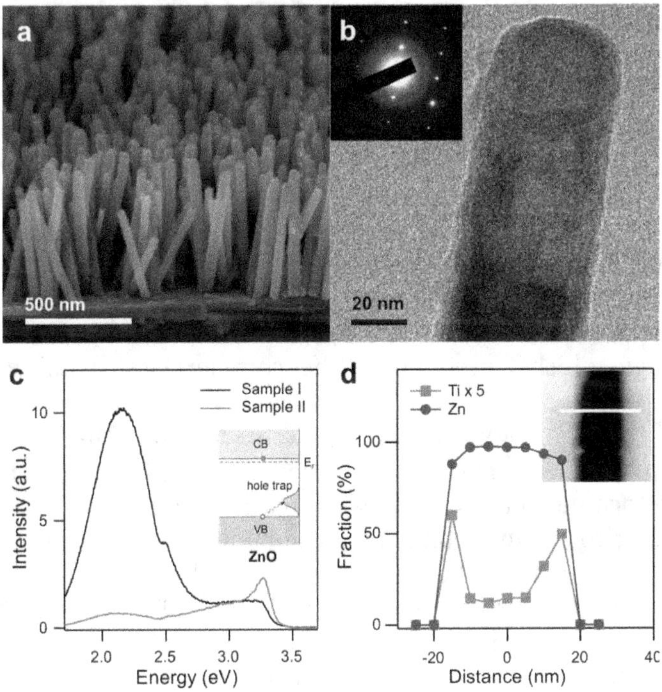

Figure 7.8 (a) SEM image of a ZnO nanowire array; (b) TEM image of a ZnO nanowire; (c) PL spectra of the ZnO nanowire array (sample I, higher intensity curve) and the ZnO/TiO_2 core/shell nanowire array (sample II, lower intensity curve) and (d) EDX elemental scans of a ZnO/TiO_2 core/shell nanowire. Reproduced from Reference 93 with permission from American Chemical Society.

Enhancing visible light absorption characteristics can improve the efficiency of the photo-catalytic reactor. Engineering of nano-structured materials to produce different nano-architectures is a promising strategy to improve the anti-reflection and absorption properties. Wei *et al.* [94] synthesized low cost, ordered hetero-nano-structure bismuth vanadate ($BiVO_4$) nano-pyramid arrays through aqueous deposition (60 °C for 6 hours) and annealed (400 °C for 30 min). The tips of the nano-pyramid were capped with CoPi nanoparticles (electro-catalyst) through photo-deposition at room temperature. The resulted structure exhibited low-cost solar water splitting without the use of a sacrificial agent [94].

Biomass Gasification

Energy from biomass gasification (BG), a flow based system, depends on continuous supply of sustainable feedstock. It is the most technically established fields of biofuel synthesis. During the thermochemical process, solid biomass such as lignocellulosic convert into combustible gases. The major application of BG process could be cogeneration of heat, power, bio-fuel, bio-tar, electricity and fertilizers. On the other hand, there are deficiencies and bottlenecks in the current technology, which are main hurdles in BG commercialization such as economically unviability and the presence of condensates, tar, dust and carbon burnout in the product streams. Catalysis research in these areas is requisite for appropriate inexpensive and innovative technology which can improve the commercial feasibility of BG [22,95].

Gasification of biomass usually occurs through many reactions such as pyrolysis, thermal cracking, dehydration, water-gas shift, and steam reforming. Supercritical water gasification (SWG) is a beneficial process to produce compressed and unpolluted gas with high H_2. One of the important advantages of using catalyst along SWG process is the lowering down of the temperature of the BG process, thus, resulting in reduced cost. Major catalysts systems used for SWG processes are usually alkaline salts (homogenous) and bimetallic, transitional metals, activated carbon (heterogeneous), etc. Kang *et al.* [96] catalyzed cellulose and lignin from feedstock of wheat straw, canola meal, and timothy grass. The gasification was performed by using by K_2CO_3 and $20Ni\text{-}0.36Ce/Al_2O_3$ (homo/heterogeneous) catalyst system. The study concluded that higher catalyst loading encouraged the hydrogen production [96]. It was also reported that impregnation of biomass with catalyst systems is another suitable and innovative method to improve the selectivity and activity of the reaction. Similarly, in another attempt, lignocellulosic feedstocks from wheat straw and pinewood were impregnated with Ni nanoparticles for SWG [97]. Hydrothermal gasification of impregnated biomass exhibited significant gas yields (hydrogen, carbon dioxide, and methane, etc.) as compared to non-catalytic gasification. Enhanced catalysis of biomass was due to the active sites present on nickel nanoparticles improving H_2 production [97].

Marine biomass like algal, known as third generation of biofuels, is another kind of biomass which is rich in C and H_2 and can be used as alternative to agricultural waste for the production of green fuel. Around 28 million tons of marine biomass presents enormous potential to produce renewable and environmentally pleasant biofuels and chemicals, which will also lower the burden on agricultural feed. The structure of algal biomass presents an ability to hydrolyze easily due to the low content of hemicelluloses and lignin. Norouzi *et al.* [98] investigated the potential of biofuel production using SWG of algal biomass from southern coast of Caspian Sea in a batch reactor. The nano-catalyst system used in this study was composed of un-promoted and Ru promoted Ni-Fe/γ-Al_2O_3. Nano-catalysts were synthesized via reverse micro-emulsion technique using 12 wt% of Ni, 6 wt% of Fe and 0.5-2 wt% of Ru. The best performance was depicted by promoted Ni–Fe/c-Al_2O_3/2 wt% Ru, during SWG of Enteromorpha intestinalis (algal) [98]. Another main bottleneck which obstructs the commercialization of BG process is the formation of tar with the gas products. Many studies have been carried out to alleviate the problem of the tar production in biomass conversion. In this respect too, nano-catalysis has played a vital role in improving the cracking behavior of tar. In a recent study, nanocomposites comprising of Ni/SiO_2 and Ni-Ce/SiO_2 were generated by deposition-precipitation method and applied as a nano-catalyst in the cracking of tar [99]. It was reported that the utilization of nanocomposites for tar cracking increased the catalytic cracking by 93% and 98.5% for Ni/SiO_2 and Ni-Ce/SiO_2, respectively. Catalytic cracking of tar also affected the quality of the product gas by increasing the carbon monoxide and hydrogen contents, whereas carbon dioxide and methane contents were decreased.

7.2.3 Other Systems

Among other functional uses of nano-catalysts for energy, Hussain *et al.* [100] reported heterogeneous ruthenium–molybdenum disulfide nano-catalyst for the selective aerobic oxidation of amines (Figure 7.9). The generated catalyst yielded with high selectivity nitriles from different aromatic, aliphatic, allylic, and heteroatomic amines. The advanced nano-catalyst reported in the study can be also potentially applied in conventional and new processes for energy applications. The authors also reported that the oxidation reactions could be achieved with a very small amount of the catalyst, which indicated the environmentally friendly and economically efficient process. In another study, Liu *et al.* [101] reported ultrafine metal oxide nano-catalysts for high-performance lithium-oxygen batteries. An electro-chemical prelithiation process was used for the catalyst synthesis, which reduces the size of $NiCo_2O_4$ (NCO) particles from 20-30 nm to a uniformly distributed domain of ~2 nm, thereby significantly improving the catalytic activity. A Li-O_2 battery employing the developed

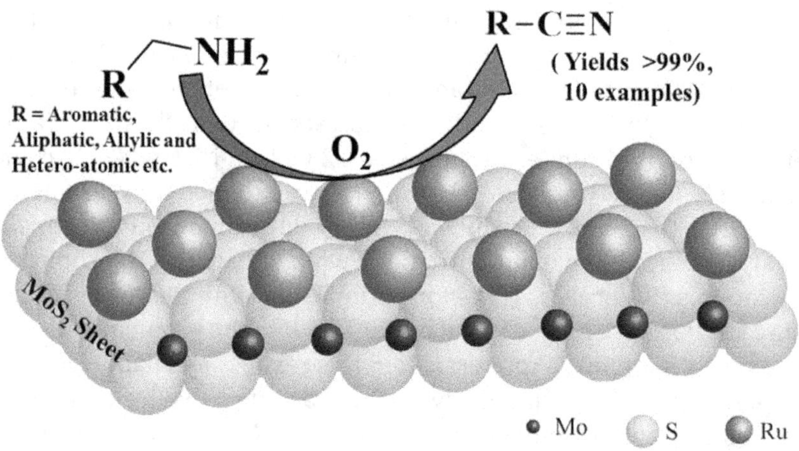

Figure 7.9 Schematic representation of the amine oxidation process. Reproduced from Reference 100 with permission from American Chemical Society.

catalyst was observed to have an initial capacity of 29,280 mAh g^{-1}. In addition, the battery retained the capacity of >1000 mAh g^{-1} after 100 cycles. Figure 7.10 also demonstrates the morphology of prelithiated (PL)-NCO nanowires (NW)/carbon fabric (CF) electrodes after prelithiation process.

7.3 Conclusion

Catalyst systems based on nano-structured materials have created interesting research areas for efficient catalysis of fossil and renewable resources. Nano-structured nano-catalyst materials present promising progress avenues in the current and future energy areas due to their higher activity and improved selectivity. Due to many current experimental, technological and environmental constraints, there is a need to improve the quality and quantity of the products generated from fossil or renewable resources from the perspective of both environment and energy requirements. In this respect, efficient and selective nano-catalysts are attractive alternatives to the conventional catalyst systems for the sustainable energy, clean environment, efficient use of resources and cost effectiveness.

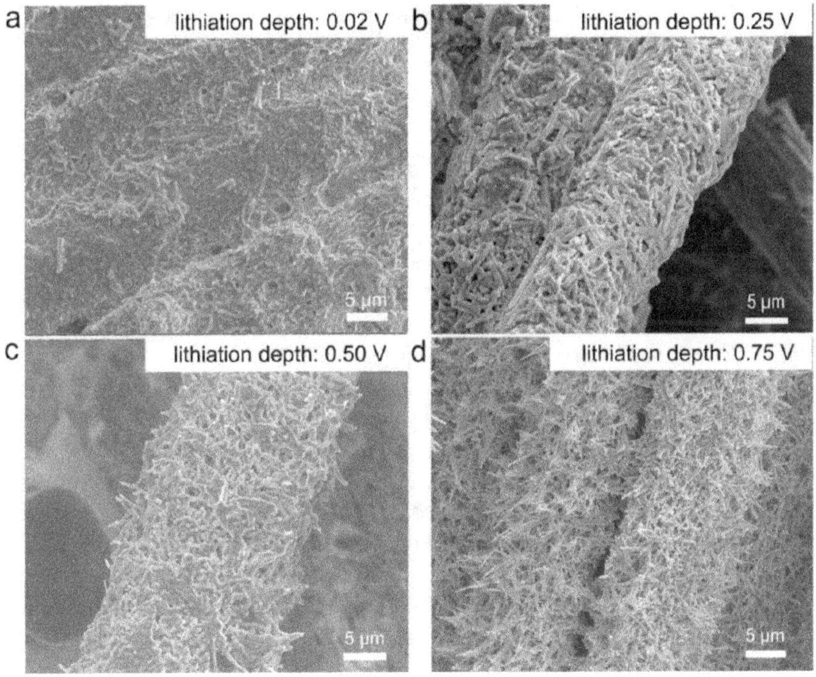

Figure 7.10 SEM images depicting the morphology of PL-NCO NWs/CF electrodes after prelithiation process at depth of (a) 0.02, (b) 0.25, (c) 0.50, and (d) 0.75 V. Reproduced from Reference 101 with permission from American Chemical Society.

References

1. Shiju, N. R., and Guliants, V. V. (2009) Recent developments in catalysis using nanostructured materials. *Applied Catalysis A: General*, **356**(1), 1-17.
2. Philippot, K., and Serp, P. (2013) Concepts in nanocatalysis. In: *Nanomaterials in Catalysis*, Serp, P., and Philippot, K. (eds.), Wiley-VCH, Germany, pp. 1-54.
3. Machado, R. M., Heier, K. R., and Broekhuis, R. R. (2001) Developments in hydrogenation technology for fine-chemical and pharmaceutical applications. *Current Opinion in Drug Discovery And Development*, **4**(6), 745-755.
4. Lovestam, G., Rauscher, H., Roebben, G., Kluttgen, B. S., Gibson, N., Putaud, J.-P., and Stamm, H. (2010) Considerations on a Definition of Nanomaterial for Regulatory Purposes. *Joint Research Centre (JRC) Reference Reports*, European Union. Online: https://ec.europa.eu/jrc/sites/jrcsh/files/jrc_reference_report_201007_nanomaterials.pdf [assessed 29th April 2017].
5. Kaur, R., and Badea, I. (2013) Nanodiamonds as novel nanomaterials for biomedical applications: drug delivery and imaging systems. *International Journal of Nanomedicine*, **8**, 203-220.
6. Gilmore, J. L., Yi, X., Quan, L., and Kabanov, A. V. (2008) Novel nanomaterials for clinical neuroscience. *Journal of Neuroimmune Pharmacology*, **3**(2), 83-94.
7. Cordier, S., Dorson, F., Grasset, F., Molard, Y., Fabre, B., Haneda, H., Sasaki, T., Mortier, M., Ababou-Girard, S., Perrin, C. (2009) Novel nanomaterials based on inorganic molybdenum octahedral clusters. *Journal of Cluster Science*, **20**(1), 9-21.
8. Garbarczyk, J. E., Wasiucionek, M., Jozwiak, P., Nowinski, J. L., and Julien, C. M. (2009) Novel nanomaterials based on electronic and mixed conductive glasses. *Solid State Ionics*, **180**(6-8), 531-536.
9. Li, Q., Mahendra, S., Lyon, D. Y., Brunet, L., Liga, M. V., Li, D., and Alvarez, P. J. J. (2008) Antimicrobial nanomaterials for water disinfection and microbial control: Potential applications and implications. *Water Research*, **42**(18), 4591-4602.
10. Zhang, L., and Webster, T. J. (2009) Nanotechnology and nanomaterials: Promises for improved tissue regeneration. *Nano Today*, **4**(1), 66-80.
11. Bonnemann, H., Khelashvili, G., Hormes, J., Kuhn, T. J., and Richter, W.-J. (2013) Nanostructured metal particles for catalysts and energy-related materials. In: *Nanomaterials in Catalysis*, Serp, P., and Philippot, K. (eds.), Wiley-VCH, Germany, pp. 123-201.
12. Schlogl, R., and Abd Hamid, S. B. (2004) Nanocatalysis: Mature science revisited or something really new? *Angewandte Chemie, International Edition*, **43**(13), 1628-1637.

13. Li, Y., and Somorjai, G. A. (2010) Nanoscale advances in catalysis and energy applications. *Nano Letters*, **10**(7), 2289-2295.
14. Chaturvedi, S., Dave, P. N., and Shah, N. K. (2012) Applications of nano-catalyst in new era. *Journal of Saudi Chemical Society*, **16**(3), 307-325.
15. Mahmoud, M. A., Tabor, C. E., El-Sayed, M. A., Ding, Y., and Wang, Z. L. (2008) A new catalytically active colloidal platinum nanocatalyst: The multiarmed nanostar single crystal. *Journal of the American Chemical Society*, **130**, 4590-4591.
16. Basic Research Needs: Catalysis for Energy (2008) *Report from the U.S. Department of Energy, Office of Basic Energy Sciences Workshop*, USA. Online: http://www.iact.anl.gov/publications/CAT_rpt.pdf [assessed 23rd April 2017].
17. Thompson, L. (2011) Applications: Energy from fossil resources. In *International Assessment of Research and Development in Catalysis by Nanostructured Materials*, Davis, R. (ed.), Imperial College Press, UK, pp. 151-183.
18. Hu, E. L., Davis, S. M., Davis, R., and Scher, E. (2011) Applications: Catalysis by nanostructured materials. In: *Nanotechnology Research Directions for Societal Needs in 2020: Retrospective and Outlook*, Roco, M. C., Mirkin, C. A., and Hersam, M. C. (eds.), Springer, Netherlands, pp. 445-466.
19. *Handbook of Heterogeneous Catalysis: 8 Volumes*, Ertl, G., Knozinger, H., Schuth, F., and Weitkamp, J. (eds.), 2nd edition, Wiley-VCH, Germany (2008).
20. Guliants, V. V. (2011) Applications: Chemicals from fossil resources. In: *International Assessment of Research and Development in Catalysis by Nanostructured Materials*, Davis, R. (ed.), Imperial College Press, UK, pp. 185-237.
21. Huber, G. (2011) Applications: Renewable fuels and chemicals. In: *International Assessment of Research and Development in Catalysis by Nanostructured Materials*, Davis, R. (ed.), Imperial College Press, UK, pp. 239-262.
22. Davis, R. J. (2011) Overview of catalysis by nanostructured materials. In: *International Assessment of Research and Development in Catalysis by Nanostructured Materials*, Davis, R. (ed.), Imperial College Press, UK, pp. 1-24.
23. Cheng, W.-C., Habib, Jr., E. T., Rajagopalan, K., Roberie, T. G., Wormsbecher, R. F., and Ziebarth, M. S. (2008) Fluid catalytic cracking. In: *Handbook of Heterogeneous Catalysis*, Ertl, G., Knozinger, H., Schuth, F., and Weitkamp, J. (eds.), 2nd Edition, Wiley-VCH, Germany, pp. 2741-2775.
24. Moser, M. D., and Bogdan, P. L. (2008) Catalytic reforming. In: *Handbook of Heterogeneous Catalysis*, Ertl, G., Knozinger, H., Schuth, F., and Weitkamp, J. (eds.), 2nd Edition, Wiley-VCH, Germany, pp. 2728-

2740.

25. Prins, R. (2008) Hydrotreating. In: *Handbook of Heterogeneous Catalysis*, Ertl, G., Knozinger, H., Schuth, F., and Weitkamp, J. (eds.), 2nd Edition, Wiley-VCH, Germany, pp. 2695-2716.
26. Sie, S. T. (2008) Isomerization. In: *Handbook of Heterogeneous Catalysis*, Ertl, G., Knozinger, H., Schuth, F., and Weitkamp, J. (eds.), 2nd Edition, Wiley-VCH, Germany, pp. 2809-2828.
27. van Veen, J. A. R., Minderhoud, J. K., Huve, L. G., and Stork, W. H. J. (2008) Hydrocracking and catalytic dewaxing. In: *Handbook of Heterogeneous Catalysis*, Ertl, G., Knozinger, H., Schuth, F., and Weitkamp, J. (eds.), 2nd Edition, Wiley-VCH, Germany, pp. 2778-2805.
28. Wei, J. (2008) Hydrodemetalation. In: *Handbook of Heterogeneous Catalysis*, Ertl, G., Knozinger, H., Schuth, F., and Weitkamp, J. (eds.), 2nd Edition, Wiley-VCH, Germany, pp. 2718-2727.
29. Rostrup-Nielsen, J. R. (2008) Steam reforming. In: *Handbook of Heterogeneous Catalysis*, Ertl, G., Knozinger, H., Schuth, F., and Weitkamp, J. (eds.), 2nd Edition, Wiley-VCH, Germany, pp. 2882-2903.
30. Hansen, J. B., and Hojlund Nielsen, P. E. (2008) Methanol synthesis. In: *Handbook of Heterogeneous Catalysis*, Ertl, G., Knozinger, H., Schuth, F., and Weitkamp, J. (eds.), 2nd Edition, Wiley-VCH, Germany, pp. 2920-2949.
31. Krause, A. O. I., and Keskinen, K. I. (2008) Etherification. In: *Handbook of Heterogeneous Catalysis*, Ertl, G., Knozinger, H., Schuth, F., and Weitkamp, J. (eds.), 2nd Edition, Wiley-VCH, Germany, pp. 2864-2879.
32. Peters, R. (2008) Fuel processors. In: *Handbook of Heterogeneous Catalysis*, Ertl, G., Knozinger, H., Schuth, F., and Weitkamp, J. (eds.), 2nd Edition, Wiley-VCH, Germany, pp. 3045-3076.
33. Gasteiger, H. A., and Garche, J. (2008) Fuel cells. In: *Handbook of Heterogeneous Catalysis*, Ertl, G., Knozinger, H., Schuth, F., and Weitkamp, J. (eds.), 2nd Edition, Wiley-VCH, Germany, pp. 3081-3117.
34. Moqadam, S. I., and Mahmoudi, M. (2013) Advent of nanocatalysts in hydrotreating process: Benefits and developements. *American Journal of Oil and Chemical Technologies*, **1**, 13-21.
35. Nejad, D. M., Rahemi, N., and Allahyari, S. (2017) Effect of tungsten loading on the physiochemical properties of nanocatalysts of Ni–Mo–W/carbon nanotubes for the hydrodesulfurization of thiophene. *Reaction Kinetics, Mechanisms and Catalysis*, **120**(1), 279-294.
36. Danforth, S. J., Liyanage, D. R., Hitihami-Mudiyanselage, A., Ilic, B., Brock, S. L., and Bussell, M. E. (2016) Probing hydrodesulfurization over bimetallic phosphides using monodisperse $Ni_{2-x}M_xP$ nanopart-

icles encapsulated in mesoporous silica. *Surface Science*, **648**, 126-135.

37. Jenifer, A. C., Sharon, P., Prakash, A., and Sande, P. C. (2015) A review of the unconventional methods used for the demetallization of petroleum fractions over the past decade. *Energy & Fuels*, **29**(12), 7743-7752.
38. Rao, S. M., and Coppens, M.-O. (2012) Increasing robustness of deactivating nanoporous catalysts by optimizing the pore network - application to hydrodemetalation. *Chemical Engineering Science*, **83**, 66-76.
39. Visaliev, M. Y., Shpirt, M. Y., Kadiev, K. H., Dvorkin, V. I., Magomadov, E. E., and Khadzhiev, S. N. (2012) Integrated conversion of extra-heavy crude oil and petroleum residue with the recovery of vanadium, nickel, and molybdenum. *Solid Fuel Chemistry*, **46**(2), 100-107.
40. *Catalysis: An Integrated Approach to Homogeneous, Heterogeneous and Industrial Catalysis*, Moulijn, J. A., van Leeuwen, P. W. N. M., and van Santen, R. A. (eds.), Volume 79, 1st edition, Elsevier, USA (1993).
41. Park, J. Y., and Somorjai, G. A. (2014) Bridging materials and pressure gaps in surface science and heterogeneous catalysis. In: *Current Trends of Surface Science and Catalysis*, Park, J. Y. (ed.), Springer, USA, pp. 3-17.
42. Musselwhite, N., Alayoglu, S., Melaet, G., Pushkarev, V. V., Lindeman, A. E., An, K., and Somorjai, G. A. (2013) Isomerization of n-hexane catalyzed by supported monodisperse PtRh bimetallic nanoparticles. *Catalysis Letters*, **143**(9). 907-911.
43. An, K., Zhang, Q., Alayoglu, S., Musselwhite, N., Shin, J.-Y., and Somorjai, G. A. (2014) High-temperature catalytic reforming of n-hexane over supported and core–shell Pt nanoparticle catalysts: Role of oxide-metal interface and thermal stability. *Nano Letters*, **14**(8), 4907-4912.
44. Madsen, C., Madsen, C., and Jacobsen, C. J. H. (1999) Nanosized zeolite crystals-convenient control of crystal size distribution by confined space synthesis. *Chemical Communications*, **1999**(8), 673-674.
45. Fan, W., Snyder, M. A., Kumar, S., Lee, P.-S., Yoo, W. C., McCornick, A. V., Penn, R. L., Stein, A., and Tsapatsis, M. (2008) Hierarchical nanofabrication of microporous crystals with ordered mesoporosity. *Nature Materials*, **7**(12), 984-991.
46. Tosheva, L., and Valtchev, V. P. (2005) Nanozeolites: Synthesis, crystallization mechanism, and applications. *Chemistry of Materials*, **17**(10), 2494-2513.
47. Koros, W. J., and Fleming, G. K. (1993) Membrane-based gas separation. *Journal of Membrane Science*, **83**(1), 1-80.
48. Koros, W. J., and Chern, R. T. (1987) Separation of gaseous mixtures

using polymer membranes. In: *Handbook of Separation Process Technology*, Rousseau, R. W. (ed.), Wiley, USA, pp. 862-953.

49. Koros, W. J., Fleming, G. K., Jordan, S. M., Kim, T. H., and Hoehn, H. H. (1988) Polymeric membrane materials for solution-diffusion based permeation separations. *Progress in Polymer Science*, **13**(4), 339-401.
50. Murase, K., Fujiwara, T., Umemura, Y., Suzuki, K., Iino, R., Yamashita, H., Saito, M., Murakoshi, H., Ritchie, K., and Kusumi, A. (2004) Ultrafine membrane compartments for molecular diffusion as revealed by single molecule techniques. *Biophysical Journal*, **86**(6), 4075-4093.
51. Garcia-Martinez, J., Johnson, M., Valla, J., Li, K., and Ying, J. Y. (2012) Mesostructured zeolite Y-high hydrothermal stability and superior FCC catalytic performance. *Catalysis Science & Technology*, **2**(5), 987-994.
52. Yang, X. Y., Tian, G., Chen, L. H., Li, Y., Rooke, J. C., Wei, Y. X., Liu, Z. M., Deng, Z., Van Tendeloo, G., Su, B. L. (2011) Well-organized zeolite nanocrystal aggregates with interconnected hierarchically micro–meso–macropore systems showing enhanced catalytic performance. *Chemistry - A European Journal*, **17**(52), 14987-14995.
53. Corma, A., Fornes, V., Pergher, S. B., Maesen, T. L. M., and Buglass, J. G. (1998) Delaminated zeolite precursors as selective acidic catalysts. *Nature*, **396**(6709), 353-356.
54. Dejhosseini, M., Aida, T., Watanabe, M., Takami, S., Hojo, D., Aoki, N., Arita, T., Kishita, A., and Adschiri, T. (2013) Catalytic cracking reaction of heavy oil in the presence of cerium oxide nanoparticles in supercritical water. *Energy & Fuels*, **27**(8), 4624-4631.
55. Lu, C. L., Dao, T. K. T., Nguyen, T., Bui, T. H., Dang, T. N. Y., Hoang, M. N., and Ho, S. T. (2013) Effect of carriers on physico-chemical properties and activity of Pd nano-catalyst in n-hexane isomerization. *Advances in Natural Sciences: Nanoscience and Nanotechnology*, **4**(4), 045001.
56. Aghdam, N. C., Ejtemaei, M., Babaluo, A. A., Tavakoli, A., Bayati, B., and Bayat, Y. (2016) Enhanced i-C5 production by isomerization of C5 isomers in BZSM-5 membrane reactor packed with Pt/ZSM-5 nanocatalyst. *Chemical Engineering Journal*, **305**, 2-11.
57. An, K., Alayoglu, S., Musselwhite, N., Na, K., and Somorjai, G. A. (2014) Designed catalysts from Pt nanoparticles supported on macroporous oxides for selective isomerization of n-hexane. *Journal of the American Chemical Society*, **136**(19), 6830-6833.
58. Behrens, M., and Armbruster, M. (2012) Methanol steam reforming. In: *Catalysis for Alternative Energy Generation*, Guczi, L., and Erdohelyi, A. (eds.), Springer, USA, pp. 175-235.
59. Koo. K. Y., Eom, H. J., Jung, U. H., and Yoon, W. L. (2016) Ni nanosheet-coated monolith catalyst with high performance for hyd-

rogen production via natural gas steam reforming. *Applied Catalysis A: General*, **525**, 103-109.
60. Purnama, H., Girgsdies, F., Ressler, T., Schattka, J. H., Caruso, R. A., Schomacker, R., and Schlogl, R. (2004) Activity and selectivity of a nanostructured CuO/ZrO_2 catalyst in the steam reforming of methanol. *Catalysis Letters*, **94**(1), 61-68.
61. Kuld, S., Thorhauge, M., Falsig, H., Elkjaer, C. F., Helveg, S., Chorkendorff, I., and Sehested, J. (2016) Quantifying the promotion of Cu catalysts by ZnO for methanol synthesis. *Science*, **352**(6288), 969-974.
62. van den Berg, R., Zecevic, J., Sehested, J., Helveg, S., de Jongh. P. E., and de Jong, K. P. (2016) Impact of the synthesis route of supported copper catalysts on the performance in the methanol synthesis reaction. *Catalysis Today*, **272**, 87-93.
63. van den Berg, R., Parmentier, T. E., Elkjaer, C. F., Gommes, C. J., Sehested, J., Helveg, S., de Jongh, P. E., and de Jong, K. P. (2015) Support functionalization to retard Ostwald ripening in copper methanol synthesis catalysts. *ACS Catalysis*, **5**(7), 4439-4448.
64. van den Berg, R., Prieto, G., Korpershoek, G., van der Wal, L. I., van Bunningen, A. J., Laegsgaard-Jorgensen, S., de Jongh, P. E., and de Jong, K. P. (2016) Structure sensitivity of Cu and CuZn catalysts relevant to industrial methanol synthesis. *Nature Communications*, **7**, 13057.
65. Rungtaweevoranit, B., Baek, J., Araujo, J. R., Archanjo, B. S., Choi, K. M., Yaghi, O. M., and Somorjai, G. A. (2016) Copper nanocrystals encapsulated in Zr-based metal-organic frameworks for highly selective CO_2 hydrogenation to methanol. *Nano Letters*, **16**(12), 7645-7649.
66. Bahmani, M., Vasheghani Farahani, B., and Sahebdelfar, S. (2016) Preparation of high performance nano-sized $Cu/ZnO/Al_2O_3$ methanol synthesis catalyst via aluminum hydrous oxide sol. *Applied Catalysis A: General*, **520**, 178-187.
67. Dry, M. E. (2008) The Fischer–Tropsch (FT) synthesis processes. In: *Handbook of Heterogeneous Catalysis*, Ertl, G., Knozinger, H., Schuth, F., and Weitkamp, J. (eds.), 2nd Edition, Wiley-VCH, Germany, pp. 2965-2992.
68. den Otter, J. H., Yoshida, H., Ladesma, C., Chen, D., and de Jong, K. P. (2016) On the superior activity and selectivity of $PtCo/Nb_2O_5$ Fischer Tropsch catalysts. *Journal of Catalysis*, **340**, 270-275.
69. Eschemann, T. O., Oenema, J., and de Jong, K. P. (2016) Effects of noble metal promotion for Co/TiO_2 Fischer-Tropsch catalysts. *Catalysis Today*, **261**, 60-66.
70. den Otter, J. H., Nijveld, S. R., and de Jong, K. P. (2016) Synergistic promotion of Co/SiO_2 Fischer–Tropsch catalysts by niobia and platinum. *ACS Catalysis*, **6**(3), 1616-1623.
71. Wang, H., Zhou, W., Liu, J.-X., Si, R., Sun, G., Zhong, M.-Q., Su, H.-Y.,

Zhao, H.-B., Rodriguez, J. A., Pennycook, S. J., Idrobo, J.-C., Li, W.-X., Kou, Y., and Ma, D. (2013) Platinum-modulated cobalt nanocatalysts for low-temperature aqueous-phase Fischer.Tropsch synthesis. *Journal of the American Chemical Society*, **135**, 4149-4158.

72. Oschatz, M., Lamme, W. S., Xie, J., Dugulan, A. I., and de Jong, K. P. (2016) Ordered mesoporous materials as supports for stable iron catalysts in the Fischer-Tropsch synthesis of lower olefins. *ChemCatChem*, **8**(17), 2846-2852.
73. Yee, K. F., Ng, E.-P., Mohamed, A. R., Adam, F., and Tan, S. H. (2016) Functionalized multi-walled carbon nanotubes as heterogeneous Lewis acid catalysts in the etherification reaction of tert-butyl alcohol and ethanol. *Chemical Engineering Communications*, **203**(10), 1385-1394.
74. Badia, J. H., Fite, C., Bringue, R., Ramirez, E., and Iborra, M. (2016) Relevant properties for catalytic activity of sulfonic ion-exchange resins in etherification of isobutene with linear primary alcohols. *Journal of Industrial and Engineering Chemistry*, **42**, 36-45.
75. Yan, X., Ke, M., Song, Z., Jiang, Q., and Yu, P. (2016) Etherification over H[small beta] zeolite modified by lanthanum ion exchange combined with low-temperature steam treatment. *RSC Advances*, **6**(18), 14799-14808.
76. Jaworski, M. A., Vega, S. R., Siri, G. J., Casella, M. L., Salvador, A. R., and Lopez, A. S. (2015) Glycerol etherification with benzyl alcohol over sulfated zirconia catalysts. *Applied Catalysis A: General*, **505**, 36-43.
77. World Energy Outlook (2009), The International Energy Agency, France. Online: http://www.worldenergyoutlook.org/media/weowebsite/2009/WEO2009.pdf [assessed 30th April 2017].
78. Centi, G., Lanzafame, P., and Perathoner, S. (2012) Introduction and general overview. In: *Catalysis for Alternative Energy Generation*, Guczi, L., and Erdohelyi, A. (eds.), Springer Science & Business Media, USA, pp. 1-28.
79. Bhandary, N., Basu, S., and Ingole, P. P. (2016) Rudimentary simple, single step fabrication of nano-flakes like AgCd alloy electro-catalyst for oxygen reduction reaction in alkaline fuel cell. *Electrochimica Acta*, **212**, 122-129.
80. Nagai, T., Yamazaki, S.-i., Fujiwara, N., Asahi, M., Siroma, Z., and Ioroi, T. (2016) Non-precious metal catalyst that combines perovskite-type oxide and iron phthalocyanine for use as a cathode catalyst in an alkaline fuel cell. *Journal of The Electrochemical Society*, **163**(5), F347-F352.
81. Devanathan, R. (2008) Recent developments in proton exchange membranes for fuel cells. *Energy & Environmental Science*, **1**(1), 101-119.
82. Kheradmandinia, S., Khandan, N., and Eikani, M. H. (2016) Synthesis and evaluation of CO electro-oxidation activity of carbon supported

SnO_2, CoO and Ni nano catalysts for a PEM fuel cell anode. *International Journal of Hydrogen Energy*, **41**(42), 19070-19080.

83. Van Schalkwyk, F. Pattrick, G., Olivier, J., Conrad, O., and Blair. S. (2016) Development and scale up of enhanced ORR Pt-based catalysts for PEMFCs. *Fuel Cells*, **16**(4), 414-427.
84. Golmohammadi, F., Gharibi, H., and Sadeghi, S. (2016) Synthesis and electrochemical characterization of binary carbon supported Pd_3Co nanocatalyst for oxygen reduction reaction in direct methanol fuel cells. *International Journal of Hydrogen Energy*, **41**(18), 7373-7387.
85. Wang, D., Yu, Y., Xin, H. L., Hovden, R., Ercius, P., Mundy, J. A., Chen, H., Richard, J. H., Muller, D. A., DiSalvo, F. J., and Abruna, H. D. (2012) Tuning oxygen reduction reaction activity via controllable dealloying: A model study of ordered Cu_3Pt/C intermetallic nanocatalysts. *Nano Letters*, **12**, 5230-5238.
86. Ahmad, M., Farhana, R., Rahman, A. A. A., and Bhargava, S. K. (2016) Synthesis and activity evaluation of heterometallic nano oxides integrated ZSM-5 catalysts for palm oil cracking to produce biogasoline. *Energy Conversion and Management*, **119**, 352-360.
87. Wang, L., and Yang, J. (2007) Transesterification of soybean oil with nano-MgO or not in supercritical and subcritical methanol. *Fuel*, **86**(3), 328-333.
88. Boz, N., Degirmenbasi, N., and Kalyon, D. M. (2009) Conversion of biomass to fuel: Transesterification of vegetable oil to biodiesel using KF loaded nano-γ-Al_2O_3 as catalyst. *Applied Catalysis B: Environmental*, **89**(3-4), 590-596.
89. Verziu, M., Cojocaru, B., Hu, J., Richards, R., Ciuculescu, C., Filip, P., and Parvulescu, V. I. (2008) Sunflower and rapeseed oil transesterification to biodiesel over different nanocrystalline MgO catalysts. *Green Chemistry*, **10**(4), 373-381.
90. Bet-Moushoul, E., Farhadi, K., Mansourpanah, Y., Nikbakht, A. M., Molaei, R., and Forough, M. (2016) Application of CaO-based/Au nanoparticles as heterogeneous nanocatalysts in biodiesel production. *Fuel*, **164**, 119-127.
91. Ahmad, H., Kamarudin, S. K., Minggu, L. J., and Kassim, M. (2015) Hydrogen from photo-catalytic water splitting process: A review. *Renewable and Sustainable Energy Reviews*, **43**, 599-610.
92. Iwase, A., Ng, Y. H., Amal, R., and Kudo, A. (2015) Solar hydrogen evolution using a $CuGaS_2$ photocathode improved by incorporating reduced graphene oxide. *Journal of Materials Chemistry A*, **3**(16), 8566-8570.
93. Liu, M., Nam, C.-Y., Black, C. T., Kamcev, J., and Zhang, L. (2013) Enhancing water splitting activity and chemical stability of zinc oxide nanowire photoanodes with ultrathin titania shells. *The Journal of Physical Chemistry C*, **117**(26), 13396-13402.
94. Wei, Y., Su, J., Wan, X., Guo, L., and Vayssieres, L. (2016) Spontaneous

photoelectric field-enhancement effect prompts the low cost hierarchical growth of highly ordered heteronanostructures for solar water splitting. *Nano Research*, **9**(6), 1561-1569.

95. Ahrenfeldt, J., Thomsen, T. P., Henriksen, U., and Clausen, L. R. (2013) Biomass gasification cogeneration - A review of state of the art technology and near future perspectives. *Applied Thermal Engineering*, **50**(2), 1407-1417.
96. Kang, K., Azargohar, R., Dalai, A. K., and Wang, H. (2016) Hydrogen production from lignin, cellulose and waste biomass via supercritical water gasification: Catalyst activity and process optimization study. *Energy Conversion and Management*, **117**, 528-537.
97. Nanda, S., Reddy, S. N., Dalai, A. K., and Kozinski, J. A. (2016) Subcritical and supercritical water gasification of lignocellulosic biomass impregnated with nickel nanocatalyst for hydrogen production. *International Journal of Hydrogen Energy*, **41**(9), 4907-4921.
98. Norouzi, O., Safari, F., Jafarian, S., Tavasoli, A., and Karimi, A. (2017) Hydrothermal gasification performance of Enteromorpha intestinalis as an algal biomass for hydrogen-rich gas production using Ru promoted Fe–Ni/γ-Al_2O_3 nanocatalysts. *Energy Conversion and Management*, **141**, 63-71.
99. Shanmuganandam, K., and Ramanan, M. V. (2016) Ni-Ce/SiO_2 nanocomposite: Characterization and catalytic activity in the cracking of tar in biomass gasifiers. *Energy Sources, Part A: Recovery, Utilization, and Environmental Effects*, **38**(16), 2418-2425.
100. Hussain, M. A., Yang, M.-H., Jang, H.-S., Hwang, S.-Y., Um, B.-H., Choi, B. G., and Kim, J. W. (2016) Two-dimensional heterogeneous ruthenium-molybdenum disulfide nanocatalyst for the selective aerobic oxidation of amines. *Industrial & Engineering Chemistry Research*, **55**(25), 7043-7047.
101. Liu, B., Yan, P., Xu, W., Zheng, J., He, Y., Luo, L., Bowden, M. E., Wang, C.-M., and Zhang, J.-G. (2016) Electrochemically formed ultrafine metal oxide nanocatalysts for high-performance lithium-oxygen batteries. *Nano Letters*, 16(8), 4932-4939.

8

Polymers and Copolymers Utilized in Photovoltaic Applications

8.1 Introduction

In early 1990s, organic photovoltaics used to exhibit short life duration and relatively low efficiency as compared to the silicon-based photovoltaics [1]. However, in past few years, organic photovoltaics have exhibited significant enhancement in efficiency and operational life duration (to 20,000 h) [2-4]. Additionally, organic photovoltaics exhibit benefits associated with operations at low-light levels like indoor applications as compared to the inorganic silicon-based photovoltaics. The efficiency of organic photovoltaic materials is affected by many components such as polymer structure, film morphology, electron acceptor choice and the ratio of the accepter and polymer donor [5].

8.2 Factors Influencing Power Conversion Efficiency of Polymer Solar Cell Devices

Factors like uncomplicated synthesis and fabrication, cost effectiveness, low weight and potential to be assembled into sizeable area flexible devices have helped polymer solar cells to exhibit promising performance in recent years [6-30]. Combination of an acceptor and photoactive blend layer of a conjugated polymer donor constitutes the polymer solar cell, sandwiched between a cathode and an anode. Either cathode or anode should have transparency for constructing the polymer solar cell. The research effort is mainly focused on gaining enhanced power conversion efficiency (PCE) of polymer solar cells. Power conversion efficiency of polymer solar cells depends on the magnitude of short-circuit current density (J_{sc}), fill factor (FF) and open-circuit voltage (V_{oc}). The charge carrier mobility of the acceptor and donor materials and absorption band wavelength control J_{sc}, whereas V_{oc} is proportionally related to the difference between the lowest unoccupied molecular orbital (LUMO) of the acceptor and the

Naman Arora and Vikas Mittal
Khalifa University of Science and Technology, Abu Dhabi, UAE

highest occupied molecular orbital (HOMO) of the donor [31-34]. Device fabrication conditions and charge carrier mobility decide the fill factor of the device. Therefore, for increasing PCE of the polymer solar cells, it is significant to gain to high value of J_{sc}, higher hole mobility for high J_{sc} and larger FF, and a low-lying HOMO energy level for obtaining higher open-circuit voltage by designing and synthesizing advanced conjugated polymer donor materials containing narrower band gap (broader absorption).

8.3 Importance of Generating Low Band Gap Polymers

The requirement behind preparing low band gap polymers is driven be the facts: i) band gap of around 1.4 eV is indicated by the Shockley-Quiesser equation which is considered to be an ideal for a single junction solar cell device [35], ii) tandem based polymer solar cells need an active material containing a band gap <1.5 eV in combination with the broad band gap material containing a band gap of 1.9 eV [26,36-39], iii) visible-transparent polymer solar cells is one of the applications which requires an active material that absorbs more near-infrared and ultra-violet light and absorbs minute visible light [40-43]. For fulfilling the requirements to achieve these objectives, various synthetic approaches have shown promising results with respect to narrowing the band gap of organic polymeric materials [44-46]. It should be noted that high PCE of devices cannot be guaranteed solely by a small band gap polymer. Appropriate arrangement of the HOMO and LUMO energy levels is significant for effective charge transfer to the electron acceptor material and for obtaining device's high open circuit voltage. There is also a requirement for enhancing J_{sc} and FF of the device through blending of the high charge carrier mobility and morphology with the acceptor material [15,31,47]. As of now, single junction devices exhibiting power conversion efficiency of around 7% have been developed by utilizing active material containing band gaps in the range 1.8-1.6 eV (mid-band gap polymer) [48-53]. Several attempts have been made for obtaining high performance polymers (band gap <1.5 eV), however, mid band gap polymers like thionothiophene are largely the best available candidates to date for photovoltaic applications.

8.4 Various Polymers and Copolymers for Photovoltaic Devices

As mentioned earlier, polymer solar cell devices have exhibited fast development in recent years with enhanced PCE [27]. The V_{oc} of the device and the harvesting of solar light are mainly determined by the active layer of polymer solar cell. Various donor polymers have been synthesized due to their potential properties of optical absorption and low band gaps [54]. For instance, Figure 8.1 shows the synthesis of such functional polymers by the combination of indacenodithiophene (IDT) and two quinoxaline derivatives to form new PIDT-diphQ and PIDT-phanQ [54]. Usually, donor-acceptor (D-A) copolymers are also introduced with the units of acceptor which affects

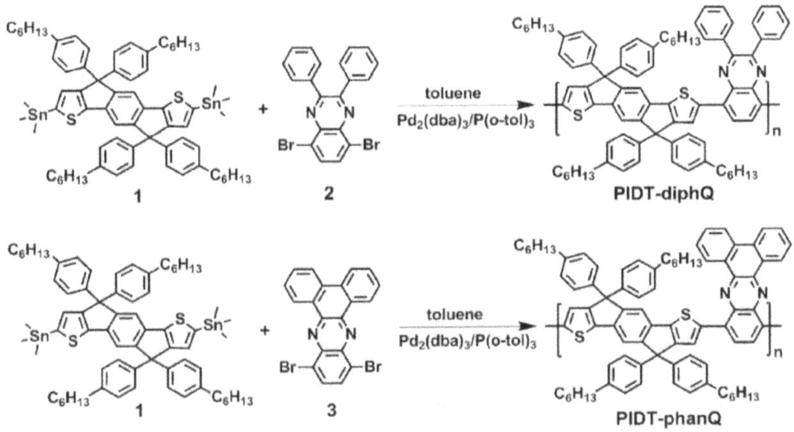

Figure 8.1 Synthesis of PIDT-diphQ and PIDT-phanQ polymers. Reproduced from Reference 54 with permission from American Chemical Society.

the HOMO level and subsequently influences the V_{oc} of polymer solar cell devices. Hence, new materials are being developed with low-lying HOMO levels. In one such study, Ye *et al.* [55] designed two-dimensional conjugated polymer acceptor (PBDTNDI-T) with naphthalene diimide (NDI) and benzo(1,2-b:4,5-b)dithiophene (BDT). In recent years, units of two dimensional conjugated BDT have been utilized as basic blocks in efficient donor polymers. Herein, the authors successfully developed two dimensional-conjugated BDT-based polymer acceptor (PBDTNDI-T), as shown in Figure 8.2. It was observed that nearly 3% of power conversion efficiency was exhibited by the polymer solar cell devices based on PBDTNDI-T. Hence, it could be concluded that affixation of side chains (alkylthiothiophene) to the π-conjugated polymer resulted in the successful conversion of polymer donor to polymer acceptor. Overall, this type of two-dimensional conjugated polymer, being a potential semiconductor, exhibits potential of significant performance for lithium batteries and transistors, along with solar cells.

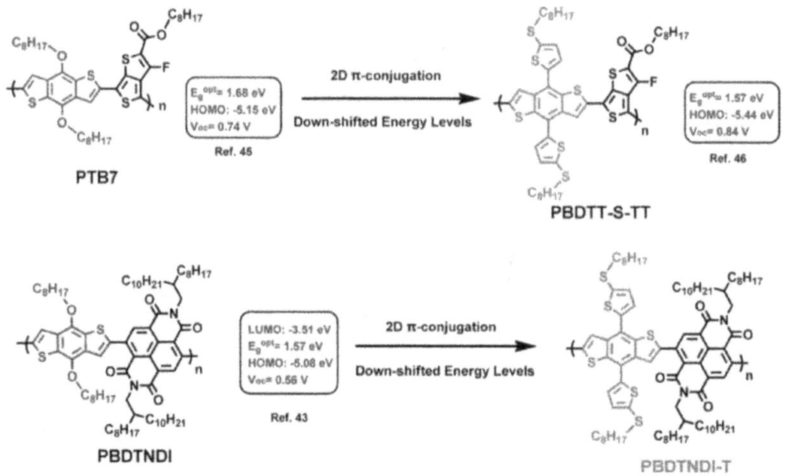

Figure 8.2 Molecular structures *of* the π-conjugated polymer donors (PTB7, PBDTT-S-TT, PBDTNDI) and the target polymer acceptor (PBDTNDI-T). Reproduced from Reference 55 with permission from American Chemical Society.

Zhang *et al.* [56] reported the preparation of a conjugated polymer with large band gap exhibiting significant performance for photovoltaic applications. The authors prepared PM6 copolymer based on 4,8-bis(5-(2-ethylhexyl)-4-fluorothiophen-2-yl)benzo[1,2-b:4,5-b′]-dithiophene (BDT-F) and 1,3-bis(thiophen-2-yl)-5,7-bis(2-ethylhexyl) benzo-[1,2-c:4,5-c′]dithiophene-4,8-dione (BDD). In this copolymer, BDT-F and BDD contributed as donor and acceptor materials, respectively. The polymer exhibited a large band gap of 1.80 eV and contained a deep level of HOMO energy, i.e. -5.45 eV. High power conversion efficiency (9.2%) and open-circuit voltage (0.98 V) with a film thickness around 75 nm were exhibited by PM6 based inverted PSC. Additionally, high average transmission (67%) in the visible range and PCE of 5.6% were exhibited by PM6 constituted semitransparent device. For the development of semitransparent single-junction polymer solar cell devices, PM6 was confirmed to be a suitable material. In another study, Liu *et al.* [57] reported the synthesis of conjugated crossed-benzodithiophene (electron-rich moiety) to generate donor-acceptor (D-A) copolymers. PCE of 3.74% and V_{oc} of 0.9 V were exhibited by the crossed-benzodithiophene and 4,7-di(thiophen-2-yl)-2,1,3-benzothiadiazole (DTBT) based polymer. The authors modified the BDT units (acceptor) with fluorinated and non-fluorinated moieties. The crossed-BDT and 4,7-di(thiophen-2-yl)-5,6-difluoro-2,1,3-benzothiadiazole (DTffBT) based polymers exhibited V_{oc} of around 1 V, which was gained due to low-lying HOMO energy level induced by fluorination. The authors also observed the blue-shift absorption and decrement in HOMO level in both fluorinated and non-fluorinated polymers.

Tandem structure is a potential route for enhancing the efficiency of polymer solar cells. Till few years back, it was not possible to develop tandem solar cells with high performance due to limited low-band gap polymers with high performance. You *et al.* [27] demonstrated the preparation of poly[2,7-(5,5-bis-(3,7-dimethyloctyl)-5H-dithieno[3,2-b:20,30-d]pyran)-alt-4,7-(5,6-difluoro-2,1,3-benzothia, which is a low band gap polymer with high performance exhibiting band gap of 1.38 eV. The polymer also exhibited deep HOMO level and high mobility due to which power conversion efficiency of 7.9% was exhibited by single-junction devices. Tandem solar cells with PCE of 10.6% were achieved by utilizing this polymer. Dou *et al.* [58] studied various low band gap polymers with alternating units of diketopyrrolopyrrole (DPP) and thienylbenzodithiophene (BDTT). The authors copolymerized BDTT unit with DPP unit containing furan (FDPP) (Figure 8.3), and PCE of 5% was achieved in a single junction solar cell by employing resultant polymer (PBDTT-FDPP) [59]. By converting furan into thiopene moiety, enhanced J_{sc} and FF were exhibited by PBDTT-DPP which further led to the enhanced PCE of 6.5% [26,59]. PCE of 8.6% was demonstrated by tandem polymer solar cells by applying PBDTT-DPP. The polymer was also utilized in visible-transparent polymer solar cells which led to a gain of 4% in power conversion efficiency with transparency of 60% in the visible region [26,43]. However, the power conversion efficiencies were limited due to the diminished range of absorption (≤ 850 nm) and poor external quantum efficiency (less than 50 %) in the near infrared region (NIR). Enhanced power conversion efficiencies could be achieved by both type of devices by reducing the band gap for harvesting large number of photons in the region of NIR and by enhancing the mobility of charge carrier of low band gap polymer. Subsequently, it was observed that replacement of atom of sulphur in the thiophene part of the DPP unit with an atom of selenium for making the selenophene-based DPP (SeDPP) unit could lead to decrement in band gap and increased the characteristics of charge transport in photovoltaic devices [60-63]. The authors reported the preparation of selenium-replaced polymer PBDTT-SeDPP, which exhibited efficiency of 7.2% in a polymer solar cell.

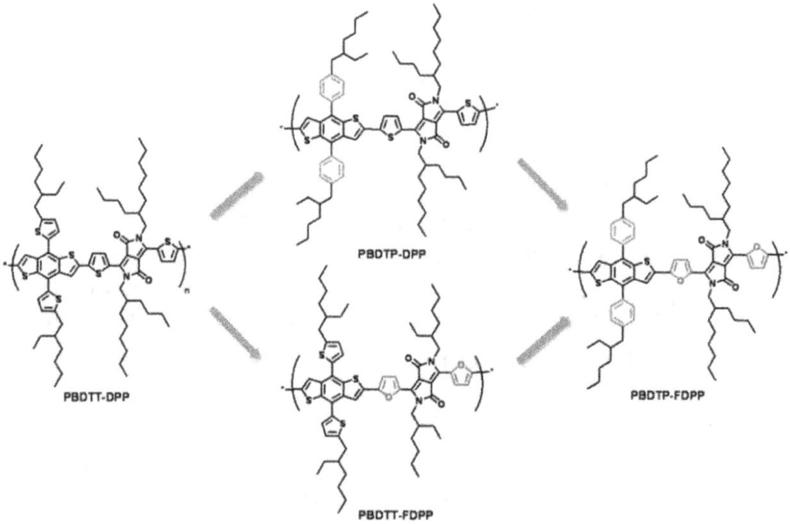

Figure 8.3 Chemical structures of PBDTT-DPP, PBDTP-DPP, PBDTT-FDPP, and PBDTP-FDPP. Reproduced from Reference 59 with permission from American Chemical Society.

Zhang *et al.* [64] designed integrated silole having silicon-bridged heptacyclic arene (ARSi) and naphthobisoxadiazole moieties for enhanced efficiency of polymer solar cells, thereby exhibiting PCE of 8.37%. In another study, Liu *et al.* [65] reported the preparation of conjugated polymers constituted of BDT and DTQx (2,3-diphenyl-5,8-di(thiophen-2-yl)quinoxaline) based units (Figure 8.4). The photovoltaic properties of polymer solar cells were characterized by using the developed polymers. It was observed that circuit current density and fill factor got affected to a small extent by fluorination process, but it was significanlty beneficial for enhancing device V_{oc}. Due to the synergistic effect of fluorination, one of the developed polymers achieved a high open-circuit voltage of 0.90 V. The results revealed that power conversion efficiencies of the polymer solar cells using the developed polymers were in the range of 5.63-8.55%, whereas the V_{oc} values ranged between 0.64-0.90 V.

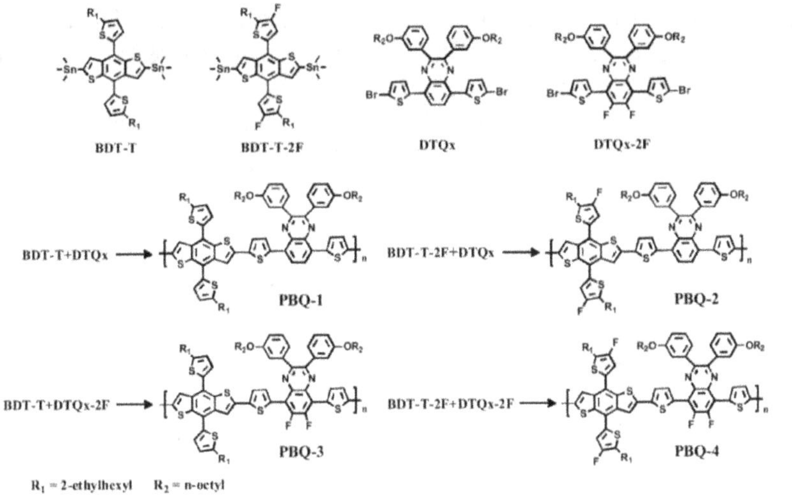

Figure 8.4 Molecular structures and synthesis strategies of the developed polymers. Reproduced from Reference 65 with permission from American Chemical Society.

Next-generation photovoltaic technology, where the polymer solar cells contain an active layer consisting of semiconducting polymer and derivative of fullerene acting as electron donor and electron acceptor, respectively, has also gained significant attention [6,66]. Duan *et al.* [67] utilized BDT and two 5,6-difluorobenzo(2,1,3)thiadiazole-based monomers for synthesizing state of the art semiconductor copolymers to develop solar cells (Figure 8.5). It was concluded that around 8% of power conversion efficiency was exhibited by random copolymers, which was higher than their binary parent polymers. Also, even after deposition of active layer from non-halogenated solvents, power conversion efficiency of 8% was achieved. In this study, the developed copolymers, thus, represented functional photovoltaic polymers for the generation of polymer solar cells. In another study, Nakano *et al.* [68] reported the synthesis of π-conjugated naphtho-[2,3-b:6,7-b]dithiophene-4,5,9,10-diimide (NDTI) based copolymers with units of different thiophene molecules. The copolymers were applied in field effect transistor and organic solar cells. The authors observed that NDTI could be promising for ambipolar organic semiconductors. Intramolecular donor-acceptor interaction could be induced by the integration of strong electron affinity of naphtho-[2,3-b:6,7-b]dithiophene-4,5,9,10-diimide with the electron rich thiophenes, which resulted in narrow energy gaps. Due to this, photo-response could be generated in the near infrared region in organic solar cells. Combination of low-lying LUMO and high-lying HOMO levels could lead to effective carrier injection of electrons and holes, enabling ambipolar operation possible in field effect transistor devices (Figure 8.6).

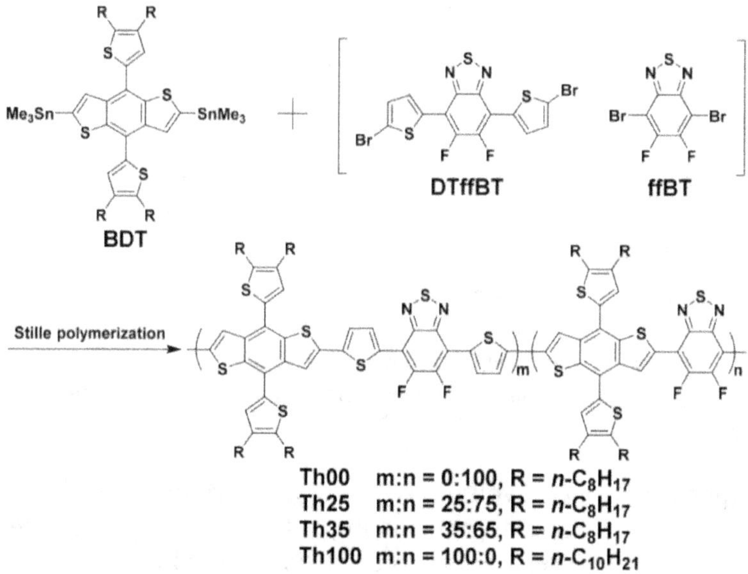

Figure 8.5 Synthesis schematic for the generation of polymers. Reproduced from Reference 67 with permission from American Chemical Society.

Liu *et al.* [69] reported the preparation and characterization of different wide band gap copolymers (PBT1-MP, PBT1-EH, PBT1-BO and PBT1-EH) with different alkyl side chains on benzodithiophene and benzodithiophene-4,8-dione as electron rich and electron deficient units, respectively. It was observed that photovoltaic, electrical and optical properties were affected due to the alkyl side chains. Additionally, these alkyl side chains had influence on molecular packing as well as crystallinity of polymers. The authors observed that long alkyl substituents on benzodithiophene-4,8-dione unit could lead to steric hindrance and twisted molecular geometry. As a result, high PCE of 10% was exhibited by PBT1-EH with bulky side chain on benzodithiophene-4,8-dione unit in fullerene as well as non-fullerene acceptor based solar cells. In another study, Zhou *et al.* [70] also reported the polymer solar cells based on naphthodithiophene diimide

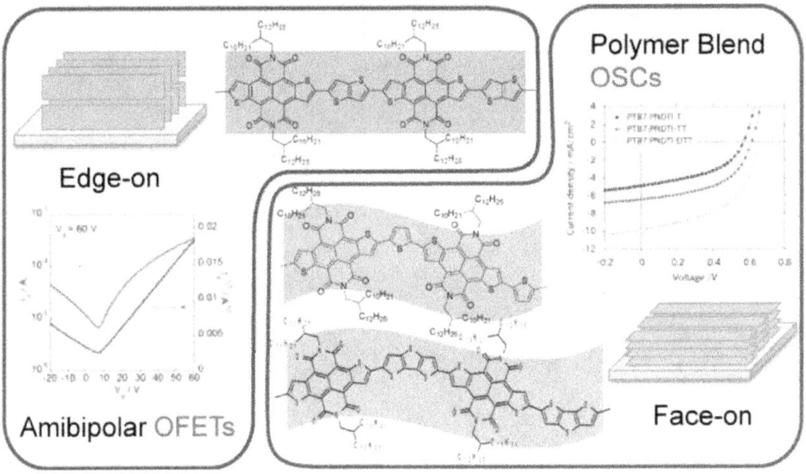

Figure 8.6 Ambipolar semiconductors in field-effect transistors and electron acceptors in polymer blend solar cells. Reproduced from Reference 68 with permission from American Chemical Society.

and bithiophene (PNDTI-BT-DT) copolymers integrated with poly[[4,8-bis[(2-ethylhexyl)oxy]benzo[1,2-b:4,5-b]-bithiophene-2,6-diyl][3-fluoro-2-[(2-ethylhexyl)carbonyl]-thieno[3,4-b]thiophenediyl]] (PTB7) as an electron donor polymer which has been utilized for many polymer/fullerene polymer solar cells [49] and other polymer solar cells [71]. The authors reported that higher photo-response was exhibited by all polymer solar cells in the near infrared region, as compared to visible region. As shown in Figure 8.7, strong absorption band was exhibited by PTB7 films in the 500-750 nm region and PNDTI-BT-DT films exhibited absorption band in the region of 600-900 nm. More sunlight could be utilized by this absorption behavior which was comparatively more beneficial than polymer-fullerene system. Under the irradiation of solar light, maximum PCE was recorded as 2.59%. Crystalline structure was formed by PNDTI-BT-DT and PTB7 blend films and was almost similar to the structure of the pristine films.

Marchiori *et al.* [72] reported the application of poly[2,7-(9,9-bis(2-ethylhexyl)-dibenzosilole)-alt-4,7-bis(thiophen-2-yl)benzo-2,1,3-thiadiazole] copolymer (PSiF-DBT) for photovoltaic devices which exhibited PCE of 4%. The PSiF-DBT copolymer was utilized as active layer in bilayer solar cell with C60, exposed to thermal annealing at 100 °C. However, it was observed that even without annealing exposure, PCE of 2% was exhibited by PSiF-DBT-based devices. This

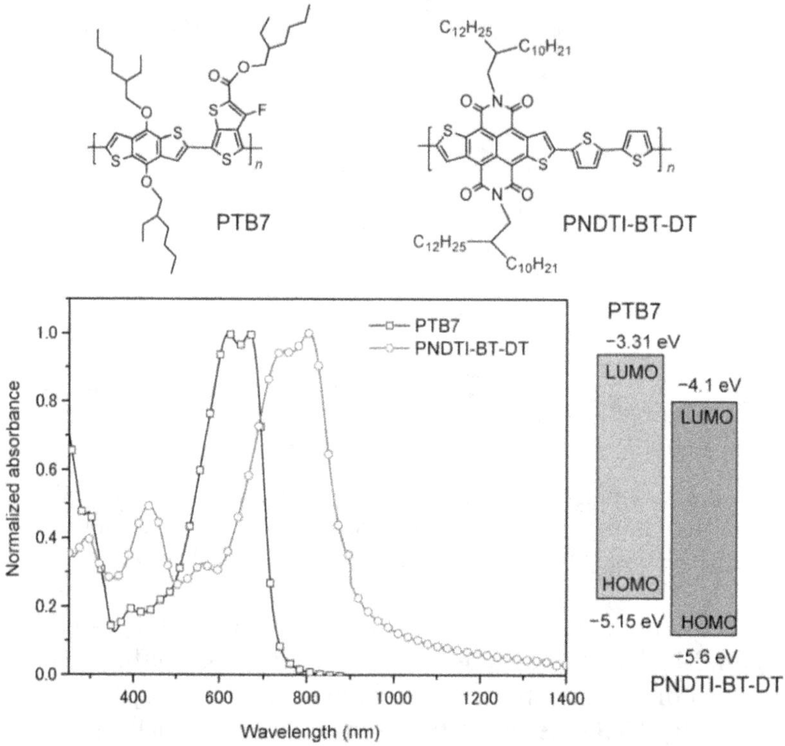

Figure 8.7 Chemical structures of PTB7 and PNDTI-BT-DT, along with absorption spectra and energy level diagrams. Reproduced from Reference 70 with permission from American Chemical Society.

was due to the reason that the cast film of PSiF-DBT contained well stacked morphology after deposition which reduced the defects concentration, thus, behaving as charge carrier traps. Consequently, there was not much improvement achieved by thermal annealing in the transportation of holes of films of PSiF-DBT. The power conversion efficiency was basically induced by the increment in roughness of the polymer surface which contributed for more efficacious interface for dissociating exciton at the heterojunction of PSiF-DBT-fullerene. Kim *et al.* [73] also prepared different π-conjugated alternating copolymers by employing cross-coupling polymerization between units of benzotriazole (electron acceptor) and benzodithiophene (electron donator). For improving the resultant polymer solubility, linear n-octyl or branched 1-octylnonyl groups were subjected to the benzotriazole part for generating 2-octyl-2H-benzo[1,2,3]triazole (LBTz) or 2-(heptadecan-9-yl)-2H-benzo[1,2,3]triazole (YBTz). High solubility was exhibited by the copolymers in organic solvents due to the presence of benzotriazole alkyl side chains. Narrow band gaps were also exhibited by the developed copolymers. It was also observed that the optical band gaps of these polymers were within the range of 1.90-2.03 eV depending on the structure of the polymer. Lower HOMO energy levels were exhibited by the as-prepared polymers as compared to poly(3-hexylthiophene) due to the presence of strong electron-accepting units of benzotriazole. It was suggested that the as-prepared polymers exhibited higher open-circuit voltages and better oxidative stability for the application in photovoltaic devices.

In another study, Saeki *et al.* [74] reported benzobisthiazole and benzothiadiazole based alternating copolymer. The copolymer contained solubilizing linear and branched alkyl chains. The authors explored the inverted cell structure where zinc oxide and MoOx/Ag were utilized as cathode buffer layer and anode respectively for improving PCE of photovoltaic devices. Enhancement of power conversion efficiency was observed from 3.8% to 6.5% after the introduction of inverted cell structure (ZnO/polymer:fullerene/MoOx/Ag). Furthermore, usage of optimized film thickness and $PC_{71}BM$ exhibited better interaction. Guo *et al.* [75] applied microphase-separated conjugated block copolymers (poly(3-hexylthiophene)–block–poly-((9,9-dioctylfluorene)-2,7-diyl-alt-[4,7-bis(thiophen-5-yl)-2,1,3-benzothiadiazole]-2′,2″-diyl) (P3HT-b-PFTBT)) for controlling donor-acceptor heterojunctions (Figure 8.8). As compared to blends of homopolymers, the use of block copolymers as active layer of cells led to higher photo-conversion. Furthermore, it was revealed by X-ray scattering that block copolymer based solar cells exhibited better performance due to the existence of self-assembly into meso-scale lamellar morphologies. Therefore, it was concluded that performance could be enhanced in excitonic solar cells through control of donor–acceptor interfaces by applying conjugated block copolymers.

For the purpose of harvesting solar energy, cost effective and flexible photovoltaic devices are required which can be prepared by employing conjugated polymers [43,76-78]. The generation of electron-donor polymers was taken as an inspiration for preparing polymer-fullerene polymer solar cells which gained significant enhancement in respect to their power conversion efficiencies recently [24,79-83]. Donor-acceptor (D-A) copolymers could provide properties of charge transportation, absorption bands with fine-tuning, HOMO/LUMO levels through the tuning of donor-acceptor interactions and π-electron

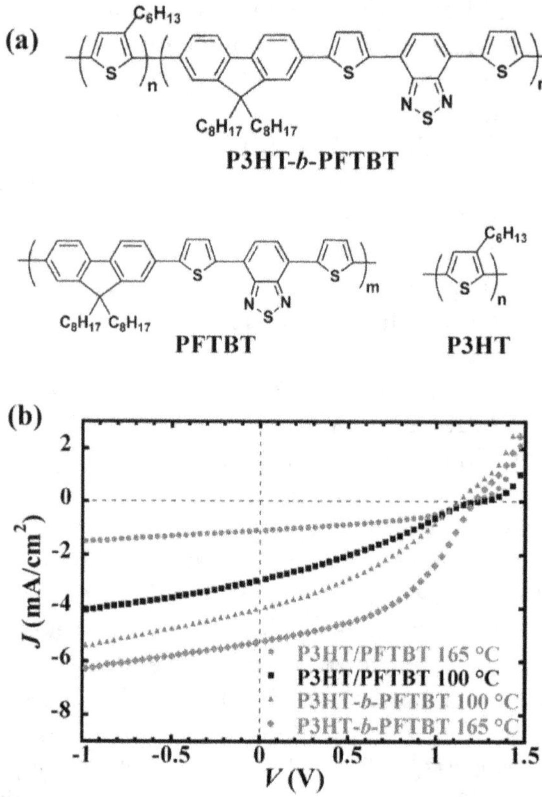

Figure 8.8 Performances of block copolymer P3HT-b-PFTBT and polymer blend P3HT/PFTBT solar cells: (a) chemical structures of P3HT-b- PFTBT, PFTBT and P3HT; (b) current–voltage characteristics of P3HT/PFTBT and P3HT-b-PFTBT photovoltaic devices. Reproduced from Reference 75 with permission from American Chemical Society.

delocalization. Researchers have also studied various electron-withdrawing units such as quinoxaline, thieno-[3,4-c]pyrrole-4,6-dione (TPD), 2,1,3-benzothiadiazole (BT) and isoindigo [28,84-90]. Tao *et al.* [91] prepared and characterized a variety of D–A1–D–A2 copolymers for generating effective polymer based solar cells. The copolymers (Figure 8.9) contained donor segments and were synthesized depending on quinoxaline-fluorinated quinoxaline and isoindigo with different electron-donating segment lengths. For the purpose of gaining increased open circuit voltage and low-lying HOMO level, fluorine atoms were subjected onto the weaker quinoxaline acceptor. It was observed that the introduction of fluorine in the qunioxaline unit

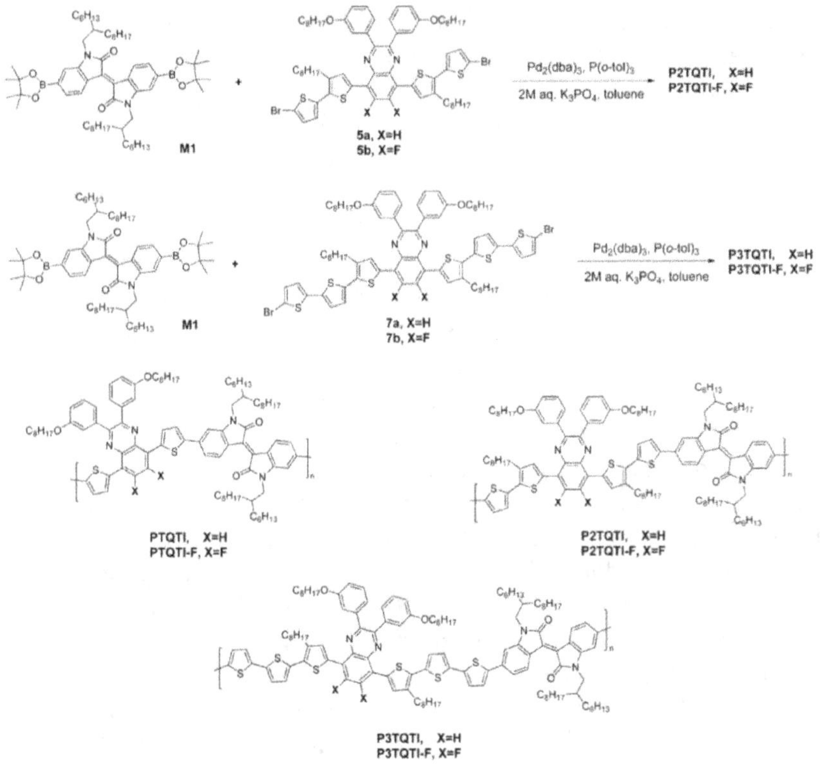

Figure 8.9 Synthetic routes of P2TQTI(F) and P3TQTI(F). Reproduced from Reference 91 with permission from American Chemical Society.

resulted in enhanced absorption coefficients, leading to high short-circuit current density and reduced HOMO levels, thus, exhibiting higher open-circuit voltage. It was also noticed that lower open-circuit voltage was exhibited by polymer solar cells which were based on P2TQTI(F) and P3TQTI(F) polymers containing extended donor segments. In contrast, polymer solar cells based on PTQTI(F) polymer containing short donor segments exhibited higher open-circuit voltage. PCE of 7% and J_{sc} of 15.5 mAcm^{-2} were exhibited by P3TQTI-F polymer containing high absorption efficient and an optimized nanostructure. In another study, Liu *et al.* [92] prepared a variety of benzodithiophene-thienothiophene copolymers containing varying fluorine content. It was observed that deep HOMO energy levels could be achieved with incorporation of fluorine in the units of thienothiophene with no absolute influence on polymer light harvesting properties and band gap. Consequently, an enhanced open-circuit voltage was gained in photovoltaic devices and high short-circuit currents were maintained at the same time. The charge transport properties and morphology significantly influenced the fill factors of photovoltaic devices in the bulk-heterojunction cell films.

Kim *et al.* [93] reported the synthesis and characterization of conjugated copolymers with low band gap, containing alternating difluorinated benzoselenadiazole (DFDTBSe) and ethylhexyloxy (EH)- or octyldodecyloxy (OD) substituted benzo[1,2-b:4,5-b]-dithiophene (BDT). The two conjugated copolymers were prepared by applying Stille cross-coupling polymerization and named as PEHBDT-DFDTBSe and PODBDT-DFDTBSe, respectively. Almost similar band gap energies of 1.66 eV and 1.69 eV were exhibited by PEHBDT-DFDTBSe and PODBDT-DFDTBSe, respectively. It was observed that molecular ordering in the $PC_{71}BM$ blended active layer films could be influenced by alkyloxy side chains present in the polymers. The 'face-on' crystal structure having a short intermolecular π–π stacking distance and an 'edge-on' structure were exhibited by PEHBDT-DFDTBSe:$PC_{71}BM$ and PODBDT-DFDTBSe:$PC_{71}BM$, respectively. In case of inverted bulk heterojunction organic photovoltaic cells, power conversion efficiency of 5.74% was achieved by fabricating PEHBDT-DFDTBSe:$PC_{71}BM$, while in tandem organic photovoltaic cells, PCE of 7.15% was achieved by fabricating PEHBDT-DFDTBSe:$PC_{71}BM$ and P3HT:IC60BA (poly(3-hexylthiophene):indene-C60-bisadduct) as the top cell and bottom cell, respectively. Figure 8.10 also shows the schematic morphological models of these materials.

Chen *et al.* [94] also reported the preparation of donor-π-acceptor copolymer which consisted of alkylthienylbenzodithiophene (BDTT) (electron-donating moiety) and fluorinated thienyl-quinoxaline (TTFQ) (electron-accepting segment) containing units of thiophene π-bridge. The authors applied Stille coupling polymerization technique for generating the copolymers and subsequently utilized them as p-type semiconducting polymer. A wide visible-near-infrared absorption band with optical band gap of 1.67 eV was exhibited by PBDTT-TTFQ copolymer and possessed a HOMO level at 5.34 eV. Additionally, high dense fibrillar nanostructure was exhibited by the films of PBDTT-TTFQ which suggested self-assembly of the segments of PBDTT and TTFQ at nanoscale. The V_{oc}, J_{sc} and FF of the bulk-heterojunction polymer solar cell were recorded as 0.75 V, 14.6 $mA.cm^2$ and 56.1%, respectively. The polymer solar cell comprised of blend of PBDTT-TTFQ : PC71BM (1 : 1) and exhibited the power conversion

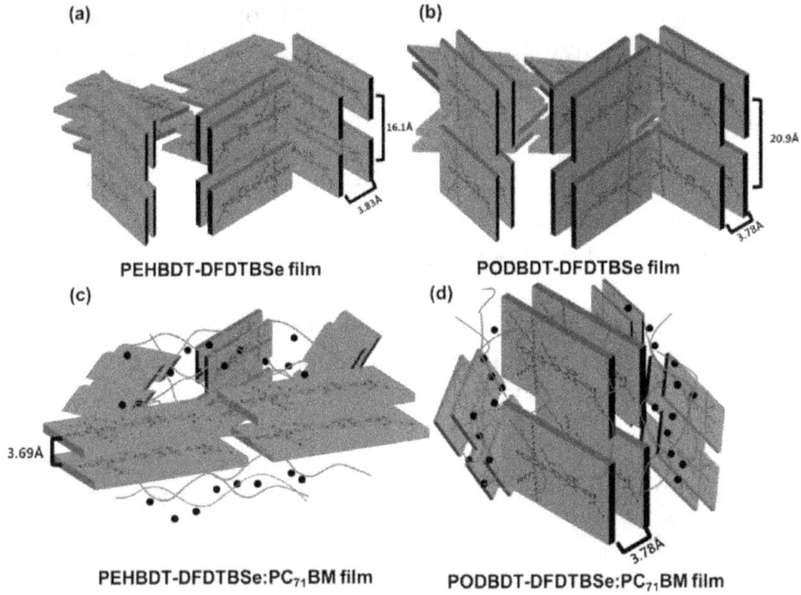

Figure 8.10 Schematic morphological models of (a, b) pure PEHBDT-DFDTBSe and PODBDT-DFDTBSe and (c, d) PEHBDT-DFDTBSe and PODBDT-DFDTBSe:PC71BM blend films. Reproduced from Reference 93 with permission from American Chemical Society.

efficiency of 6.1%. The authors concluded that the copolymers could be utilized for next-generation solar cells. Also, TTFQ unit was observed to be an ideal building block for developing polymer solar cells with enhanced performance. In another study, Dang *et al.* [80] reported the preparation of different copolymers containing a donor unit of indacenodithiophene (IDT), which is considered to be one of the most potential electron rich units because of its high degree of planarity which provides increased delocalization of an electron and therefore, a high charge-carrier mobility [54,95,96]. The copolymers were designed with either meta-hexyl-phenyl or para-hexylphenyl side chains and thiophene-quinoxalinethiophene acceptor unit. The authors utilized different type of acceptors, i.e. fluorinated unit (TQF) and non-fluorinated unit (TQ) where fluorinated copolymer exhibited little lower HOMO level as compared to non-fluorinated copolymers because of fluorine atom, which has electron-withdrawing nature [97,98]. Enhanced photovoltaic performance, higher molecular weight and improved solubility were exhibited by meta-substituted copolymers in both non-fluorinated and fluorinated copolymers. In addition, enhanced V_{oc} of 0.96 V and PCE of around 6.6% were achieved by bulk-heterojunction polymer solar cell based on the blends of fluorinated IDT copolymer (PIDTTQF-m) and $PC_{71}BM$, thereby making the material compatible for the generation of tandem devices. In case of devices based on non-fluorinated PIDTTQ-m: $PC_{71}BM$, enhanced photocurrent and slightly lower photo-voltage were achieved along with high power conversion efficiency (7.8%), which was one of the highest solar cell efficiencies in terms of IDT-based copolymers.

Ye *et al.* [99] reported the synthesis of polymer solar cell based on PBDT-TS1:PC71BM, exhibiting PCE of 9.48%. The authors designed conjugated (BDT) and thieno[3,4-b]thiophene (TT)-based copolymer (PBDTTS1, Figure 8.11a) through the replacement of 2-ethylhexyl with octyl, which was similar to the other reported polymers structures, i.e. PBT7-Th [100] and PBDTT-S-TT [82]. Linear alkylthio

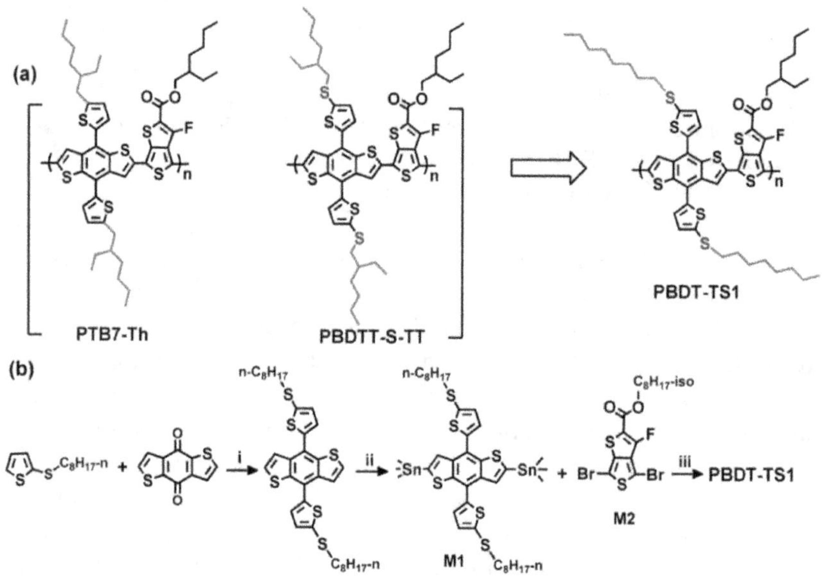

Figure 8.11 (a) Molecular structure of two reference polymers and target polymer; (b) synthesis procedure of PBDT-TS1. Reproduced from Reference 99 with permission from American Chemical Society.

chains were introduced for synthesizing PBDT-TS1 and its preparation route is demonstrated in Figure 8.11b. It is well known that alkyl chains are utilized as basic units to design organic semiconductors for obtaining well balanced intermolecular packing, high crystallinity and enhanced performance of device [101-105]. Moreover, since high starting point could be gained in conventional device, power conversion efficiency of 10% or more could be achieved in single-junction polymer solar cells through the integration of effective structures of devices [58,92,100,106,107]. For instance, Figure 8.12 shows the field-effect hole mobility through control of nanoscale polymer aggregation [102].

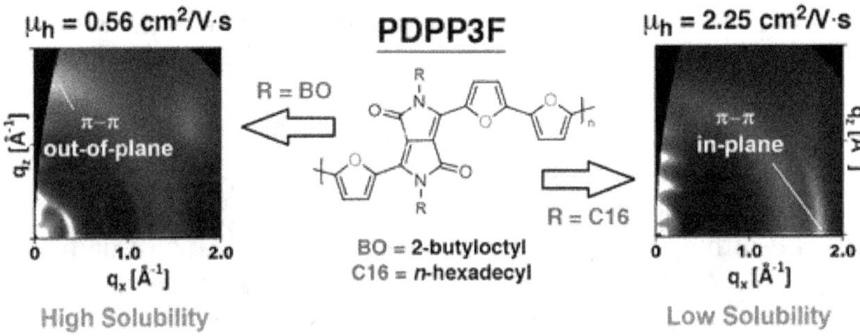

Figure 8.12 Field-effect hole mobility through control of nanoscale polymer aggregation. Reproduced from Reference 102 with permission from American Chemical Society.

Cui *et al.* [82] also reported the preparation of a low band gap two-dimensional conjugated polymer (PBDTT-S-TT). The authors subjected an alkylthio substituent onto the thiophene side chain in the polymer. The polymer's hole mobility was observed to be increased to 4.08 x $10^{-3}cm^2V^{-1}s^{-1}$ due to the alkylthio substituent and HOMO energy level was shifted downwards by 0.11 eV. It was noticed that improved V_{oc} of 0.84 eV was exhibited by polymer solar cell device based on PBDTT-S-TT which was recorded as highest value among all BDT and TT units based copolymers. Moreover, enhanced power conversion efficiency of around 8.42% was exhibited by PBDTT-S-TT based device. It could be concluded that molecular modification achieved through the introduction of an alkylthio side chain was a beneficial approach for broadening the absorption, down shifting the HOMO energy level and increasing the hole mobility of the two-dimensional conjugated polymer containing low band gap in order to enhance the photovoltaic performance of the polymer solar cells. Zhao *et al.* [108] reported the preparation of the semiconducting conjugated polymer, (BTT-BTz) poly{benzo(1,2-b:3,4-b':5,6-d") trithiophene-alt-4,4'-dihexyl-2,2'-bithiazole} containing a broad band gap and higher open circuit voltage based on benzotrithiophene and bithiazole units. The power conversion efficiency and open circuit voltage exhibited by BTT-BTz copolymer/$PC_{71}BM$ photoactive layer based device were recorded as 2.49% and 0.83 V, respectively. After treatment with 1,8-diiodoctane, appropriate morphology of the blend film was achieved and the PCE value reached 5.06% because of an increment in the values of J_{sc} and FF. So far, the observed efficiency of device has been the highest among all reported BTT or BTz derivatives based conjugated copolymers. It could be concluded that BTT-BTz based semiconducting copolymer can act as one of the potential active layers in the tandem cell structure for capturing various parts of the solar spectrum.

In another study, Keshtov *et al.* [109] applied Stille reaction conditions for the preparation of two donor-acceptor functional copolymers 4,8-bis-(5-bromothiophene-2-yl)-benzo [1,2,5]thiadiazole (P1) and 4,8-dithiophene-2-yl-benzo[1,2-c;4,5-c']-bis-[1,2,5]thiadiazole (P2) (acceptor segment) with BDT (donor segment). Broad absorption bands were exhibited by the copolymers. With regards to optical band gaps, P1 and P2 exhibited the values of 1.62 eV and 1.16 eV, respectively. For the fabrication of polymer solar cells, P1 and P2 along with fullerene derivatives (PC60BM and PC70BM) were taken as electron donors and acceptor, respectively. The power conversion efficiency of around 3.45%, 1.19%, 2.17% and 0.80% were exhibited by the bulk heterojunction solar cells based on P1:$PC_{70}BM$, P2:$PC_{70}BM$, P1:$PC_{60}BM$ and P2:$PC_{60}BM$ blends (casted from THF solvent), respectively. P1 as electron donor exhibited enhanced power conversion efficiency due to its high V_{oc} which was attributed to the deep lying HOMO level, higher short circuit current density attributed to more force for exciton dissociation at the donor–acceptor joint into the layer of the bulk heterojunction solar cell and its relatively high mobility. Subsequently, the authors utilized DIO/THF (1,8-diiodooctane/tetrahydrofuran) solvent for casting blends and improved power conversion efficiencies of P1:$PC_{70}BM$ and P2:$PC_{70}BM$ blends were observed as 5.30% and 1.58%, respectively. The observed increment in power conversion efficiency was due to the larger donor-acceptor interface in the active layer for exciton dissociation, enhanced crystallinity of copolymer phase and balanced transportation of charge in the device.

Gedefaw *et al.* [110] prepared p-type copolymers P(1)-PFQBDT-OR, P(2)-PFQBDT-TROR, P(3)-PFQBDT-TR and P(4)-PFQTT, termed as P(1), P(2), P(3), and P(4), respectively, which involved fluorinated quinoxaline (FQ) acceptor unit with substituted benzodithiophene (BDT) or an unsubstituted thieno[3,2-b]thiophene (TT). The copolymers (P(1), P(2), P(3), and P(4)) in combination with PC61BM were utilized for the fabrication of blade-coated bulk heterojunction solar cells. The power conversion efficiencies of the devices based on P(1), P(2), P(3) and P(4) were recorded as 4.1%, 3.8%, 5.2% and 3.8%, respectively. Ho *et al.* [111] also reported the preparation of isoindigo-based low band gap copolymers (PnTI) with an expanded thiophene unit in the donor segment of the polymer. Isoindigo, being an acceptor unit, has gained much attention because of its sustainability (as it is extracted from plants) as well as its strong electron-withdrawing capability [112]. Absorption range of polymer was expanded by the increment in the thiophene unit length, while the HOMO and LUMO levels remained within an appropriate range for polymer solar cells. After the introduction of bithiophene into the donor segment with centrosymmetric conformation, the polymers exhibited improved crystallinity and consequently, enhanced hole mobility. Eventually, polymer P6TI exhibited the power conversion efficiency of around 7.25% as well as high J_{sc} of 16.24 mAcm^{-2}. So far, this PCE has been observed to be highest among all other isoindigo-based low band gap copolymers.

In a recent study, Andersen *et al.* [113] demonstrated the synthesis of slot-die coated photovoltaic device based on semi-random copolymer of poly-3-hexylthiophene and diketopyrrolopyrrole (P3HTT-DPP-10%) mixed with phenyl-C61-butyric acid methyl ester ($PC_{60}BM$). The device was fabricated by utilizing small roll-coater (slot-die coating only) and flexographic printing under an appropriate environment onto an indium tin oxide (ITO)-free substrate. For overcoming the low values of J_{sc} and FF achieved for single junction devices, the devices were generated by following tandem geometry, enabling the viability for employing thinner junction films. Single junction geometry based device exhibited power conversion efficiency of 1.31%, whereas the tandem geometry based device exhibited an efficiency of 1.36%. Earmme *et al.* [71] prepared naphthalene diimide-selenophene (PNDIS-HD) copolymer acceptor and a thiazolothiazole-dithienosilole (PSEHTT) copolymer donor based bulk heterojunction solar cells (Figure 8.13). The solar cells exhibited power conversion efficiency of around 3.3%. External quantum efficiency and J_{sc} were recorded as 47% and 7.78 mA cm^{-2} respectively. These parameters could be compared to the performance of similarly examined [6,6]-Phenyl-C60-butyric acid methyl ester

(PC60BM)/PSEHTT devices. Appropriate electron and hole transport in the polymer-polymer blend solar cells was attributed to the PNDIS-HD morphology, i.e. lamellar crystalline which confirmed its potential photovoltaic properties.

Yuan *et al.* [114] reported the preparation and characterization of narrow band gap conjugated polymers based on a high mobility template for generating high performance visibly transparent photovoltaic devices. Nguyen *et al.* [81] reported the synthesis of a series of dialkoxyphenylene-BT which contained semi-crystalline low band gap polymers. The polymers were prepared by non-covalent conformational locking for enhancing intermolecular ordering, chain planarity and thermal stability without disturbing solution processability. Uniformly-distributed interpenetrating nano-fibrillar network

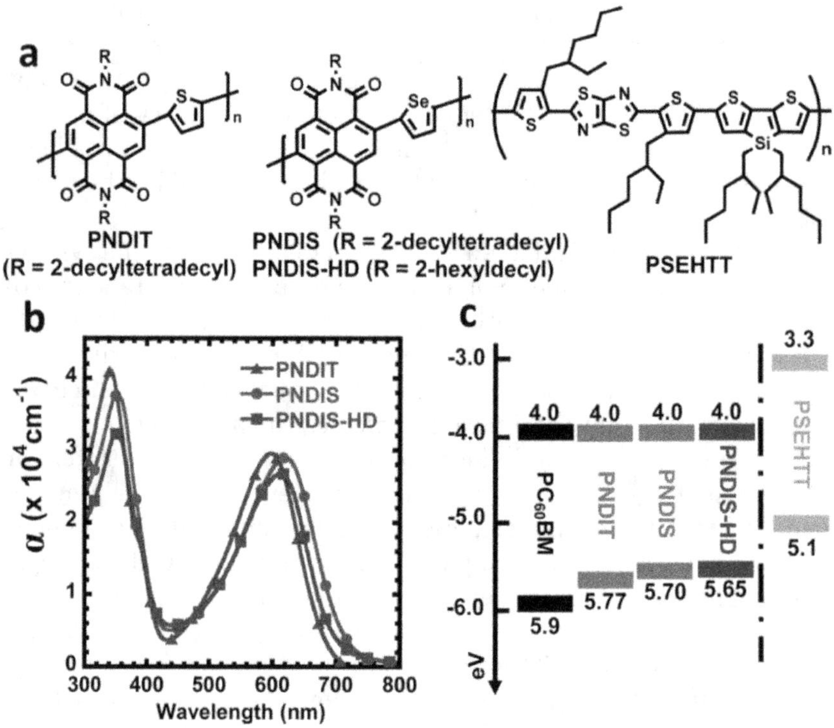

Figure 8.13 (a) Molecular structures of acceptor (PNDIT, PNDIS and PNDIS-HD) and donor (PSEHTT) polymers, (b) UV-vis absorption spectra of PNDIT, PNDIS and PNDIS-HD, (c) LUMO/HOMO energy levels of PNDIT, PNDIS, PNDIS-HD, PC60BM and PSEHTT. Reproduced from Reference 71 with permission from American Chemical Society.

morphologies with $PC_{70}BM$ were obtained by the polymers, exhibiting properly-balanced electron and hole mobilities. Power conversion efficiency of around 9.39% was exhibited by polymer solar cells based on these polymers in a structure of 290 nm thick single-cell device without using an extra interfacial layer. In another study, Liao *et al.* [115] employed dual doped zinc oxide nano-film as the cathode interlayer for generating single junction inverted polymer solar cell which exhibited PCE of around 10.31%.

Chen *et al.* [116] reported another single-junction polymer solar cell which exceeded PCE of around 10%. Huang *et al.* [117] also employed Stille coupling reaction for the preparation of two conjugated copolymers (named as PSBDTeFBT and PSBDTeTID) based on asymmetric 4-(2-ethylhexyloxy)-8-(2-ethylhexylthio)benzo(1,2-b:4,5-b')dithiophene (SBDT) unit and different conjugated side chain group DTffBT and TID (5,6-difluoro-4,7-bis(4-(2-ethylhexyl)thiophen-2-yl)benzo[c][1,2,5]thiadiazole and 6-(thiophen-2-yl)-di(2-ethylhexy)isoindigo). The authors characterized the polymers by using cyclic voltammetry, UV-Vis absorption spectroscopy and thermal gravimetric analysis for analyzing electrochemical, optical and thermal properties, respectively. Good solubility in common organic solvents and enhanced thermal stability were exhibited by the copolymers. The values of deep HOMO energy levels for 5,6-difluoro-benzodiathiazole and isoindigo-based copolymers were observed as -5.56 eV and -5.48 eV, respectively. The power conversion efficiency of 5% and high open circuit voltage of 0.92 V were exhibited by the photovoltaic device based on 5,6-difluoro-benzodiathiazole-based polymer/phenyl-C71-butyric acid methyl ester.

Li *et al.* [118] reported the synthesis of hybrid rod-coil diblock copolymers with coordinatively attached nanocrystals of cadmium selenide (CdSe) and studied their optical properties and applicability for photovoltaic devices.

Lu *et al.* [119] reported ternary blend polymer solar cell having poly-3-oxothieno[3,4-d]isothiazole-1,1-dioxide/benzodithiophene (PID2), polythieno[3,4-b]-thiophene/benzodithiophene (PTB7) as donors and [6,6]-phenyl C71 butyric acid methyl ester (PC71BM) as an acceptor. The values of FF, PCE, V_{oc} and J_{sc} of the as prepared ternary solar cell were observed as 68.7%, 8.22%, 0.72 V and 16.8 mA cm^{-2}, respectively.

In another study, Najari *et al.* [120] employed direct heteroarylation polymerization for the synthesis of a series of push-pull conjugated polymers based on thieno[3,4-c]pyrrole-4,6-dione (TPD), furo[3,4-c]pyrrole-4,6-dione (FPD) and selenopheno[3,4-c]-pyrrole-4,6-dione (SePD).

Further, Squeo *et al.* [121] employed Stille cross-coupling technique for the preparation of ultra-low band gap a,b-unsubstituted 4,4-(difluoro-4-bora-3a,4a-diaza-s-indacene)-based conjugated polymer.

8.5 Conclusions

In the past few years, organic photovoltaics have exhibited immense development with respect to improvements in both lifetime and efficiency.

In this chapter, the necessity of low band gap polymers as well as other requirements for the fabrication of polymer solar cells have been discussed. The review presents thorough compilation of recent studies reporting the design, synthesis and property evaluation of various polymers and copolymers for utilization in photovoltaic applications. This chapter works as a benchmark for developing the polymer systems for photovoltaic applications in future.

References

1. Brabec, C. J. (2004) Organic photovoltaics: technology and market. *Solar Energy Materials and Solar Cells*, **83**, 273-292.
2. Krebs, F. C., and Spanggaard, H. (2005) Significant improvement of polymer solar cell stability. *Chemistry of Materials*, **17**, 5235-5237.
3. Reyes-Reyes, M., Kim, K., Dewald, J., Lopez-Sandoval, R., Avadhanula, A., Curran, S., and Carroll, D. L. (2005) Meso-structure formation for enhanced organic photovoltaic cells. *Organic Letters*, **7**, 5749-5752.
4. Ma, W., Yang, C., Gong, X., Lee, K., and Heeger, A. J. (2005) Thermally stable, efficient polymer solar cells with nanoscale control of the interpenetrating network morphology. *Advanced Functional Materials*, **15**, 1617-1622.
5. Hoppe, H., and Sariciftci, N. S. (2006) Morphology of polymer/fullerene bulk heterojunction solar cells. *Journal of Materials Chemistry*, **16**, 45-61.
6. Yu, G., Gao, J., Hummelen, J. C., Wudl, F., and Heeger, A. J. (1995) Polymer photovoltiac cells: Enhanced efficiencies via a network of internal donor-acceptor heterojunctions. *Science*, **270**, 1789-1791.
7. Arias, A. C., MacKenzie, J. D., McCulloch, I., Rivnay, J., and Salleo, A. (2010) Materials and applications for large area electronics: soluti-

on-based approaches. *Chemical Reviews*, **110**, 3-24.
8. Chen, J., and Cao, Y. (2009) Development of novel conjugated donor polymers for high-efficiency bulk-heterojunction photovoltaic devices. *Accounts of Chemical Research*, **42**, 1709-1718.
9. Li, Y. (2012) Molecular design of photovoltaic materials for polymer solar cells: toward suitable electronic energy levels and broad absorption. *Accounts of Chemical Research*, **45**, 723-733.
10. Su, Y.-W., Lan, S.-C., and Wei, K.-H. (2012) Organic photovoltaics. *Materials Today*, **15**, 554-562.
11. Po, R., Carbonera, C., Bernardi, A., and Camaioni, N. (2011) The role of buffer layers in polymer solar cells. *Energy & Environmental Science*, **4**, 285-310.
12. Kim, J.-H., Song, C.E., Kang, I.-N., Shin, W.S., and Hwang, D.-H. (2013) A highly crystalline low band-gap polymer consisting of perylene and diketopyrrolopyrrole for organic photovoltaic cells. *Chemical Communications*, **49**, 3248-3250.
13. Kim, J.-H., Song, C.E., Kim, H. U., Grimsdale, A. C., Moon, S.-J., Shin, W. S., Choi, S. K., and Hwang, D.-H. (2013) High open circuit voltage solution-processed tandem organic photovoltaic cells employing a bottom cell using a new medium band gap semiconducting polymer. *Chemistry of Materials*, **25**, 2722-2732.
14. Wienk, M. M., Kroon, J. M., Verhees, W. J., Knol, J., Hummelen, J. C., van Hal, P. A., and Janssen, R. A. (2003) Efficient methano [70] fullerene/MDMO-PPV bulk heterojunction photovoltaic cells. *Angewandte Chemie, International Edition*, **115**, 3493-3497.
15. Li, G., Zhu, R., and Yang, Y. (2012) Polymer solar cells. *Nature Photonics*, **6**, 153-161.
16. Cheng, Y.-J., Yang, S.-H., and Hsu, C.-S. (2009) Synthesis of conjugated polymers for organic solar cell applications. *Chemical Reviews*, **109**, 5868-5923.
17. Li, Y., and Zou, Y. (2008) Conjugated polymer photovoltaic materials with broad absorption band and high charge carrier mobility. *Advanced Materials*, **20**, 2952-2958.
18. Gendron, D., and Leclerc, M. (2011) New conjugated polymers for plastic solar cells. *Energy & Environmental Science*, **4**, 1225-1237.
19. Zhou, H., Yang, L., and You, W. (2012) Rational design of high performance conjugated polymers for organic solar cells. *Macromolecules*, **45**, 607-632.
20. Mihailetchi, V. D., van Duren, J. K., Blom, P. W., Hummelen, J. C., Janssen, R. A., Kroon, J. M., Rispens, M. T., Verhees, W. J. H., and Wienk, M. M. (2003) Electron transport in a methanofullerene. *Advanced Functional Materials*, **13**, 43-46.
21. Dou, L., You, J., Hong, Z., Xu, Z., Li, G., Street, R. A., and Yang, Y. (2013) A decade of organic/polymeric photovoltaic research. *Advanced Materials*, **25**, 6642-6671.

22. Dhanabalan, A., Van Duren, J., Van Hal, P., Van Dongen, J., and Janssen, R. (2001) Synthesis and characterization of a low bandgap conjugated polymer for bulk heterojunction photovoltaic cells. *Advanced Functional Materials*, **11**, 255-262.
23. Li, W., Furlan, A., Hendriks, K. H., Wienk, M. M., and Janssen, R. A. (2013) Efficient tandem and triple-junction polymer solar cells. *Journal of the American Chemical Society*, **135**, 5529-5532.
24. He, Z., Zhong, C., Su, S., Xu, M., Wu, H., and Cao, Y. (2012) Enhanced power-conversion efficiency in polymer solar cells using an inverted device structure. *Nature Photonics*, **6**, 591-595.
25. Liu, S., Zhang, K., Lu, J., Zhang, J., Yip, H.-L., Huang, F., and Cao, Y. (2013) High-efficiency polymer solar cells via the incorporation of an amino-functionalized conjugated metallopolymer as a cathode interlayer. *Journal of the American Chemical Society*, **135**, 15326-15329.
26. Dou, L., You, J., Yang, J., Chen, C.-C., He, Y., Murase, S., Moriarty, T., Emery, K., Li, G., and Yang, Y. (2012) Tandem polymer solar cells featuring a spectrally matched low-bandgap polymer. *Nature Photonics*, **6**, 180-185.
27. You, J., Dou, L., Yoshimura, K., Kato, T., Ohya, K., Moriarty, T., Emery, K., Chen, C.-C., Gao, J., and Li, G. (2013) A polymer tandem solar cell with 10.6% power conversion efficiency. *Nature Communications*, **4**, 1446.
28. Wang, E., Ma, Z., Zhang, Z., Vandewal, K., Henriksson, P., Inganas, O., Zhang, F., and Andersson, M. R. (2011) An easily accessible isoindigo-based polymer for high-performance polymer solar cells. *Journal of the American Chemical Society*, **133**, 14244-14247.
29. Guo, X., Zhang, M., Tan, J., Zhang, S., Huo, L., Hu, W., Li, Y., and Hou, J. (2012) Influence of D/A ratio on photovoltaic performance of a highly efficient polymer solar cell system. *Advanced Materials*, **24**, 6536-6541.
30. Zhang, M., Guo, X., Zhang, S., and Hou, J. (2014) Synergistic effect of fluorination on molecular energy level modulation in highly efficient photovoltaic polymers. *Advanced Materials*, **26**, 1118-1123.
31. Brabec, C. J., Sariciftci, N. S., and Hummelen, J. C. (2001) Plastic solar cells. *Advanced Functional Materials*, **11**, 15-26.
32. He, Y., and Li, Y. (2011) Fullerene derivative acceptors for high performance polymer solar cells. *Physical Chemistry and Chemical Physics*, **13**, 1970-1983.
33. He, Y., Chen, H.-Y., Hou, J., and Li, Y. (2010) Indene–C60 bisadduct: a new acceptor for high-performance polymer solar cells. *Journal of the American Chemical Society*, **132**, 1377-1382.
34. Guo, X., Cui, C., Zhang, M., Huo, L., Huang, Y., Hou, J., and Li, Y. (2012) High efficiency polymer solar cells based on poly (3-hexylthiophene)/indene-C 70 bisadduct with solvent additive. *Energy & Envi-*

ronmental Science, **5**, 7943-7949.

35. Shockley, W., and Queisser, H. J. (1961) Detailed balance limit of efficiency of p-n junction solar cells. *Journal of Applied Physics*, **32**, 510-519.
36. Hadipour, A., de Boer, B., Wildeman, J., Kooistra, F. B., Hummelen, J. C., Turbiez, M. G., Wienk, M. M., Janssen, R. A., and Blom, P. W. (2006) Solution-Processed Organic Tandem Solar Cells. *Advanced Functional Materials*, **16**, 1897-1903.
37. Kim, J. Y., Lee, K., Coates, N. E., Moses, D., Nguyen, T.-Q., Dante, M., and Heeger, A. J. (2007) Efficient tandem polymer solar cells fabricated by all-solution processing. *Science*, **317**, 222-225.
38. Chou, C. H., Kwan, W. L., Hong, Z., Chen, L. M., and Yang, Y. (2011) A metal-oxide interconnection layer for polymer tandem solar cells with an inverted architecture. *Advanced Materials*, **23**, 1282-1286.
39. Gevaerts, V. S., Furlan, A., Wienk, M. M., Turbiez, M., and Janssen, R. A. (2012) Solution processed polymer tandem solar cell using efficient small and wide bandgap polymer: fullerene blends. *Advanced Materials*, **24**, 2130-2134.
40. Bailey-Salzman, R. F., Rand, B. P., and Forrest, S. R. (2006) Semitransparent organic photovoltaic cells. *Applied Physics Letters*, **88**, 233502.
41. Ng, G.-M., Kietzke, E. L., Kietzke, T., Tan, L.-W., Liew, P.-K., and Zhu, F. (2007) Optical enhancement in semitransparent polymer photovoltaic cells. *Applied Physics Letters*, **90**, 103505.
42. Meiss, J., Holzmueller, F., Gresser, R., Leo, K., and Riede, M. (2011) Near-infrared absorbing semitransparent organic solar cells. *Applied Physics Letters*, **99**, 193307.
43. Chen, C.-C., Dou, L., Zhu, R., Chung, C.-H., Song, T.-B., Zheng, Y. B., Hawks, S., Li, G., Weiss, P. S., and Yang, Y. (2012) Visibly transparent polymer solar cells produced by solution processing. *ACS Nano*, **6**, 7185-7190.
44. Thompson, B. C., and Frechet, J. M. (2008) Polymer–fullerene composite solar cells. *Angewandte Chemie, International Edition*, **47**, 58-77.
45. Boudreault, P.-L. T., Najari, A., and Leclerc, M. (2010) Processable low-bandgap polymers for photovoltaic applications. *Chemistry of Materials*, **23**, 456-469.
46. Beaujuge, P. M., and Frechet, J. M. (2011) Molecular design and ordering effects in π-functional materials for transistor and solar cell applications. *Journal of the American Chemical Society*, **133**, 20009-20029.
47. Li, G., Shrotriya, V., Huang, J., Yao, Y., Moriarty, T., Emery, K., and Yang, Y. (2005) High-efficiency solution processable polymer photovoltaic cells by self-organization of polymer blends. *Nature Materials*, **4**, 864-868.

48. Chen, H.-Y., Hou, J., Zhang, S., Liang, Y., Yang, G., Yang, Y., Yu, L., Wu, Y., and Li, G. (2009) Polymer solar cells with enhanced open-circuit voltage and efficiency. *Nature Photonics*, **3**, 649-653.
49. Liang, Y., Xu, Z., Xia, J., Tsai, S.T., Wu, Y., Li, G., Ray, C., and Yu, L. (2010) For the bright future-bulk heterojunction polymer solar cells with power conversion efficiency of 7.4%. *Advanced Materials*, **22**(20), E135-E138.
50. He, Z., Zhong, C., Huang, X., Wong, W.Y., Wu, H., Chen, L., Su, S., and Cao, Y. (2011) Simultaneous enhancement of open-circuit voltage, short-circuit current density, and fill factor in polymer solar cells. *Advanced Materials*, **23**, 4636-4643.
51. Amb, C. M., Chen, S., Graham, K. R., Subbiah, J., Small, C. E., So, F., and Reynolds, J. R. (2011) Dithienogermole as a fused electron donor in bulk heterojunction solar cells. *Journal of the American Chemical Society*, **133**, 10062-10065.
52. Chu, T.-Y., Lu, J., Beaupre, S., Zhang, Y., Pouliot, J.-R., Wakim, S., Zhou, J., Leclerc, M., Li, Z., and Ding, J. (2011) Bulk heterojunction solar cells using thieno [3, 4-c] pyrrole-4, 6-dione and dithieno [3, 2-b: 2′, 3′-d] silole copolymer with a power conversion efficiency of 7.3%. *Journal of the American Chemical Society*, **133**, 4250-4253.
53. Huo, L., Zhang, S., Guo, X., Xu, F., Li, Y., and Hou, J. (2011) Replacing alkoxy groups with alkylthienyl groups: a feasible approach to improve the properties of photovoltaic polymers. *Angewandte Chemie, International Edition*, **123**, 9871-9876.
54. Zhang, Y., Zou, J., Yip, H.-L., Chen, K.-S., Zeigler, D.F., Sun, Y., and Jen, A. K.-Y. (2011) Indacenodithiophene and quinoxaline-based conjugated polymers for highly efficient polymer solar cells. *Chemistry of Materials*, **23**, 2289-2291.
55. Ye, L., Jiao, X., Zhang, H., Li, S., Yao, H., Ade, H., and Hou, J. (2015) 2D-conjugated benzodithiophene-based polymer acceptor: design, synthesis, nanomorphology, and photovoltaic performance. *Macromolecules*, **48**, 7156-7163.
56. Zhang, M., Guo, X., Ma, W., Ade, H., and Hou, J. (2015) A Large-Bandgap Conjugated Polymer for Versatile Photovoltaic Applications with High Performance. *Advanced Materials*, **27**, 4655-4660.
57. Liu, D., Gu, C., Xiao, M., Qiu, M., Sun, M., and Yang, R. (2015) A new highly conjugated crossed benzodithiophene and its donor–acceptor copolymers for high open circuit voltages polymer solar cells. *Polymer Chemistry*, **6**, 3398-3406.
58. Dou, L., Chang, W. H., Gao, J., Chen, C. C., You, J., and Yang, Y. (2013) A selenium-substituted low-bandgap polymer with versatile photovoltaic applications. *Advanced Materials*, **25**, 825-831.
59. Dou, L., Gao, J., Richard, E., You, J., Chen, C.-C., Cha, K. C., He, Y., Li, G., and Yang, Y. (2012) Systematic investigation of benzodithiophene- and diketopyrrolopyrrole-based low-bandgap polymers designed

for single junction and tandem polymer solar cells. *Journal of the American Chemical Society*, **134**, 10071-10079.

60. Kronemeijer, A. J., Gili, E., Shahid, M., Rivnay, J., Salleo, A., Heeney, M., and Sirringhaus, H. (2012) A selenophene-based low-bandgap donor–acceptor polymer leading to fast ambipolar logic. *Advanced Materials*, **24**, 1558-1565.
61. Shahid, M., McCarthy-Ward, T., Labram, J., Rossbauer, S., Domingo, E. B., Watkins, S. E., Stingelin, N., Anthopoulos, T. D., and Heeney, M. (2012) Low band gap selenophene–diketopyrrolopyrrole polymers exhibiting high and balanced ambipolar performance in bottom-gate transistors. *Chemical Science*, **3**, 181-185.
62. Ballantyne, A. M., Chen, L., Nelson, J., Bradley, D. D., Astuti, Y., Maurano, A., Shuttle, C. G., Durrant, J. R., Heeney, M., and Duffy, W. (2007) Studies of highly regioregular poly (3-hexylselenophene) for photovoltaic applications. *Advanced Materials*, **19**, 4544-4547.
63. Klein, M. F., Pfaff, M., Muller, E., Czolk, J., Reinhard, M., Valouch, S., Lemmer, U., Colsmann, A., and Gerthsen, D. (2012) Poly (3-hexylselenophene) solar cells: Correlating the optoelectronic device performance and nanomorphology imaged by low-energy scanning Transmission electron microscopy. *Journal of Polymer Science, Part B: Polymer Physics*, **50**, 198-206.
64. Zhang, Z., Lin, F., Chen, H.-C., Wu, H.-C., Chung, C.-L., Lu, C., Liu, S.-H., Tung, S.-H., Chen, W.-C., and Wong, K.-T. (2015) A silole copolymer containing a ladder-type heptacylic arene and naphthobisoxadiazole moieties for highly efficient polymer solar cells. *Energy & Environmental Science*, **8**, 552-557.
65. Liu, D., Zhao, W., Zhang, S., Ye, L., Zheng, Z., Cui, Y., Chen, Y., and Hou, J. (2015) Highly efficient photovoltaic polymers based on benzodithiophene and quinoxaline with deeper HOMO levels. *Macromolecules*, **48**, 5172-5178.
66. Lu, L., Zheng, T., Wu, Q., Schneider, A. M., Zhao, D., and Yu, L. (2015) Recent advances in bulk heterojunction polymer solar cells. *Chemical Reviews*, **115**, 12666-12731.
67. Duan, C., Gao, K., van Franeker, J. J., Liu, F., Wienk, M. M., and Janssen, R. A. (2016) Toward practical useful polymers for highly efficient solar cells via a random copolymer approach. *Journal of the American Chemical Society*, **138**, 10782-10785.
68. Nakano, K., Nakano, M., Xiao, B., Zhou, E., Suzuki, K., Osaka, I., Takimiya, K., and Tajima, K. (2016) Naphthodithiophene diimide-based copolymers: Ambipolar semiconductors in field-effect transistors and electron acceptors with near-infrared response in polymer blend solar cells. *Macromolecules*, **49**, 1752-1760.
69. Liu, T., Pan, X., Meng, X., Liu, Y., Wei, D., Ma, W., Huo, L., Sun, X., Lee, T. H., and Huang, M. (2017) Alkyl side-chain engineering in wide-bandgap copolymers leading to power conversion efficiencies over

10%. *Advanced Materials*, **29**(6), 1604251.

70. Zhou, E., Nakano, M., Izawa, S., Cong, J., Osaka, I., Takimiya, K., and Tajima, K. (2014) All-polymer solar cell with high near-infrared response based on a Naphthodithiophene Diimide (NDTI) copolymer. *ACS Macro Letters*, **3**, 872-875.
71. Earmme, T., Hwang, Y.-J., Murari, N.M., Subramaniyan, S., and Jenekhe, S. A. (2013) All-polymer solar cells with 3.3% efficiency based on naphthalene diimide-selenophene copolymer acceptor. *Journal of the American Chemical Society*, **135**, 14960-14963.
72. Marchiori, C. F., Yamamoto, N. A., Matos, C. F., Kujala, J., Macedo, A. G., Tuomisto, F., Zarbin, A. J., Koehler, M., and Roman, L. S. (2015) Annealing effect on donor-acceptor interface and its impact on the performance of organic photovoltaic devices based on PSiF-DBT copolymer and C60. *Applied Physics Letters*, **106**, 133301.
73. Kim, J.-H., Kim, H. U., Song, C. E., Kang, I.-N., Lee, J.-K., Shin, W. S., and Hwang, D.-H. (2013) Benzotriazole-based donor–acceptor type semiconducting polymers with different alkyl side chains for photovoltaic devices. *Solar Energy Materials and Solar Cells*, **108**, 113-125.
74. Saeki, A., Tsuji, M., Yoshikawa, S., Gopal, A., and Seki, S. (2014) Boosting photovoltaic performance of a benzobisthiazole based copolymer: a device approach using a zinc oxide electron transport layer. *Journal of Materials Chemistry A*, **2**, 6075-6080.
75. Guo, C., Lin, Y.-H., Witman, M. D., Smith, K. A., Wang, C., Hexemer, A., Strzalka, J., Gomez, E. D., and Verduzco, R. (2013) Conjugated block copolymer photovoltaics with near 3% efficiency through microphase separation. *Nano Letters*, **13**, 2957-2963.
76. Krebs, F. C., Espinosa, N., Hosel, M., Sondergaard, R. R., and Jorgensen, M. (2014) Rise to power-OPV-based solar parks. *Advanced Materials*, **26**, 29-39.
77. Espinosa, N., Hosel, M., Jorgensen, M., and Krebs, F. C. (2014) Large scale deployment of polymer solar cells on land, on sea and in the air. *Energy & Environmental Science*, **7**, 855-866.
78. Zhan, X., and Zhu, D. (2010) Conjugated polymers for high-efficiency organic photovoltaics. *Polymer Chemistry*, **1**, 409-419.
79. Dou, L., Chen, C.-C., Yoshimura, K., Ohya, K., Chang, W.-H., Gao, J., Liu, Y., Richard, E., and Yang, Y. (2013) Synthesis of 5 H-dithieno [3, 2-b: 2′, 3′-d] pyran as an electron-rich building block for donor-acceptor type low-bandgap polymers. *Macromolecules*, **46**, 3384-3390.
80. Dang, D., Chen, W., Himmelberger, S., Tao, Q., Lundin, A., Yang, R., Zhu, W., Salleo, A., Muller, C., and Wang, E. (2014) Enhanced photovoltaic performance of indacenodithiophene-quinoxaline copolymers by side-chain modulation. *Advanced Energy Materials*, **4**(15), 1400680.
81. Nguyen, T. L., Choi, H., Ko, S.-J., Uddin, M. A., Walker, B., Yum, S.,

Jeong, J.-E., Yun, M., Shin, T., and Hwang, S. (2014) Semi-crystalline photovoltaic polymers with efficiency exceeding 9% in a ~300 nm thick conventional single-cell device. *Energy & Environmental Science*, **7**, 3040-3051.

82. Cui, C., Wong, W.-Y., and Li, Y. (2014) Improvement of open-circuit voltage and photovoltaic properties of 2D-conjugated polymers by alkylthio substitution. *Energy & Environmental Science*, **7**, 2276-2284.
83. Chen, Z., Cai, P., Chen, J., Liu, X., Zhang, L., Lan, L., Peng, J., Ma, Y., and Cao, Y. (2014) Low band-gap conjugated polymers with strong interchain aggregation and very high hole mobility towards highly efficient thick-film polymer solar cells. *Advanced Materials*, **26**, 2586-2591.
84. Stalder, R., Mei, J., and Reynolds, J.R. (2010) Isoindigo-based donor-acceptor conjugated polymers. *Macromolecules*, **43**, 8348-8352.
85. Lei, T., Wang, J.-Y., and Pei, J. (2014) Design, synthesis, and structure–property relationships of isoindigo-based conjugated polymers. *Accounts of Chemical Research*, **47**, 1117-1126.
86. Robb, M., Ku, S., Brunetti, F., and Hawker, C. (2013) A renaissance of color: New structures and building blocks for organic electronics. *Journal of Polymer Science, Part A: Polymer Chemistry*, **51**(6), 1263-1271.
87. Chen, M. S., Niskala, J. R., Unruh, D. A., Chu, C. K., Lee, O. P., and Frechet, J. M. (2013) Control of polymer-packing orientation in thin films through synthetic tailoring of backbone coplanarity. *Chemistry of Materials*, **25**, 4088-4096.
88. Wang, E., Mammo, W., and Andersson, M. R. (2014) Isoindigo-based polymers and small molecules for bulk heterojunction solar cells and field effect transistors. *Advanced Materials*, **26**, 1801-1826.
89. Wang, E., Ma, Z., Zhang, Z., Henriksson, P., Inganas, O., Zhang, F., and Andersson, M. R. (2011) An isoindigo-based low band gap polymer for efficient polymer solar cells with high photo-voltage. *Chemical Communications*, **47**, 4908-4910.
90. Ma, Z., Wang, E., Vandewal, K., Andersson, M. R., and Zhang, F. (2011) Enhance performance of organic solar cells based on an isoindigo-based copolymer by balancing absorption and miscibility of electron acceptor. *Applied Physics Letters*, **99**, 143302.
91. Tao, Q., Xia, Y., Xu, X., Hedstrom, S., Backe, O., James, D. I., Persson, P., Olsson, E., Inganas, O., and Hou, L. (2015) D–A1–D–A2 copolymers with extended donor segments for efficient polymer solar cells. *Macromolecules*, **48**, 1009-1016.
92. Liu, P., Zhang, K., Liu, F., Jin, Y., Liu, S., Russell, T. P., Yip, H.-L., Huang, F., and Cao, Y. (2014) Effect of fluorine content in thienothiophene-benzodithiophene copolymers on the morphology and performance of polymer solar cells. *Chemistry of Materials*, **26**, 3009-3017.

93. Kim, J.-H., Shin, S. A., Park, J. B., Song, C. E., Shin, W. S., Yang, H., Li, Y., and Hwang, D.-H. (2014) Fluorinated benzoselenadiazole-based low-band-gap polymers for high efficiency inverted single and tandem organic photovoltaic cells. *Macromolecules*, **47**, 1613-1622.
94. Chen, H.-C., Chen, Y.-H., Liu, C.-H., Hsu, Y.-H., Chien, Y.-C., Chuang, W.-T., Cheng, C.-Y., Liu, C.-L., Chou, S.-W., and Tung, S.-H. (2013) Fluorinated thienyl-quinoxaline-based D–π–A-type copolymer toward efficient polymer solar cells: synthesis, characterization, and photovoltaic properties. *Polymer Chemistry*, **4**, 3411-3418.
95. McCulloch, I., Ashraf, R. S., Biniek, L., Bronstein, H., Combe, C., Donaghey, J. E., James, D. I., Nielsen, C. B., Schroeder, B. C., and Zhang, W. (2012) Design of semiconducting indacenodithiophene polymers for high performance transistors and solar cells. *Accounts of Chemical Research*, **45**, 714-722.
96. Chen, Y.-C., Yu, C.-Y., Fan, Y.-L., Hung, L.-I., Chen, C.-P., and Ting, C. (2010) Low-bandgap conjugated polymer for high efficient photovoltaic applications. *Chemical Communications*, **46**, 6503-6505.
97. Dang, D., Chen, W., Yang, R., Zhu, W., Mammo, W., and Wang, E. (2013) Fluorine substitution enhanced photovoltaic performance of a D–A 1–D–A 2 copolymer. *Chemical Communications*, **49**, 9335-9337.
98. Albrecht, S., Janietz, S., Schindler, W., Frisch, J., Kurpiers, J., Kniepert, J., Inal, S., Pingel, P., Fostiropoulos, K., and Koch, N. (2012) Fluorinated copolymer PCPDTBT with enhanced open-circuit voltage and reduced recombination for highly efficient polymer solar cells. *Journal of the American Chemical Society*, **134**, 14932-14944.
99. Ye, L., Zhang, S., Zhao, W., Yao, H., and Hou, J. (2014) Highly efficient 2D-conjugated benzodithiophene-based photovoltaic polymer with linear alkylthio side chain. *Chemistry of Materials*, **26**, 3603-3605.
100. Liao, S. H., Jhuo, H. J., Cheng, Y. S., and Chen, S. A. (2013) Fullerene derivative-doped zinc oxide nanofilm as the cathode of inverted polymer solar cells with low-bandgap polymer (PTB7-Th) for high performance. *Advanced Materials*, **25**, 4766-4771.
101. Mei, J., and Bao, Z. (2013) Side chain engineering in solution-processable conjugated polymers. *Chemistry of Materials*, **26**, 604-615.
102. Chen, M. S., Lee, O. P., Niskala, J. R., Yiu, A.T., Tassone, C. J., Schmidt, K., Beaujuge, P. M., Onishi, S. S., Toney, M. F., and Zettl, A. (2013) Enhanced solid-state order and field-effect hole mobility through control of nanoscale polymer aggregation. *Journal of the American Chemical Society*, **135**, 19229-19236.
103. Zhang, F., Hu, Y., Schuettfort, T., Di, C.-a., Gao, X., McNeill, C. R., Thomsen, L., Mannsfeld, S. C., Yuan, W., and Sirringhaus, H. (2013) Critical role of alkyl chain branching of organic semiconductors in enabling solution-processed n-channel organic thin-film transistors with mobility of up to 3.50 $cm^2 V^{-1} s^{-1}$. *Journal of the American Chemical*

Society, **135**, 2338-2349.

104. Lei, T., Dou, J. H., and Pei, J. (2012) Influence of alkyl chain branching positions on the hole mobilities of polymer thin-film transistors. *Advanced Materials*, **24**, 6457-6461.
105. Meager, I., Ashraf, R. S., Mollinger, S., Schroeder, B. C., Bronstein, H., Beatrup, D., Vezie, M. S., Kirchartz, T., Salleo, A., and Nelson, J. (2013) Photocurrent enhancement from diketopyrrolopyrrole polymer solar cells through alkyl-chain branching point manipulation. *Journal of the American Chemical Society*, **135**, 11537-11540.
106. Guo, X., Zhang, M., Ma, W., Ye, L., Zhang, S., Liu, S., Ade, H., Huang, F., and Hou, J. (2014) Enhanced photovoltaic performance by modulating surface composition in bulk heterojunction polymer solar cells based on PBDTTT-C-T/PC71BM. *Advanced Materials*, **26**, 4043-4049.
107. Chen, K. S., Yip, H. L., Salinas, J. F., Xu, Y. X., Chueh, C. C., and Jen, A. K. Y. (2014) Strong photocurrent enhancements in highly efficient flexible organic solar cells by adopting a microcavity configuration. *Advanced Materials*, **26**, 3349-3354.
108. Zhao, X., Yang, D., Lv, H., Yin, L., and Yang, X. (2013) New benzotrithiophene derivative with a broad band gap for high performance polymer solar cells. *Polymer Chemistry*, **4**, 57-60.
109. Keshtov, M., Marochkin, D., Kochurov, V., Khokhlov, A., Koukaras, E., and Sharma, G. (2014) New conjugated alternating benzodithiophene-containing copolymers with different acceptor units: synthesis and photovoltaic application. *Journal of Materials Chemistry A*, **2**, 155-171.
110. Gedefaw, D., Tessarolo, M., Prosa, M., Bolognesi, M., Henriksson, P., Zhuang, W., Seri, M., Muccini, M., and Andersson, M. R. (2016) Induced photodegradation of quinoxaline based copolymers for photovoltaic applications. *Solar Energy Materials and Solar Cells*, **144**, 150-158.
111. Ho, C.-C., Chen, C.-A., Chang, C.-Y., Darling, S. B., and Su, W.-F. (2014) Isoindigo-based copolymers for polymer solar cells with efficiency over 7%. *Journal of Materials Chemistry A*, **2**, 8026-8032.
112. Puchalska, M., Polec-Pawlak, K., Zadrozna, I., Hryszko, H., and Jarosz, M. (2004) Identification of indigoid dyes in natural organic pigments used in historical art objects by high-performance liquid chromatography coupled to electrospray ionization mass spectrometry. *Journal of Mass Spectrometry*, **39**, 1441-1449.
113. Andersen, T. R., Dam, H. F., Burkhart, B., Angmo, D., Corazza, M., Thompson, B. C., and Krebs, F. C. (2014) Medium area, flexible single and tandem junction solar cells based on roll coated semi-random copolymers. *Journal of Materials Chemistry C*, **2**, 9412-9415.
114. Yuan, J., Ford, M., Ding, G., Dong, H., Wang, M., Han, L., Li, Y., Bazan, G. C., and Ma, W. (2016) Narrow bandgap conjugated polymers

based on a high-mobility polymer template for visibly transparent photovoltaic devices. *Journal of Materials Chemistry A*, **4**, 17333-17343.

115. Liao, S.-H., Jhuo, H.-J., Yeh, P.-N., Cheng, Y.-S., Li, Y.-L., Lee, Y.-H., Sharma, S., and Chen, S.-A. (2014) Single junction inverted polymer solar cell reaching power conversion efficiency 10.31% by employing dual-doped zinc oxide nano-film as cathode interlayer. *Scientific Reports*, **4**, 6813.
116. Chen, J. D., Cui, C., Li, Y.Q., Zhou, L., Ou, Q. D., Li, C., Li, Y., and Tang, J. X. (2015) Single-junction polymer solar cells exceeding 10% power conversion efficiency. *Advanced Materials*, **27**, 1035-1041.
117. Huang, Y., Wu, F., Zhang, M., Mei, S., Shen, P., and Tan, S. (2015) Synthesis and photovoltaic properties of conjugated polymers with an asymmetric 4-(2-ethylhexyloxy)-8-(2-ethylhexylthio) benzo [1, 2-b: 4, 5-b′] dithiophene unit. *Dyes and Pigments*, **115**, 58-66.
118. Li, S., Li, Y., Wisner, C. A., Jin, L., Leventis, N., and Peng, Z. (2014) Synthesis, optical properties and photovoltaic applications of hybrid rod–coil diblock copolymers with coordinatively attached CdSe nanocrystals. *RSC Advances*, **4**, 35823-35832.
119. Lu, L., Xu, T., Chen, W., Landry, E. S., and Yu, L. (2014) Ternary blend polymer solar cells with enhanced power conversion efficiency. *Nature Photonics*, **8**, 716-722.
120. Najari, A., Beaupre, S., Allard, N., Ouattara, M., Pouliot, J. R., Charest, P., Besner, S., Simoneau, M., and Leclerc, M. (2015) Thieno, furo, and selenopheno [3, 4-c] pyrrole-4, 6-dione copolymers: Air-processed polymer solar cells with power conversion efficiency up to 7.1%. *Advanced Energy Materials*, **5**(23), 1501213.
121. Squeo, B. M., Gasparini, N., Ameri, T., Palma-Cando, A., Allard, S., Gregoriou, V. G., Brabec, C. J., Scherf, U., and Chochos, C. L. (2015) Ultra low band gap α, β-unsubstituted BODIPY-based copolymer synthesized by palladium catalyzed cross-coupling polymerization for near infrared organic photovoltaics. *Journal of Materials Chemistry A*, **3**, 16279-16286.

9

Thermally Phase Changing Materials: Synthesis, Properties and Applications

9.1 Introduction

Materials that undergo phase change at particular temperature are defined as thermally phase changing materials (PCMs). These materials can store and release heat in the form of energy, as required by specific applications. Such a release of energy is facilitated by their phase change owing to the change in temperature. They allow relatively large amount of energy to be stored in comparatively small volumes. The energy released/absorbed by phase transition from solid to liquid (heat of fusion) is generally much higher than the sensible heat. Due to their attractive and controllable phase change properties, these materials are useful in many applications. For instance, PCMs have been used as smart building materials, energy storage devices, heat recovering agents, etc. Water/ice is a very useful phase change material, which has been used to cool buildings in summer for many decades. By virtue of their versatile and moldable properties, these materials have also found applications in smart textiles and aerospace industry [1]. In these applications, PCMs have been able to provide solutions to the thermal problems, which are otherwise not possible to resolve using the conventional systems. Their attractive characteristic to maintain the temperature at a constant level during the phase change is of particular relevance.

Materials undergo different types of phase changes, viz. solid-solid, solid-liquid and liquid-gas. Solid-solid and solid-liquid phase change materials have garnered widespread research and commercial attention due to superior textural and thermal properties. In addition, materials with phase change properties in between solid-solid and solid-liquid phase changes are known as shape stabilized or form stable PCMs. Their structures comprise of supporting materials and working substances. They can physically stay in solid state even if the working materials undergo phase change from solid to liquid, which makes it very easy to use them in many practical applications [2].

Liyamol Jacob and Vikas Mittal
Khalifa University of Science and Technology, Abu Dhabi, UAE

Various methods are utilized for the synthesis of PCM materials. Blending, adsorption, impregnation, sol-gel method and copolymerization are some of the widely used methods for this purpose. These methods are effective in producing PCMs with different physical, thermal and textural properties. Additionally, these methods help to tune the structure, properties and morphological attributes of the PCMs to make them suitable for particular applications. Thus, the ability of the phase change materials to be molded according to the intended use makes them one of the most advanced and efficient materials [3].

9.2 Different Phase Change Materials (PCMs)

9.2.1 Nanomaterial PCMs

Melting temperature, thermal conductivity and energy storage density are the main aspects that decide the thermal performance of a PCM. The enhancement in the properties of the PCMs by composite formation leads to the expansion of their spectrum of applications [4]. Specifically, low thermal conductivity of the phase change materials can be manipulated by incorporating them with different kinds of nanomaterials. Nanomaterials such as nanoparticles, nanofibers, nanotubes and nanosheets have gained significant research attention for this purpose [5-8]. Many studies have also reported nano-encapsulated PCM systems.

Vanadium dioxide encapsulated with gold nanoparticles shows excellent surface plasmon resonance combined with phase change properties. Of specific intertest are its quantum phase changing properties, which interact effectively with the electrical and optical properties of the material. Gold nanoparticles in the nanocomposite structure act like photocathodes, which cause very fast phase transition in vanadium dioxide nanostructure, when illuminated by a spectrally resonant femtosecond laser pulse [9].

Paraffin wax incorporated with carbon nanotubes (CNTs) is a multifunctional phase change composite where energy storage can be driven by small voltages [10]. The CNTs act as flexible encapsulation scaffold for the wax and also retain a highly conducting network over it during the phase change process. The phase change enthalpy and thermal conductivity of the paraffin wax is improved due to the uniform interpenetration of the nanotube network and paraffin wax with high affinity towards each other. The developed material can be effectively utilized in sunlight absorption and conventional lithium ion batteries [10].

N-octadecane nanocapsules as organic PCMs have been used for thermal energy storage [11]. These nanocapsules were developed by trapping n-octadecane as a core with St (styrene)-MMA (methyl-methacrylate) copolymer shell using miniemulsion *in-situ* polymerization method. Nanocapsules were observed to have a melting point of 29.5 °C and a crystallization temperature of 24.6 °C, measured by differential scanning calorimetry (DSC). The enthalpy of melting and crystallization of the PCMs was found to be 107.9 and 104.9 $J.g^{-1}$ respectively.

Other studies on different polymer nanocomposite PCMs have also confirmed their potential of application in various fields. Sol-gel method was used to form paraffin and TiO_2 composite, whereas *in-situ* polymerization was reported to form urea-formaldehyde for nano-encapsulation of n-tetradecane [12,13]. Some of the PCMs can also unavoidably form formaldehyde which can pose threat to the environment and animal life [14].

As an example, Figure 9.1 shows the transmission electron micrographs (TEM) and high-resolution TEM (HRTEM) of phase changing ZnS nanostructures [15].

9.2.2 Carbon PCMs

Carbon materials have been employed in many studies to enhance the property profiles of different PCMs. Activated carbon, CNTs and graphene have been extensively used in the synthesis of advanced PCMs [16].

To achieve stable phase change composites, polyethylene glycol (PEG) was coupled with activated carbon and silica [17]. A prominent effect on the crystallinity of PEG by the use of porous carbon mesh and silica sieves was observed. The material showed improved latent heat, low melting point, low supercooling and high heat storage efficiency.

Phase changing materials with carbon nanotubes have obtained special interest owing to the potential for many applications. Composite systems formed by incorporating single walled CNTs (SWCNTs), multi-walled CNTs (MWCNTs) and carbon nanofibers have been used to analyze the improvement in the latent heat energy storage systems using PCMs [18]. The DSC studies revealed a remarkable improvement in the latent heat as compared to pure wax. Among the different

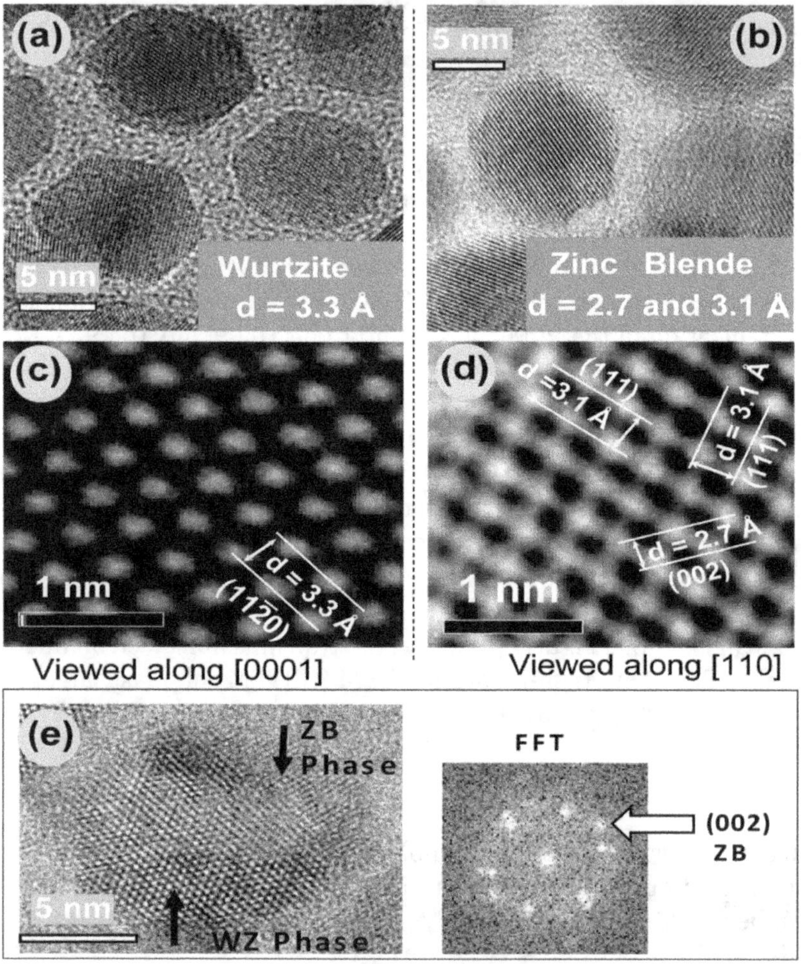

Figure 9.1 HRTEM images of (a) WZ and (b) ZB ZnS platelets; (c,d) the corresponding magnified HRTEM images. (e) An intermediate TEM image showing the existence of both WZ and ZB phases in a single platelet. Right, the FFT with identification of the (002) plane of ZB ZnS. Reproduced from Reference 15 with permission from American Chemical Society.

composites, the SWCNTs showed the best improvement of 13% for 1% loading owing to higher surface area and molecular density. The study was combined with a theoretical model which was found to be in good agreement with the experimental studies. Paraffin wax modified with CNTs showed shift in the melting point to a lower temperature [19]. This was further enhanced by increasing the carbon nanotubes content. The addition of CNTs reduced the latent heat capacity as well. The thermal conductivity was observed to improve in the case of both liquid and solid phases.

Elgalfy and Lafdi [20] studied the behavior of paraffin wax with carbon nanofibers. The cooling rate was predicted theoretically by analyzing the changing thermal response of the composite during solidification process. The cooling rate was observed to show improvement owing to the change in the thermal conductivity with regard to the pure paraffin wax. Based on these results, the authors introduced an analytical model which effectively predicted the thermal conductivity. The model could be further modified to analyze and predict the effect of surface modification of carbon nanofillers on phase change materials.

SWNTs/poly(methylmethacrylate) composite has been reported for studying the improvement in the thermal properties of the polymer by adding different percentage of SWCNTs [21]. Both thermal and electrical conductivity improved by the introduction of SWCNTs to the polymer matrix. For instance, improvements of 57% in the thermal conductivity and several magnitudes in the electrical conductivity were observed by incorporating 7 wt% SWCNTs in the polymer matrix. The percolation of SWCNTs in the polymer network was observed to have influenced the thermal properties of the polymer matrix manifold.

Another study used the ability of carbon nanotubes to confine the materials inside the tubular structure to create materials that undergo dimensionally confined phase changes [22]. Inside the carbon nanotubes, the material can show a solid-liquid critical point beyond the point where solid-liquid distinction disappears, unlike its bulk counterpart. This feature can be used for the direct and continuous transformation from sloid to liquid phases. The study developed numerical simulations where the ice transitions showed different phases which could not be found in the bulk transition, along with different critical temperature points. Water, inside CNTs with 1.1 nm to 1.4 nm dimensions, was shown to undergo phase change from hexagonal cubic phase to heptagonal cubic phase, and a continuous phase change was observed from solid-like square or pentagonal ice nanotubes.

Nano-architectured phase change material composite made of graphene and 1-octadeconal was reported to demonstrate improved thermal conductivity without much change in the heat of fusion. Figure 9.2 shows the improvement in the conductivity of the composite as a function of graphene content [23].

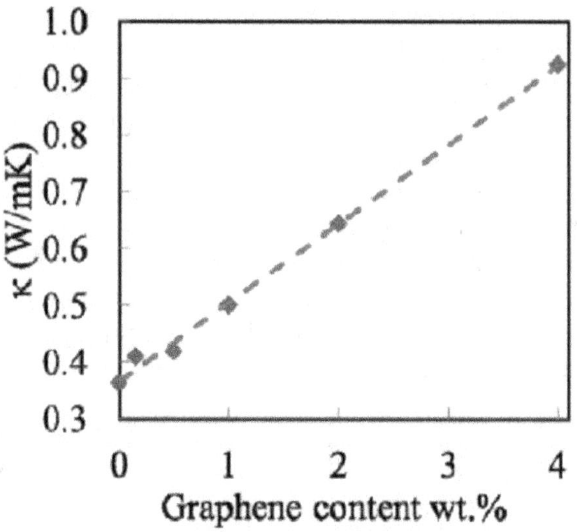

Figure 9.2 Measured thermal conductivity of the sample vs filler content for graphene/PCM composites. Reproduced from Reference 23 with permission from American Chemical Society.

9.2.3 Metal Foam PCMs

Phase changing materials combined with metal foams have attracted recent attention due to the improvement in the thermal properties of the base materials. Thermal transport, effective thermal conductivity, forced and natural convection, thermal radiation as well as pool and flow boiling heat transfer in the material can be increased by incorporating it with metal foams [24]. Use of aluminum foam embedded paraffin wax was studied for heat transfer enhancement [25]. The experiments showed that the metal foam was able to improve solid liquid phase change heat transfer, due to the enhanced thermal conduction inside the metal foam. It was also evident that the structure of the metal foams affected the overall heat conduction behavior of the material.

Use of copper metal foam for passive thermal management using paraffin wax was reported in another study [26]. The apparatus consisted of a hybrid heat sink with parallel fins attached to it and copper foam-paraffin composite material saturated in its hollow basement. For comparison, improvement of thermal properties using pure paraffin wax and solid copper basement were also analyzed. The use of metal foam in the paraffin matrix reduced the time required for the wax to reach its melting point and also effectively reduced the surface temperature of the system. Surface temperature could be further reduced by decreasing the pore density and porosity of the foam. Enhancement in thermal conduction by the metal foam surpassed the natural convection in the pure paraffin wax which caused a linear increase in the temperature during the melting process.

Embedding ultrathin graphite foams into PCMs can enhance the thermal conductivity by almost 18 times without much effect on the melting temperature or mass specific heat of fusion [27]. Wide applicability of such foams to a range of PCMs can help in achieving hybrids with high thermal cycling stability.

In another study, melting phase change heat transfer on open celled metallic foams with paraffin was analyzed both theoretically and experimentally [28]. Highly porous copper metal foam was used as the nanofiller in the paraffin matrix. The inner temperature distribution as well as wall temperature distribution of the copper foam were analyzed during the melting of the paraffin wax. Local natural convection inside the wax was suppressed by the presence of the metal foam even though the thermal conductivity was increased. Numerical study was effective in predicting the velocity, interface location during phase change and thermal field of the synthesized composites. Another study also utilized PCMs stabilized with copper metal foam for safety applications inside Li ion batteries [29]. The latent heat of absorption and natural convection inside the PCMs could be used for the surface temperature reduction and maintenance of constant temperature range in the battery material. The increase in porosity and porous density, however, caused an enhancement in the surface temperature of the battery.

9.3 Classification of PCMs Based on Morphology

PCMs show different morphologies based on the synthesis method, base material and filler used during composite formation. As mentioned earlier, numerous synthesis methods have been developed to generate PCMs with different morphologies, such as electrospinning, microencapsulation, vacuum impregnation, simple submersion, etc. [30-32]. The morphologies include film, core-shell, fibrous, porous, capsules, etc. Morphology of the PCMs is an important factor, especially when these are integrated into different systems. PCMs with fibrous and porous morphology are analyzed further in this section.

9.3.1 Fibrous Structure

PCMs with fibrous morphology are synthesized mainly using electrospinning, microencapsulation and sol-gel methods. McCann *et al.* [33] used melt coaxial electrospinning, which is an innovative method for the synthesis of PCM nanofibers. This method has been successful in electrospinning organic materials like paraffin wax and creating fibers with long chain hydrocarbon in the core, in a single step. The authors focused on materials that underwent phase change near room temperature as well as body temperature, thus, making them more attractive for practical applications. Rapid solidification of hydrocarbons by the evaporative cooling of carrier solution resulted in segmented morphology inside the core of the composite material. Figure 9.3 shows scanning (SEM) and transmission images of electrospun octadecane@TiO_2-PVP nanofibers using melt coaxial spinneret.

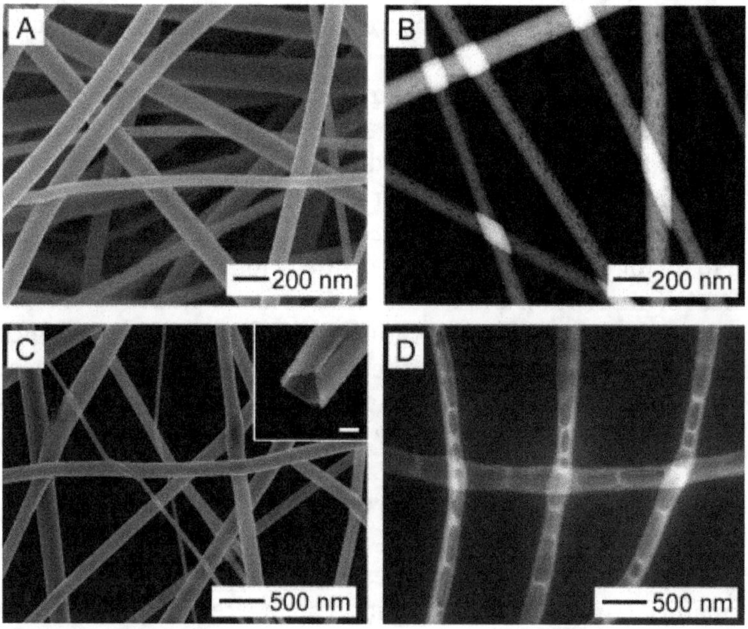

Figure 9.3 SEM image of the nanofibers with 7% octadecane loading, (B) TEM image of the fibers with the octadecane having been removed by soaking in hexane for 24 h, (C) SEM image of octadecane@TiO_2-PVP nanofibers with 45% octadecane loading and (D) TEM image of these fibers after the octadecane had been removed by soaking in hexane for 24 h. Reproduced from Reference 33 with permission from American Chemical Society.

Chen *et al.* [34] also reported electrospun fatty acids/polyethylene terephthalate (PET) composite fibers as novel form-stable phase change materials. The latent heat and phase transition temperature of the fatty acid/PET composite fibers were observed to vary with the type of the fatty acids employed in the study. In addition, these characteristics were maintained after multiple heating-cooling cycles.

Phase changing thermal materials with fibrous structure have been used in a wide range of applications, as also obvious from a number of patents on this subject [35-39]. For example, phase changing paraffinic hydrocarbons have been incorporated with different types of coatings for application in thermal insulations [38]. In another study, a phase change material from a mixture of plant oils was used for incorporation into electrospun poly(vinyl alcohol) nanofibers [40]. The authors used an emulsion electrospinning technique for this purpose.

9.3.2 Porous Structure

Vacuum impregnation method has been widely used for the synthesis of PCMs with porous structure or porous filler material. This method is (a) convenient to use, (b) economical and (c) able to load the porous fillers into the matrix effectively. The granular filler material can be anything from fibrous graphene, clay materials, fly ash, etc., and the phase changing materials can be paraffin fax, fatty acids, thickened latex, etc. [41].

Polyethylene glycol was impregnated into the porous structure of diatomite for thermal energy storage applications [42]. PEG exhibited good retention capability into the porous diatomite without any apparent leakage from the composite material. The composite showed good thermal reliability, cycling stability and melting temperature around ambient conditions. Thermal conductivity of the composite could be further improved by adding expanded graphite to the porous structure in varying mass fractions. In another study, erythritol was impregnated in different porous materials like perlite, gamma alumina and diatom earth for heat recovery applications [43]. Porosity of the materials affected the melting temperature of the phase change composite. Materials with low porosity showed lower melting point. Expanded perlite was found to be the best filler material which exhibited 83% retention of latent heat of erythritol, its complete impregnation into the pores and good cycling retention ability. Numerical investigation of the melting process of a phase change material inside a porous matrix also confirmed that the use of materials with high pore density and thermal conductivity improved the properties of PCMs for energy storage applications [44,45]. Isothermal phase changes in metallic systems using the melting of pure gallium in a porous cavity has also been numerically analyzed [46].

Zhang *et al.* [47] studied the phase changing behavior of capric acid and paraffin with three different porous building materials. It was observed that different phase change materials acted differently in the porous matrices. A number of other studies have reported phase change materials with porous expanded graphite and porous carbon materials [48,49].

9.4 Applications of PCMs

PCMs have been used in a wide range of applications like energy storage [50,51], construction [52-55], textile industry, heat storage [56-60], random access memory technology [61], data storage [62], non-volatile memories [63], etc. The properties of the PCMs to store heat energy in its latent form and to undergo changes in the phase by virtue of heat release and heat storage are made use of in most of their applications. When the temperature is supplied, it causes breaking of chemical bonds, thereby, causing a change from solid to liquid form. There are specific chemical, kinetic, thermal and physical properties that are desired for specific applications. The physical properties correspond to the attainment of desired phase equilibrium which prevents phase segregation. The thermal property is the phase change in a suitable temperature range, depending on external conditions. From the kinetic point of view, there should not be any super-cooling. From the chemical perspective, the materials should have chemical stability, non-toxicity, fire safety, compatibility and environment suitability, even after multiple cycles of use [64].

9.4.1 Energy Storage

Renewable energy is one of the major topics of discussion in 20th century. Direct solar radiation is one of the upcoming solutions for the energy crises we are facing these days with greenhouse gas emissions and other environmentally polluting energy usage. Appropriate conversion of energy has been a major research focus in the recent years. New advanced energy storage systems are also needed to be reliable, environment friendly and efficient in performance. The PCMs with improved properties are one of the attractive candidates for new generation green energy store houses. Their large scale production and application in real life situations is one of the major concerns that needs to be tackled [65].

Thermal energy storage has been mainly achieved using fatty acids and their composite PCMs. It can be utilized in space heating and other practical applications [66]. Polymer based PCMs have been used extensively used in energy storage applications. In one such study, the system consisting of polyethylene/wood flour mixed with micro-encapsulated paraffin (MEP) was developed [67]. Thermal conductivity of the system was improved by adding micro mist graphite (MMG). It was observed from the SEM images that the MEP particles remain undamaged, and form-stable PCMs had homogenous composition. Also, the leakage of molten paraffin was prevented by using the matrix and shell of MEP. It was noted that as-prepared form-stable PCMs had freezing and melting temperatures as well as latent heat suitable for latent heat thermal energy storage (LHTES) applications. It was also observed that after subjecting to 100 melt-freeze cycles, the resultant material had a better thermal stability. The form-stable PCM showed an enhancement of 17% in the thermal conductivity with the addition of 8.8 wt% MMG. The incorporation of MMG had no negative impact on the mechanical behavior of form-stable composite PCMs.

As mentioned earlier, Li *et al.* [29,68] also reported a passive thermal management system for lithium ion batteries using porous metal foam saturated with phase change materials. The composite reduced the surface temperature of the battery and correspondingly enhanced the temperature distribution owing to the increased effective thermal conductivity.

9.4.2 Smart Buildings

Construction and building materials with smart properties compared to the conservative building materials are in increasing demand now a days. Many smart building materials with energy storage capabilities have been developed using suitable phase change materials. This is achieved by incorporating PCMs into materials used for floors, walls and ceilings [69-75].

Microencapsulation is one of the methods used for the synthesis of materials for application in construction industry. For this purpose, PCMs have been incorporated with gypsum, concrete, slurries, etc. [76]. A mixture of PCMs and fluids like water or slurries can act as a two phase latent heat storage fluid. The slurries combined with PCMs are of immense use in civil applications for maintaining temperature inside the buildings [77].

Good performance stability, suitable phase transformation temperature range and high latent heat were exhibited by composites generated by blending dodecanol, butyl stearate and RT20 with organically modified montmorillonite (MMT) [78]. The composite materials showed high compatibility with gypsum. The composites were effective in decreasing the energy consumption by the buildings by decreasing the variation in temperature inside the building. RT20/MMT composite was found to the most suitable material, and the performance increased with enhancing the mass ratio of the composite. In another study, wall boards made from PCM copolymer were observed to enhance the temperature comfort inside the buildings [79]. The wall boards were successful in promoting natural convection inside the room. A numerical method was also suggested by the authors for future building applications.

9.4.3 Textile Applications

Another area where PCMs have found wide range of applications is the production of smart textiles. In these applications, these materials provide appropriate environment for the body and ensure that the physical conditions around the human body are suitable for survival or comfort [80-82].

The phase changing materials can create a small micro-climate around the body which can protect it from the changes in outside temperature. Thus, these materials help in maintaining the thermal balance inside the body. The functions of a PCM material inside a fabric range from thermal conduction as well as heat transfer between the body and the environment to the thermal isolation of the body. Thermal insulation can be achieved by using active or passive insulating materials [83]. In a representative study, Xiao *et al.* [84] developed a silk fabric with thermal controlling properties, using temperature-controlled water vapor annealing.

In-situ polymerization is a common method for the microencapsulation of PCMs for textile applications. Sarier and Onder [85] reported polyurea-formaldehyde microcapsules containing different waxes, using *in-situ* polymerization method. In another study, Sanchez *et al.* [86] reported the preparation of microcapsules with a high amount of PRS paraffin wax, prepared using Shirasu porous glass (SPG) emulsification method and simultaneous suspension polymerization, for fabric applications. The microcapsule material exhibited a high latent heat of 102.42 J/g, confirming its suitability for textile applications. Figure 9.4 shows the schematic of the microencapsulation system.

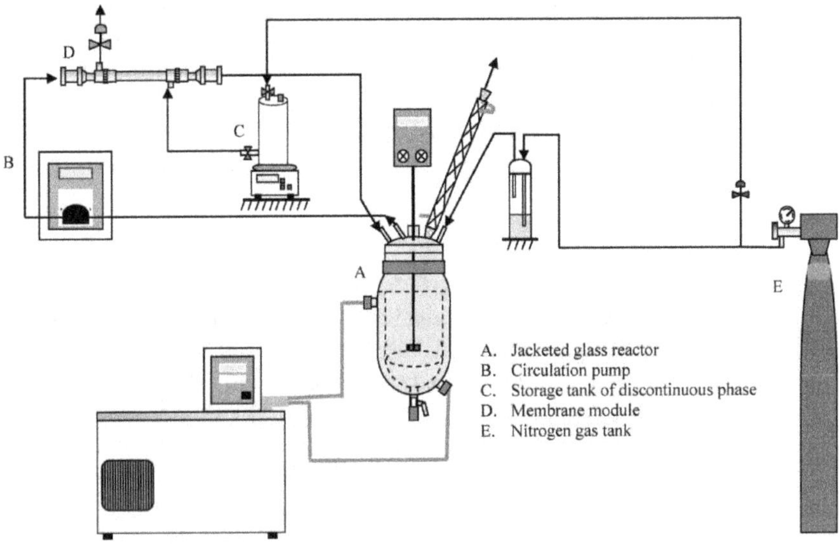

Figure 9.4 Schematic diagram of the microencapsulation system. Reproduced from Reference 86 with permission from American Chemical Society.

9.5 Conclusions

This chapter reviews the properties and applications of different phase changing materials. A number of methods have been employed to synthesize composite assemblies incorporating PCMs, however, economical, environmental friendly and bulk production methods need to be explored further. In addition, the morphology of the phase change materials needs to be taken into consideration while using these synthesis methods as it affects the material's integration for different applications. Thermal and chemical stability as well as conductivity of PCMs also need to be optimized for their practical use. Overall, PCMs are effective in taking human comfort to the next level by contributing to the development of smart buildings, energy storage media, textiles, etc.

References

1. Wang, X., Zhang, Y.-P., Xiao, W., Zeng, R.-L., Zhang, Q.-L., and Di, H.-F. (2009) Review on thermal performance of phase change energy storage building envelope. *Chinese Science Bulletin*, **54**, 920-928.
2. Shi, J.-N., Ger, M.-D., Liu, Y-.-M., Fan, Y.-C., Wen, N.-T., Lin, C.-K., and Pu, N.-W. (2013) Improving the thermal conductivity and shape-stabilization of phase change materials using nanographite additives. *Carbon*, **51**, 365-372.
3. Jankowski, N. R., and McCluskey, F. P. (2014) A review of phase change materials for vehicle component thermal buffering. *Applied Energy*, **113**, 1525-1561.
4. Ma, Z., Lin, W., and Sohel, M. I. (2016) Nano-enhanced phase change materials for improved building performance. *Renewable and Sustainable Energy Reviews*, **58**, 1256-1268.
5. Parameshwaran, R., and Kalaiselvam, S. (2014) Energy conservative air conditioning system using silver nano-based PCM thermal storage for modern buildings. *Energy and Buildings*, **69**, 202-212.
6. Khodadadi, J. M., Fan, L., and Babaei, H. (2013) Thermal conductivity enhancement of nanostructure-based colloidal suspensions utilized as phase change materials for thermal energy storage: A review. *Renewable and Sustainable Energy* Reviews, **24**, 418-444.
7. Kashani, S., Ranjbar, A. A., Madani, M. M., Mastiani, M., and Jalaly, H. (2013) Numerical study of solidification of a nano-enhanced phase change material (NEPCM) in a thermal storage system. *Journal of Applied Mechanics and Technical Physics*, **54**, 702-712.
8. Wu, S. Y., Wang, H., Xiao, S., and Zhu, D. S. (2013) An investigation of melting/freezing characteristics of nanoparticle-enhanced phase change materials. *Journal of Thermal Analysis and Calorimetry*, **110**, 1127-1131.
9. Appavoo, K., Wang, B., Brady, N. F., Seo, M., Nag J., Prasankumar R. P., Hilton D. J., Pantelides, S. T., and Haglund, Jr. R. F., (2014) Ultrafast phase transition via catastrophic phonon collapse driven by plasmonic hot-electron injection. *Nano Letters*, **14**(3), 1127-1133.
10. Chen, L., Zou, R., Xia, W., Liu, Z., Shang, Y., Zhu, J., Wang, Y., Lin, J., Xia, D., and Cao, A. (2012) Electro- and photodriven phase changec based on wax-infiltrated carbon nanotube sponges. *ACS Nano*, **6**(12), 10884-10892.
11. Tumirah, K., Hussein, M. Z., Zulkarnain, Z., and Rafeadah, R. (2014) Nano-encapsulated organic phase change material based on copolymer nanocomposites for thermal energy storage. *Energy*, **66**, 881-890.
12. Fang, G., Chen, Z., and Li, H. (2010) Synthesis and properties of microencapsulated paraffin composites with SiO_2 shell as thermal en-

ergy storage materials. *Chemical Engineering Journal*, **163**, 154-159.

13. Fang, G., Li, H., Yang, F., Liu, X., and Wu, S. (2009) Preparation and characterization of nano-encapsulated n-tertradecane as phase change material for thermal energy storage. *Chemical Engineering Journal*, **153**, 217-221.
14. Wu, W., Bostanci, H., Chow, L. C., Ding, S. J., Hong, Y., Su, M., Kizito, J. P., Gschwender, L., and Snyder, C. E. (2011) Jet impregnation and spray cooling using slurry of nano-encapsulated phase change materials. *International Journal of Heat and Mass Transfer*, **54**, 2715-2723.
15. Karan, N. S., Suresh, S., Sarma, D. D., Paromita, K., Ravishankar, N., and Pradhan, N. (2011) Thermally controlled cyclic insertion/ejection of dopant ions and reversible zinc blende/wurtzite phase changes in ZnS nanostructures. *Journal of The American Chemical Society*, **133**, 1666-1669.
16. Ettouney, H., El-Dessouky, H., and Al-Kandari, E. (2004) Heat transfer characteristics during melting and solidification of phase change energy storage process. *Industrial and Engineering Chemistry Research*, **43,** 5350-5357.
17. Feng, L., Zhao, W., Zheng, J., Frisco, S., Song, P., and Li, X. (2011) The shape-stabilized phase change materials composed of polyethylene glycol and various mesoporous matrices (AC, SBA-15 and MCM-41). *Solar Energy Materials and Solar Cells*, **95**(12), 3550-3556.
18. Shaikh, S., Lafdi, K., and Hallinan, K. (2008) Carbon nano-additives to enhance latent energy storage of phase change materials. *Journal of Applied Physics*, **103**, 094302.
19. Wang, J., Xie, H., and Xin, Z. (2009) Thermal properties of paraffin based composites containing multi-walled carbon nanotubes. *Thermochimica Acta*, **488**(1-2), 39-42.
20. Elgafy, A., and Lafdi K. (2005) Effect of carbon nanofiber additives on thermal behavior of phase change materials. *Carbon*, **43**, 15, 3067-3074.
21. Bonnet, P., Sireude, D., Garnier, B., and Chauvet, O. (2007) Thermal properties and percolation in carbon nanotube-polymer composites. *Applied Physics Letters*, **91**, 201910.
22. Koga, K., Gao, G. T., Tanaka, H., and Zeng, X. C. (2001) Formation of ordered ice nanotubes inside carbon nanotubes. *Nature*, **412**, 802-805.
23. Yavari, F., Fard, H. R., Pashayi, K., Rafiee, M. A., Zamiri, A., Yu, Z., Ozisik, R., Borca-Tasciuc, T., and Koratkat, N. (2011) Enhanced thermal conductivity in a nanostructured phase change composite due to low concentration graphene additives. *The Journal of Physical Chemistry C*, **115**, 8753-8758.
24. Zhao, C. Y. (2012) Review on thermal transport in high porosity cellular metal foams with open cells. *International Journal of Heat and*

Mass Transfer, **55**(13), 3618-3632.

25. Chen, Z., Gao, D., and Shi, J. (2014) Experimental and numerical study on melting of phase change materials in metal foams at pore scale. *International Journal of Heat and Mass Transfer*, **72**, 646-655.
26. Qu, Z. G., Li, W. Q., Wang, J. L., and Tao, W. Q. (2012) Passive thermal management using metal foam saturated with phase change material in a heat sink. *International Communications in Heat and Mass Transfer*, **39**(10), 1546-1549.
27. Ji, H., Sellan, D. P., Pettes, M. T., Kong, X., Ji, J., Shi, L., and Ruoff, R. S. (2014) Enhanced thermal conductivity of phase change materials with ultrathin-graphite foams for thermal energy storage. *Energy and Environmental Science*, **7**, 1185-1192.
28. Li, W. Q., Qu, Z. G., He, Y. L., and Tao, Y. B. (2012) Experimental and numerical studies on melting phase change heat transfer in open-cell metallic foams filled with paraffin. *Applied Thermal Engineering*, **37**, 1-9.
29. Li, W. Q., Qu, Z. G., He, Y. L., and Tao, Y. B. (2014) Experimental study of a passive thermal management system for high-powered lithium ion batteries using porous metal foam saturated with phase change materials. *Journal of Power Sources*, **255**(1), 9-15.
30. Fukai, J., Hamada, Y., Morozumi, Y., and Miyatake, O. (2003) Improvement of thermal characteristics of latent heat thermal energy storage units using carbon-fiber brushes: experiments and modeling. *International Communications in Heat and Mass Transfer*, **46**, 4513-4525.
31. Chow, L. C., Zhong, J. K., and Beam, J. E. (1996) Thermal conductivity enhancement for phase change storage media. *International Communications in Heat and Mass Transfer*, **23**(1), 91-100.
32. Cui, Y., Liu, C., Hu, S., and Yu, X., (2011) The experimental exploration of carbon nanofiber and carbon nanotube additives on thermal behavior of phase change materials. *Solar Energy Materials and Solar Cells*, **95**(4), 1208-1212.
33. McCann, J. T., Marquez, M., and Xia, Y. (2006) Melt coaxial electrospinning: A versatile method for the encapsulation of solid materials and fabrication of phase change nanofibers. *Nano Letters*, **6**(12), 2868-2872.
34. Chen, C., Wang, L., and Huang, Y. (2008) Morphology and thermal properties of electrospun fatty acids/polyethylene terephthalate composite fibers as novel form-stable phase change materials. *Solar Energy Materials and Solar Cells*, **92**(11), 1382-1387.
35. Shannon, T. (2003) Fibrous Materials Exhibiting Thermal Change During Use, patent US 20050136765A1.
36. Pause, B. (1997) Interactive Thermal Insulating System having a Layer Treated with a Coating of Energy Absorbing Phase Change Material Adjacent a Layer of Fibers Containing Energy Absorbing

Phase Change Material, patent US 6077597A.

37. Buckley, T. M. (1992) Flexible Composite Material with Phase Change Thermal Storage, patent US 6004662A.
38. Bryant, Y. G., Colvin, D. P., Driscoll, J. C., and Mulligan, J. C. (1995) Thermal Insulating Coating Employing Microencapsulated Phase Change Material and Method, patent US 5804297A.
39. Bryant, Y. G., and Colvin, D. P. (1987) Fiber with Reversible Enhanced Thermal Storage Properties and Fabrics Made Therefrom, patent US 4756958A.
40. Zdraveva, E., Fang, J., Mijovic, B., and Lin, T. (2015) Electrospun Poly (vinyl alcohol)/ phase change material fibers: morphology, heat properties, and stability. *Industrial and Engineering Chemistry Research,* **54**, 8706-8712.
41. Zhang, D., Zhou, J., Wu, K., and Li, Z. (2005) Granular phase changing composites for thermal energy storage. *Solar Energy,* **78**(3), 471-480.
42. Karaman, S., Karaipekli, A., Sarı, A., and Biçer, A. (2011) Polyethylene glycol (PEG)/diatomite composite as a novel form-stable phase change material for thermal energy storage. *Solar Energy Materials and Solar Cells,* **95**(7), 1647-1653.
43. Nomura, T., Okinaka, N., and Akiyama, T. (2009) Impregnation of porous material with phase change material for thermal energy storage. *Materials Chemistry and Physics,* **115**(2), 846-850.
44. Mesalhy, O., Lafdi, K., Elgafy, A., and Bowman, K. (2005) Numerical study for enhancing the thermal conductivity of phase change material (PCM) storage using high thermal conductivity porous matrix. *Energy Conversion and Management,* **46**(6), 847-867.
45. Lafdi, K., Mesalhy, O., and Elgafy, A. (2008) Graphite foams infiltrated with phase change materials as alternative materials for space and terrestrial thermal energy storage applications. *Carbon,* **46**(1), 159-168.
46. Brent, A. D., Voller, V. R., and Reid, K. J. (1988) Enthalpy-porosity technique for modeling convection-diffusion phase change: application to the melting of a pure metal. *Numerical Heat Transfer,* **13**(3), 297-318.
47. Zhang, D., Tian, S., and Xiao, D. (2007) Experimental study on the phase change behavior of phase change material confined in pores. *Solar Energy,* **81**(5), 653-660.
48. Zhang, Z., and Fang, X. (2006) Study on paraffin/expanded graphite composite phase change thermal energy storage material. *Energy Conversion and Management,* **47**(3), 303-310.
49. Zhou, D., and Zhao, C. Y. (2011) Experimental investigations on heat transfer in phase change materials (PCMs) embedded in porous materials. *Applied Thermal Engineering,* **31**(5), 970-977.
50. Farid, M., and Kong, W. J. (2001) Underfloor heating with latent heat

storage. *Proceedings of the Institution of Mechanical Engineers, Part A, Journal of Power and Energy*, **215**, 601-609.

51. Farkas, D., and Birchenall, C. E. (1985) New eutectic alloys and their heats of transformation. *Metallurgical and Materials Transactions A*, **16**, 323-328.
52. Vafai, K., and Sarkar, S. (1986) Condensation effects in a fibrous insulation slab. *Journal of Heat Transfer*, **108** (8), 667-675.
53. Kedl, R. J. (1990) Conventional Wallboard with Latent Heat Storage for Passive Solar Applications. *Proceedings of the 25th Intersociety Energy Conversion Engineering Conference*, USA, doi: 10.1109/IECEC.1990.716492.
54. Feldman, D., Banu, D., Hawes, D., and Ghanbari, E. (1991) Obtaining an energy storing building material by direct incorporation of an organic phase change material in gypsum wallboard. *Solar Energy Materials*, **22**, 231-242.
55. Drake, J. B. (1987) *A study of the Optimal Transition Temperature of PCM Wallboard for Solar Energy Storage*. Oak Ridge National Laboratory, USA. Online: https://www.osti.gov/servlets/purl/6100261 [accessed 21st December 2018].
56. Charach, C., Zarmi, Y., and Zemel, A. (1987) Simple method for assessing the thermal performance of PCM panels. *Proceedings of the Biennial Congress of the International Solar Energy Society*, Germany, pp. 1212-1216.
57. Pincemin, S., Olives, R., Py, X, and Christ M. (2008) Highly conductive composites made of phase change materials and graphite for thermal storage. *Solar Energy Materials & Solar Cells*, **92**, 603-613.
58. Ait Adine, H., and El Qarnia, H. (2009) Numerical analysis of the thermal behaviour of a shell-and-tube heat storage unit using phase change materials. *Applied Mathematical Modelling*, **33**, 2132-2144.
59. Lacroix, M. (1993), Numerical simulation of a shell and tube latent heat thermal energy storage unit. *Solar Energy*, **50**(4), 357-367.
60. Dincer, I., and Rosen, M. A. (2002) *Thermal Energy Storage, Systems and Applications*, John Wiley & Sons, USA.
61. Raoux, S., Shelby, R. M., Jordan-Sweet, J., Munoz, B., Salinga, M., Chen, Y.-C., Shih, Y.-H., Lai, E.-K., and Lee, M.-H. (2008) Phase change materials and their application to random access memory technology. *Microelectronic Engineering*, **85**(12), 2330-2333.
62. Lencer, D., Salinga, M., and Wuttig, M. (2011) Design rules for phase-change materials in data storage applications. *Advanced Materials*, **23**, 2030-2058.
63. Raoux, S., Welnic, W., and Ielmini, D. (2010), Phase change materials and their application to nonvolatile memories. *Chemical Reviews*, **110** (1), 240-267.
64. Baetens, R., Jelle, B. P., and Gustavsen, A. (2010) Phase change materials for building applications: A state-of-the-art review. *Energy*

and Buildings, **42**(9), 1361-1368.

65. Akhiani, A. R., Mehrali, M., Latibari, S. T., Mehrali, M., Mahlia, T. M. I., Sadeghinezhad, E., and Metsellar, H. S. C. (2015) One-step Preparation of form-stable phase change material through self-assembly of fatty acid and graphene. *The Journal of Physical Chemistry C*, **119**(40), 22787-22796.
66. Sharma, A, Tyagi, V. V., Chen, C. R, and Buddhi, D. (2009) Review on thermal energy storage with phase change materials and applications. *Renewable and Sustainable Energy Reviews*, **13**(2), 318-345.
67. Li, J., Xue, P., Ding, W., Han, J., and Sun, G. (2009) Micro-encapsulated paraffin/high-density polyethylene/wood flour composite as form-stable phase change material for thermal energy storage. *Solar Energy Materials and Solar Cells*, **93**(10), 1761-1767.
68. Li, W. Q., Wan, H., Jing, T. T., Li, Y. B., Liu, P. J., He, G. Q., and Qin, F. (2019) Microencapsulated phase change material (MEPCM) saturated in metal foam as an efficient hybrid PCM for passive thermal management: A numerical and experimental study. *Applied Thermal Engineering*, **146**, 413-421.
69. Khudhair, A. M, and Farid, M. M. (2004) A review on energy conservation in building applications with thermal storage by latent heat using phase change materials. *Energy Conversion and Management*, **45**(2), 263-275.
70. De Gracia, A., and Cabeza, L. F. (2015) Phase change materials and thermal energy storage for buildings. *Energy and Buildings*, **103**, 414-419.
71. Heim, D., and Clarke, J. A. (2003) Numerical Modeling and Thermal Simulation of Phase Change Material with ESP-R. E*ighth international IBPSA Conference*, The Netherlands, pp. 459-466.
72. Tyagi, V. V., Kaushik, S. C., Tyagi, S. K., and Akiyama, T. (2011) Development of phase change materials based microencapsulated technology for buildings: a review. *Renewable and Sustainable Energy Reviews*,**15**, 1373-1391.
73. Ibanez, M., Lazaro, A., Zalba, B., and Cabeza, L. F. (2005) An approach to the simulation of PCMs in building applications using TRNSYS. *Applied Thermal Engineering*, **25**, 1796-807.
74. Boh, B., and Sumiga, B. (2008) Microencapsulation technology and its applications in building construction materials. *RMZ - Materials and Geo environment*, **55**(3), 329-344.
75. Chandra Sekhar, G., Venkatesu, P., Hofman, T., and Prabhakara Rao, M. V. (2002) Solid-liquid equilibria of long chain n-alkanes (C18-C28) in n,n-dimethylacetamide. *Fluid Phase Equilibrium*, **201** 219-231.
76. Morikawa, J., and Hashimoto, T. (2000) Simultaneous measurement of heat capacity and thermal diffusivity in solid-solid and solid-liquid phase transitions of n- alkane. *Thermochimica Acta*, **352-353**,

291-296.

77. Delgado, M., Lazaro, A., Mazo, J., and Zalba, B. (2012) Review on phase change material emulsions and microencapsulated phase change material slurries: Materials, heat transfer studies and applications. *Renewable and Sustainable Energy Reviews*, **16**(1), 253-273.
78. Fang, X., Zhang, Z., and Chen, Z. (2008) Study on preparation of montmorillonite-based composite phase change materials and their applications in thermal storage building materials. *Energy Conversion and Management*, **49**(4), 718-723.
79. Kuznik, F., and Virgone, J. (2009) Experimental assessment of a phase change material for wall building use. *Applied Energy*, **86**(10), 2038-2046.
80. Alvarado, J. L., Marsh, C., Sohn, C., Vilceus, M., Hock, V., Phetteplace, G, and Newell, T. (2006) Characterization of supercooling suppression of microencapsulated phase change material by using DSC. *Journal of Thermal Analysis and Calorimetry*, **86** (2) 505-509.
81. Quanying, Y., Chen, L., and Lin, Z. (2008) Experimental study on the thermal storage performance and preparation of paraffin mixtures used in the phase change wall. *Solar Energy Materials and Solar Cells*, **92,** 1526-1532.
82. Van Miltenburg, J. C. (2000) Fitting the heat capacity of liquid n-alkanes: new measurements of n-heptadecane and n-octadecane. *Thermochimica Acta*, **343,** 57-62.
83. Mondal, S. (2008) Phase change materials for smart textiles - An overview. *Applied Thermal Engineering*, **28,** 1536-1550.
84. Hu, X., Shmelev, K., Sun, L., Gil, E.-S., Park, S.-H., Cebe, P., and Kaplan, D. L. (2011) Regulation of silk material structure by temperature-controlled water vapor annealing. *Biomacromolecules*, **12**, 1686-1696.
85. Sarier, N., and Onder, E. The manufacture of microencapsulated phase change materials suitable for the design of thermally enhanced fabrics. *Thermochimica Acta*, **452**(2), 149-160.
86. Sanchez, L., Lacasa, E., Carmona, M., Rodrigues, J. F., and Sanchez, P. (2008) Applying an experimental design to improve the characteristics of microcapsules containing phase change materials for fabric uses. *Industrial and Engineering Chemistry Research*, **47**(23), 9783-9790.

10

Thermally Conducting Polymer Nanocomposites: Synthesis, Properties and Applications

10.1 Introduction

Polymers are best known for their insulating behavior and relatively higher specific heat capacities. As most of the polymers (e.g. polyethylene, polypropylene, epoxy) have low thermal conductivities in the range of 0.11-0.44 $W.m^{-1}.K^{-1}$, optimization of their thermal conductivities is vital for melt processing and specific applications. During polymer melt processing, thermal conductivity plays a key role due to the requirement of efficient heat transfer for processing cycle times. Similarly, certain applications of polymers also require enhanced thermal conduction such as electronic packaging, assembly, batteries, solar cells, satellite devices, etc. Generally, in perfect solids, atoms are in constant modes of coupled and coordinated vibrations with high frequency and comparatively small amplitudes (shorter wavelengths) which can be assumed as elastic waves. Analogous to photon, quantum of vibrational energy is called phonon. Contrary to metals, the dominant heat conduction mechanism in most of the polymers is through phonons [1]. Specifically, heat conduction in polymers is accomplished through combination of phonons and motion of chains molecules i.e. vibrational and rotational motion. Heat transfer in polymers mainly depends on the extent of crystallinity of the material, thus, the polymer with high crystallinity has higher conductivity due to ordered structure as compared to the polymer with amorphous or disordered structure. Thermal conductivity can be defined as analogous to Fick's first law:

$$q = kA\frac{\Delta T}{L}$$

where q represents the heat flux or rate of heat transfer, A is the cross-sectional area, k is the thermal conductivity, ΔT is the temper-

Ali U. Chaudhry and Vikas Mittal*
Khalifa University of Science and Technology, Abu Dhabi, UAE
**Current address: Texas A&M University, Qatar*

ature difference and L is the conduction path length [2]. Enhancement in the thermal conductivity of polymeric materials can be achieved by fabricating polymer composites and alloys by the addition of organic or inorganic fillers.

Nano-scale materials have at least one external dimension or internal feature in the size range of 1-100 nanometers [3]. Nanocomposites generated using nano-fillers such as layered silicates, graphene nano-sheets, graphite nanoparticles, carbon nanotubes (CNTs) and metal oxide nanoparticles have attracted much research attention [4]. Using nano-fillers in polymer matrices is advantageous due to high surface to volume ratio and improved properties at much lower filler loading such as low density, mechanical strength as well as enhanced electrical, thermal, optical, and rheological properties. Same approach has also been adopted to improve the thermal conductivity of polymers by the addition of heat conductive fillers in the matrices. In literature, various thermally conductive fillers such as aluminum nitride, aluminum oxide (Al_2O_3), silica (SiO_2), graphite, boron nitride (BN), graphene, metallic particles (Cu), diamond, etc. have been reported. In conducting polymer composites, the improvement of thermal conductivity is thus mainly attributed to the incorporated fillers due to very low heat conduction ability of polymers. Filler properties like shape, size, concentration, interaction with polymer, alignment and orientation have strong effect on the conductivity of the composite systems (Figure 10.1) [1,5]. This review discusses recent developments in enhancing thermal conductivity of polymer based nanocomposites. Various polymer and filler types as well as techniques used to evaluate the thermal conductivity and microstructure of thermally conducting polymer nanocomposites are discussed, followed by the applications of these composites.

10.2 Modeling of Thermal Conductivity

Designing of polymer composites for improved thermal conductivity depends on many factors such as type of polymer and filler, concentration of filler (loading), processing conditions, combinations of different reinforcements, etc. Prediction of thermal conductivity through modeling is vital for designing effective polymer composites with optimum set of properties. Among various available approaches, two mainly employed models for thermal conductivity prediction are analytical micro-mechanical models (MME) (constitutive equations) and finite element (FC) simulations. Modeling of pure material is comparatively easier than the composite system due to evolution of complex morphology during processing i.e. varying degrees of filler dispersion, network formation, and interfacial resistance. Thus, the

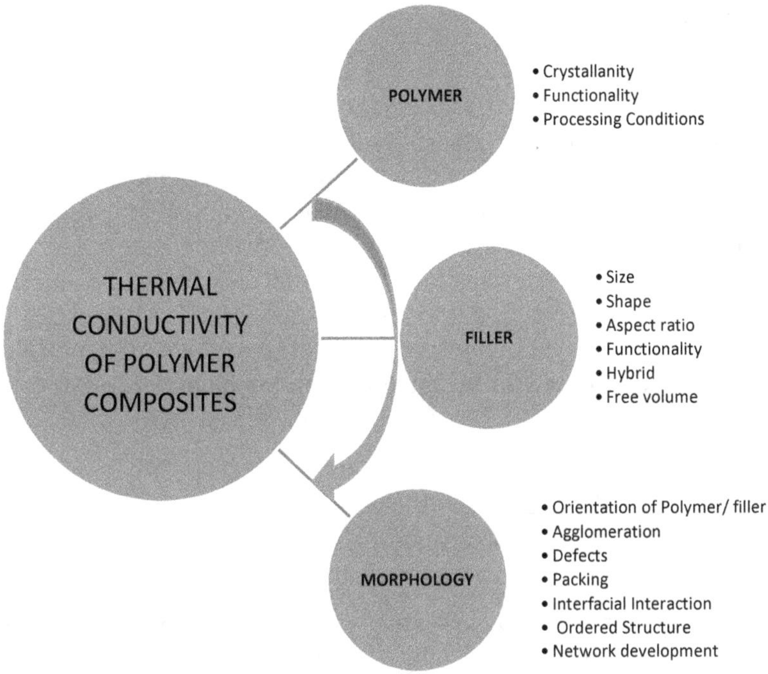

Figure 10.1 Design parameters for thermal conductivity of polymer composites.

predicted thermal conductivity values for polymer composites from these models are generally qualitative or semi-quantitative. MME models are very useful for rapid evaluation of thermal conductivity for composite system on the basis of composition and properties of the host polymers and fillers. However, these models are often limited to certain composite systems like homogenous matrix incorporated with mono-disperse sphere-shaped or perfectly aligned ellipsoidal fillers. In this regard, two main approaches are adopted to predict the thermal conductivity of the composites i.e. upper bound (parallel model, linear mixing rule) and lower bound (series model, inverse mixing rule). In basic parallel model, the overall conductivity is predicted based on the assumption that the heat flux is the weighted sum of heat fluxes through domain of each composite phase, whereas temperature gradient is uniform. Parallel model also assumes that the particles are in complete contact forming percolating network. According to the parallel model, thermal conductivity can be calculated in terms of volume fractions of each phase as

$$k_c = k_p \Phi_p + k_m \Phi_m$$

where k_c, k_p and k_m are the thermal conductivities of the composite, polymer phase and filler phase, whereas ϕ_p and ϕ_m are the volume fractions of the polymer and filler respectively. On the contrary, the basic series model assumes that the overall temperature gradient is based on the weighted sum of temperature gradients through the domain of each phase, whereas heat flux is uniform. It also assumes that filler particles are completely out of contact with each other and percolation is impossible to achieve. According to the series model, the thermal conductivity can be calculated in terms of volume fractions as [1,6]

$$k_c = \frac{1}{(k_p + \Phi_p) + (k_m + \Phi_m)}$$

Among these models, inverse mixing rule (series model) is more common due to the ability to make predictions matching the experimental data. Series model has also formed the basis of many models built on complex weighted averages of k_c, k_p, ϕ_p and ϕ_m. These complex models frequently consider semi-theoretical fitting parameters and depend on effective medium approximations (EMA) or effective medium theories (EMT). Based on EMA approach, many common models such as Maxwell model, Bruggeman model, Hasselman-Johnson model, Rayleigh models, Lewis-Nielsen model and percolation model have been developed [7,8]. Maxwell model considers spherical non-interacting particles embedded in a continuous matrix, where effective thermal conductivity of the composite k_{eff} is given as

$$k_{eff} = k_m \left[1 + \frac{3\Phi_f}{\frac{(k_f + 2k_m)}{k_f - k_m} - \Phi_f} \right]$$

where k_f and ϕ_f represent the thermal conductivity and volume fraction of the filler phase. The validity of Maxwell model is traditionally for composite systems under 25% of filler fraction, which led to other modifications for particle shape and different phases of filler particles [9,10]. Similarly, Rayleigh introduced another model based on the thermal interactions between the filler particles. The model considers the filler particles as cubically arranged spherical shape inclusions in the matrix and can be used for higher volume fraction of filler [11]. According to Rayleigh model, k_{eff} can be calculated as

$$k_{eff} = k_m \left[1 + \frac{3\Phi_f}{\left(\frac{k_f + 2k_m}{k_f - k_m}\right) - \Phi_f + 1.569\left(\frac{k_f - k_m}{3k_f - 4k_m}\right)\Phi_f^{\frac{10}{3}} + \ldots\ldots} \right]$$

Using Maxwell and Rayleigh models, Hasselman-Johnson developed another model considering filler volume fraction, interfacial gaps, thermal resistance and particle size in order to develop the relation for k_{eff}. According to Hasselman-Johnson model, equations of k_{eff} for spherical, cylindrical, and flat plate filler geometry are given as

$$k_{eff} = k_m \left[\frac{2\left(\frac{k_f}{k_m} - \frac{k_f}{ah_c}\right)\Phi_f + \frac{k_f}{k_m} + \frac{2k_f}{ah_c} + 2}{\left(1 - \frac{k_f}{k_m} + \frac{k_f}{ah_c}\right)\Phi_f + \frac{k_f}{k_m} + \frac{2k_f}{ah_c} + 2} \right] ----Spherical$$

$$k_{eff} = k_m \left[\frac{\left(\frac{k_f}{k_m} - \frac{k_f}{ah_c} - 1\right)\Phi_f + \left(1 + \frac{k_f}{k_m} + \frac{2k_f}{ah_c}\right)}{\left(1 - \frac{k_f}{k_m} + \frac{k_f}{ah_c}\right)\Phi_f + \left(1 - \frac{k_f}{k_m} + \frac{k_f}{ah_c}\right)} \right] ----Cylindrical$$

$$k_{eff} = \left[\frac{k_f}{\left(1 - \frac{k_f}{k_m} + \frac{2k_f}{ah_c}\right)\Phi_f + \left(\frac{k_f}{ah_c}\right)} \right] ----Flat\ plate$$

where a and h_c represent the particle radius and boundary conductivity respectively [12,13]. Further, Bruggeman used differential equations to calculate infinitesimal changes in incrementally constructed composite system. This approach is usually called differential effective medium theory (DEM) and can be used for different systems and high filler volume fractions. Using Bruggeman approach, many research studies obtained k_{eff} for different systems given as

$$(1-\Phi_f)^3 = \left(\frac{k_m}{k_{eff}}\right)^{(1+2\alpha)/(1-\alpha)} \left[\frac{k_{eff}-k_f(1-\alpha)}{k_m-k_f(1-\alpha)}\right]^{\frac{3}{(1-\alpha)}} ----Particulate-Composite$$

$$\frac{k_f}{k_m} = \frac{1}{(1-\Phi)^{3(1-\alpha)/(1+2\alpha)}} ----ZnS-Diamond$$

$$(1-\Phi_f) = \left(\frac{k_m}{k_{eff}}\right)^{1/3} \left[\frac{k_{eff}k_fR_{int}+ak_{eff}-\alpha k_1)}{k_mk_fR_{int}+ak_m-\alpha k_1}\right] ----Composite$$

where α is a dimensionless parameter, which depends on the interfacial thermal resistance (R_{int}) between the filler and matrix and $\alpha = a_k/a$, where a is the particle size, a_k is the Kapitaz radius ($a_k = R_{int}k_m$) [14,15]. Lewis-Nielsen model was another simple and popular model reported in literature for moderate filler fractions (<40%). The benefits of this model are its applicability for a broad range of particle shapes and arrangements. The k_{eff} of a composite according to this model is given as

$$k_{eff} = \frac{1+AB\Phi_f}{1-B\psi\Phi_f},\ B = \left(\frac{\left(k_f/k_m\right)-1}{\left(k_f/k_m\right)+A}\right),\ \psi = 1+\left(\frac{1-\Phi_m}{\Phi_m^2}\right)$$

where ϕ_m and A represent maximum filler volume fraction and shape coefficient for the filler particles respectively [8,16,17].

10.3 Measurement Techniques

Many techniques with different precision levels have been introduced for determining the thermal conductivity of materials. In general, the measurement techniques can be divided into two basic groups i.e. steady state and transient methods. The techniques in the former group perform the measurement when system has attained stability, whereas the techniques in the latter group take the measurement during heating or cooling process. The ability of the measuring techniques in both groups primarily depends on factors such as temperature range, material type and range of thermal conductivity [18]. Steady state techniques deal with the measurements of heat flux and temperature gradient through the materials and generally take longer time to achieve equilibrium. Steady state techniques include methods like guarded hot plate, axial flow, heat flow meter and pipe method [1]. Guarded hot plate method is usually employed for materials with low thermal conductivity and requires large sample size. The advantage of this method is its high accuracy (2%) and its usability in broad range of temperature of 80-1500 K. In this method, the sample is placed between the cold and hot plates attached with heaters and thermal insulation. The heat is allowed to pass through the specimen and the temperature is measured at both sides when the system attains steady state. Thermal conductivity is calculated on the basis of heat flux and temperature gradient across the sample, thickness and surface area of specimen [19]. Axial flow method is a longitudinal method to measure wide range of thermal conductivity of the materials using low temperature. In this method, the conductivity is usually measured by creating a temperature difference across the sample which is placed between two suitable references of known conductivity and attached with heater and heat sink [20]. Heat flow meter is similar to guarded hot plate method except the presence of heat flux sensor instead of main heater. It is simple in construction and has the ability to operate between −173-473 K, 90-1300 K, 298-2600 K for normal, axial and radial heat flow respectively [21]. Pipe method is another kind of radial heat flow method with wide range of operating temperature (293-2700 K) and targeted thermal conductivity range (0.02-2 W/(m.K)). It uses the principle of temperature gradient where cylindrical sample contains a core heater surrounded by a heat sink [18,22]. As mentioned earlier, transient methods perform the thermal conductivity measurements during transient addition of heat to the sample. Various transient methods include laser flash method, transient hot wire method and transient plane source method. In these methods, the thermal diffusivity is recorded as a function of time which is further used to calculate the thermal conductivity. In laser flash method, a sample disk is heated by laser pulse and the temperature is monitored by a detector. Transient hot wire method is similar to the pipe method where a hot wire is embedded in the test sample which acts as a sensor for the heater. The heating in transient plane source method is achieved with a known electrical current pulse. In this method, the heating element is placed between two test samples of same material, while temperature resistance is measured as a function of time [23-26].

10.4 Polymer and Fillers for Thermally Conducting Nanocomposites

As mentioned earlier, most of the polymers are insulator at macroscale and usually have thermal conductivity below 0.5 W/(m.K). Generally, the polymers with higher degree of crystalline domains have higher thermal conductivity than the amorphous polymers. Similarly, presence of side chains in the main chain also result in lower thermal conductivity of the polymer. Specifically, in thermoplastic polymers, thermal conductivity is affected by the factors such as chain structure and orientation, crystallinity and interchain interactions. On the contrary, the factors which affect the thermal conductivity in thermoset polymers are mainly liquid crystal domain size and content, curing conditions and orientation [1].

Recent studies have indicated that the polymers with stiff backbone, such as conducting polymers, can have improved thermal conductivity due to suppressed chain conformation. Similarly, polymers with strong interchain interactions (e.g. Teflon) can also suppress the chain conformation [27,28]. In similar manner, polymers containing more crystalline phases have higher thermal conductivity than the amorphous polymers due to the fact that amorphous phases cause phonon scattering, which is not observed in crystalline phase (e.g. poly (ether ether ketone)). In crystalline phases, thermal conductivity further depends on the extent of extended crystal structure and thickness of lamellae in the polymer morphology [29]. Moreover, the thermal conductivity is also affected with the direction of chains i.e. transverse direction can have significantly higher thermal conductivity. Similarly, orientation of polymer chains through mechanical stretching can also enhance thermal conductivity [30]. It was also reported that the thermal conductivity of a H-bond-accepting amorphous polymer ((poly(N-acryloyl piperidine)) was improved due to engineered interchain interaction by blending with H-bond donating polymers such as (poly(acrylic acid), poly(vinyl alcohol), or poly(4-vinyl phenol)) [31].

Thermosets matrices are formed by crosslinking of either liquid monomers (epoxy resin) or solid polymers (rubber, polyethylene). In most of the cases, the mechanical and thermal properties of thermosets are superior to those of thermoplastic polymers. Similar to thermoplastics, thermosets also have amorphous (epoxy) and crystalline phases (liquid crystals) with low and high thermal conductivity, respectively. It was reported that the thermal conductivity of thermosets from liquid crystals strongly depends on the liquid crystal domain size and content. It usually increases with the volume percentage of anisotropic structure and liquid crystal domain size. As mentioned earlier, other factors like curing conditions have also significant great influence on the thermal conductivity of the polymer. For instance, in case of liquid crystal processing, higher temperature and slow cooling reduce the crystalline phase by changing from highly oriented smectic state, then to weakly oriented nematic state and finally to isotropic amorphous state [32-34]. Similar to thermoplastics, the thermal conductivity also varies with the orientation of polymer chains and significantly increases in the orientation direction, e. g. in highly oriented liquid crystalline thermosets [35].

It is also vital to discuss the factors associated with various fillers which significantly determine the thermal conductivity of the whole composite system due to low thermal conductivity of the polymer matrices. Different types of filler materials based on ceramic (BeO, Al_2O_3), metal (Cu, Al) and carbon (graphite) categories have been used with polymers for improved thermal conductivity. Generally, the materials which conduct heat through phonons are less efficient than those through electrons due to the resistance to scattering and higher speed of electrons [36]. Metallic fillers are quite effective in this process and are widely used in order to improve the thermal conductivity of polymers. The major disadvantage associated with metallic fillers is their high density and inherited electrical conductivity which can increase the weight and reduce the dielectric breakdown voltage, respectively [37]. On the other hand, carbon based materials belong to the class of conductive materials which have very high thermal conductivity and comparatively lower weight. As a result, materials like graphite, exfoliated graphite, carbon nanotubes, carbon fibers, and conductive carbon black have been extensively considered as conductive fillers for improved thermal conductivity [38]. Contrary to metal and carbon bases fillers, ceramic fillers are intrinsically electrically insulating materials. Transport of heat through ceramic fillers is mostly through phonons and this category usually comprises of both oxides (BeO, SiO_2) and non-oxides (AlN, BN, Si_3N_4) based fillers. Some shortcomings associated with ceramic fillers were also reported in the literature such as anisotropic behavior, crystal structure, crystallinity, presence of surface impurities, grain boundaries, etc. [39,40]. In addition to the inherited properties of fillers, the thermal conductivity of the system also depends on other factors such as filler loading level, shape and size of particles and interfacial adhesion. Usually, the thermal conductivity of the composites exhibits non-linear behavior with increasing filler loading and requires higher loading of fillers to achieve filler-to-filler connections [41]. Moreover, fillers with low aspect ratio exhibit a lower thermal conductivity as compared to the fillers with high aspect ratio (fibers, tubes). Usually, one dimensional fillers (1-D) develop well connected long conductive pathways in composite systems which result in significantly improved thermal conductivity. Examples of 1-D fillers are carbon nanotubes, carbon fibers, Si_3N_4 nanowires, boron nitride nanotubes, silver nanowires, copper nanowires, etc. [42-44]. Similarly, two dimensional (2-D) fillers, such as platelet fillers, are also considered as high aspect ratio fillers and can exhibit very high in-plane thermal conductivity, examples of such fillers with plate like morphology are boron nitride, graphene, Al_2O_3, TiB_2 and SiC [45]. Further, filler size is also vital for improving the thermal conductivity of the polymer composite systems as smaller size filler particles provide more interfacial area which results in more phonon scattering. Larger sized filler particles create effective network in the polymer composite system which reduces the thermal interfacial resistance. It has also been reported that the thermal conductivity does not depend on the particle size as nano-sized fillers were also observed to exhibit similar enhancement in thermal conductivity as the micro-sized fillers. Overall, the dependence of the thermal conductivity is generally linked to network formation by conductive fillers regardless of the size [46-48]. Further, the thermal conductivity can also be improved by using other techniques such as the use of hybrid fillers, different filler surface treatments, etc. [49-51]. Microstructure of the composite also affects the thermal conductivity of the system. Development of the composite microstructure occurs during processing of the composite and depends on many factors such as processing conditions and ability of filler to orientate, agglomerate and form network in the polymer. During processing, filler particles can also be oriented using externally applied fields and shear or extensional flow. Similarly, formation of network by filler particles in polymers can be achieved by self-assembly of the filler particles in the polymer, molding of the mixture of filler and polymer powders, in-situ polymerization or double percolation. Agglomeration of filler particles is also occasionally required in order to achieve composite systems with isotropic thermal conductivity [52-56].

Before moving to the applications of various thermally conducting polymer nanocomposite systems, few general examples of thermally conducting nanocomposites are presented here in order to underline the above-mentioned factors. Shahil and Balandin reported epoxy nanocomposites using graphene - multilayer graphene hybrid filler [57]. Significant increase in the thermal conductivity was of the composites was observed as a function of filler volume fraction, as shown in Figure 10.2. As observed in Figure 10.2a, an enhancement of nearly 25 times in the thermal conductivity of the nanocomposites was achieved, which was significantly higher than corresponding composites with graphite and carbon black. In another study, Yao *et al.* [58] reported the nanocomposites of polyaniline with single-wall nanotubes. The nanocomposites exhibited small increase in thermal conductivity, however, the enhancement was much less as compared to

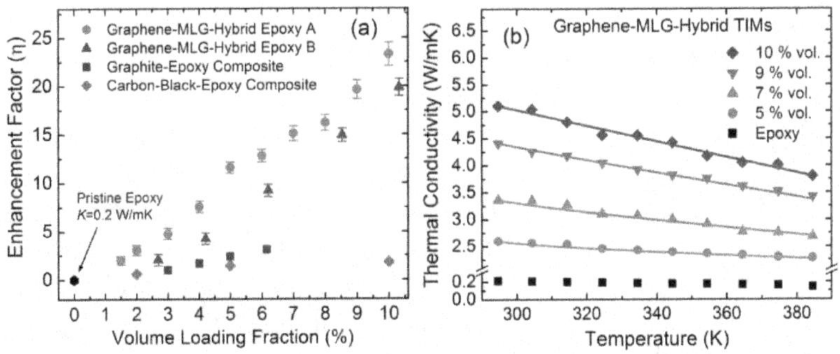

Figure 10.2 (a) Thermal conductivity enhancement factor as a fraction of the filler volume fraction; (b) thermal conductivity of the thermal interface materials as a function of filler fraction and temperature. Reproduced from Reference 57 with permission of American Chemical Society.

the predictions, as shown in Figure 10.3. Even with filler fraction of 41 wt%, the thermal conductivity of the nanocomposite was 1.5 W/(mK). The authors attributed this phenomenon to the phonon scattering effect of the nano-interfaces generated by polyaniline/nanotubes nanocomposite structure. Thus, varying degrees of thermal conductivity enhancements have been achieved in literature.

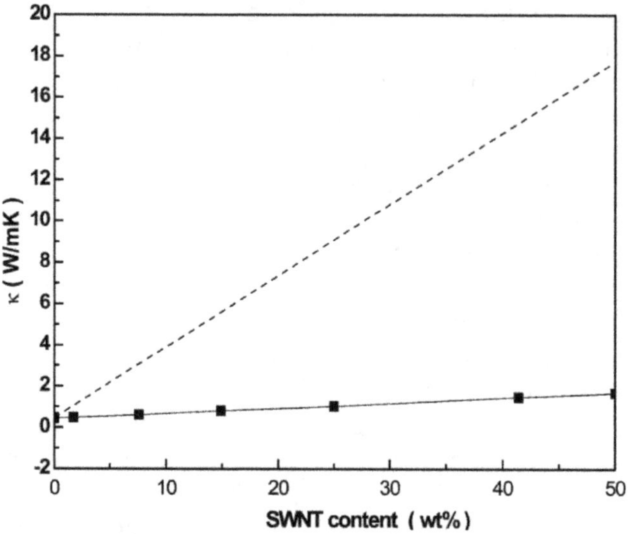

Figure 10.3 Thermal conductivity of the polyaniline-single walled nanotubes nanocomposites as a function of filler content. Dotted line indicates the prediction of thermal conductivity development with filler amount. Reproduced from Reference 58 with permission from American Chemical Society.

10.5 Applications

As mentioned earlier, polymer nanocomposites have advantages over conventional composite systems due to significant improvements in mechanical properties, gas barrier, thermal stability, chemical resistance, electrical and thermal conductivity. The improved properties of polymer nanocomposites at low volume fraction are due to the incorporated fillers with high surface to volume ratio and large interfacial area between the polymer and filler phases. With increasing demand in thermal management applications, where heat removal has become tremendously significant, the field of thermally conductive nanocomposites is expanding rapidly. These materials represent new potential for replacing metal, ceramic and pure polymer components in numerous applications owing to recyclability, electrical insulation, manufacturability, compactness, easy processing, light weight, anti-corrosion performance, etc. [59]. The following sections discuss selected evolving application areas of thermally conducting nanocomposites.

10.5.1 Heat Exchange Materials

Heat exchangers plays an important role in refrigeration, air conditioning as well as energy recovery applications in industries, especially oil and gas industries. The thermal performance in heat exchangers is driven by efficiency, cost, size, and quality of the heat exchanging system. Use of conventional materials such as metals has encountered limitations in many applications. The advantages of using polymers as heat exchange materials over conventional metallic alloys are ease of processing and handling, vast range of mechanical properties, high volume to weight ratio, light weight, low cost, resistance against fouling and absence of corrosion. Examples of such polymers useful as heat exchange materials after modification of thermal conductivity by the addition of thermally conducting fillers are fluoro-plastics, epoxy liquid crystal polymer, phenolic resins, polyamide, polyester resins, polyethylene terephthalate, polycarbonate, low density polyethylene, polyetheretherketone, polypropylene, polysulfone, polyvinyl chloride, polyvinylidene difluoride, polyphenylsulfone, etc. [60]. For instance, in a study underlining the impact of synergism between the polymer and filler phases to enhance thermal conductivity, Xu and Buehler [61] studied heat transfer performance at interfaces of carbon nanotubes. By employing the technique of polymer wrapping, interfacial thermal conductivity of carbon nanotube junctions was significantly improved. As shown in Figure 10.4, optimum polymer density led to more than 40% increase in the interfacial thermal conductivity.

Fouling of heat exchanger surfaces in process industries is one of the major problems which considerably increases the process costs. This problem may be overcome by introducing hydrophobicity on the internal surfaces of the heat exchanger. Rungraeng *et al.* [62] introduced thermally conducting hydrophobic anti-fouling coating comprising of carbon nanotubes (MWCNTs) and polytetrafluoroethylene on the heat exchanger internal surface. The nanocomposite coating exhibited a reduction of 97% in the surface energy as compared to the stainless steel surface. Superhydrophobic nature of the nanocomposite coating (5 wt% MWCNTs) was revealed by increased contact angle from 71.2° to 141.1° due to increased micro/nanoscale surface roughness. The thermal conductivity of the nanocomposite coating was observed to be 10.25 W/m.K, which was close to the thermal conductivity of stainless steel (16.3 W/m.K) [62]. Similar technique was adopted by Kananeh *et al.* [63] where thermoset polymers (epoxy, polyurethane and polyamide) were reinforced with nano-powders of Al_2O_3, SiO_2, etc., and coated on stainless steel. Cleaning in place (CIP) time was reduced by 70% in the case of plates coated with thermally conducting nanocomposite [63].

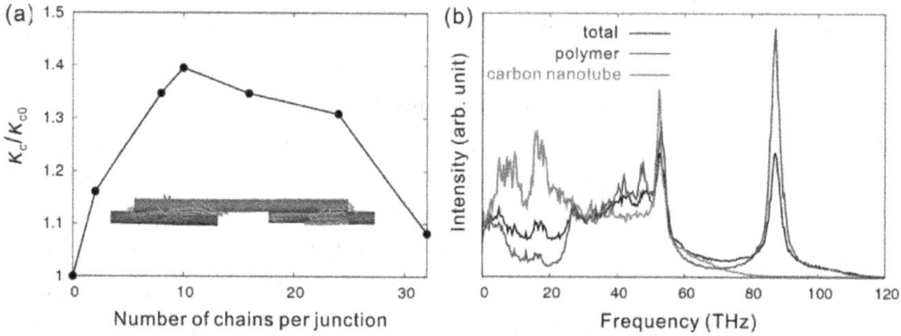

Figure 10.4 (a) Enhancement of the interfacial thermal conductivity by polyethylene chain wrapping as a function of number of chains per junction; (b) phonon spectrum of the hybrid system containing carbon nanotubes and polyethylene chains. Reproduced from Reference 61 with permission from American Chemical Society.

Scale deposition in heat exchanger tubes of process plants is also a key concern which causes fouling, corrosion and erosion, thus, using a tube surface with properties like high thermal conductivity and super hydrophobicity may reduce the cleaning and maintenance efforts. Hwang *et al.* [64] used adherent and protective coatings consisting of conducting nanocomposites of polytetrafluoroethylene (PTFE) and polyphenylenesulfide (PPS) blend with different concentrations of carbon nanotubes (MWNTs) for high-performance heat exchanger used in geothermal binary cycle power plants. It was reported that using a blend of PTFE/PPS (3:2) helped to overcome the drawbacks associated with PTFE and PPS. The blend components also complemented each other as the poor thermal conductivity, abrasion resistance and adhesion of PTFE were improved by the addition of PPS, whereas low glass transition temperature (T_g), low strength and brittleness of PPS were improved by blending with PTFE. Addition of surface modified MCWNTs (up to 5 wt%) to the blend further improved the hydrophobicity, electrical, thermal and mechanical properties. The contact angle measurements revealed that the addition of MWCNTs (5 wt%) to the PTFE/PPS blend increased the water contact angle of blend from 131° to 171° due to surface roughness. Further, the addition of functionalized MWCNTs changed the morphology of the nanocomposites from smooth to nano-pin-like [64].

In another study, Marconnet *et al.* [65] developed epoxy nanocomposites reinforced with aligned carbon nanotubes. At 1 vol% of filler fraction in epoxy, the thermal conductivity of the polymer increased by 100%. Increasing the filler content to 17 vol% led to an increase in the thermal conductivity by a factor of 18, as shown in Figure 10.5. The developed nanocomposites exhibit high potential for heat exchange applications. Similarly, for heat exchanger material, the thermal conductivity of poly(vinylidene fluoride) (PVDF) was improved with the help of reduced graphene oxide (rGO) [66]. The nanocomposites were prepared by solvent casting in the absence as well as presence of an external electric field. It was observed that with the

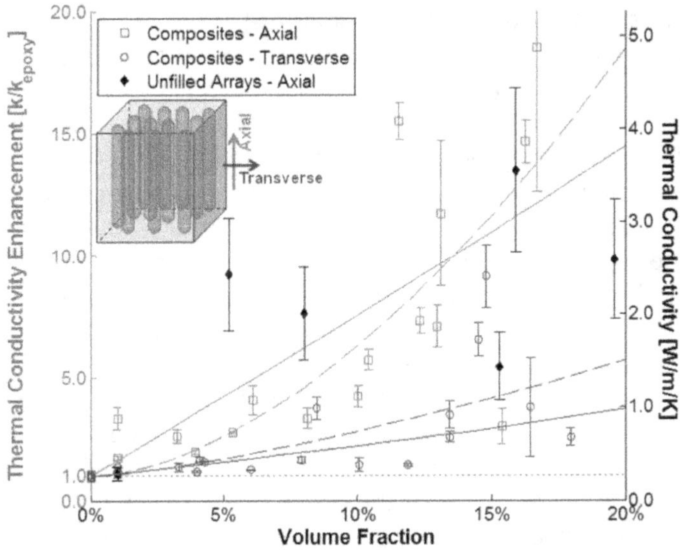

Figure 10.5 The axial and transverse thermal conductivity of CNT nanocomposites and unfilled arrays as a function of volume fraction. Reproduced from Reference 65 with permission from American Chemical Society.

addition of up to 10 wt% rGO to PVDF, the thermal conductivity of the nanocomposites increased due to uniform filler dispersion. Beyond 10 wt% rGO in PVDF, the conductivity increased at a faster rate due to effective contact and connection between the rGO sheets. At 20 wt% filler content, the thermal conductivity of the system was observed to be maximum i.e. 0.562 W/m.K, which was 212% as compared to PVDF. Beyond 20 wt% of rGO content in PVDF, the thermal conductivity of the composite decreased owing to filler agglomeration, which led to scattering of phonons and increased ITR between PVDF and agglomerated graphene. The study also demonstrated that the application of external electric field during solution processing of nanocomposites was advantageous to align the rGO sheets in the polymer. In case of applied electric field, the thermal conductivity of the composite with 7 wt% filler content was significantly higher due to the alignment of the rGO sheets in the composite. Application of electric field reduced the filler percolation threshold owing to the better dispersion in the composite, thus, leading to enhanced thermal conductivity at lower filler concentration. As a comparison, rGO at 1 wt% level exhibited 23% increased thermal conductivity as compared to the composite without filler alignment [66].

Hagenmueller *et al.* [67] studied the thermal conductivity of the high density and low density polyethylene nanocomposites with single wall carbon nanotubes. The evolution of thermal conductivity with filler loading, degree of polymer crystallinity as well as polyethylene alignment. High density polyethylene composite with a filler fraction of 0.2 had twice the thermal conductivity as low density polyethylene composite. It indicated that the polymer with high crystallinity had lower interfacial thermal resistance. The thermal conductivity along the alignment direction was observed to increase with polyethylene alignment irrespective of the filler loading (Figure 10.6). The thermal conductivity of the composite was observed to evolve from both polymer and filler phases, thus, underlining the importance of polymer matrix in defining the thermal conductivity and corresponding applications of the composites. As expected, the electrical conductivity of the composites was also increased with addition of carbon nanotubes.

10.5.2 Materials for Harnessing Solar Energy

The importance of energy from alternative resources is becoming progressively vital owing to shrinking crude oil reserves and environmental hazards of liberated gases. Bringing renewable energy into the main stream is also a significant challenge of the current century. Alternative or renewable energy refers to the energy which is obtained from the sources other than earth's natural resources and which do not damage the environment. In this category, main technologies can be wind, hydro, solar, geothermal, biomass, etc. Among these different technologies providing renewable and sustainable energy, usage of solar energy has increased rapidly in recent years. Usually harnessing of solar energy is achieved through photovoltaic cells,

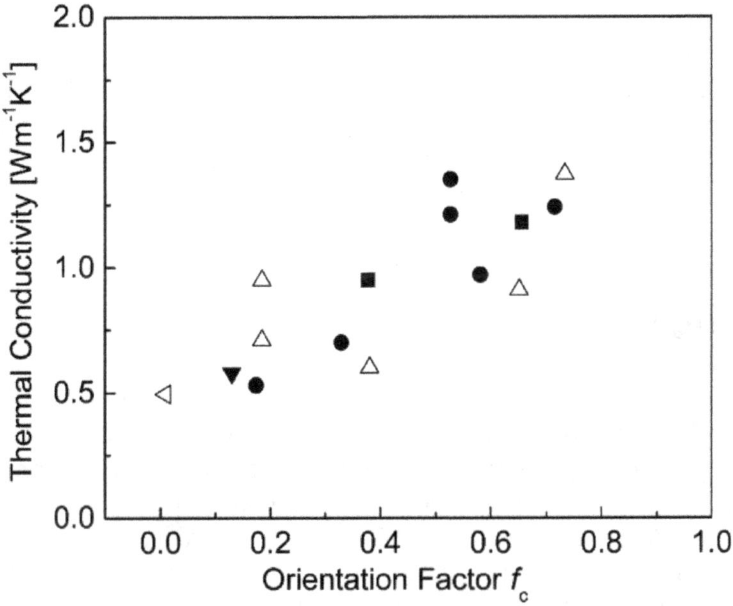

Figure 10.6 Thermal conductivity of various HDPE nanocomposites as a function of the chain orientation; (◁) isotropic HDPE, (▼) nominally isotropic SWNT/HDPE with φ ~ 0.006, (△) aligned HDPE fibers, (•) aligned SWNT/HDPE with φ ~ 0.006, and (■) aligned SWNT/ HDPE composites with φ ~ 0.012. Reproduced from Reference 67 with permission from American Chemical Society.

solar heating, molten salt power plants, solar architecture, and artificial photosynthesis. Photovoltaic cells have the characteristic of zero emission and represent clean and efficient route to convert sunlight into electric energy. However, inefficient heat dissipation from the cells is one of the major drawbacks in these cells. It was also reported that with an increase of 1 °C in the operating temperature leads to approximately 0.4–0.5% and 0.25% decline of the relative conversion efficiency for crystalline and amorphous silicon based cells, respectively [68]. In order to improve the heat dissipation ability of the photovoltaic cells, the thermal conductivity of the electrically insulating ethylene-vinyl acetate (EVA) layer was improved using thermally conductive fillers [69,70]. Cross-linked EVA composites were fabricated with different thermally conductive fillers such as Al_2O_3, SiC, ZnO, BN, AlN or MgO. Thermal conductivity of the composites containing SiC, ZnO and BN system was observed to be 2.85, 2.26 and 2.08 W/m·K respectively at filler content of 60 vol%. Thermally conductive layers produced from these composites were able to successfully improve the heat dissipation ability of the photovoltaic cells [69]. In another attempt, the thermal conductivity of EVA was improved by incorporating BN particles in varying concentrations from 0-60 wt% [71]. The thermal conductivity increased from 0.24W/m·K to 0.80W/m·K for the 60 wt% composite. It was also observed that the performance of the solar cell was increased by 0.3% owing to 6% more cooling than the standard laminate [71].

10.5.3 Batteries

Similar to solar cells, thermal management in batteries has pivotal role in battery's performance and service life. Higher temperature (>30°C) affects the battery life, whereas slow heat dissipation rate may result in internal pressure accumulation which could lead to catastrophic destruction of the cell. In order to overcome the overheating and safety challenges related to battery operation, highly thermally conductive materials are desirable in batteries which are capable to increase the rate of heat dissipation. In a related study, composites of graphite with polyvinylidene fluoride and carbon-black were studied for generating negative electrodes [72]. Introduction of a phase change material (PCM) in Li-ion battery packs for thermal management has also become a common practice. PCM takes excessive heat from the battery pack and stores it as latent heat, followed by phase change over a small temperature range. Usually PCMs have very low thermal conductivity which can be further improved by the addition of thermally conductive fillers. Paraffin wax is extensively applied as thermal PCM owing to its availability, stability, and heat storing and phase change ability. Shirazi *et al.* [73] conducted theoretical studies on nanocomposites of paraffin with fullerene, graphene, and CNTs with different fractions as phase change materials for lithium ion batteries. Computational investigations indicated that with increasing the concentration of fillers in the paraffin matrix, the rate of heat dissipation of the battery pack to the environment increased based on charging/discharging rates (C-rates). Similarly, Goli *et al.* [74] used nanocomposites comprising of paraffinic hydrocarbons and graphene sheets of different dimensions. The thermal conductivity of the Li-ion batteries was improved dramatically on incorporating graphene sheets, thus, leading to significant decrease in the temperature rise. The improved thermal conductivity of the nanocomposites was attributed to the attachment of hydrocarbon molecules to the graphene sheets which reduced the thermal interface resistance between the paraffin and graphene, as shown by Raman spectroscopy. In another study, graphene was used as an additive in paraffin wax from 0% up to 15%, where graphene/paraffin showed an enhanced thermal performance due to the graphene additive [75]. Similar experimental studies using different fillers such as carbon nano-fibers, SWCNTs, MWCNTs, SiO_2, Al_2O_3, Fe_2O_3, ZnO with paraffin have also been reported [76,77].

10.5.4 Light-Emitting Diode Devices

Light-emitting diode (LED) is a semiconductor p-n junction diode which is used to convert electrical energy into light energy, commonly known as electroluminescence phenomenon. Some of the significant applications of LEDs are illuminated displays, lighting, signaling, data communication, etc. Thermal management in LED devices is important as generated heat can affect the performance and service life. Consequently, efficient thermal dissipation system for LEDs is vital to maintain the luminous efficiency and lifetime of a LED. Generally, for thermal interface materials (TIM), PCMs like grease, gels (low conductivity), etc., are used whereas Cu and Al alloys (high weight and cost) are used as heat spreader and heat sink. Thermally conducting nanocomposites with improved thermal conductivity can be used as alternative materials for TIM, heat spreader and heat sink purposes in LEDs. Hu *et al.* [78] studied nanocomposite of silicone matrix with varying volume fractions of CNTs and micro-scale metallic (Ni) particles/spheres for TIM. The inclusion of CNTs (2%) helped to improve the thermal conductivity of the composite by 2.25 times (at 40 vol% Ni). From the percolation theory, improved thermal conductivity of the composite was attributed to ultra-high conductivity of CNTs and formation of conductive paths through interconnection of CNTs and nickel particles [78]. Similarly, nanocomposites of epoxy resin with functionalized SiC particles, MWNTs and hexagonal boron nitride (h-BN) were fabricated and applied as TIM with significant performance [79, 80].

10.5.5 Electronic Packaging

Heat dissipation in electronic packaging is a challenge due to continuous development of higher power density and miniaturization. The purpose of thermal management in electronic packaging is to guarantee the performance and service life of devices by maintaining the temperature in the working range. Similar to LEDs and batteries, applications of electronic packaging such as single and stack chip packages and electronic control units (ECU) also need TIM and the heat sink with improved thermal conductivity [81]. Along with the improved thermal conductivity for TIM for chip packages, the material should also have good adhesion and rheological properties. Recently, Sun *et al.* [82] reported novel nanocomposites as TIM for integrated circuits and electronic packaging. The nanocomposite consisted of Ag coated polyimide network and indium matrix. Nanocomposites were fabricated in three steps. During the first step, electro-spinning technique was used to produce polyimide nano-fibers followed by surface modification for imide-ring cleavage. The infiltration of Ag was achieved by placing the fibrous network in $AgNO_3$ aqueous solution for ion exchange reaction. Further, liquid indium was infiltrated in the nanocomposite by using an infiltration device. Thermal cycling under extreme environments (−40 to 115 °C at 2 cycles/h) indicated slight increase of thermal resistance after 1000 cycles. The TIM nanocomposite had good adhesion, comparably unchanging reliability and consistency due to indium [82]. Similarly, nano-structured polyurethane and indium tin bismuth (InSnBi, melting point of 60°C) composite was fabricated by injecting metal into the nano-fibers. Mechanical testing indicated that shear strength of the sample decreased to 50% after 500 cycles due to overflow of the alloy [83]. Similarly, Xu *et al.* [84] prepared nanocomposites from Ag nanowires embedded in polycarbonate templates as TIM for electronic cooling. Nanocomposites exhibited improved intrinsic thermal conductivity of 30.3 W/m.K. Further, as a replacement of metals for TIM, high-performance TIM was prepared from composites of few-layer graphene (FLG) and epoxy resin [85]. In this research, FLG was prepared by interlayer catalytic exfoliation which is a cost-effective method, whereas the nanocomposites were prepared by solution mixing followed by molding (Figure 10.7). Nanocomposites exhibited improved thermal and mechanical reliability within ranges of operating temperature and pressure. Thermal interface resistance between FLG TIM and copper was observed in the range of 3.2 and 4.3 mm^2K/W

for 5 vol% and 10 vol% at 330 K respectively, which was lower than that of commercially available TIMs. Similar studies were also reported where nanocomposites were fabricated using graphene nanoplatelets and epoxy resin [86,87].

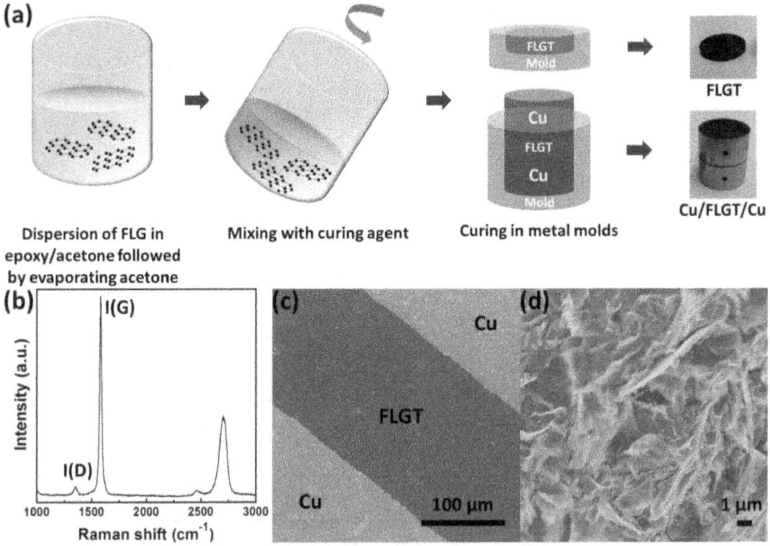

Figure 10.7 (a) Schematic of the FLG-TIM and Cu/FLG-TIM/Cu preparation; (b) Raman spectrum of a FLG flake; and (c) SEM image of the interface. Reproduced from Reference 85 with permission from American Chemical Society.

Further, in order to generate synergist effect of graphene and Ag for TIM, nanocomposites from hybrid fillers of thiophenol modified silver nanowires (mAgNWs) and acrylamide-modified reduced graphene oxide (AA-RGO) were fabricated with silicone rubber [88]. Improved thermal conductivity (1.152 W/m.K) of the composite was observed owing to synergetic effects and development of three-dimensional network structure in silicone rubber due to π-π interactions. It was observed that mAgNWs aided the graphene sheets to avoid re-stacking and aggregation in the composites, thus, leading to new heat conduction networks [88]. The full use of superior thermal properties of graphene is sometimes restricted due to high thermal resistance at the polymer-filler interface [89]. Interfacial thermal resistance mainly depends on factors like covalent and noncovalent functionalization, weak interactions between polymer chains and graphene surface, graphene doping, acetylenic linkage in graphene, etc. Molecular dynamics simulations on graphene-epoxy nanocomposites indicated that previously mentioned factors reduce the interfacial thermal resistance, as shown in Figure 10.8 for covalent functional groups [89].

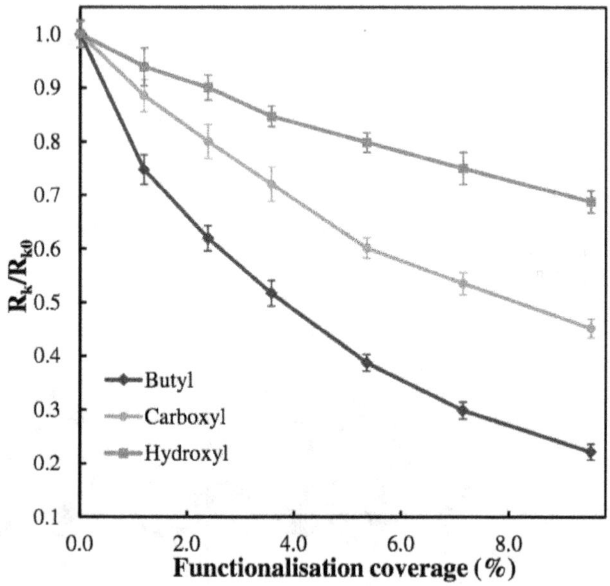

Figure 10.8 Relative interfacial thermal resistance with respect to the coverage of different covalent functional groups. Reproduced from Reference 89 with permission from American Chemical Society.

Electrically insulated hexagonal boron nitride (h-BN) has excellent in-plane (~600 W/m.K) and poor out-of-plane thermal conductivity. Nanocomposites prepared from randomly oriented h-BN nanoparticles generate materials with low thermal conductivity and limit its application for electronic packaging materials. Research studies have exhibited that directing the orientation of h-BN nanoparticles in the polymer matrix can produce nanocomposites with enhanced thermal conductivity. For this purpose, magnetically activated h-BN particles were synthesized by surface modification of h-BN nanoparticles with super paramagnetic Fe_2O_3 nanoparticles [90]. The surface modification was achieved by electrostatic interaction of $(Fe_2O_3)^+$ nanoparticles with $(h\text{-}BN)^-$ nano-platelets, which generated magnetically reactive h-BN nanoparticles (mh-BN). Nanocomposites were generated for electronic encapsulation by solution mixing of epoxy resin and mh-BN followed by placing between two rare-earth magnets, whereas curing was performed in the presence of magnets. The thermal conductivity ($\kappa_{composite}/\kappa_{matrix} = 5.7$) of mh-BN/epoxy nanocomposites at 20 wt% filler content was significantly enhanced by vertically aligned filler as compared to the thermal conductivity ($\kappa_{composite}/\kappa_{matrix}=2.8$) of randomly oriented rh-BN/epoxy nanocomposites. Finite element analysis of nanocomposites as underfill material in a flip-chip package exhibited promising material performance for advanced micro-electronic packaging [90].

10.5.6 Energy Storage

Depleting fossil fuel resources and rising awareness of environmental concerns indicate that there is a need to reassess our energy needs by discovering alternative sources of energy and effectively utilizing the existing resources. Use of PCMs is a convenient way to store latent heat through melting (store) and solidifying (emit). For energy storage purposes, mainly paraffins or paraffinic hydrocarbons are employed due to their abundance from natural and synthetic resources, chemically inert nature, high latent heat, etc. Various methods of energy storage reported in the literature are storage through mechanical, electrical, thermal and thermochemical processes [91]. Many energy storage applications using PCMs have been discussed in literature such as thermal energy storage, conditioning of buildings [92], solar cooking, solar power plants [93], cold energy battery [94], medical applications [95], turbine inlet chilling with thermal energy storage, cooling of heat and electrical engines, waste heat recovery [96], spacecraft thermal systems [97], textiles [98], etc. [99]. A great deal of research focus has been devoted to improve the thermal conductivity of thermoplastics, thermosets and paraffin waxes for energy storage by the incorporation of different inorganic and organic thermally conductive fillers. Some of the thermal management applications require shape stabilization and structure integrity during energy storage operation. Polymers with rigid backbone can provide support and protection for the PCM materials from the leakage issues. For this purpose, grafting of PCM on to the rigid chain can be beneficial owing to distinctive thermal management. In this way, the attached PCM molecules can act as energy storage units whereas rigid chains perform the supporting role.

In order to achieve the unique characteristics of PCM materials, Pielichowska *et al.* [101] developed shape stabilized PCM by grafting octadecanol on poly(styrene-co-maleic anhydride) and graphene oxide nano-platelets (GONPs) were added to the grafted polymer through solution blending method [100]. The purpose of adding GONPs was to improve thermal stability of the PCM at higher temperature due to adsorption and shielding behavior of GONPs sheet-like structure. Similarly, in order to improve the leakage and low thermal conductivity of PCM, thermosets incorporated with thermally conductive fillers were also used as carrier of PCM. Poly(ethylene glycol) (PEG) is a semi-crystalline promising PCM due to its properties such as non-toxicity, inertness, low super cooling and cost effectiveness. In order to avoid the leakage above its melting point, chemical linking with polyisocyanates was performed to produce polyurethane using one-step bulk polymerization. Moreover, graphite nano-platelets GNPs were also incorporated into PU/PEG to improve thermal conductivity and shape stability [102]. Three-dimensional (3-D) microcrystalline cellulose aerogels (CAG) is a light weight and porous material and can be used as support for PCM for thermal energy storage applications. Use of PEG with 3-D CAG can also be beneficial owing to inter-molecular interactions between PEG and cellulose molecules. It was reported that thermal conductivity of PEG/3-DCAG blends was improved by 463% with the incorporation of 5.3 wt% of GNPs, along with retaining of the shape upon compressing at melting point of PEG [103]. Nano-fiber nanocomposites produced from electro-spinning of nylon (polyamide 6, PA6), PEG and SiO_2, Fe_2O_3, ZnO, Al_2O_3 nanoparticles were also reported as PCM with improved thermal stability and stability [77,104].

Similarly, direct nanocomposites of PCM materials with different thermally conductive fillers were reported for energy storage applications using solar energy. In a related study, nanocomposites of different waxes like paraffin wax and microcrystalline wax with α nano-alumina (α-Al_2O_3) (0.5, 1.0 and 2.0 wt%) were prepared using melt mixing and intensive sonication. The results exhibited that the thermal conductivity of the nanocomposite improved, thus, leading to rise in latent heat and reduction in melting temperature [105]. Xia *et al.* [106] prepared nanocomposites of acetamide and expanded graphite (EG) (10 wt%) for energy storage application in the active solar systems. It was reported that with the addition of EG, the thermal conductivity of the nanocomposites increased by five-fold. Along with the thermal conductivity, other thermal properties of were also improved such as shifting of melting/freezing points 66.95/42.46 °C to 65.91/65.52 °C and latent heat from 194.92 to 163.71 kJ.kg^{-1} for pure acetamide and acetamide/EG respectively. The use of acetamide/EG in a latent thermal energy storage unit performed better by reducing heat storage and retrieval periods to 45% and 78% than that of acetamide alone [106]. In a similar work, nanocomposites of paraffin/HDPE with 3 wt% of EG were generated for enhanced thermal conductivity. The material exhibited improved thermal conductivity by 24% due to thermally conductive networks and was used as PCM for thermal energy storage [107]. In another study, attempt to improve the shape stability and thermal storage performance of PCM was made by generating nanocomposites of paraffin/halloysite nanotube (P-HNT) by absorbing P into the pores of HNT and applied for solar energy storage. The nanocomposite materials were proved to be effective for solar energy storage after 50 melt–freeze cycles [108]. For medium range temperature solar energy storage applications, sebacic acid/expanded graphite nanocomposites were synthesized in order to avoid the low extent of sub-cooling of sebacic acid. The PCM exhibited negligible sub-cooling along with excellent thermal properties, thermal reliability, stability, and formability [109]. Use of cellulosic matrix loaded with stable microcapsules of PCM (25 and 50 wt%) was reported for cold storage of perishable products. Using approaches of experiments and computations, it was observed that PCM/cellulose composite served as effective cold storage material [110].

Recently, it was estimated that 40% of entire energy is usually spent in industrial and residential building activities like lighting, heating and cooling. In accordance with the global policies of effective and efficient consumption of energy, it is necessary to pay attention to the thermal management of buildings. Research on energy efficiency in buildings has been expanding rapidly to generate improved systems and design approaches to ensure efficient use of building activities, especially in many energy-intensive industries. Fang *et al.* [111] introduced a novel method of incorporating organic PCMs into building materials. Firstly, nanocomposites were prepared by adding organically modified montmorillonite (OMM) in organic PCMs followed by incorporation in gypsum. The aim of OMM addition in PCM was to improve the thermal conductivity and generate compatibility with gypsum. Nanocomposites generated with 20 wt% of OMM performed much better in decreasing the building energy consumption due to higher latent heat, appropriate phase change temperature and solidity [111]. Similar studies have also been reported on the generation of nanocomposites with PCMs [112,113]. Moreover, the direct incorporation of octadecane (C18) and tetracosane (C24) (eutectic mixture) as PCM into gypsum was also reported. The composite as PCM exhibited outstanding thermal regulation properties even after 1000 melting and freezing cycles. In addition, the composite did not lose its chemical structure and performed as favorable material to regulate room temperature [114]. Warzoha *et al.* [115] studied the effect of different volume fractions of randomly oriented herringbone style graphite nanofibers (HGNF) on thermal properties of paraffin used as PCM. It was observed that as the percolation approached, the thermal conductivity of the nanocomposites increased exponentially. Also, the thermal boundary resistance analysis demonstrated that the thermal resistance was higher at HGNF–PCM interface as compared to HGNF–HGNF interfaces in solid and liquid phases [115].

10.6 Conclusion

In this chapter, different factors affecting the thermal conductivity of the polymers and their nanocomposites have been discussed. Theoretical calculations and modeling of nanocomposites may help to select the polymer, filler and parameters of nanocomposites fabrication so as to achieve suitable thermally conducting polymer nanocomposite. Generally, the thermal conductivity of the polymer nanocomposites increases with increasing filler loading. Improvement in the thermal conductivity can be achieved at lower loading of fillers by using crystalline and oriented polymers. The interfacial thermal resistance can be decreased using fillers with large sized particles. Further, the use of hybrid fillers and modification of filler surface are other techniques to develop conducting networks between different fillers in a matrix. Finally, different applications of thermally conducting polymer nanocomposites in alternative and sustainable energy resources along with efficient use of existing energy have been discussed so as to achieve effective use of these materials in a large number of industries, especially in energy intensive oil and gas industries.

References

1. Chen, H., Ginzburg, V. V., Yang, J., Yang, Y., Liu, W., Huanf, Y., Du, L., and Chen, B. (2016) Thermal conductivity of polymer-based composites: Fundamentals and applications. *Progress in Polymer Science*, **59**, 41-85.

2. Callister, W. D., and Rethwisch, D. G. (2014) *Thermal Properties*, in *Materials Science and Engineering: An Introduction*, Wiley, USA.
3. Lovestam, G., Rauscher, H., Roebben, G., Kluttgn, B. S., Gibson, N., Putaud, J.-P., and Stamm, H. (2010) *Considerations on a definition of nanomaterial for regulatory purposes.* JRC Reference Reports. Online: https://ec.europa.eu/jrc/sites/jrcsh/files/jrc_reference_report_201007_nanomaterials.pdf (assessed 24th February 2017).
4. Mittal, V. (2014) Functional polymer nanocomposites with graphene: A review. *Macromolecular Materials and Engineering*, **299**, 906-931.
5. Idumah C. I., and Hassan, A. (2016) Recently emerging trends in thermal conductivity of polymer nanocomposites. *Reviews in Chemical Engineering*, **32**, 413-457.
6. Ebadi-Dehaghani, H., and Nazempour, M. (2012) Thermal conductivity of nanoparticles filled polymers. In: *Smart Nanopraticles Technology*, Hashim, A. (ed.), INTECH Open. Online: http://www.intechopen.com/books/smart-nanoparticles-technology/thermal-conductivity-of-nanoparticles-filled-polymers- (assessed 10th February 2017).
7. Kumar, P. M., Kumar, J., Tamilarasan, R., Sendhilnathan, S., and Suresh, S. (2015) Review on nanofluids theoretical thermal conductivity models. *Engineering Journal*, **19**(1). Online: http://citeseerx.ist.psu.edu/viewdoc/download;jsessionid=2F215B408291AC43A147C40DF7276C83?doi=10.1.1.676.9543&rep=rep1&type=pdf (assessed 31st January 2017).
8. Pietrak, K., and Wisniewski, T. S. (2015) A review of models for effective thermal conductivity of composite materials. *Journal of Power Technologies*, **95**(1), 14-24.
9. Burger, H. (1915) Das lertvermogen verdummter mischristallfreier lonsungen. *Phys. Zs.*, **20**, 73-76.
10. Hamilton, R., and Crosser, O. (1962) Thermal conductivity of heterogeneous two-component systems. *Industrial & Engineering Chemistry Fundamentals*, **1**(3), 187-191.
11. Rayleigh, L. (1892) LVI. On the influence of obstacles arranged in rectangular order upon the properties of a medium. *Philosophical Magazine Series 5*, **34**(211), 481-502.
12. Swartz, E. T., and Pohl, R. O. (1989) Thermal boundary resistance. *Reviews of Modern Physics*, **61**(3), 605-668.
13. Powell, Jr., B. R., Youngblood, G. E., Hasselman, D. P., and Bentsen, L. D. (1980) Effect of thermal expansion mismatch on the thermal diffusivity of glass-Ni composites. *Journal of the American Ceramic Society*, **63**(9-10), 581-586.
14. Benveniste, Y. (1987) Effective thermal conductivity of composites with a thermal contact resistance between the constituents: Nondilute case. *Journal of Applied Physics*, **61**(8), 2840-2843.

15. Landauer, R. (1952) The electrical resistance of binary metallic mixtures. *Journal of Applied Physics*, **23**(7), 779-784.
16. Conway, J., and Sloane, N. J. A. (2013) *Sphere Packings, Lattices and Groups*, Springer, USA.
17. Devpura, A., Phelan, P. E., and Prasher, R. S. (2000) Percolation Theory Applied to the Analysis of Thermal Interface Materials in Flip-Chip Technology. In: *ITHERM 2000. The Seventh Intersociety Conference on Thermal and Thermomechanical Phenomena in Electronic Systems*, Cat. No.00CH37069.
18. Yuksel, N. (2016) The review of some commonly used methods and techniques to measure the thermal conductivity of insulation materials. In: Insulation Materials in Context of Sustainability, Almusard, A., and Almassad, A. (eds.), INTECH Open. Online: http://www.intechopen.com/books/insulation-materials-in-context-of-sustainability/the-review-of-some-commonly-used-methods-and-techniques-to-measure-the-thermal-conductivity-of-insul (assessed 25th February 2017).
19. Tong, X.C. (2011) Characterization methodologies of thermal management materials. In: *Advanced Materials for Thermal Management of Electronic Packaging*, Springer, USA, pp. 59-129.
20. Corsan, J. M. (1992) Axial heat flow methods of thermal conductivity measurement for good conducting materials. In: *Compendium of Thermophysical Property Measurement Methods: Volume 2, Recommended Measurement Techniques and Practices*, Maglic, K. D., Cezairliyan, A., and Peletsky, V. E. (eds.), Springer, USA, pp. 3-31.
21. Czichos, H., Saito, T., and Smith, L. (2006) *Springer Handbook of Materials Measurement Methods*, vol. 978, Springer, Germany.
22. Yesilata, B., and Turgut, P. (2007) A simple dynamic measurement technique for comparing thermal insulation performances of anisotropic building materials. *Energy and Buildings*, **39**(9), 1027-1034.
23. Vozar, L. (1996) A computer-controlled apparatus for thermal conductivity measurement by the transient hot wire method. *Journal of Thermal Analysis and Calorimetry*, **46**(2), 495-505.
24. Kwon, S. Y., and Lee, S. (2012) Precise measurement of thermal conductivity of liquid over a wide temperature range using a transient hot-wire technique by uncertainty analysis. *Thermochimica Acta*, **542**, 18-23.
25. Solorzano, E., Rodriguez-Perez, M. A., and de Saja, J. A. (2008) Thermal conductivity of cellular metals measured by the transient plane sour method. *Advanced Engineering Materials*, **10**(4), 371-377.
26. Min, S., Blumm, J., and Lindemann, A. (2007) A new laser flash system for measurement of the thermophysical properties. *Thermochimica Acta*, **455**(1–2), 46-49.
27. Zhang, T., Wu, X., and Luo, T. (2014) Polymer nanofibers with outstanding thermal conductivity and thermal stability: fundamental

linkage between molecular characteristics and macroscopic thermal properties. *The Journal of Physical Chemistry C*, **118**(36), 21148-21159.

28. Zhang, T., and Luo, T. (2012) Morphology-influenced thermal conductivity of polyethylene single chains and crystalline fibers. *Journal of Applied Physics*, **112**, 094304.
29. Hansen, D., and Bernier, G. A. (1972) Thermal conductivity of polyethylene: The effects of crystal size, density and orientation on the thermal conductivity. *Polymer Engineering & Science*, **12**(3), 204-208.
30. Liu, J., and Yang, R. (2010) Tuning the thermal conductivity of polymers with mechanical strains. *Physical Review B*, **81**(17), 174122.
31. Kim, G.-H., Le, D., Shanker, A., Shao, L., Kwon, M. S., Gidley, D., Kim, J., and Pipe, K. P. (2015) High thermal conductivity in amorphous polymer blends by engineered interchain interactions. *Nature Materials*, **14**(3), 295-300.
32. Giang, T., Park, J., Cho, I., Ko, Y., and Kim, J. (2013) Effect of backbone moiety in epoxies on thermal conductivity of epoxy/alumina composite. *Polymer Composites*, **34**(4), 468-476.
33. Harada, M., Hamaura, N., Ochi, M., and Agari, Y. Thermal conductivity of liquid crystalline epoxy/BN filler composites having ordered network structure. *Composites, Part B: Engineering*, **55**, 306-313.
34. Akatsuka, M., Takazawa, Y., and Sugawara, K. (2006) Thermosetting Resin Compounds, US Patent 20060276568 A1.
35. Hammerschmidt, A., Geibel, K., and Strohmer, F. (1993 In situ photopolymerized, oriented liquid-crystalline diacrylates with high thermal conductivities. *Advanced Materials*, **5**(2), 107-109.
36. Wypych, G. (2000) *Handbook of Fillers*. Chem Tech Publishing, Canada.
37. Mamunya, Y. P., Davydenko, V. V., Pissis, P., and Lebedev, E. V. (2002) Electrical and thermal conductivity of polymers filled with metal powders. *European Polymer Journal*, **38**, 1887-1897.
38. Stoller, M. D., Park, S., Zhu, Y., An, J., and Ruoff, R. S. (2008) Graphene-based ultracapacitors. *Nano Letters*, **8**(10), 3498-3502.
39. Liu, Z., Wu, B., and Gu, M. (2007) Effect of hydrolysis of AlN particulates on corrosion behavior of Al/AlNp composite in neutral chloride solution. *Composites, Part A: Applied Science and Manufacturing*, **38**(1), 94-99.
40. Yokota, H., and Ibukiyama, M. (2003) Microstructure tailoring for high thermal conductivity of β-Si_3N_4 ceramics. *Journal of the American Ceramic Society*, **86**(1), 197-199.
41. Bhattacharya, S. K. (1986) *Metal Filled Polymers*, Taylor & Francis, USA.
42. Tavman, I. H., and Akinci, H. (2000) Transverse thermal conductivity of fiber reinforced polymer composites. *International Communi-*

cations in Heat and Mass Transfer, **27**(2), 253-261.

43. Kusunose, T., Yagi, T., Firoz, S. H., and Sekino, T. (2013) Fabrication of epoxy/silicon nitride nanowire composites and evaluation of their thermal conductivity. *Journal of Materials Chemistry A*, **1**(10), 3440-3445.
44. Yu, J., Chen, Y., Wuhrer, R., Liu, Z., and Ringer, S. P. (2005) In situ formation of BN nanotubes during nitriding reactions. *Chemistry of Materials*, **17**(20), 5172-5176.
45. Hill, R. F., and Supancic, P. H. (2002) Thermal conductivity of platelet-filled polymer composites. *Journal of the American Ceramic Society*, **85**(4), 851-857.
46. Pashayi, K., Fard, H. R., Lai, F., Iruvanti, S., Plawsky, J., and Borca-Tasciuc, T. (2012) High thermal conductivity epoxy-silver composites based on self-constructed nanostructured metallic networks. *Journal of Applied Physics*, **111**(10), 104310.
47. Li, T.-L., and Hsu, S.L.-C. (2010) Enhanced thermal conductivity of polyimide films via a hybrid of micro- and nano-sized boron nitride. *The Journal of Physical Chemistry B*, **114**(20), 6825-6829.
48. Zhou, W., Qi, S., Tu, C., Zhao, H., Wang, C., and Kou, J. (2007) Effect of the particle size of Al2O3 on the properties of filled heat-conductive silicone rubber. *Journal of Applied Polymer Science*, **104**(2), 1312-1318.
49. Leung, S. N., Khan. M. O., Chan, E., Naguib, H. E., Dawson, F., Adinkrah, V., and Lakatos-Hayward, L. (2013) Synergistic effects of hybrid fillers on the development of thermally conductive polyphenylene sulfide composites. *Journal of Applied Polymer Science*, **127**(5), 3293-3301.
50. Pak, S. Y., Kim, H. M., Kim, S. Y., and Youn, J. R. (2012) Synergistic improvement of thermal conductivity of thermoplastic composites with mixed boron nitride and multi-walled carbon nanotube fillers. *Carbon*, **50**(13), 4830-4838.
51. Zhu, B. L., Zheng, H., Wang, J., Ma, J., Wu, J., and Wu, R. (2014) Tailoring of thermal and dielectric properties of LDPE-matrix composites by the volume fraction, density, and surface modification of hollow glass microsphere filler. *Composites, Part B: Engineering*, **58**, 91-102.
52. Raman, C. (2008) Boron Nitride in Thermoplastics: Effect of Loading, Particle Morphology and Processing Conditions. in *Proceedings of the NATAS Annual Conference on Thermal Analysis and Applications*, USA.
53. Choy, C. L., Leung, W. P., Kowk, K. W., and Lau, F. P. (1992) Elastic moduli and thermal conductivity of injection-molded short-fiber–reinforced thermoplastics. *Polymer Composites*, **13**(2), 69-80.
54. Choy, C. L., Wong, Y. W., Yang, G. W., and Kanamoto, T. (1999) Elastic modulus and thermal conductivity of ultradrawn polyethylene.

Journal of Polymer Science, Part B: Polymer Physics, **37**(23), 3359-3367.

55. Choy, C. L., Luk, W. H., and Chen, F. C. (1978) Thermal conductivity of highly oriented polyethylene. *Polymer*, **19**, 155-162.
56. Han, S., Lin, J. T., Yamada, Y., and Chung, D. D. L. (2008) Enhancing the thermal conductivity and compressive modulus of carbon fiber polymer–matrix composites in the through-thickness direction by nanostructuring the interlaminar interface with carbon black. *Carbon*, **46**(7), 1060-1071.
57. Shahil, K. M. F., and Balandin, A. A. (2012) Graphene–multilayer graphene nanocomposites as highly efficient thermal interface materials. *Nano Letters*, **12**(2), 861-867.
58. Yao, Q., Chen, L., Zhang, W., Liufu, S., and Chen, X. (2010) Enhanced thermoelectric performance of single-walled carbon nanotubes/polyaniline hybrid nanocomposites. *ACS Nano*, **4**(4), 2445-2451.
59. Han, Z., and Fina, A. (2011) Thermal conductivity of carbon nanotubes and their polymer nanocomposites: A review. *Progress in Polymer Science*, **36**(7), 914-944.
60. T'Joen, C., Park, Y., Wang, Q., Sommers, A., Han, X., and Jacobi, A. (2009) A review on polymer heat exchangers for HVAC&R applications. *International Journal of Refrigeration*, **32**(5), 763-779.
61. Xu, Z., and Buehler, M. J. (2009) Nanoengineering heat transfer performance at carbon nanotube interfaces. *ACS Nano*, **3**(9), 2767-2775.
62. Rungraeng, N., Cho, Y.-C., Yoon, S. H., and Jun, S. (2012) Carbon nanotube-polytetrafluoroethylene nanocomposite coating for milk fouling reduction in plate heat exchanger. *Journal of Food Engineering*, **111**(2), 218-224.
63. Kananeh, A. B., Scharnbeck, E., Kuck, U. D., and Rabiger, N. (2010) Reduction of milk fouling inside gasketed plate heat exchanger using nano-coatings. *Food and Bioproducts Processing*, **88**(4), 349-356.
64. Yang, E., Hwang, T., Kumar, A., and Kim, K. J. (2016) Anti-biofouling, thermal, and electrical performance of nanocomposite coating with multiwall carbon nanotube and polytetrafluoroethylene-blended polyphenylenesulfide. *Advances in Polymer Technology*, DOI: 10.1002/adv.21728.
65. Marconnet, A. M., Yamamoto, N., Panzer, M. A., Wardle, B. L., and Goodson, K. E. (2011) Thermal conduction in aligned carbon nanotube-polymer nanocomposites with high packing density. *ACS Nano*, **5**(6), 4818-4825.
66. Guo, H., Li, X., Li, B., Wang, J., and Wang, S. (2017) Thermal conductivity of graphene/poly(vinylidene fluoride) nanocomposite membrane. *Materials & Design*, **114**:, 355-363.

67. Haggenmueller, R., Guthy, C., Lukes, J. R., Fischer, J. E., and Winey, K. I. (2007) Single wall carbon nanotune/polyethylene nanocomposites: thermal and electrical conductivity. *Macromolecules*, **40**, 2417-2421.
68. Krauter, S., Araujo, R. G., Schroer, S., Hanitsch, R., Salhi, M. J., Triebel, C., and Lemoine, R. (1999) Combined photovoltaic and solar thermal systems for facade integration and building insulation. *Solar Energy*, **67**(4–6), 239-248.
69. Lee, B., Liu, J. Z., Sun, B., Shn, C. Y., and Dai, G. C. (2008) Thermally conductive and electrically insulating EVA composite encapsulants for solar photovoltaic (PV) cell. *eXPRESS Polymer Letters*, **2**(5), 357-363.
70. Kim, N., Kim, D., Kang, H., and Park, Y.-G. (2016) Improved heat dissipation in a crystalline silicon PV module for better performance by using a highly thermal conducting backsheet. *Energy*, **113**, 515-520.
71. Allan, J., Pinder, H., and Dehouche, Z. (2016) Enhancing the thermal conductivity of ethylene-vinyl acetate (EVA) in a photovoltaic thermal collector. *AIP Advances*, **6**(3), 035011.
72. Maleki, H., Selman, J. R., Dinwiddie, R. B., and Wang, H. (2001) High thermal conductivity negative electrode material for lithium-ion batteries. *Journal of Power Sources*, **94**(1), 26-35.
73. Shirazi, A. H. N., Mohebbi, F., Kakavand, M. R. A., He, B., and Rabczuk, T. (2016) Paraffin nanocomposites for heat management of lithium-ion batteries: A computational investigation. *Journal of Nanomaterials*, 2131946.
74. Goli, P., Legedza, S., Dhar, A., Salgado, R., Renteria, J., and Balandin, A. A. (2014) Graphene-enhanced hybrid phase change materials for thermal management of Li-ion batteries. *Journal of Power Sources*, **248**, 37-43.
75. Zhang, Y., Yue, W., Zhang, S., Huang, S., and Liu, J. (2016) Experimental Investigation of Paraffin Wax with Graphene Enhancement as Thermal Management Materials for Batteries. In: *17th International Conference on Electronic Packaging Technology (ICEPT)*, China, DOI: 10.1109/ICEPT.2016.7583385.
76. Shaikh, S., Lafdi, K., and Hallinan, K. (2008) Carbon nanoadditives to enhance latent energy storage of phase change materials. *Journal of Applied Physics*, **103**(9), 094302.
77. Babapoor, A., Karimi, G., and Sabbaghi, S. (2016) Thermal characteristic of nanocomposite phase change materials during solidification process. *Journal of Energy Storage*, **7**, 74-81.
78. Xuejiao, H., Linan, J., and Goodson. K. E. (204) Thermal Conductance Enhancement of Particle-Filled Thermal Interface Materials Using Carbon Nanotube Inclusions. *The Ninth Intersociety Conference on Thermal and Thermomechanical Phenomena In Electronic Systems,*

USA, IEEE Cat. No. 04CH37543.

79. Liang, Q., Xiu, Y., Lin, W., Moon, K.-S., and Wong, C. P. (2009) Epoxy/h-BN Composites for Thermally Conductive Underfill Material. *59th Electronic Components and Technology Conference*, USA.
80. Liang, Q., Moon, K.-S., Jiang, H., and Wong, C. P. (2012) Thermal conductivity enhancement of epoxy composites by interfacial covalent bonding for underfill and thermal interfacial materials in Cu/low-K application. *IEEE Transactions on Components, Packaging and Manufacturing Technology*, **2**(10), 1571-1579.
81. Prasher, R. (2006) Thermal interface materials: historical perspective, status, and future directions. *Proceedings of the IEEE*, **94**(8), 1571-1586.
82. Sun, S., Chen, S., Luo, X., Fu, Y., Ye, L., and Liu, J. (2016) Mechanical and thermal characterization of a novel nanocomposite thermal interface material for electronic packaging. *Microelectronics Reliability*, **56**, 129-135.
83. Lu, X., Zhuang, M., Zhang, L., Ye, L., and Liu, J. (2012) Environmental Reliability of Nano-structured Polymer-Metal Composite Thermal Interface Material. *13th International Conference on Electronic Packaging Technology & High Density Packaging*, China.
84. Xu, J., Munari, A., Dalton, E., Mathewson, A., and Razeeb, K. M. (2009) Silver nanowire array-polymer composite as thermal interface material. *Journal of Applied Physics*, **106**(12), 124310.
85. Park, W., Guo, Y., Li, X., Hu, J., Liu, L., Ruan, X., and Chen, Y. P. (2015) High-performance thermal interface material based on few-layer graphene composite. *The Journal of Physical Chemistry C*, **119**(47), 26753-26759.
86. Jimenez-Suarez, A., Moriche, R., Prolongo, S. G., and Urena, A. (2016) GNPs reinforced epoxy nanocomposites used as thermal interface materials. *Journal of Nano Research*, **38**, 18-25.
87. Gu, J., Yang, X., Lv, Z., Li, N., Liang, C., anf Zhang, Q. (2016) Functionalized graphite nanoplatelets/epoxy resin nanocomposites with high thermal conductivity. *International Journal of Heat and Mass Transfer*, **92**, 15-22.
88. Lin, S.-C., Ma, C.-C. M., Liao, W.-H., Wang, J.-A., Zeng, S.-J., Hsu, S.-Y., Chen, Y.-H., Hsiao, S.-T., Cheng, T.-Y., Lin, C.-W., and Hsiao, P.-Y. (2016) Preparation of a graphene–silver nanowire hybrid/silicone rubber composite for thermal interface materials. *Journal of the Taiwan Institute of Chemical Engineers*, **68**, 396-406.
89. Wang, Y., Yang, C., Pei, Q.-X., and Zhang, Y. (2016) Some aspects of thermal transport across the interface between graphene and epoxy in nanocomposites. *ACS Applied Materials & Interfaces*, **8**(12), 8272-8279.
90. Lin, Z., Liu, Y., Raghavan, S., Moon, K.-s., Sitaraman, S. K., and Wong, C.-p. (2013) Magnetic alignment of hexagonal boron nitride plate-

lets in polymer matrix: toward high performance anisotropic polymer composites for electronic encapsulation. *ACS Applied Materials & Interfaces*, **5**(15), 7633-7640.

91. Sharma, A., Tyagi, V. V., Chen, C. R., and Buddhi, D. (2009) Review on thermal energy storage with phase change materials and applications. *Renewable and Sustainable Energy Reviews*, **13**(2), 318-345.
92. Liu, L., and Khodadadi, J. M. (2016) Thermal conductivity enhancement of phase change materials for thermal energy storage: A review. *Renewable and Sustainable Energy Reviews*, **62**, 305-317.
93. Kenisarin, M., and Mahkamov, K. (2007) Solar energy storage using phase change materials. *Renewable and Sustainable Energy Reviews*, **11**(9), 1913-1965.
94. Huo, Y., and Rao, Z. (2017) Investigation of phase change material based battery thermal management at cold temperature using lattice Boltzmann method. *Energy Conversion and Management*, **133**, 204-215.
95. Grim, T. E., and Haines, J. R. (1990) Orthopedic Device Having Gel Pad with Phase Change Material, US Patent 4964402 A.
96. Kaizawa, A., Maruoka, N., Kawai, A., Kamano, H., Jozuka, T., Senda, T., and Akiyama, T. (2008) Thermophysical and heat transfer properties of phase change material candidate for waste heat transportation system. *Heat and Mass Transfer*, **44**(7), 763-769.
97. Mulligan, J. C., Colvin, D. P., and Bryant, Y. G. (1996) Microencapsulated phase-change material suspensions for heat transfer in spacecraft thermal systems. *Journal of Spacecraft and Rockets*, **33**(2), 278-284.
98. Mondal, S. (2008) Phase change materials for smart textiles – An overview. *Applied Thermal Engineering*, **28**(11-12), 1536-1550.
99. Farid, M. M., Khudhair, A. M., Razack, S. A. K., and Al-Hallaj, S. (2004) A review on phase change energy storage: materials and applications. *Energy Conversion and Management*, **45**(9–10), 1597-1615.
100. Pielichowska, K., Nowak, M., Szatkowski, P., and Macherzynska, B. (2016) The influence of chain extender on properties of polyurethane-based phase change materials modified with graphene. *Applied Energy*, **162**, 1024-1033.
101. Liu, L., Kong, L., Wang, H., Niu, R., and Shi, H. (2016) Effect of graphene oxide nanoplatelets on the thermal characteristics and shape-stabilized performance of poly(styrene-co-maleic anhydride)-g-octadecanol comb-like polymeric phase change materials. *Solar Energy Materials and Solar Cells*, **149**, 40-48.
102. Pielichowska, K., Bieda, J., and Szatkowski, P. (2016) Polyurethane/graphite nano-platelet composites for thermal energy storage. *Renewable Energy*, **91**, 456-465.
103. Yang, J., Zhang, E., Li, X., Zhang, Y., Qu, J., and Yu, Z.-Z. (2016) Cellulose/graphene aerogel supported phase change composites with

high thermal conductivity and good shape stability for thermal energy storage. *Carbon*, **98**, 50-57.

104. Babapoor, A., Karimi, G., and Khorram, M. (2016) Fabrication and characterization of nanofiber-nanoparticle-composites with phase change materials by electrospinning. *Applied Thermal Engineering*, **99**, 1225-1235.
105. Mohamed, N. H., Soliman, F. S., El Maghraby, H., and Moustfa, Y. M. (2017) Thermal conductivity enhancement of treated petroleum waxes, as phase change material, by α nano alumina: Energy storage. *Renewable and Sustainable Energy Reviews*, **70**, 1052-1058.
106. Xia, L., and Zhang, P. (2011) Thermal property measurement and heat transfer analysis of acetamide and acetamide/expanded graphite composite phase change material for solar heat storage. *Solar Energy Materials and Solar Cells*, **95**(8), 2246-2254.
107. Zhang, R., Moon, K.-s., Lin, W., and Wong, C. P. (2010) Preparation of highly conductive polymer nanocomposites by low temperature sintering of silver nanoparticles. Journal of Materials Chemistry, **20**, 2018-2023.
108. Zhang, J., Zhang, X., Wan, Y., Mei, D., and Zhang, B. (2012) Preparation and thermal energy properties of paraffin/halloysite nanotube composite as form-stable phase change material. *Solar Energy*, **86**(5), 1142-1148.
109. Wang, S., Qin, P., Fang, X., Zhang, Z., Wang, S., and Liu, X. (2014) A novel sebacic acid/expanded graphite composite phase change material for solar thermal medium-temperature applications. *Solar Energy*, **99**, 283-290.
110. Melone, L., Altomare, L., Cigada, A., and De Nardo, L. (2012) Phase change material cellulosic composites for the cold storage of perishable products: From material preparation to computational evaluation. *Applied Energy*, **89**(1), 339-346.
111. Fang, X., Zhang, Z., and Chen, Z. (2008) Study on preparation of montmorillonite-based composite phase change materials and their applications in thermal storage building materials. *Energy Conversion and Management*, **49**(4), 718-723.
112. Fang, X., and Zhang, Z. (2006) A novel montmorillonite-based composite phase change material and its applications in thermal storage building materials. *Energy and Buildings*, **38**(4), 377-380.
113. Zhang, Z., and Fang, X. (2006) Study on paraffin/expanded graphite composite phase change thermal energy storage material. *Energy Conversion and Management*, **47**(3), 303-310.
114. Karaipekli, A., Sarı, A., and Bicer, A. (2016) Thermal regulating performance of gypsum/(C18–C24) composite phase change material (CPCM) for building energy storage applications. *Applied Thermal Engineering*, **107**, 55-62.
115. Warzoha, R. J., Weigand, R. M., and Fleischer, A. S. (2015) Tempera-

ture-dependent thermal properties of a paraffin phase change material embedded with herringbone style graphite nanofibers. *Applied Energy*, **137**, 716-725.

Index

A

abrasion resistance, 394
absorbency, 44
adhesion, 44, 55, 133, 138-140, 143, 146, 160, 190, 204, 394, 400
adsorbed water, 207
adsorbents, 265
adsorption, 48, 51, 77, 178, 267, 304, 310, 362, 404
aerogels, 225, 228, 265, 285, 404
affinity, 234, 335, 362
AFM, 38
agglomeration, 297, 383, 391
aggregation, 37, 210, 345, 357-358, 401
alcoholysis, 312
alumino-silicate, 86
American Society for Testing and Materials, 13
amine functionalized, 142, 185, 239
amorphous, 34, 39, 48, 231, 257, 267, 272, 280, 288, 296, 388-389, 397, 409
anisotropic, 40, 52, 64, 101, 136, 164, 389-390, 414
anisotropic phase, 40
anti-fouling, 393
antibacterial, 62-63, 66, 69-71, 73-74, 78-79, 83-84, 88, 99, 102-110, 113-117, 119-122, 172
antibacterial activity, 63, 99, 104, 106-107, 109-110, 115-117, 120-121
antimicrobial, 59, 66-67, 70-71, 73-74, 90, 94, 97, 104, 106-109, 113, 115, 117,

121, 124, 127, 320
aspect ratio, 39-40, 151, 171, 179, 274, 383, 390
atomization, 37

B

ball milling, 153, 155
barrier properties, 47, 50, 57
BET, 193, 313, 327
biocompatibility, 59, 64, 66, 86, 88, 90, 92, 98, 128, 133, 136, 140-142, 144-147, 152, 167, 171-173
biodegradability, 44, 312
biodegradable, 1, 3, 47, 57, 159-160
biofouling, 69, 80, 84, 92, 94, 129, 411
biofuels, 1-7, 9, 11, 13-15, 17-19, 21-27, 29, 31, 317
biological systems, 95
blending, 7, 14, 46, 49, 169, 186, 195, 204, 207, 330, 362, 372, 388, 394, 404
bridging, 323
brittleness, 394

C

cancer therapy, 62-63, 66, 68, 79, 87, 89, 91, 123, 129
capacitance retention, 223, 229, 231, 233, 235, 237, 248, 250, 253, 256-257, 260
capacitors, 221-222, 230, 265-266, 271-272, 274-275, 279
catalysis, 24, 28, 59, 66, 113, 213, 293-294, 296, 302-304, 306-308, 314, 316-317, 319-327
catalyst poisoning, 177, 293, 310
cell adhesion, 138-140, 146, 160
cell proliferation, 127, 133, 136, 139, 146
chain length, 307
chemical bonds, 370
chemical doping, 225
chemical stability, 203, 205, 245, 327, 370, 373
chitin, 55
clay, 53, 95, 179-180, 299, 369
coatings, 34, 37, 71, 102, 121, 166, 168, 215, 242, 315, 369, 394, 411
combustion, 7, 112, 175, 272
composition, 18, 22, 133-134,

138, 142, 145, 270, 295, 298, 359, 371, 383
conducting polymers, 132, 141, 159, 161, 165, 227, 276, 279, 388
contact angle, 393-394
core-shell, 96, 102, 113, 232, 259, 263, 269, 272, 284, 290-291, 298, 311, 323, 367
corrosion, 7, 10, 67, 91, 249, 314-315, 392-394, 409
Coulombic efficiency, 243, 245-246, 255-256, 259-260, 280
counter ion, 41
cracking, 3, 293, 296, 299-300, 312-313, 316-317, 321, 324, 327-328
crosslinking, 132-134, 136-137, 147, 149-150, 163, 166, 389
crystal growth, 268
crystal structure, 148-149, 342, 388
crystallinity, 37, 48-49, 148, 151, 153, 336, 345-347, 363, 381, 388, 390, 396
crystallization, 7, 33, 46, 57, 172, 363
crystallization temperature, 363
current density, 186, 200, 203-205, 230, 329, 334, 341, 346, 354
CVD, 230
cycle life permanence, 226
cytotoxicity, 64, 68, 88-89, 107, 120, 151, 160

D

degree of crystallinity, 37
dehydration, 287, 296, 316
delamination, 35-36
differential scanning calorimetry, 363
diffractograms, 303
diffusion, 154, 193, 205, 213, 288, 298-300, 302, 311, 324, 377
diffusivity, 157, 207, 379, 387
dimensional stability, 193
dissociation, 305-306, 310, 346
DMFC, 175, 188-189, 191-192, 194-195, 202, 205-207, 210, 212, 309-311
dopant, 375
drug delivery, 59, 61, 86, 90, 93-95, 100, 117, 123, 129, 131-132, 159-160, 320
ductility, 155
durability, 20, 44, 176, 193, 203, 311

E

eco-friendly, 59, 61-62, 68, 74, 77-79, 84-90, 94, 99, 102, 104-105, 121, 124
EDX, 315
elastic modulus, 410
elasticity, 44
electric double layer, 221
electrical insulation, 392
electrochemical properties, 121, 224, 229, 236, 238, 249, 253-254, 256-258, 260, 273, 284
electrodeposition, 109, 272
electron microscopy, 16, 38, 252, 355
electron transfer, 307
electrostatic interactions, 21, 200, 205, 209
embedding, 220, 367
encapsulation, 362-363, 376, 402, 414
enzymatic, 8, 11, 26, 34, 40, 51-53, 164, 293, 309
etherification, 296, 307-308, 322, 326
exfoliation, 224, 264, 400

F

fermentation, 2-4, 8-9, 14, 17, 23, 25-26, 293, 312
field emission scanning electron microscopy, 38
filler dispersion, 383, 395
filler fraction, 385, 391, 395-396
Fischer-Tropsch, 3, 296, 305, 325-326
flakes, 233-234, 237, 248, 309, 326
flexibility, 34, 37, 44, 147, 192, 221, 225, 238
flocculation, 21
fluorescence, 91, 124
fracture resistance, 242
fracture toughness, 151, 153
free volume, 383
friction coefficient, 151, 153, 155-156
FTIR, 48
functionalized graphene, 165, 195-197, 218-219, 224-225, 228, 239, 264

G

gas sensing, 117
gasification, 296, 316-317, 328
glass transition temperature, 394
graphitic carbon, 213, 229, 250, 267, 284

H

hardness, 145, 150, 152-153, 155
hazardous gases, 294-295, 312
hierarchical arrangement, 227
hollow particles, 249, 284
humidity, 46, 93, 177, 180, 188-189, 196, 203, 211-212, 218, 220
husk, 74
hybrid nanoparticles, 67, 72, 249
hydrocarbons, 4-6, 8, 25, 112, 296, 302, 305, 314, 369, 403
hydrodemetalation, 296-298, 322-323
hydrodesulfurization, 297, 322
hydrogels, 86, 123, 126, 132-136, 138-139, 142, 147, 156-160, 162-165, 168-169, 173
hydrogen bonding, 34, 36, 56, 157, 195, 200, 203, 206-208
hydrogenation, 18-19, 30, 72, 110, 278, 296, 305, 320, 325
hydrothermal, 60, 264, 266, 274-275, 283, 289, 300, 317, 324, 328
hydrotreating, 296-297, 322

I

immobilization, 11, 77, 113, 166, 265
impact strength, 148
impedance, 133, 219
in-situ polymerization, 372, 391
interface stability, 244
interfacial interaction, 383
interlayers, 288
ionic dyes, 41

isomerization, 296, 300, 302, 322-324

J

joint implants, 170

L

lamellar structure, 171
layer-by-layer, 186
layered double hydroxides, 226, 261, 265
layered silicates, 382
leaching, 188, 308
LIB, 242-255
life cycle assessment, 21
liquefaction, 296
LSB, 257-260

M

macroporous, 224, 257, 264, 288, 301, 324
mass spectroscopy, 18
MEA, 311
mechanical strength, 44, 48, 54, 176, 178-180, 185-187, 191-192, 201, 212, 238, 382
melt mixing, 144, 149, 180, 404
melt processed, 166-167
melting temperature, 46, 362, 367, 369, 404
membrane electrode assembly, 190, 193, 311
mesoporosity, 229, 300
metal ions, 36
metal oxides, 10, 84, 223, 250, 260, 313
metal surface, 7
microcapsules, 372, 380, 405
migration, 25, 200, 207, 209, 304, 314
miscibility, 357
mixing, 144, 149, 180, 184, 383-384, 400, 402, 404
mobility, 93, 200, 207, 210, 315, 329-330, 333, 343, 345, 347-348, 351, 357-358, 360
modeling, 37, 377, 379, 382,

406
MOF, 245, 247, 249, 259, 280, 284, 290, 304-305
molecular sieves, 307-308
monolith, 302, 324
montmorillonite, 53, 124, 372, 380, 405, 415
multilayers, 279

N

nanocapsules, 285, 363
nanoflowers, 282, 288
nanofluids, 407
nanowire array, 249, 262, 315, 413
network structure, 229, 236, 401, 409
NMR, 19
nucleation, 77, 112, 128

O

oil and gas, 393, 406
oil-water separation, 113
optical properties, 39, 114, 121, 336, 349, 360
oxide layer, 286
oxygen reduction reaction, 188, 289, 309, 311, 326-327

P

PAFC, 175, 309
particle size, 89, 282, 300, 385-386, 390, 410
PEDOT, 132, 142, 161, 241, 253, 279
PEMFC, 182, 185, 187, 189, 192, 195, 197-198, 206-207, 215-216, 309-310
penetration, 242
percolation, 19, 45, 47, 180, 365, 375, 384, 391, 396, 399, 406, 408
permeability, 166, 177, 180, 189, 191-196, 201, 204, 207-208, 210-211
petrochemicals, 294
phase changing materials, 361, 363, 365-367, 369, 371-373, 375, 377, 379
phase equilibrium, 41, 370, 379
phase separation, 42, 53,

186, 209
phase transformation, 372
PLA, 46, 56-57, 129, 132, 140-141, 144, 166, 348
plasticization, 93
polydispersity index, 39
pores, 16, 156, 233, 299, 369, 377, 405
porosity, 85-86, 131, 136, 223, 225, 234, 254, 290, 302, 305, 367, 369, 375, 377
porous carbon mesh, 363
porous network structure, 236
pre-treatment, 5, 8, 11, 20
precipitation, 60, 209, 297, 302-303, 313, 317
precursor, 19, 30, 66, 73, 84-85, 87-88, 90, 93-94, 244, 251-252, 302
protective coatings, 215, 394
proton exchange, 175-176, 179, 194, 207, 211-220, 326
pseudocapacitors, 221-222, 268, 271, 275
pyrolysis, 3, 24, 280, 296, 316

R

Raman spectroscopy, 399
rate capability, 82, 225, 231, 235, 242-244, 246, 249-250, 252-255, 257-260, 280, 283-285
recyclability, 392
redox characteristics, 235
reforming, 293, 296, 298-299, 302, 314, 316, 321-325
regeneration, 120, 133-136, 138-139, 141-142, 145, 158, 165, 307, 320
relaxation, 243
reversible storage capacity, 82, 242, 246, 248, 250, 253, 255-257, 259-260
roughness, 198, 338, 393-394

S

scanning electron microscopy, 16, 38
self-cleaning, 117, 293
self-healing, 93, 129
SEM, 16, 314-315, 319, 368, 371, 401

semi-crystalline, 34, 348, 357, 404
semiconductor, 113, 116, 332, 335, 399
shear thinning, 35, 37
SIB, 254-257
silica, 16, 29, 85-86, 93, 96, 123, 213, 297-298, 303, 305-306, 323, 363, 382
sol-gel, 153, 179, 190, 362-363, 368
solubility, 9, 12, 307, 313, 338-339, 343, 349
solution casting, 47-48, 185
sonication, 37, 48, 121, 152-153, 404
specific capacitance, 222-231, 233-243, 246, 248-249, 253, 270
specific heat, 367, 381
specific surface area, 306
spinel-type metal oxide, 253
spinning, 273, 368, 376, 400, 404
splitting, 39, 121, 296, 314-316, 327-328
starch, 2, 26, 45, 56-57, 80, 88, 126, 132
steam reforming, 296, 302, 314, 316, 322, 324-325
stiffness, 49, 52, 59, 147, 205
storage modulus, 47-48, 205
sulfite pulps, 35
supercapacitor, 125, 223-238, 240-241, 262-279
supercritical water, 300, 316, 324, 328
surface charge, 39
surface chemistry, 101
surface energy, 393
surface grafting, 164
surface modification, 56, 64, 123, 365, 400, 402, 410
surface modifiers, 147
surface roughness, 198, 393-394
surfactants, 42, 54

T

temperature programmed reduction, 305
templating, 290, 302
tensile strength, 38, 44, 46-50, 90, 140, 148, 151, 181-182, 189, 192-193, 204-205, 208
ternary composite framework, 231
thermal conductivity, 57, 90, 362, 365-367, 369-371, 374, 376-377, 381-415

thermal degradation, 37, 53
thermal expansion, 407
thermal resistance, 37, 385-386, 396, 400-402, 406
thermally stable, 216, 350
thermogravimetric analysis, 42
thermoset, 388, 393
tissue engineering, 84, 88-89, 123, 131-135, 137-147, 149, 151, 153, 155, 157-167, 169, 171, 173, 293
titanium dioxide, 83, 119-120, 250
toughness, 59, 90, 147, 149, 151-153, 181, 189
toxicity, 3, 44, 66, 68-69, 90, 108, 120, 161, 370, 404
transesterification reactions, 11
transmission electron microscopy, 252, 355
tribological properties, 167-168, 171-172
triglycerides, 12, 312
turnover frequency, 297-298

U

ultimate tensile strength, 48
ultracapacitors, 221, 264, 409

V

vegetable oil, 1, 23, 28, 296, 311, 327

W

wear, 147-156, 166-170, 172-173
wettability, 6, 49, 150, 155, 171

X

X-ray, 303-304, 339
X-ray diffraction, 304
xerogels, 273

Y

Young's modulus, 38, 46, 48, 50, 138, 144, 155, 157, 181-182, 192, 204-205

Z

zeolites, 10, 296, 299-300, 307-308, 313

www.ingramcontent.com/pod-product-compliance
Lightning Source LLC
LaVergne TN
LVHW011531200926
848484LV00001B/231